Ethnic Landscapes of America

John A. Cross

Ethnic Landscapes of America

Springer

John A. Cross
University of Wisconsin Oshkosh
Oshkosh, WI, USA

ISBN 978-3-319-54008-5   ISBN 978-3-319-54009-2 (eBook)
DOI 10.1007/978-3-319-54009-2

Library of Congress Control Number: 2017942774

Printed on acid-free paper

This Springer imprint is published by Springer Nature
The registered company is Springer International Publishing AG
The registered company address is: Gewerbestrasse 11, 6330 Cham, Switzerland

Acknowledgments

Appreciation must be given to those scholarly presses that have not only published the works of the many leading scholars of ethnicity and landscape, but which have graciously granted permission to include the longer quotations that pepper the manuscript, enabling the quoted authors to express their findings and observations in their own words.

The longest quotations in *Ethnic Landscapes of America* are appropriately from that pioneering scholar of landscapes, John Brinkerhoff Jackson. These come from Jackson's *A Sense of Place, a Sense of Time*, New Haven: Yale University Press, 1994, Copyright © 1994 by Yale University, with quotations reprinted by permission of Yale University Press. Not only did Allen G. Noble's edited collection *To Build in a New Land: Ethnic Landscapes in North America*, 1992, Copyright © 1992 by Johns Hopkins University Press, serve as the textbook for the Ethnic Landscapes of America class that I have taught for over two decades, but Noble's contributors include some of the most prominent landscape scholars of the past half century, and their quotes are reprinted by permission of Johns Hopkins University Press. Quotations have also been taken from Dell Upton's edited collection *America's Architectural Roots: Ethnic Groups that Built America*, Washington: The Preservation Press, 1986, Copyright © 1986 by National Trust for Historic Preservation. These are reprinted by permission of John Wiley and Sons, which has acquired the rights to Upton's publication.

David Hackett Fischer graciously gave permission to quote from his *Albion's Seed: Four British Folkways in America*, Copyright © 1989 by David Hackett Fisher, whose copyright has been assigned to Oxford University Press. These quotations are reprinted by permission of Oxford University Press, Inc. Quotations from the many volumes of the *Handbook of North American Indians*, published between 1978 and 2008 by the Smithsonian Institution Scholarly Press, are reprinted with permission of the Department of Anthropology, National Museum of Natural History. Selections from *Florida's Colonial Architectural Heritage* by Elsbeth K. Gordon, Gainesville: University Press of Florida, 2002, are reprinted with permission of the University Press of Florida.

Permission was granted by the American Association of Geographers, formerly the Association of American Geographers, to include quotations from the *Annals of the Association of American Geographers*, *The Professional Geographer*, the *Yearbook of the Association of Pacific Coast Geographers*, as well as the guidebook printed for one of its annual conferences. The American Geographical Society graciously granted permission to quote from articles within *The Geographical Review* and the *American Geographical Society's Focus on Geography*. Similarly, the *Journal of the Southwest* courteously extended permission to quote from several articles. Not only did the *Journal of Cultural Geography* previously publish some of my work that is quoted in this book, but assurance was given that my desired quotes fell within the fair use doctrine.

The University of Chicago Press provides guidelines for the fair use of its publications, and it notes that authors, whose uses are "eligible," need not seek permission. Such openness is greatly appreciated, and it is something this author wishes other presses would emulate. Thus, knowing their definition of fair use, the quotes from one of their works are longer than those from several other publishers from which no information regarding their fair use policies could be obtained.

All of the color photographs included in this book were taken by the author. The author expresses appreciation to the Library of Congress for making available several of the historical photographs included in Chap. 3. Many of the maps included in this book were created by the author, while others have been credited to and reprinted from several U.S. Bureau of the Census publications. Others are from the U.S. Geological Survey and the National Agricultural Statistics Service's *Census of Agriculture Atlas*.

The author wishes to express appreciation to the many individuals who assisted in this endeavor. In 1995 the University of Wisconsin Oshkosh Faculty Development Board provided funding for the development of the author's Ethnic Landscapes of America course, the first course approved on this campus to meet the University's newly established ethnic studies graduation requirement. Over the subsequent years, many students made helpful suggestions, including sharing of information regarding how their families displayed their ethnicity in the structures at their farmsteads or around their homes as well as how various ethnicities were exhibited in their home communities.

The author appreciates the encouragement provided by Zachary Romano of Springer Nature as the manuscript has been prepared for publication. In particular, the author wishes to thank his wife, Joann Noe Cross, for her patience, encouragement, and assistance. Not only did she accompany me on many of the journeys to visit various ethnic communities, but she read the manuscript, making many helpful suggestions. It is to her that I wish to dedicate this book.

Contents

1 Introduction to Ethnic Landscapes

Every day the people of the United States see reminders that numerous ethnic groups played pivotal roles in shaping the nation, and that ethnic Americans and recently arrived immigrants continue to remold the country. Some of the landscape features are conspicuous, such as the magnificent ethnic architecture displayed in cathedrals and other houses of worship, while others are subtle, being displayed in the way that fields are shaped and by the crops that are being grown. While there is much that the rest of the world would identify as American, many of these features were shaped by individuals who settled the country from specific locales or have evolved by the blending of traditions from several nations, adapting them to meet local needs and values. We are constantly reminded that waves of immigrants have left their indelible marks on the American cultural landscape.

1.1 Ethnicity Defined

While precise definitions of "ethnic" are muddled (Zelinsky 2001), many writers have utilized the word as a practical classification "identifying a distinctive people with a common culture evident in their shared history, language and other characteristics" (Banton 2015:100). Although some social scientists have argued about the social relevance of ethnic origins (Banton 2015) and some would argue that ethnics are defined by comparison with the dominant group, it could be argued that all groups having their own cultural heritage or ancestry are ethnic, including the dominant group when viewed from the perspective of peoples of another ancestry.

Sociologists have long argued about the meaning of ethnicity and the efficacy of the melting pot, in which the cultural values of the numerous ethnic groups that formed the contemporary American culture were blended together (Sowell 1981). Immigration historian Roger Daniels (2002:95) explains, "Although the notion of a melting pot is a myth, some groups did in fact, melt, and all groups, even those who persisted the longest, changed over time through contact with the new environment and new peoples." Others have noted how distinct cultural values have remained, a century or more following immigration from the old country (Salamon 1992). Rather than identifying with the American melting pot, it may be more appropriate to suggest a stew pot, where the separately identifiable vegetables and pieces of meat are akin to the distinct ethnic groups that settled America, which still maintain aspects of visibly distinct cultures, yet the flavors of all the groups merged to form the stew's rich gravy.

There are many items in the American diet that immediately remind the diner—particularly if that consumer is a visitor from overseas—that the food dish is associated with the United States. Yet numerous Americans, including many who are descendents of individuals who immigrated two or three centuries ago, seek out ethnic eateries. Frequently these eating places serve foods that are totally unrelated to the patron's ethnic heritage, thus the popularity of Chinese restaurants among people of European descent or Mexican cantinas by persons of Northern European ancestry who live far from the Mexican border. Indeed, ethnic restaurants are often one of the most conspicuous aspects of the cultural landscape that outsiders often see—and interact with—when they visit ethnic neighborhoods. Just as many Americans appreciate the culinary talents contributed by immigrants from disparate backgrounds, they also appreciate a diversity of music, literature, and architecture.

The architecture that we see in the churches, schools and houses, particularly within the older historic districts of American cities or those rural areas that have been bypassed by progress, and in the barns and outbuildings on its farms, often contain many features that are clearly identified with specific ethnic groups. Indeed, immigrants typically built their structures using designs and materials with which they had familiarity in the Old Country, yet often substituting materials given local resources. When possible, many immigrant groups also sought to settle on lands that had climates and vegetation that were reminiscent of home. Some immigrant groups sought to live near other ethnic groups with which they had some familiarity or linguistic association.

J.A. Cross, *Ethnic Landscapes of America*, DOI 10.1007/978-3-319-54009-2_1

Unlike the populations of Africa, Asia, and Europe, virtually every American is the descendent of immigrants who arrived in the last few millennia—and most are descendents of far more recent immigrants. The earliest immigrants to North America south of Beringia, the area of the Bering land bridge connecting Asia with Alaska, arrived about 15,000 years ago, travelling south along a coastal ice-free corridor opened with the waning of the Pleistocene Ice Age (Goebel et al. 2008). Later waves subsequently arrived from Siberia, while the earliest authenticated European settlement in North America, that of the Vikings in Newfoundland, was established just over 1000 years ago, and it was doomed for quick abandonment. Five hundred years later Europeans began their unprecedented expansion across the Atlantic Ocean.

Beginning with the Native Americans, who themselves came in several distinct waves, each subsequent wave of immigrants reshaped the land of the continent as they decided how it should be used, building upon some of the material culture from their predecessors and adding some of their own. Thus study of the resulting landscapes reveals not only the heritage of those who currently occupy the land, but it also uncovers the history and ethnic heritage of those who previously occupied the area.

1.2 Geographic Definitions of Cultural Landscapes

Geographers have long been interested in the human use of the earth's surface. The British geographer Alan Baker notes that "description and interpretation of landscapes has been a long and honourable tradition within geography" (Baker 1992:6). The word "landscape" evokes many notions in the mind of the public, ranging from the famous landscape paintings hanging on the walls of museums to highly manicured gardens and the accomplishments of landscape architects. As John Stilgoe (2015:198) quips, even academic administrators "routinely ask what landscape studies 'actually means.'" One of the most succinct definitions of landscape is provided by a geographer. John Fraser Hart, in providing his 'simple vernacular definition of landscape', notes it is "the things we see" (Hart 1995:23). However, what we see is also culturally conditioned, and not all societies are "landscape-aware" or appreciative of landscape (Baker 2003:112).

Many geographers want to see more in the concept of a landscape. Donald Meinig argues that a landscape is more than formal parks, gardens, or the grassy yard in front of a house and that a landscape is more than an attractive countryside. Thus he notes that for many persons "it may require a considerable stretching of their usual sense of the key term: landscape" (Meinig 1990:xv). One of the best explanatory definitions of landscape is provided by John Stilgoe, author of *Common Landscape of America, 1580 to 1845.* He writes,

> Landscape is a slippery word. It means more than scenery painting, a pleasant rural vista, or ornamental planting around a country house. It means shaped land, land modified for permanent human occupation, for dwelling, agriculture, manufacturing, government, worship, and for pleasure. A landscape happens not by chance but by contrivance, by premeditation (Stilgoe 1982:3).

Stilgoe explains that the word landscape has German origins, noting that Landschaft describes "a collection of dwellings and other structures crowded together within a circle of pasture, meadow, and planting fields and surrounded by unimproved forest or marsh" (Stilgoe 1982:12), but he also notes how Germans distorted the definition "in the name of nation-building politics" (Stilgoe 2015:199). While there has been some ambiguity regarding the precise definition of landscape, geographers clearly understand "landscapes as being cultural constructions" (Baker 2003:112).

John Brinckerhoff Jackson, who founded *Landscape Magazine* and edited it for nearly two decades, was a preeminent commentator on landscape study for the last half of the twentieth century. In his *Discovering the Vernacular Landscape*, Jackson defines his term, "landscape: a composition of man-made or man-modified spaces to serve as infrastructure or background for our collective existence" (Jackson 1984:8). He continued, writing "my interest in the topic is confined to trying to see how certain organizations of space can be identified with certain social and religious attitudes" (Jackson 1984:8). Near the end of his long and productive career, when he was in his mid-eighties, Jackson reworked his definition of landscape. In 1995 he wrote "landscape is more than an area of attractive rural or natural scenery. It is a space or collection of spaces made by a group of people who modify the natural environment to survive, to create order, and to produce a just and lasting society" (Jackson 1995:43).

While the greatest modification of the environment is found within cities, much of the geographic study of landscapes is rural. Indeed, Stilgoe (2005:11) argues that "landscape is not cityscape but essentially rural, essentially the product of tradition." Many scholars in studying landscapes focus upon the vernacular landscape, which Stilgoe (2015:211) defines as "one made and used by people most of the time, not the great buildings and gardens beloved of historians of architecture and landscape architecture." As Wilbur Zelinsky (2011:218) explains, this "oversight in dealing with a country where more than three-quarters of the population is urban clearly reflects an essentially subconscious rural bias, a Saurerian mind-set shared with many of my contemporary cultural geographers." While many material features of American cities, even in those that can be identified as genuine culture areas into themselves, evolved to meet the generic social and economic needs of their growing multicultural populations that were more focused upon being American than ethnic, it is within cities today that geographers can best appreciate the visual impacts that today's newly arrived immigrants are making (Airriess 2015).

Through these definitions it is clear that a landscape is not only highly cultural and comprehensive, but it includes features that have been deliberately shaped to reflect the needs of a place's inhabitants. The way the land is shaped and the cultural values that inspire its molding are strongly influenced by the particular group of people who are making this modification. Donald Meinig, the renowned historical geographer, writes that the landscape includes "virtually all that can be seen has been created or altered by human intervention" (Meinig 1990:xv). The landscape includes evidence about the contemporary society that occupies an area, but also the past. Furthermore, Meinig (1990:xv) notes that "landscape…is part of a vast, cluttered, complex repository of society, an archive of tangible evidence about our character and experience as a people through all our history." Indeed, Brian Short, a British historical geographer, writes that "[l]andscapes are also cumulative, and older landscapes may never quite die" (Short 2006:110). Thus, there are almost always traces on the landscapes of historical settlements. Yet, at the same time it must be understood that the historic landscape of a particular area at a specific time may be quite different from what we see from the past in that landscape today. Thus, historical preservation efforts are of tremendous value in providing glimpses of how our predecessors shaped the landscape, even though the structures preserved are not necessarily representative of their entire communities.

There are also ethnic and gender biases in the way that landscape is defined, described, interpreted, and appreciated. Don Mitchell explains:

> Landscape was more than a way of seeing, more than a representation, more than ideology—though it was very deeply all of these. It was a substantive, material reality, a place lived, a world produced and transformed, a commingling of nature and society that is struggled over and in (Mitchell 2003:792).

That there is a close relationship between landscape and law is clear to some landscape scholars, inasmuch as the law determines who shall occupy the land and how they should do so. Given both the legal traditions of private landownership in the United States and the cultural values brought by many immigrants, the development and interpretation of the landscape of America is strongly shaped by the nation's long and unique history of interaction of many peoples bringing their traditions with them.

1.3 Geographic Study of Migration

Geographers have long been interested in the movement and migration of peoples, an important element of cultural geography. Furthermore, as Karl Raitz eloquently explains,

> The transfer of cultural elements from the Old World and their subsequent patters has been a major theme in geographical studies. Many immigrant groups attempted to establish a way of life similar to that of their homeland, but with the exception of a few sectarian groups, such as the Hasidic Jews, the Amish, or the Hutterites, most failed. A number of ethnic groups transferred individual elements of their culture to the New World. In many settlements material artifacts or social institutions have been preserved through several generations, and across the country are large areas that still bear the hallmarks of the original settlers (Raitz 1979:88, Reprinted with permission of the American Geographical Society).

It is thus the visible consequences of the migration of a wide number of disparate immigrant groups that comprise the ethnic landscapes that are the topic of this book. Yet to study the results of this migration requires an understanding of the complex interplay of a wide variety of human groups and their environment over considerable periods of time to shape the ethnic landscapes that are seen today.

1.4 Landscape Components

Given that a landscape includes all that we see, a landscape has both physical and human components (Hart 1998). Nature determined the shape of the land, whether flat or mountainous, whether wet or dry, and whether hot or cold. Nature determined what resources were available, whether the soil was fertile, whether desirable minerals occurred near the surface, and the location of rivers and the seacoast. Nature determined whether forests or prairies grew in the wilderness and the vegetative species that were best suited for the environmental conditions. Nature determined what crops could be grown on the soils and what animals moved across the land.

Human action determined how the land would be used, what crops would actually be grown, how those crops would be cultivated and how the fields should be arranged. Humans decided whether the forests should be cleared so the lands could be used for growing their crops and whether animals should be grazed, both actions that changed the botanical composition of the meadows or forests. Humans decided which mineral resources would be used and how they would be used, which river waters would be floated upon and whether these waters would be diverted to irrigate crops. Changes in technology over time influenced the human decisions regarding what natural materials were actually useful resources and what actions might be taken to reshape or modify the landscape. Humans chose where bridges should be built over rivers and how those bridges would be constructed, while previously decisions might have been made about the best location for a ford. Humans decided whether they should live in dispersed groups or concentrate in villages. Humans selected the sites where villages would be established and what types of structures their settlements should contain.

The structures added by people include not only their houses, but also structures associated with their economy and their system of land division. The economic uses of the land reflected not only the resources and technology available to the population, but also their political system regulating such activity and their religious beliefs. Furthermore, as John Fraser Hart (1995:28) explains, cultural landscape reflects the character of their "agglomerations of houses into villages, town, and cities, with all of their associated features."

Human shaping of the landscape did not occur overnight. As Michael Conzen has noted, "cultural landscape is, in truth, a composite of the historical interaction between nature and human action" (Conzen 1990:2). Therefore, in viewing the cultural landscape, we see the results of both contemporary and previous decisions as to what structures best serve the needs of the population. Many structures that we see today, whether currently in use or collapsing into ruin, were built a century or more ago and may have fulfilled functions that are obsolete in today's economy or society. Indeed, as Paul Oliver (1990:10) explains, "Few of mankind's artefacts have the longevity of houses…their survival to the present being a record of responsiveness to altering life-styles and societies in change." Thus, "[s]tudies of landscape…necessitate an historical perspective. Because landscapes have histories they possess a compelling human significance" (Baker 1992:7).

During the 1990s the study of American ethnic geography gained considerable stature. Michael Conzen (1990) published *The Shaping of the American Landscape*, in which nearly half of the contributors described how various ethnic groups had contributed to the nation's cultural landscape. Allen Noble (1992) edited *To Build in a New Land: Ethnic Landscapes in North America*, which summarized the work of the pre-eminent pioneers looking at examples of the landscape imprints of over a dozen ethnic groups. At nearly the same time an American Ethnic Geography Specialty Group was established in the Association of American Geographers (now the American Association of Geographers), signifying greater recognition of the evolving field. Scholarly attention to the contributions of multiple ethnic groups had also attracted attention of historians, such as Takaki (1993) and Woodard (2011), among other social scientists, such as Sowell (1981 and 1996). Likewise, the history of immigration by various groups has received study (Daniels 2002).

Early this century ethnic geographers had not only sought "a deeper understanding of sense of place and place-making" (Estaville et al. 2003:589), but had noted the importance of the "homeland" concept when describing the attachment of certain ethnic groups to a specific place with which they identified (Nostrand and Estaville 2001). Those groups that have clearly defined homelands in the United States, such as the Hispanos and Navajos of the Southwest, have created some of the most distinctive ethnic landscapes in the nation, ones that have long attracted the attention of geographers studying ethnic landscapes. Building upon the work of these and many other pioneers within the field, whose works helped inform this book, cultural geographers can readily identify the influence of ethnicity upon the landscape.

1.5 Ethnicity and Landscape

Just as ethnic groups can be distinguished by their different languages, their distinctive preparation of food, and their choice of music, ethnic groups have their own attitudes towards land. These include notions about what is the best way to use or to divide their land. Should their land be held in common by village members or should private ownership by specific persons be the norm? If the land is to be held privately, how are lands traditionally distributed? Some groups have customarily had a large number of free-holding farmers, each whom can demonstrate those lands to which he or she has title, while in other societies feudal ownership with nobility dominated, in which most farmers were tied by various obligations to a nobleman who had actual ownership of the fields that the farmers tilled, paying rent, a share of their harvest, or fulfilling other obligations for the use of the land. In other societies lands were held by a tribal group and not owned by individuals, or lands were government owned, as occurred throughout the formerly Communist countries of Eastern Europe and Russia.

Likewise, various populations have their own ideas about what are the best shapes for farmlands and how the land should be surveyed. How were the boundaries between various lands marked? Are they to be fenced, marked with a hedgerow, or otherwise indicated? These decisions leave a profoundly long-lived mark on the landscape. Indeed, Edward Price, the geographer who spent his career studying patterns of dividing the land, wrote "landownership (cadastral) patterns are among the most persistent features of the human landscape" (Price 1995:6). Furthermore, should the homes of the farmers be located by their fields, or should the farmers live in a nearby village? If they live in a village, is this community a tight clustering of buildings, possibly enclosed by a wall, or do the residents live in a narrow string of houses along a road or irrigation canal? Or, are the dwellings arranged around a church?

1.5.1 Religious Structures

Ethnic groups vary in the role that religion plays in their everyday lives. Of primary significance are the questions: do they practice a religion? And, if so, what is their religious affiliation? Likewise, does this group practice its religion in private or in public? Are religious institutions or buildings a central focus for a community? Do members make a public display of their devotion, and if they do, how do these manifest themselves? Some groups that are very devout have built churches and monasteries that conspicuously tower over the community, while other groups that may be at least as devout, such as the Amish, refrain from building any religious structures. Likewise, are questions about what the religious traditions of a group instruct about how and whether its dead should be laid to rest in cemeteries. Similarly, does a group erect carved stone grave markers that are likely to withstand the ravages of time or use simple wooden markers that are gone a generation later? In addition, we need to consider the way the religious teachings of a particular ethnic group instruct its members about the use of their lands or whether there are natural landscapes that are to be held sacred. If certain lands or structures are sacred, what are they, and are these held sacred by the entire society?

Where ethnic groups have erected conspicuous churches, synagogues, temples, or mosques, it is often these structures that best display and distinguish the group's ethnicity. The immigrant church played a prominent role in the maintenance of the native language and the culture of the immigrants to the Great Plains, while also functioning as a community center. As Robert Ostergren (1981:228) explains, the "church is the dominant structure on the rural landscape." By providing "a visual link between the new church in America and the mother church in Europe…the church was physically symbolic of its role as the keeper of culture and of continuity with the past" (Ostergren 1981:230).

Certain religious groups, such as Roman Catholics, erect large amounts of religious statuary, while to others such items are considered blasphemous. Many ethnic groups are strongly linked to certain religious denominations and have distinct notions of how their houses of worship should look. Whereas a German Catholic church might have a towering steeple and be adorned by elaborately carved stone statues of several saints, a Protestant Welsh chapel may often be plain, without the conspicuous steeple and the statuary. Yet there are often similarities in architecture of churches among certain ethnic groups. The ethnic heritage of a population is typically displayed by the language and wording of the religious inscriptions on tombstones and above the church entrance, the presence or absence of stained glass windows, the type of images where windows are displayed, and the character of statues of saints (Williams 1997). Thus, while sometimes ethnic groups may erect religious structures that are quite similar to those of nearby ethnic groups—and the differences are visible by only examining some of the more subtle detailing or inscriptions—at other times, or among very different ethnic groups, significant architectural differences are displayed.

The architectural differences among churches of various denominations, often ethnically associated, have deep meanings. Richard Wolniewicz explains that the "architecture, decoration, and spatial arrangement" of churches "reflect basic differences in the value systems of the believers" (Wolniewicz 1997:73). These differences in values may influence not only how the church is built and located, but the world view of the community and how it shapes its use of the land. Thus, even for groups where the church functions as the social center of the community, one community may tend to cluster its dwellings near the church and a parochial school, while the residents of another are more widely dispersed.

Cemeteries often outlast the church building, whether or not it was located nearby, and may function, as Chris Post (2013: 329) clarifies, as "relic landscapes—those that reinforce the historic heritage of those regions…[and] play a major role in the formation of place identity." In addition, because of the nature of the gravestones, their inscriptions and artwork, naming and location, "cemeteries work as powerful elements of the cultural landscape…[and] they are also infused with religious and cultural meaning" (Post 2013: 335). Not only can cemeteries display cultural or ethnic shifts that may occur within a community, but they may contain the oldest material artifacts remaining to mark the legacy of certain ethnic groups.

1.5.2 Housing

Ethnic groups often differ in their attitudes about how their houses should be built. As Fred Kniffen (1965:549) explains, housing "reflects cultural heritage, current fashion, functional needs, and the positive and negative aspects of noncultural environment." Drawing upon the work of J.B. Jackson, Alan R. H. Baker (2003:118) points out that "[t]he elementary unit of landscape is the *individual dwelling*, for Jackson the oldest and by far the most significant 'man-made' component of a landscape." Far more numerous than religious buildings that often required huge investments of community time, talent, and treasure, houses reflected the vernacular, or everyday, traditions of their builders. Yet do people share a building with many others, or do small numbers of individuals occupy free-standing structures?

Even more basic than the preferred architectural styles are questions about how the family is organized. Are typical families extended or nuclear? Thus, how many generations and kin are usually found living together? Likewise, how many children are found in the typical family? What cultural controls influence gender-related activities outside of the home? How permanent or temporary is the home? What other buildings are found near the home? Are dwellings used exclusively for living, or is space shared with farm animals, or is part of the building devoted to a workspace?

Of considerable importance in guiding the immigrants' construction of their homes is their knowledge of how houses were built in the Old Country. However, as the number of generations separating a group from its Old Country increases, typically the influence of these Old Country architectural styles decrease, and those that are retained are often at increasing deviance from contemporary ways in the Old Country. Where several ethnic groups share a community or a later arriving population displaces a previous dominant group, the introduction of new styles or ways of building may result in an evolving style of architecture, reflecting several ethnic influences, or it may result in a new hybrid vernacular.

The building styles brought from the Old Country were often a response to the climate there, the local availability of building materials, and to that group's traditions of employment and social class. Depending upon how the environmental conditions of the Old Country match those of the new settlement in America, houses and other buildings were modeled following the traditional styles of the former homeland. Thus, in downtown Honolulu, Hawaii, one can see several wood-frame houses built in the Cape Cod style to house Congregational missionaries who arrived from New England in the early 1800s. While this style was inappropriate for Hawaii, the building knowledge brought by early Spanish settlers into northern New Mexico was far better adapted. In summary, as John Fraser Hart notes,

> The cultural baggage of a group may affect the form and appearance of its structures, because different groups of people have their own distinctive ideas about how particular types of buildings should look. Some cultural geographers, in fact, have argued that house types are among the best indicators of cultural diffusion (Hart 1995:34).

Houses are a particularly useful conveyor of ethnic heritage when they are constructed by their owners or occupants, or by others within the community, following traditional or vernacular styles.

At other times later arriving settlers have had little opportunity to materially influence their style of housing. If they must rent houses or apartments already built, or acquire tract housing, their ethnic imprint may be limited to "the surficial and ephemeral décor of house façades" (Lehr 1990:264) or the colors used in the painting of their houses. In such cases, as John Lehr (1990:264) explains, "It was only in the building of his church, the icon of national identity in the new land, that the urban immigrant was effectively able to cast his personality into the urban landscape." Indeed, many later arriving ethnic immigrants occupy buildings designed and built by earlier arriving ethnic groups, and thus the ethnic flavor of their neighborhood may be strongly influenced by ethnic groups that may no longer be resident.

Modern houses are often constructed following formal architectural plans and using factory assembled trusses and framing. Thus, they are far less useful in conveying ethnic heritage. This is also true for many modern churches. Within many suburban areas and newly established neighborhoods of tract houses the overall imprint is strongly a generic American landscape. While the observer may be able to identify a few ethnic detailings on some of the houses or in the yards that surround them, such as the display of a national flag or the placement of a religious statue linked to a specific ethnic identity in the yard, the overwhelming majority of modern American homes lack such ethnic identifiers. While such neighborhoods would be clearly identifiable to a visitor from overseas as being American, to best view landscapes that clearly display ethnic elements linked to the various groups that together shaped the American identity, one must often visit historic districts, neighborhoods that have been the foci of recent immigrants, or areas that have been occupied by a single locally dominant ethnic group for a long time. Yet, contemporary folk housing varies locally, and "material-geographic examination of vernacular building continues to reveal location-associated relationships between manufacturers, buildings, their material supplies, production processes, products, and perspectives" (Peters, et al. 2015:73–74).

1.6 Old and New World Experience of Ethnic Groups

Ethnic groups varied greatly in their Old World and New World experiences. Some immigrants relocated to similar environments where long-held traditions gave them an advantage, while others needed to quickly adjust to accommodate unfamiliar conditions (Fig. 1.1). For example, the similarity of the climate and the suitability of customary agricultural crops differed depending upon their homeland. Scandinavians had experience as lumbermen, having worked in the forest for centuries, experience that served them well when they settled in forested areas of Wisconsin, Minnesota, and Washington. Likewise, Ukrainians knew how to break the sod of prairie grasslands from their long experience with growing wheat in the Steppes. Thus they successfully settled areas of the cold wind swept prairies of the Dakotas and Saskatchewan, where their traditional varieties of wheat were particularly well adapted. Because of their experience in growing rice in coastal West Africa, many slaves from that area were brought to coastal Georgia and South Carolina, where they were put to work utilizing their agricultural skills.

The similarity or differences in employment opportunities and economic activities of particular ethnic groups between their Old and New World settlements influenced the transference of many cultural traditions. Germans leaving their farming villages in Europe to cultivate lands and raise livestock in the American Midwest could put their knowledge about how to build a barn to good use in their new home. On the other hand, immigrants who had to find new occupations upon arrival, such as the Irish tenant farmers displaced by the Potato Famine of the 1840s who relocated to New York City

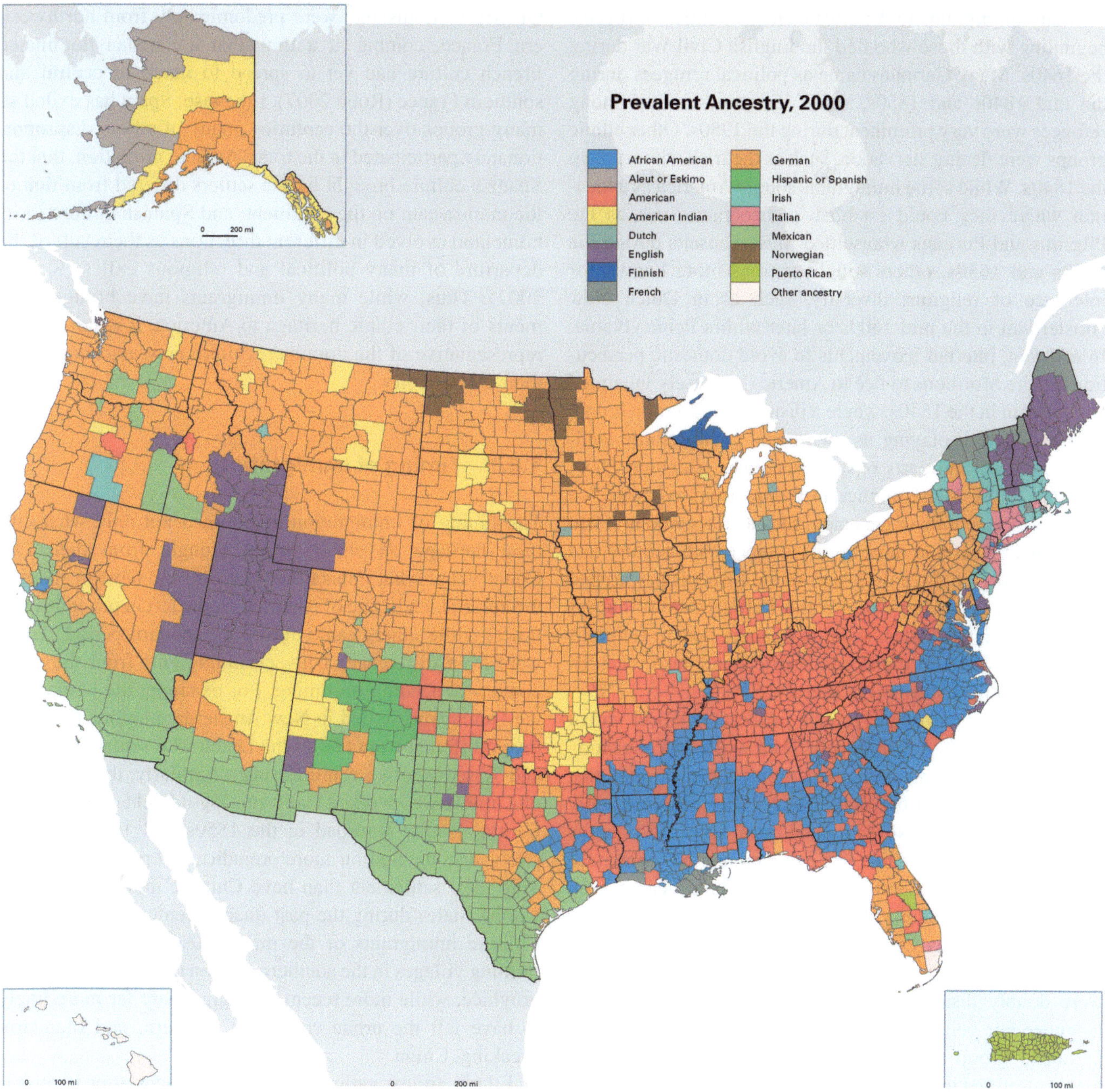

Fig. 1.1 Prevalent ancestry per county in the United States, 2000. The ethnic group that is most numerous in each county is indicated. While in some counties the most prevalent group comprised well over half of the population, within other counties where many groups are represented, the prevalent ancestry group was a much smaller fraction of the total population. In some counties the ethnic landscape was strongly influenced by a group that is no longer a prominent component of the contemporary population (Reproduced from Suchan et al. (2007), *Census Atlas of the United States,* page 141)

and Boston, lacked the same type of opportunity to showcase and maintain their rural building traditions and farming practices in their new urban setting.

It is important to remember that there was tremendous variation in the social, economic, and political standing among different ethnic immigrants. Some individuals came from prominent families in the Old World and brought trappings of their wealth with them and could buy property upon arrival, yet a far greater number lacked such standing. Some, such as the tens of thousands brought during the eighteenth century from the Guinea Coast of Africa as slaves, had no standing at all, significantly limiting, but not eliminating, their opportunity to shape America's cultural landscape. At the same time, America's indigenous population faced warfare, epidemic diseases, slavery, and genocide, greatly diminishing its role in continuing to shape the landscape.

Many immigrants to America came as political refugees, beginning with those who fled the English Civil War during the 1640s. Many Germans came as political refugees during the mid 1840s and 1850s, while Vietnamese and Hmong refugees were very prominent during the 1980s. Other ethnic groups were fleeing starvation, such as the Irish who came in the 1840s. While some immigrants sought America as a location where they could establish a theocracy, such as the Pilgrims and Puritans who settled Massachusetts during the 1620s and 1630s, others sought regions more known for tolerance of religious diversity, such as in Dutch New Amsterdam in the mid 1600s or later within Pennsylvania. In addition, internal movements to avoid domestic persecution led the Mormons to flee to America's sparsely inhabited Great Basin in the 1840s, where a distinctive Mormon homeland evolved, displaying its own home-grown ethnic landscape in Utah and parts of adjacent states (Bennion 2001, Jackson 2003). Many immigrant groups from overseas fled religious persecution or genocide, such as Jews who first fled pogroms in Russia at the end of the nineteenth century and the Holocaust a half-century later or the Armenians who fled from Turkey following World War I. Refugees are still arriving in the United States, such as Somalis fleeing religious and political strife in the Horn of Africa, and most recently, those escaping the turmoil in Iraq and Syria.

The standing of ethnic groups in the Old World varied from prominent to downtrodden. Their skills and education varied greatly among immigrant groups, even groups that shared citizenship in the same country. For example, early English settlers into New England came from East Anglia and had high levels of literacy, even among the women. In contrast, the English settlers going to Virginia came from England's West Country and were much more likely to be illiterate, and many came as indentured servants and were treated as poorly as slaves. Yet many later immigrant groups were doubly disadvantaged, neither speaking the English language nor being literate in any language. Even within a particular group today of immigrants today, "migration is highly stratified in terms of class, evidencing bifurcation and differential treatment of the skilled and less skilled…. [and] unequal treatment of immigrants of color is amply evident in most immigrant destinations" (Kaplan and Chako 2015:135). How that group shapes and adapts to their new home is not only dependent upon the characteristics of the immigrants, but also the attributes of the community which receives them.

Ethnic immigrants to North America often represented only small segments of the Old Country society. Not only were certain social and economic groups far more likely to immigrate than others, but certain regions of many Old Countries were likely to dominate the immigrant stream. For example, four counties in southern China were the source of most Chinese immigrants to the United States before the middle of the twentieth century. French settlers, whether to Quebec or Louisiana, were predominately from northwestern France, coming at a time that a Parisian dominated French culture had yet to spread to much of central and southern France (Robb 2007). Likewise, Spain has exiled so many groups over the centuries, many of whom disproportionately participated in the trans-Atlantic migration, that the Spanish culture brought by the settlers differed from that of the mainstream on the continent, and Spanish culture in its homeland evolved in different directions as the result of the departure of many political and religious exiles (Kamen 2007). Thus, while many immigrants have brought elements of their ethnic heritage to America, it is not always representative of the dominant culture now associated with their Old Country.

1.6.1 Timing of Immigration

The timing of immigration determined not only what cultural heritage an ethnic group brought from their Old Country, but what impact that group would have in shaping the cultural landscape of America. Some left in pre-industrial times, while others came during the modern period. There were even differences between immigrants from the same country depending on timing. For example, the heritage of the Dutch who settled in New Amsterdam before 1654 was much different from that of the Dutch who settled Michigan and Wisconsin two centuries later. Similarly, the thousands of Chinese who worked California's gold fields and built the Central Pacific Railroad in the 1850s and 1860s, respectively, encountered far more prejudice and prohibitions limiting their settlement than have Chinese immigrants to the United States during the past quarter century. Furthermore, Chinese immigrants of the mid-1800s largely came from farming villages in the southern Canton (or now Guangdong) province, while more recent immigrants are far more likely to have left the urban centers of northern, and Mandarin speaking, China.

Ethnic groups varied in their experiences after immigration. The timing of their arrival determined what lands were available for settlement, the amount of competition they had with other groups and with which groups they would be interacting, and what economic opportunities existed. For example, the Midwest was open to settlement when German settlers were arriving from the 1840s through the 1880s, but the frontier shifted farther west each year. Lands in the Great Plains were open for homesteading when Russians and Ukrainians arrived, as lands suitable for agricultural settlement in the Midwest had already been occupied. By the time Mexican farm workers began arriving in the American Midwest in large numbers in the last quarter of the twentieth century, no cheap farmlands were available, leaving the Mexican immigrants little choice other than to work as migrants picking fruit or vegetables, to labor cutting and packing meat in slaughterhouses, or to seek non-farm jobs in the cities. Some groups were the first settlers within a region, while others obtained lands from those who were already there, or occupied those less desirable lands that had been passed over by the earlier arrivals.

Many groups moved into areas that were already settled by other ethnic groups, yet the degree to which that landscape was already shaped varied considerably. For example, consider how German immigrants shaped the landscape of southeastern Wisconsin during the 1840s and 1850s, arriving only two or three decades after the earlier clearing of the lands by Yankees from upstate New York, who themselves occupied lands that had recently been acquired from Native American tribes. Both Caucasian groups shaped the resulting landscape, yet the imprint of the Germans remains quite prominent and German-Americans are the largest population group today. In contrast, the combined European imprint on the landscape of Wisconsin and Minnesota had nearly two centuries to develop before the arrival of thousands of Hmong refugees from Southeast Asia during the 1980s, who have simply not been able to have the same type of impact on the urban landscape, even though they are the largest minority group in many of that region's cities.

Whether an ethnic group maintained its presence within a region after its arrival, or whether it simply assimilated into the broader population, also profoundly affected the manner it shaped its landscape. While some communities have not altered their ethnic composition since their founding, others have seen complete turnovers of their populations. Consider, for example, how an Italian neighborhood in New York City became the site of that city's Chinatown and how Harlem changed from a neighborhood of prosperous Europeans into a ghetto of African-Americans and blacks from the Caribbean. In both cases the earlier group built most of the buildings that are seen today, but their uses, signage, and ornamentation reflect the desires of their current occupants.

Being the first arriving ethnic group, or at least an early arrival, provides an ethnic group with exceptional opportunities to shape the cultural landscape, particularly if the earliest arriving groups came in large enough numbers and with economic or social prominence. Thus, although the Spanish colonized part of South Carolina before the English, their numbers were insufficient to have had any lasting consequence, and indeed their settlement was abandoned after a decade. Likewise, while the Dutch first established posts along the Delaware River and named the Schuylkill River near Philadelphia, the cultural landscape that came to dominate southeastern Pennsylvania was shaped by the thousands of English Quakers and German Pietists who arrived during the eighteenth century. However, once established, the groups that made this first imprint then shaped the actions of those who followed.

Native American groups were obviously the first to occupy the lands that were to become the United States, yet their numbers were overwhelmed by the greater economic and political strength of the European immigrants. Those that came in contact with European colonists the earliest, particularly those along the Atlantic coast, suffered the greatest loss of numbers and cultural vitality. Nevertheless, these indigenous peoples had shaped the cultural landscape, and, as we will see in Chapters 2, 3, and 4, their actions influenced the behavior of those who displaced them. In many parts of the country their contributions to the cultural landscape are still visible, and in large parts of the West they remain the dominant population. The size of various American Indian groups and the timing of contact with Europeans influenced the degree to which Native American cultural influences survived.

1.6.2 Size of Ethnic Group

The size of an ethnic group determined its ability to influence landscape development. For examples, Germans ethnics are very numerous—indeed, the most commonly reported ancestry during the 2000 U.S. Census. Over half of the farmers in America's Heartland are of German ancestry (Salamon 1992) and many aspects of the nation's farm landscape reflect German values and cultural heritage. Although many architectural reminders of German heritage can be seen on both American farms and in cities, Germans could only partly shape the landscape as many institutions had already been created by earlier settlers. In contrast, those Cornish and Luxembourgers who settled in America were few, and they could only influence small areas, or were subsumed by larger groups.

Ethnic groups varied considerably in their rate of assimilation into American society. Some groups assimilated rapidly, some have yet to fully assimilate, and others were sufficiently prominent that they altered American culture as they became assimilated. The language and religion of the ethnic group determined the degree and speed by which the group became assimilated into American society. For example, Scots rapidly assimilated, and much of their culture is hard to distinguish from that of the English. They generally spoke the same language—although with a distinctly different accent, and both groups were largely Protestant—although of different denominations. Likewise, the Irish assimilated fairly quickly, as many also shared the English language, although their Roman Catholicism differed from Church of England Protestantism. In contract, the assimilation of Hispanics has been slower, given the differences in both language and religion. Furthermore, where populations can readily move back and forth between their "origins and destinations" and easily communicate with their homelands, such as is common among both Hispanic and West Indian immigrants, their maintenance of a distinct ethnicity is enhanced (Zelinsky 2001:57).

Slow rates of assimilation permits a group to have a greater influence on the cultural landscape, particularly if the group is sizable and dominates the population of a particular area. Thus, we see the perseverance of the Hispano Homeland across northern New Mexico and the Cajun region of southern Louisiana. Racial differences also slow rates of intermarriage, mixing, and assimilation. Thus, on several coastal South Carolina islands there remain descendants of African slaves who still speak Gullah and who have maintained a culture distinct not only from the Southern white culture, but also from the nation's evolving urban African-American culture.

1.6.3 Contested Landscapes and Evolving Toponyms

Early arriving populations frequently have the opportunity to affix place names to the physical features and communities they establish. The selection of these toponyms often highlights the ethnicity and values of the group that figuratively stakes its claim to territory through the naming process. Place names vary widely among ethnic groups, with some groups naming places after their patron saint or the saint for the calendrical date of the place's discovery or establishment, while others give names describing a characteristic of the place in their own language. Some settler groups show a strong affinity for naming places after the community from which they had come or for their sponsor, while others seek to honor famous personages or political figures. Names given to locations by indigenous peoples have often influenced the names written on the maps for those locations by Europeans, who often rendered their phonetic interpretations of the Native American place name and whose toponyms have not always been appreciated by the indigenous populations. Indeed, consider how several Native American groups have changed their names over the past couple decades, including the Papago who became the Tohono O'odham, the Winnebago who became the Ho-Chunk, and San Juan Pueblo became Ohkay Owingeh. As marginalized groups have regained stature, pressures grow to replace names given by outsiders with their traditional indigenous names. Recent examples include the changing the name of Alaska's Mount McKinley to Denali and the renaming of numerous locations within Canada's Inuit dominated Nunavut, as the Eskimo, now called by their preferred name, the Inuit, reestablish their cultural claim upon the territory. Indeed, as Mark Monmonier (2006:88) explains, "Restoration of native place names became a key element in reestablishing cultural sovereignty."

Throughout the United States, but particularly within those communities with sizable African-American populations, streets have been named to honor Dr. Martin Luther King, Jr. (Alderman and Inwood 2013). Similarly, streets have been renamed to commemorate Caesar Chavez in many locales with significant Mexican-American communities. Yet as Eliott Tretter (2011:52) explains, "black commemorations are still a 'black thing,' and their geographies are mostly limited to 'black places'...[and thus] they remain symbols of a separate 'black nation' that is marginalized and remain excluded from a more inclusive universal sociocultural historical geography of the United States." As we will see in a later chapter, the legacy of slavery in intricately interwoven with the landscapes that African-Americans and their plantation owners created, and how that landscape is presented—or even if it is presented—to the public remains a topic of contention.

1.7 Generic American Landscapes

When identifying distinct ethnic landscapes, reflecting upon the Old Country heritage of Americans, it is important to be able to make contrasts with what is generically American. What is typically American about our landscape? What makes our landscape different from that found in Europe, Africa, or Asia? Or how is the landscape of a rural German-American community different from the landscape surrounding a village in Germany? Why does it differ? We could ask virtually identical questions for a wide array of other ethnic groups. Our understanding would benefit by both posing and answering such questions.

It is also vital to ask why immigrants brought certain cultural traits with them, but abandoned others. The suitability of previous customs was influenced by other settlers and immigrants, as well as by the physical environment of the region. Did immigrants arrive in time to influence American institutions, or after such institutions had largely been established? For example, early French and Spanish settlements are marked by their distinctive ways of surveying their lands, with all of its implications for the layout of fields and roads, while later immigrants had little opportunity to so shape their landscape.

American institutions and political ideology has had a profound impact upon creating a landscape that contrasts with that in the nations from which settlers emigrated. Americans have long displayed an anti-urban bias—even as more Americans than ever now live in metropolitan areas. These attitudes have resulted in cities that are shaped very differently from those in Europe, even if they contain certain buildings that were constructed to scrupulously follow the architectural style of the homeland. Yet Americans, as a group, have their own notions about the use of space and land ownership that have left a distinct imprint upon much of the nation's landscape.

Fig. 1.2 Aerial photograph over southern Indiana showing grid pattern of fields and roads resulting from the U.S. Public Land Survey

1.7.1 Public Land Survey System and American Landscape

The American way of surveying its land is quite distinctive from what is found in much of the world. Indeed, an individual seeing high altitude aerial photographs taken over the American Great Plains, the Midwest (Fig. 1.2), or parts of the South could quickly tell that the view is American, and not European, Asian, or African, simply by noting the rectangular shape of the fields. Such rigid rectangular patterns are simply not found in many areas of the world. America's choice of its survey system, and how it was implemented, profoundly influenced its resulting settlement pattern, necessitating the evolution of a distinctly American landscape.

The Public Land Survey system adopted in the United States in 1785 was first utilized in eastern Ohio, and then extended to other public lands, providing a consistent means to survey lands before they were transferred into private ownership. The survey pattern for most of the United States, with the exception of those lands near certain former colonial settlements, resulted in lands being surveyed into square sections, one mile (1.61 kilometers) on the side, or into square or rectangular fractions of these squares. Thus, farm fields are rectangular in shape and property lines run parallel or perpendicular to the compass aligned to the North. Roads, both rural and urban, typically run North-South or East-West, up and down the slopes unless the topography is particularly rugged. Rural roads are typically spaced one mile apart, following the boundaries between the survey sections (Fig. 1.3).

The American rural landscape is characterized by dispersed farmsteads, initially with a farmhouse and barn located on each farm. Although farm consolidation has reduced the number of farms and resulted in the abandonment of many farmhouses, even in the heyday of homesteading farm families usually lived some distance from their nearest neighbors. Because farmers built their houses on their own lands, and indeed the Homestead Act of 1862 required such behavior as one of the conditions to receive the free land, at best only four farms cornered. Thus, if a farmer built on the corner of his property near his neighbors, only three neighbors could be nearby. Many farmers built their houses more centrally, widely dispersing the farm population across the land. Rural schools and churches were likewise widely dispersed among the farmsteads, and what rural villages developed were populated by tradesmen, not farmers. This pattern contrasts sharply with the rural farm villages that are characteristic of much of Europe, Africa, and Latin America, where farmhouses and frequently barns are located together in the village.

The U.S. Public Land Survey System necessitated changes in the ways immigrants shaped their land and fields, arranged their villages, placed their houses, and controlled their lands (Fig. 1.4). Although in a few cases European

Fig. 1.3 Aerial photograph showing section line roads and square and rectangular field patterns east of Sherwood, Wisconsin. The roads following the one mile square boundaries of the survey section are clearly visible between the snow covered fields in the winter

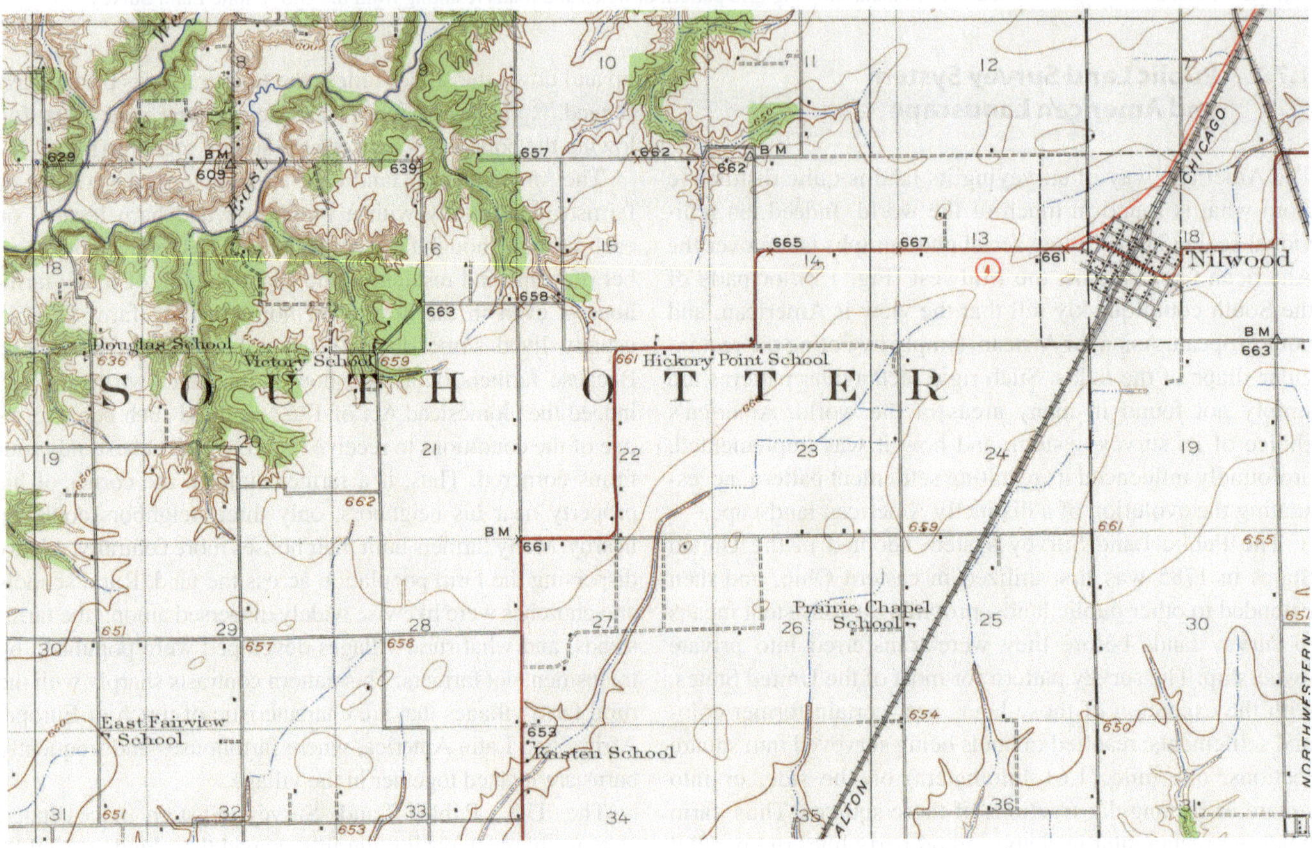

Fig. 1.4 Topographic map showing section line roads and square mile sections in the South Otter Township of Illinois. Note how the highway follows both section lines and quarter section lines, requiring many 90° turns. The patterns of forest, shown in green, also illustrate the importance of the Public Land Survey in shaping the landscape. The village of Nilwood demonstrates the influence of railroads in reorienting the grid of communities to the railroad (Reproduced from U.S. Geological Survey's 15 min Carlinville, Illinois quadrangle map)

immigrants were able to purchase large blocks of land and replicate a village pattern reminiscent of the Old Country, such as the Amana villages founded by German immigrants in Iowa and in the Texas Hill Country where several large land-grants were given to German settlers by the Republic of Texas, most Germans and other European immigrants in America were dispersed upon their own rural farms. Yet there are parts of the United States, such as in the river valleys of New Mexico, where earlier land survey systems had already divided the land, resulting in different patterns on the cultural landscape.

Americans are far more likely to number their streets (or to assign them a letter from the alphabet) than what one sees in Europe, where street names sometimes change every block. In their study of the geography of street numbering, Rose-Redwood and Kadonaga (2016:49) note that the "increase in the proportion of places with numbered streets from East to West is…partly attributable to the national survey grid that provided a template for settlement west of the Ohio River." Thus, with the conspicuous exception of New England, where numbered streets are rarely found, the use of numbered or lettered streets reduces the likelihood that street naming highlights the ethnicity of the community's settlers.

1.7.2 Railroads and the American Landscape

Much of the United States was settled at the same time as, or after, the arrival of a railroad, unlike most villages of Europe and Asia that greatly predated the invention of the locomotive. Not only did the railroad companies determine where their tracks should run and where stations would be built to serve settlers, but the railroads were granted huge acreages along their routes by the Federal government. The sale of these lands brought both profit and customers to the railroad, and many railroads sent agents to Europe to attract immigrants to settle their lands. Thus, the railroad could directly influence how the villages were located and surveyed along its route, and the railroads determined the ethnicity of those who settled within those villages and on the nearby farmlands by their decision as to where to send their agents.

The railroads had numerous landscape influences. Besides determining where communities would be located and how they should be named, the railroad companies surveyed lots and platted communities along their tracks, and sold lots to prospective residents and businesses. The plat usually had the main street of the business district either parallel to the railroad tracks or perpendicular to the rail line (Hudson 1985). This often resulted in T-shaped villages arranged in a grid, but in a grid that conformed to the routing of the railroad rather than the compass that determined the survey of the outlying farmlands. Although residents of the newly established communities often brought their own notions of how their houses, business buildings, and churches should look, the railroads also influenced the type of architecture that was utilized. Railroads brought mail-order houses to many communities, and such houses were described in catalogs, including ones issued by the mail-order giant, the Sears-Roebuck Company. All parts of the houses were delivered in a railroad freight car, including pre-cut lumber, shingles, nails, windows, and of course, the assembly instructions. Such mail order houses, of which tens of thousands were sold, contributed to a standardization of American styles.

Certain railroads, realizing the importance of promoting tourism, deliberately sought to build their stations or rail-side hotels to showcase the ethnic architecture of the region. The best example is the Atchison, Topeka and Santa Fe Railroad, which constructed many of its stations in the American Southwest to evoke the style of Spanish missions, reflecting the ethnic heritage of the region that it served (Fig. 1.5). The Great Northern Railroad constructed huge timbered resort hotels in the Northern Rockies mimicking the architecture of Alpine chalets. Such structures were not incongruous with the vernacular architecture of many of the Scandinavians who settled along the route of that railroad.

1.7.3 The Automobile and the American Landscape

America's love affair with the automobile has created an urban landscape that greatly differs from that of many of the cities of Europe and Asia. With its reliance upon private automobiles for routine transportation, huge portions of the land area of American cities are devoted to streets, expressways, and parking lots (Fig. 1.6). With the typical American family now owning at least two automobiles, the expansion of the country's highway system facilitates long-distance commutes, particularly in such rapidly growing metropolitan areas as Los Angeles, Houston, and Atlanta. The construction of homes surrounded by spacious yards in sprawling suburbs has become the norm within most American metropolitan areas, particularly within those cities of the South and West that have experienced their greatest growth in the Post-World War II period (Fig. 1.7).

Americans drive to sprawling shopping centers, strip malls, and huge stand-alone big-box department stores, such as Walmart, to purchase their groceries, clothing, and housewares (Fig. 1.8). These commercial zones have led to the demise of the downtown business districts in both America's small towns and the central business districts of many of its larger cities, reshaping the urban landscape to make it even more different from that of many traditional European and Asian cities. The American city, as it evolved, is visually and functionally distinct from those in the Old Country. As John Brinckerhoff Jackson (1997:238) eloquently noted,

Fig. 1.5 Railroad station in downtown San Diego, California. The high rise buildings and the large parking lot filled with vehicles typify the generic American city, yet the railroad station constructed by the Atchison, Topeka, and Santa Fe Railroad evokes the Hispanic ethnicity of California's and the Southwest's early settlers

Fig. 1.6 Dense grid of streets and expressways blanket the Los Angeles, California area. Single family detached housing dependent upon the automobile for access characterizes the American urban landscape

Fig. 1.7 New suburban residential development along the greens of a golf course south of Orlando, Florida. Single family housing on large lots sprawling across the landscape characterizes the modern and wealthier suburban landscape of America

Fig. 1.8 Huge parking lots surround the Ontario Mills retail outlet mall in Ontario, California, within the Los Angeles Basin. Note the ready access to expressways and the expansive cloverleaf interchange between Interstate Highways 10 and 15. Expressways are a much more prominent part of the American landscape than in most European nations

> what makes most modern American cities interesting in spite of their similarity is that they are not like cities in Europe or like American cities of the past: they are not pedestrian cities; they are not to be explored on foot. They have to be explored in a car, and even that takes time, because they stretch for miles and miles—street after street of one-story single-family houses (Jackson 1997:238).

Nevertheless, many architectural features found within American cities, even the sprawling ones, provide visual connections with the ethnic heritage of the residents.

1.7.4 Use of Space in American Cities

The sprawling of American cities, while greatly facilitated by the construction of streetcar and interurban rail lines beginning in the mid nineteenth century and expressway systems beginning a century later, is more than a consequence of modern transportation. It is also a manifestation of negative American attitudes towards cities dating from the time of the nation's founding (Conzen 1996). Thomas Jefferson and other early political leaders championed the virtues of a rural land-owning population over an urban elite.

American cities have been viewed as quite distinctive from those elsewhere. Lutz Holzner (1996:289) in explaining why the landscape of American cities differs from those of Europe and Asia, notes that American cities are less concentrated and include a large transitional region between the rural countryside and the urban core "in a pseudo urban-rural mix." He observes that American cities "are not compact and centered around one major urban nucleus, no matter how impressive the skylines of skyscrapers in the old downtowns might appear, but…form a multi-nucleated, mosaic-like landscape which is neither entirely urban nor entirely rural" (Holzner 1996:289). Holzner, like Conzen, describes the American anti-urban biases, noting that even the early seventeenth century layout of Plymouth, Massachusetts "was by design not a compact settlement European-style but a clear forerunner of suburban America, with individual houses on individual lots" (Holzner 1996:293). Furthermore, he notes that the physical arrangement of American cities displayed a more suburban orientation rather that an "'urbane' notion of cities" (Holzner 1996:293) throughout almost all of American history.

A number of criteria distinguish American cities from those elsewhere, besides the huge investment in highways and parking lots. Michael Conzen notes that in the United States cities there is a sharp contrast between

> a 'downtown' core of great building height and density that usually stands on the site of the city's historical kernel…; and on the other hand, a vast penumbra of mostly low-density land uses and building patterns, with an overwhelming bias towards single-family residential construction (Conzen 1996:277).

While sprawling suburbia marks most American cities as well as exurban developments, as we will see in a later chapter, the American penchant for detached housing surrounded by manicured grass lawns reflects the colonists' ethnic roots.

Americans have long been noted as being a highly mobile people. This has had many impacts upon the American landscape. Within many urban areas, one ethnic group has followed another into specific districts, and the expansion of racial or ethnic minorities, particularly African-Americans and Hispanics, into certain neighborhoods has prompted "white flight," as previous occupants of neighborhoods relocate to new suburban areas, where higher property costs limit such racial or ethnic mixing. Such evacuation of the inner city by Americans of European ethnicity has resulted in huge African-American ghettos, such as those on the south side of Chicago, the north side of Milwaukee, or sprawling across most of Detroit. At the same time such ghettos develop, many residential properties have become abandoned, which is particularly conspicuous in Detroit, where the city has been steadily losing population. The migration to the suburbs is only one example of American mobility, and the 2010 Census shows that many African-Americans and Hispanics have now joined the movement to the suburbs.

Americans frequently relocate, seeking better employment opportunities, or seeking locations with a more palatable climate—both physical and cultural. After the closing of the frontier over a century ago, Americans have been steadily leaving much of the Great Plains as well as rural communities throughout the nation. The West Coast and the Southern States have rapidly grown, attracting new residents from many of the Northern states, where population growth is well below the national norms. Americans often relocate at the time of their retirement, contributing to very rapid growth in Florida, Arizona, and Nevada.

The constant movement of the American population has had several important consequences upon the visible ethnic landscape. The mixing of the population leads to a more homogenized national culture and breaks down the dominance of any particular ethnic group in the development of new settlements. The movement of peoples has resulted in many rural settlements and urban neighborhoods being occupied and reshaped by ethnic groups different from the ones that first established and built the community, so that the older structures and layout of the community display the influence and values of one ethnic group, but the subsequent structures and uses of that landscape display the motivations of later groups, who may or may not have any appreciation for the architecture and cultural institutions shaped by their predecessors. Yet the exodus of population from many rural counties in the Plains and Midwest has resulted in the imprint of the initial European settlers remaining the most visible element upon the contemporary cultural landscape, which has been little altered in the century since settlement except for the abandonment of many structures. The ruined houses and barns upon the American landscape speak of two more distinctly American traits. Similarly, in urban areas one sees that "much of the building ethos in American cities is the rapid obsolescence of physical construction and the flimsiness of materials and assembly in the overwhelming majority of structures built with short life-expectancy" (Conzen 1996:278). Because many Americans invest far less in the permanence of their structures than their European forebears, not only are these buildings less likely to survive over extended periods of time, but they are more likely to be replaced by newer structures whose design and purpose is more in keeping with their contemporary needs. Such behavior is consistent with those many centripetal forces that have led to a homogenization of much of the American landscape. Yet as Wilbur Zelinsky (2011) concludes in his aptly named final book, *Not Yet a Placeless Land*, while the United States has become an increasingly homogenized land, it has not become a placeless land. Many regional differences continue to exist and new arrivals contribute to many distinctive places.

1.8 Ethnic Persistence

Even though much of what is being built on the American landscape today has little in common with the traditional architecture of Europe, Asia, and Africa, much remains, particularly within various historic districts and rural areas, that highlights the influence of numerous ethnic groups upon America's cultural landscape. Within certain areas, ethnically distinct homelands have evolved, persisted, and even grown. In other areas small ethnic enclaves, past or present, can be seen by their distinct shaping of the cultural landscape. America remains highly ethnic, and while in many areas that is best manifest by a generic American cultural imprint, one need not travel far to find commercial signs in languages other than English and buildings whose function, shape, and design conspicuously digress from the American norm.

Both the current distribution and former presence of particular ethnic populations help explain the regional patterns of ethnicity that one sees on the cultural landscape today. Obviously, geographic continuity of a particular ethnic group determines the nature of the resulting ethnic landscape. Thus, in its portrayal of the ancestry of the American people, the authors of the *Census Atlas of the United States* (Suchan et al. 2007) display maps showing both the distribution of the population reporting specific ancestries in 2000 and the proportion of the foreign-born population from that nation in 1900. They note,

> For some ancestries, continuity in geographic distribution from 1900 to 2000 is evident. For instance, in 1900, Norwegians were a large share of the foreign-born population in parts of Wisconsin, Minnesota, North Dakota, and South Dakota. One hundred years later, ancestry data from Census 2000 still indicated high percentages of Norwegian ancestry in these states' populations. The geographic distributions of Russian, Polish, and Swedish ancestries in 2000 also mirror their foreign-born distributions in 1900…. Ancestry data reveal the country's links to many heritages and illuminate our diverse roots (Suchan et al. 2007:140).

Where less correspondence between the earlier patterns of immigration and contemporary ancestry patterns exist, the viewer can still find aspects of ethnicity upon the landscape. But the landscape is more complex, with some of the older structures displaying ethnicities of one group, while more recent structures, or alterations to other older buildings, display the heritage of more recent arrivals. To appropriately explain what one sees on the cultural landscape today, one needs to understand both the contemporary and historical patterns of ancestry.

The greatest numerical growth of any ancestry group between the 1990 and 2000 Censuses were those "who reported American and no other ancestry" (Suchan et al. 2007:139), a number which had grown to 20.2 million in 2000 (Fig. 1.9). Such growth is not necessarily a reflection that Americans have lost their ancestral roots—although that is definitely the answer for some Americans—but also that many Americans have become the product of such a diversity of ancestries that reporting a specific heritage would be inappropriate. (Indeed, the Census form only allowed space for respondents to list two ancestries, an awkward situation for any individual whose four grandparents each came from a different nation.) There is a strong correspondence between the spatial pattern of individuals reporting American ancestry and those regions of the Upland South that were settled early by Scots-Irish and English ethnics, but which were largely bypassed by the waves of European immigrants in the nineteenth century. At the same time, the rise in those numbers reporting a generic American ancestry is consistent with the increasing prominence of a generic American landscape, which contrasts all the more sharply with those homelands, ethnic islands, and vestiges of ethnicity that continue to shape the landscape.

Maps illustrating the various ancestry groups discussed in the chapters are reproduced from several Census Bureau publications reporting the results of the 2000 Census (Brewer and Suchan 2001; Suchan et al. 2007) and the author has prepared maps illustrating the spatial distribution at the county-level nationwide of specific ancestries, utilizing data from the 2000 Census. While the 2010 Census has provided useful updated information regarding the spatial patterns of various racial groups—as well as Hispanics, it collected no ancestry data comparable to the 2000 Census. Although the American Community Survey has included questions regarding ancestry, small sample size bias has left most rural counties without reportable statistics, reducing the usefulness of

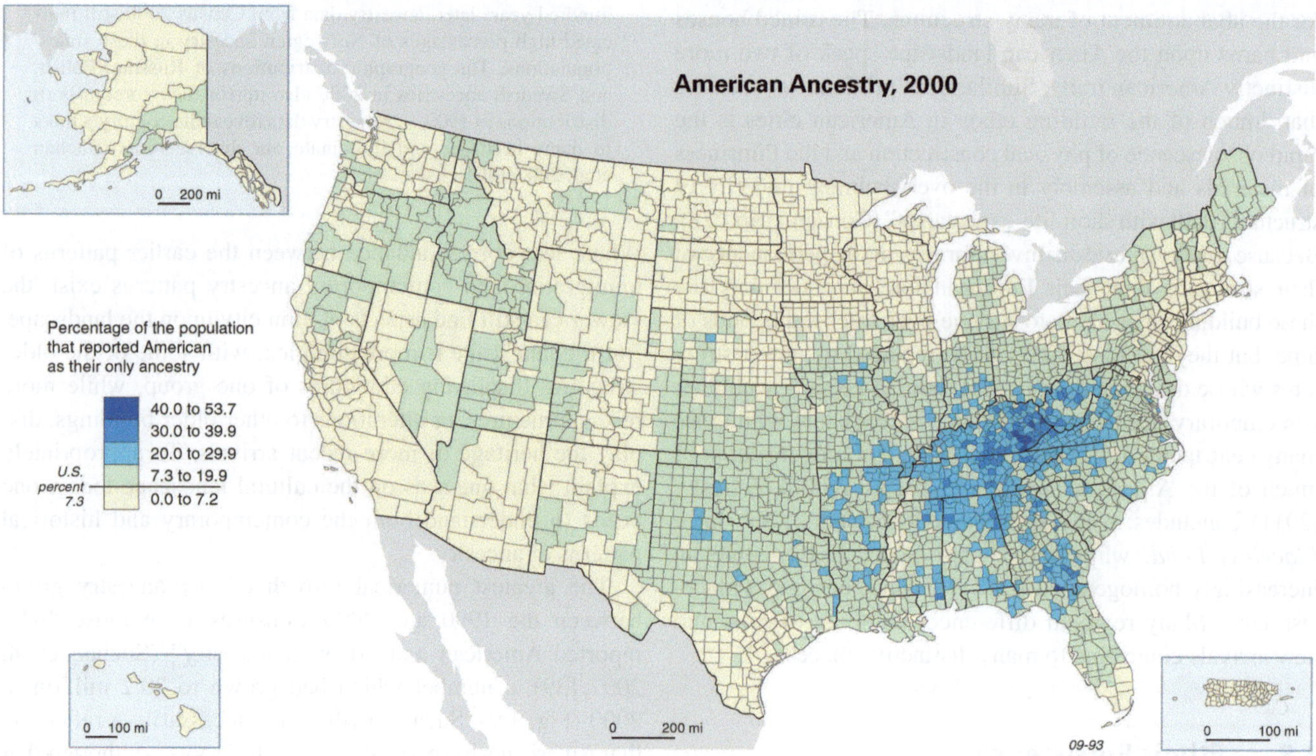

Fig. 1.9 Map showing the percentage of individuals who reported their ancestry on the 2000 Census indicating that "American" was their primary or secondary ancestry. Note the largest concentrations of those reporting American ancestry are concentrated in the Upland South, a region which was largely settled by individuals with British roots who immigrated two centuries ago, and which was largely ignored by the waves of European immigrants who arrived during the nineteenth century and Asians and Hispanics who arrived during the twentieth century. Nevertheless, even within this region elements of the early settlers' Scots, Irish, and English ethnicity still are visible on the landscape, just as the heritage of Native Americans and African-Americans can be seen in much of the South (Reproduced from Suchan et al. (2007), *Census Atlas of the United States*, page 155)

the more recent data and necessitating the consideration of data compiled and averaged over five annual surveys, thus leaving the 2000 Census data as the most recent and reliable information regarding the population and spatial distribution of most ethnic groups in the nation. However, given the time necessary for populations to shape or reshape the ethnic landscape, the 2000 Census data is likely to retain its value in helping to explain the contemporary ethnic landscape for several decades to come.

1.9 Prelude to Specific Ethnic Chapters

The ethnic landscapes explored in this book are generally described in the order that the various groups settled in America. Thus, the next three chapters of this book look at the landscape impacts of Native Americans, followed by chapters that look at the cultural landscapes created by Spanish, French, and English settlers. Beginning in colonial times, Africans were brought to America, and their impacts upon the landscape began at that time. Although a few other European groups settled during colonial times, the impacts of the huge waves of European settlers from Central and Northern Europe during the mid and late 1800s are examined next, even though it is impossible to include detailed information on every ethnic group or each area they settled. Later chapters explore the impacts of Italians, Eastern Europeans, and more recently, immigrants from the Middle East, from China, and from Southeast Asia. As America now receives, although not always welcomes, over a million immigrants a year, the ethnic imprint upon the nation's cultural landscape remains highly visible, albeit constantly evolving.

Many ethnic geographers, historical geographers, and architectural historians have described specific cultural landscapes both in detail and with eloquence. Quotations from their works have been generously provided with the hope that the interested reader will consult many of their works for greater understanding and appreciation of the contributions of the various groups described. Inevitably, not every group has received equal attention. Some are too small to have had much impact upon the cultural landscape, while others have had an impact, yet it is too regionally confined to have attracted the attention of either geographers or other ethnic scholars. Rather than view this work as a complete comprehensive guide to all of the ethnic groups that shaped the American landscape, it is better to view this book as an attempt to describe most of the prominent ethnic landscapes in the United States and to provide brief introductions to the landscape contributions of many other ethnic groups. Hopefully, additional study will enable additional ethnic groups to receive appropriate recognition for the contributions they have made in shaping or reshaping the American landscape.

References

Airriess, Christopher A., ed. 2015. *Contemporary Ethnic Geographies in America*. Lanham: Rowman & Littlefield Publishers.

Alderman, Derek H., and Joshua Inwood. 2013. Street Naming and the Politics of Belonging: Spatial Injustices in the Toponymic Commemoration of Martin Luther King Jr. *Social and Cultural Geography* 14 (2): 211–233.

Baker, Alan R.H. 1992. Introduction: On Ideology and Landscape. In *Ideology and Landscape in Historical Perspective*, ed. Alan R.H. Baker and Gideon Biger, 1–14. Cambridge: Cambridge University Press.

———. 2003. *Geography and History: Bridging the Divide*. Cambridge: Cambridge University Press.

Banton, Michael. 2015. *What We Now Know About Race and Ethnicity*. New York: Berghahn.

Bennion, Lowell C. 2001. Mormondom's Deseret Homeland. In *Homelands: A Geography of Culture and Place across America*, ed. Richard L. Nostrand and Lawrence E. Estaville, 184–209. Baltimore: Johns Hopkins University Press.

Brewer, Cynthia A., and Trudy A. Suchan. 2001. *Mapping Census 2000: The Geography of U.S. Diversity*. Washington: U.S. Bureau of the Census.

Conzen, Michael P. 1990. Introduction. In *The Making of the American Landscape*, ed. Michael P. Conzen, 1–8. Boston: Unwin Hyman.

———. 1996. The Moral Tenets of American Urban Form. In *Human Geography in North America: New Perspectives and Trends in Research Innsbrucker Geographische Studien Band 26*, ed. Klaus Krantz, 275–287. Innsbruck: Universität Innsbruck.

Daniels, Roger. 2002. *Coming to America: A History of Immigration and Ethnicity in American Life*. 2nd ed. New York: Harper Perennial.

Estaville, Lawrence E., Susan W. Hardwick, James P. Allen, and Ines M. Miyares. 2003. American Ethnic Geography. In *Geography in America at the Dawn of the 21st Century*, ed. Gary L. Gaile and Cort J. Willmont, 586–599. New York: Oxford University Press.

Goebel, Ted, Michael R. Waters, and Dennis H. O'Rourke. 2008. The Late Pleistocene Dispersal of Modern Humans in the Americas. *Science* 319: 1487–1501.

Hart, John Fraser. 1995. Reading the Landscape. In *Landscape in America*, ed. George F. Thompson, 23–42. Austin: University of Texas Press.

———. 1998. *The Rural Landscape*. Baltimore: The Johns Hopkins University Press.

Holzner, Lutz. 1996. American Ideologies and the Building of Compromise-Landscapes in Urban America. In *Human Geography in North America: New Perspectives and Trends in Research Innsbrucker Geographische Studien Band 26*, ed. Klaus Krantz, 289–299. Innsbruck: Universität Innsbruck.

Hudson, John C. 1985. *Plains Country Towns*. Minneapolis: University of Minnesota Press.

Jackson, John Brinckerhoff. 1984. *Discovering the Vernacular Landscape*. New Haven: Yale University Press.

———. 1995. In Search of the Proto-landscape. In *Landscape in America*, ed. George F. Thompson, 43–50. Austin: University of Texas Press.

———. 1997. The Vernacular City. In *Landscape in Sight: Looking at America*, ed. John Brinckerhoff Jackson and Helen Lefkowitz Horowitz, 237–247. New Haven: Yale University Press.

Jackson, Richard H. 2003. Mormon Wests: The creation and evolution of an American region. In *Western Places, American Myths: How We Think About the West*, ed. Gary J. Hausladen, 135–165. Reno: University of Nevada Press.

Kamen, Henry. 2007. *The Disinherited: Exile and the making of Spanish Culture: 1492–1975*. New York: Harper Collins Publisher.

Kaplan, David H., and Elizabeth Chako. 2015. Placing Immigrant Identities. *Journal of Cultural Geography* 32 (1): 129–138.

Kniffen, Fred. 1965. Folk Housing: Key to Diffusion. *Annals of the Association of American Geographers* 55 (4): 549–577.

Lehr, John C. 1990. Preservation of the Ethnic Landscape in Western Canada. *Prairie Forum* 15 (2): 263–276.

Meinig, Donald W. 1990. Forward. In *The Making of the American Landscape*, ed. Michael P. Conzen, xv–xvi. Boston: Unwin Hyman.

Mitchell, Don. 2003. Cultural Landscapes: Just Landscapes or Landscapes of Justice? *Progress in Human Geography* 27 (6): 787–796.

Monmonier, Mark. 2006. *From Squaw Tit to Whorehouse Meadow: How Maps Name, Claim, and Inflame*. Chicago: University of Chicago Press.

Noble, Allen G. 1992. *To Build in a New Land: Ethnic Landscapes in North America*. Baltimore: Johns Hopkins University Press.

Nostrand, Richard L., and Lawrence E. Estaville. 2001. *Homelands: A Geography of Culture and Place Across America*. Baltimore: Johns Hopkins University Press.

Oliver, Paul. 1990. *Dwellings: The House Across the World*. 2nd ed. Austin: University of Texas Press.

Ostergren, Robert C. 1981. The Immigrant Church as a Symbol of Community and Place in the Upper Midwest. *Great Plains Quarterly* 1 (Fall): 225–238.

Peters, James S., David T. Damery, and Richard W. Wilkie. 2015. Folk Housing Revisited. *Geographical Review* 105 (1): 61–77.

Post, Chris W. 2013. Heritage, Amenity, and the Changing Landscape of the American West. *Journal of Cultural Geography* 30 (3): 328–355.

Price, Edward T. 1995. Dividing the Land: Early American Beginnings of Our Private Property System. Geography Research Paper No. 238. Chicago: University of Chicago Press.

Raitz, Karl B. 1979. Themes in the Cultural Geography of European Ethnic Groups in the United States. *Geographical Review* 69 (1): 79–94.

Robb, Graham. 2007. *The Discovery of France: A Historical Geography, from the Revolution to the First World War*. New York: W. W. Norton and Company.

Rose-Redwood, Reuben, and Lisa Kadonaga. 2016. The Corner of Avenue a and Twenty-third Street': Geographies of Street Numbering in the United States. *The Professional Geographer* 68 (1): 39–52.

Salamon, Sonya. 1992. *Prairie Patrimony: Family, Farming, and Community in the Midwest*. Chapel Hill: University of North Carolina Press.

Short, Brian. 2006. *England's Landscape: the South East*. London: Collins.

Sowell, Thomas. 1981. *Ethnic America: A History*. New York: Basic Books.

———. 1996. *Migrations and Cultures: A World View*. New York: Basic Books.

Stilgoe, John R. 1982. *Common Landscape of America, 1580 to 1845*. New Haven: Yale University Press.

———. 2005. *Landscapes and Images*. Charlottesville: University of Virginia Press.

——. 2015. *What Is Landscape?* Cambridge: The MIT Press.

Suchan, Trudy A., Marc J. Perry, James D. Fitzsimmons, Anika E. Juhn, Alexander M. Tait, and Cynthia A. Brewer. 2007. *Census Atlas of the Unites States, Series CENSR-29*. Washington: U.S. Census Bureau.

Takaki, Ronald. 1993. *A Different Mirror: A History of Multicultural America*. Boston: Little, Brown and Company.

Tretter, Eliot M. 2011. The Power of Naming: The Toponymic Geographies of Commemorated African-Americans. *The Professional Geographer* 63 (1): 34–54.

Williams, Peter W. 1997. *Houses of God: Region, Religion, and Architecture in the United States*. Urbana: University of Illinois Press.

Wolniewicz, Richard. 1997. Comparative Ethnic Church Architecture. *Polish American Studies* 54: 53–73.

Woodard, Colin. 2011. *American Nations: A History of the Eleven Rival Regional Cultures of North America*. New York: Viking.

Zelinsky, Wilbur. 2001. *The Enigma of Ethnicity: Another American Dilemma*. Iowa City: University of Iowa Press.

——. 2011. *Not Yet a Placeless Land: Tracking an Evolving American Geography*. Amherst: University of Massachusetts Press.

2 Native American Landscapes in the Eastern United States

The cultural landscape of Native Americans within much of the eastern United States is best represented by their historical legacy rather than by vibrant contemporary Native American communities. While a few ethnic islands of American Indian populations do exist in the Atlantic coast states, the landscape legacy of Native Americans is best seen in the physical remnants of their prehistoric cultural landscapes. Although small isolated population centers have survived, such as of the Seminoles in the Florida Everglades, the Cherokee in the foothills of the Smoky Mountains, the Lumbees in coastal North Carolina, and the Mohawk and Seneca in upstate New York, among others, most of these groups have become highly acculturated by their long contact with European- and African-Americans.

Much of what is seen today in the small Indian reservations and enclaves in the eastern United States is no different from what one sees among their white or black neighbors. Nevertheless, recent reaffirmation of Native American treaty rights that granted sovereignty has not only begun to alleviate the poverty found in so many American Indian communities through the opening of tribal casinos and the reacquisition of some of their former lands, but it has strengthened their resolve and capacity to maintain their heritage. Indeed, some groups have reinstituted regular powwows in an effort to revitalize their cultural legacies and the casinos have brought many Americans into greater contact with Native American culture.

2.1 Native American Legacy of the Past

The Native American population of North America north of the contemporary Mexican border has been estimated around four million at the time of first contact with Europeans over five centuries ago (Denevan 1976). Their numbers in Mexico and the West Indies may have been at least ten times greater. Indeed, recent estimates of a pre-contact population of around 100 million throughout the Americas have been made (Mann 2006). Through their widespread use of fire, hunting and gathering, and agricultural practices, Native Americans had extensively shaped the pre-Columbian distribution of plants and animals (Denevan 1992, 2011; Cronon 2003), as well as created numerous settlements. Widespread and rapid Indian depopulation resulted from the European contact, such that by the time that many European colonists arrived, the Indian groups they encountered represented mere vestiges of a far more prominent recent past.

2.1.1 Disease and Depopulation

Disease often preceded the arrival of Europeans, as is clearly documented by the observations of the Spaniards led by Hernando de Soto, who journeyed from Tampa Bay across Florida, Georgia, into the Carolinas, and then Tennessee, Alabama, Mississippi, Arkansas, eastern Texas and Louisiana between 1539 and 1543. This group traveled through many areas that less than a century earlier had far larger populations, yet few remained at the sites of the former villages at the time of their journey. Some major centers, such as Etowah (in present Georgia) and Moundville (in Alabama), were still occupied but were in decline, and their populations collapsed shortly thereafter. Later European settlers found large unoccupied expanses to settle in parts of the American southeast. There was evidence that smallpox epidemics had already occurred in North America, having been spread inland by Native Americans who came in contact with earlier explorers (Smith 1989). For example, at the time of Cabeza de Vaca's exploration of the Texas Coast in 1528, an epidemic of smallpox was being spread from Indian to Indian well in advance of the European explorers who had first brought the disease. The epidemic infection rate among Native Americans exceeded 80%, but it affected far smaller numbers of the Europeans, who had greater natural immunities to this disease (Saunt 2004). Similarly, smallpox killed about 90% of the Massachusetts coastal Indians between 1617 and 1619, depopulating the area that the Pilgrims chose to settle (Thornton 1987). Having descended from a small initial

J.A. Cross, *Ethnic Landscapes of America*, DOI 10.1007/978-3-319-54009-2_2

population, Native Americans, as a group, possessed a very restricted number of human leukocyte antigens, rendering them far more vulnerable to the pandemic spread of many infectious diseases than Europeans, Africans, and Asians, who were far less vulnerable because of the much greater genetic diversity within their populations (Mann 2006).

Besides smallpox, other major epidemic diseases, including plague, tuberculosis, measles, and typhus, as well as alcoholism, which increased the vulnerability of Native Americans to a variety of diseases, exposure, and abuse by white settlers and traders, also contributed to high loss of life. As Charles Mann (2006:123) explains, "Disease turned whole societies to ash." In addition, warfare, in particular numerous conflicts between Indians and European settlers (Thornton 1987), reduced the numbers of those who managed to survive the white men's diseases. Disproportional losses from diseases by the tribes, as well as advantages that certain tribes had in obtaining firearms over others, also led to internal conflict among the various Indian tribes.

Warfare often led to genocide. Early examples of conflict include the Pequot War of 1637 in the Connecticut Valley and the Governor Kieft War of 1640–45 between the Dutch and Native Americans around New Amsterdam. During King Philip's War of 1675–76 between English settlers and the Narragansett in Rhode Island, 3000 Narragansett were killed (Washburn 1978). In total, one-quarter of all New England Indians died between 1620 and 1720 because of wars with Europeans. White settlers often massacred Indian villagers; indeed that was the cause of Governor Kieft's War. At times the Europeans deliberately introduced smallpox.

Many other examples of the calamitous consequences of warfare upon the Native American population can be cited for other regions of North America and during later centuries. It is also noteworthy that the attitude of English settlers towards Indians differed from that of the French and Spanish. The French saw the Indians as business collaborators and lived and worked among the Indians, frequently taking Indian women as their wives, resulting in a large population of mixed-race Metis. The Spanish were initially guilty of genocide on a grand scale, chronicled by Bartolome de las Casas (1992) in his *Devastation of the Indies*, first published in 1552. Indians were largely eliminated from the West Indies by the Spanish, with the number of deaths on Hispaniola alone exceeding one million within a decade of Columbus's arrival. The Spanish emphasis then changed to conversion of the Indians to Roman Catholicism and their enslavement in New Spain, centered upon Mexico. Indians were also enslaved in what became the eastern United States, with tens of thousands of Indians captured in the Carolinas as slaves during 1600s and 1700s (Saunt 2004). These Indian slaves were either put to work locally or sold to planters in the West Indies to replace indigenous Indian slaves who had perished.

The landscape consequences of this depopulation of the American countryside were significant. Such demographic collapse "led to widespread settlement discontinuity," as Karl Butzer (1992:352) explains. He notes,

> To grasp the implications of such discontinuity, one must imagine what almost total depopulation would mean…silent villages, decaying cities, fields lying waste…. Such desolation set the stage for landscape transformation, within a new colonial context (Butzer 1992:352).

Nevertheless, much remained visible on the lands taken by the European settlers to remind them that vibrant Indian communities had once flourished centuries earlier. These reminders are far more monumental than the arrowheads and potsherds that can be frequently spotted on freshly plowed fields.

2.2 Prehistoric American Indian Landscapes

Prehistoric Indian groups constructed hundreds of thousands of effigy, temple, and burial mounds throughout much of the area of the Mississippi and Ohio River Valleys and the southeastern United States (Denevan 1992). Although time and erosion have obscured many mounds and modern agricultural plowing and urban development have obliterated a significant number of those that were visible to the first white farmers, many mounds—and even a few distinctive Indian fields—still remain. The earthen mounds that were built are of many sizes and shapes. Some are simple conical dots, while others form long lines. Effigy mounds portray a variety of real and mythical figures. Mounds range from small piles of soil to towering structures. The largest rise over a hundred feet in height, and some of the earthworks are nearly a mile long.

Without doubt, the earthworks left by prehistoric Americans are a prominent landscape feature. Although many of the Native American earthworks were abandoned in pre-Columbian times, "they constituted a conspicuous element of the landscape of 1492 and some are still prominent" (Denevan 1992:377). Several distinct prehistoric cultures created the numerous earthworks that remain today. The Adena culture, centered upon the Ohio River Valley, and the roughly contemporaneous and more widespread Hopewell culture, whose influence extended from the Great Lakes to the Gulf of Mexico, were responsible for building burial mounds and other earthworks between roughly 800 BC and 600 AD. The Mississippian culture, whose core was focused upon the central Mississippi River Valley, including those areas where it was joined by the Missouri, Arkansas, and Ohio Rivers, arose later, reaching its greatest development in the fourteenth and fifteenth century. Mississippian village sites, known today for their impressive earthen temple mounds, "are largely confined to the major river valleys of the area where fertile bottom lands, oft-flooded and renewed, made intensive horticulture possible" (Jennings 1968:216). Not only are many Mississippian earthworks still visible on the cultural landscape today, but several early Spanish explorers and French traders wrote accounts of their visits to then-inhabited Mississippian villages (Bailey 2004).

2.2.1 Mississippian Temple and Platform Mounds

Among the greatest concentrations of monumental mounds (Fig. 2.1) is the complex at Cahokia, near East St. Louis, Illinois, where over a 100 mounds are spread across an area of six square miles (Fowler and Hall 1978). The largest mound, now named Monk's Mound, measures 1080 feet by 710 feet and rises in four terraces to a height of 100 feet. The base of Monk's Mound covers 14 acres, and at one time a large wooden building sat on the top terrace surrounded by a wall. In the vicinity are numerous other mounds, built to accommodate temples and council houses, as well as others for signal and burial purposes, positioned around six major plazas. The largest mounds have been preserved within Illinois' Cahokia State Park, yet a large complex of 27 mounds in nearby St. Louis, Missouri was destroyed in the mid-1800s. The Cahokia settlement was occupied between 800 and 1500 AD and reached its peak around 1100. It is representative of Mississippian or Temple Mound culture and is estimated to have once supported a population of 15,000–30,000. Indeed, "Cahokia was the largest settlement north of the Rio Grande until surpassed by New York City in 1775" (Denevan 1992:377).

Earthen platform or temple mounds similar to those at Cahokia are still visible at numerous other widely spread locations. To the north at Aztalan State Park in southern Wisconsin are three large mounds, including one with two tiers, located on a site that once had pallisaded walls (Fig. 2.2). Archaeologists speculate that this site, which was contemporary with Cahokia, was possibly an outpost or colony of it, where Mississippian and Late Woodland populations intermingled (Birmingham 2010). To the south, Emerald Mound, the second largest ceremonial mound within the United States, is located a short distance northeast of Natchez, Mississippi. Covering eight acres (Fig. 2.3), this Mississippian mound complex was occupied from around 1250 until 1600, approximately the time the first European explorers entered the area. Atop the bluff facing the Tennessee River in Shiloh National Military Park are six flat-topped temple mounds plus a burial mound, the largest and best preserved mound site within the Tennessee Valley. The 29 mounds that are preserved at Moundville Archaeological Park on a bluff along the Black Warrior River in Alabama mark what "was one of the premier Mississippian civic-ceremonial centers in the Southeast" (Brown 2004:581). In northern Georgia at Etowah Mounds State Park are several large mounds (Fig. 2.4), the largest which measures 360 by 280 feet and rises 67 feet in height. With a truncated platform and three levels, it also

Fig. 2.1 Monk's Mound rises one-hundred feet above the surrounding countryside in the American Bottoms area northeast of East St. Louis, Illinois. Over 100 Indian mounds besides Monk's Mound are preserved in Cahokia State Park

Fig. 2.2 Two terraces are clearly visible on the largest platform mound preserved in Aztalan State Park, northwest of Watertown, Wisconsin. A reconstructed wooden palisade surrounds a portion of the mound

Fig. 2.3 Emerald Mound covers eight acres, with a smaller upper mound atop the large plateau-like surface that comprises most of the Indian mound. Occupied until around 1600, this mound is located a mile from the Natchez Trace Parkway to the northeast of Natchez, Mississippi

represents the Mississippian culture. Farther south additional mounds are preserved in Ocmulgee National Monument, near Macon, Georgia; at Kolomoki Mounds Historic Park in southwestern Georgia (Fig. 2.5), which date to around 1100 years ago; and at Lake Jackson Mounds State Park and Letchworth-Love Mounds State Park, both near Tallahassee, Florida, that represent the Weeden Island culture, a early local variant of the Mississippian culture. To the west, impressive

Fig. 2.4 View of smaller truncated pyramid mound from earthen ramp leading to summit terrace of larger mound in Etowah Mounds State Park, Georgia

Fig. 2.5 Fifty-six foot high great temple mound within Kolomoki Mounds Historic Park near Blakely, Georgia. This large flat topped mound was erected between 1400 and 1700 years ago

pyramidal earthen temple mounds are found at Toltec State Park (Fig. 2.6) and Parkman State Park, both in Arkansas, just to name a few of the numerous Mississippian and the slightly earlier Woodland archaeological sites.

Throughout much of the eastern interior Southeast United States, the "similarities in the layout of Mississippian mound centers—mound summit architecture, contrast between mortuary and nonmortuary uses

Fig. 2.6 The two pyramid shaped earthen mounds were constructed near a former channel of the Arkansas River between the eighteenth and eleventh centuries. An earthen wall, portions which remain, surrounds this site, preserved in Toltec Mounds Archaeological State Park, southeast of Little Rock, Arkansas

of mounds, mound-plaza arrangement, etc.—indicate that most were constructed according to general, widely shared plans" (Hally and Mainfort 2004:280). The environmental settings of many of the Mississippian sites are also similar—a location along a river terrace of a floodplain in close proximity to the river (Griffin 1978). What distinguishes some of these sites, such as Moundville and Etowah, is the complexity of the mounds and their spatial arrangements. Far more typical are the sites of smaller Indian villages, marked by a single small pyramidal mound.

There is more to the sites than just the mounds. Angel Mounds State Historic Site near Evansville, Indiana preserves a complex that is described by James Griffen, a prominent archaeologist, as

> a classic example of a major Middle Mississippi fortified town with three platform mounds plus three to eight other mounds, a town square or plaza, house remains with evidence of much rebuilding and relocation of successive structures, burials…and evidence of agriculture and extensive exploitation of the rich floral and faunal food resources (Griffin 1978:548).

It was once surrounded by a stockade that was nearly one mile long, and its mound today covers an area of 644 by 350 feet and is 28 feet high. Yet, the mounds are the most visible landscape legacy of the prehistoric Indian village.

2.2.2 Woodland Complex or Effigy Mound Culture

Generally coincident with the time Cahokia's temple and platform mounds were built, or slightly earlier, thousands of elaborate effigy mounds were constructed along lakes and rivers within the Upper Midwest. Within Wisconsin alone over 15,000 mounds were created, of which about one-sixth were actual effigies (Birmingham and Eisenberg 2000). Built by those peoples who archaeologists classify as being of the Woodland Complex or Effigy Mound Culture, which was a late derivative of the Hopewell culture (Jennings 1968), these relatively low mounds are at times hundreds of feet long. Although best known for effigy mounds representing various animals, mounds in low conical shapes or linear arrangements (Fig. 2.7) are far more common.

Archaeologists note that the Woodland Complex or Effigy Mound Culture produced a large number of mound groups consisting of low earthen profiles, rising several feet above ground level, of a wide variety of animals, birds, and reptiles. Furthermore, these effigy mounds of animals "are usually found along a ridge crest with the feet, naturally, pointing downhill" (Spencer and Jennings 1965:71). The areal extent of these mounds varies widely. Panther mounds may be ten feet wide with a height of no more than four feet, yet with

Fig. 2.7 A line of conical earthen mounds extends towards the bluff overlooking the Mississippi River in Effigy Mounds National Monument, located near McGregor, Iowa. Hundreds of additional mounds, in shapes of animals, dots, and lines, are found on the bluffs on both sides of the Mississippi River

Fig. 2.8 Wingspan of several hundred feet and body of bird effigy mound on grounds of Mendota State Hospital near Madison, Wisconsin. The largest bird effigy at that site has a wingspan of 624 feet

their lengthy tails they may be well over a hundred feet long. Others are far smaller, yet some are gigantic. Indeed, on the grounds of Mendota State Hospital near Madison, Wisconsin is a bird effigy mound (Fig. 2.8) with a wingspan of 624 feet (Birmingham 2010). In the Madison area 1200 individual effigy mounds were built in a four lake area between Lake Mendota and Lake Kegonsa, clustered "in at least 160 different groups" (Birmingham 2010:106). Although many of the low effigy mounds were destroyed by agricultural activity, their locations can sometimes be identified by noticing differences in the soil. Rather than simply mounding the adjacent soil into the shapes of animals, the Indians typically hauled in sands to build their mounds, enhancing the distinctiveness of their effigies.

One of the best locations to see these mounds is Effigy Mounds National Monument in northeastern Iowa, just across the Mississippi River from Wisconsin. Atop the bluffs rising several hundred feet above the Mississippi River floodplain are 191 mounds, 29 of which are in the form of bears or birds. The Great Bear Mound measures 70 feet from its shoulders to its forelegs, is 137 feet long, and is over three feet high (Lenzendorf 2000). Elsewhere in the park are a series of conical mounds arranged in a line, as well as long linear mounds. Similar mounds are found across the Mississippi River in Wisconsin's Wyalusing State Park, atop the bluffs facing the river. Numerous comparable effigy mounds are found at other sites throughout eastern Iowa, northwestern Illinois, and southern Wisconsin (Fig. 2.9). Near Baraboo, Wisconsin a mound in the shape of a human that is over 200 feet long is preserved in a county park, while an unusual shallow intaglio mound of a mythical water panther is found in Fort Atkinson, Wisconsin.

One of the largest effigy mounds is Serpent Mound (Fig. 2.10), located in south central Ohio. Constructed by the Fort Ancient Culture, a Mississippian group who occupied Ohio at about the same time that effigy mounds were being constructed along the Mississippi River between Iowa and Wisconsin, Serpent Mound is 1330 feet long and about three feet high. Located atop a bluff, the sinuous mound represents an uncoiling snake, located near several much older burial mounds. While Ohio has many other mounds, it has few effigy mounds.

2.2.3 Hopewell Period Earthworks and Mounds

Nearly a 1000 years before the Mississippian peoples constructed their mounds at Cahokia and elsewhere throughout the Ohio and Mississippi Valleys, people of the Adena and Hopewell cultures built a wide variety of earthworks (Fig. 2.11). Many of the finest remaining examples of their cultural landscapes can be found in Ohio, where over a hundred Hopewell sites have been identified. One of the best concentrations is found in Hopewell Culture National Historical Park, formerly Mound City Group National Monument (Fig. 2.12), located outside of Chillicothe, Ohio. This village and burial site was occupied about 0–200 AD. Although 23 mounds are found in an area of 13 acres that is surrounded by a low earthen wall, other earthworks and causeways connected a four square mile area.

Fig. 2.9 Part of an effigy mound in shape of a panther preserved at Lizard Mound County Park northeast of West Bend, Wisconsin. Note the legs on both sides of the sign, and the tail extending out of view to the left. Two linear mounds are visible in the distance

Fig. 2.10 Serpent Mound displays the sinuous winding body of a snake effigy, which leads to a tightly coiled body to the left of the area shown in this photograph. Located at Serpent Mound State Historic Site in south central Ohio, the snake effigy is 1330 feet long

Fig. 2.11 A large ditch and earthen wall surround the Great Mound in Mounds State Park near Anderson, Indiana. This earthwork, which was built by Adena peoples approximately 2250 years ago, surrounded a central platform which was reached by a gateway through the earthen embankment. Archaeologists think that dips in the embankment were aligned to signify astronomical events

Fig. 2.12 Low earthen walls surround large areas of small mounds and other earthworks in Hopewell Culture National Historical Park, located near Chillicothe, Ohio

The largest assemblage of Hopewell earthworks is located in Newark, Ohio. While the complex once covered over four square miles, unfortunately farming and urban expansion obliterated much of the remains. As the archaeologist Bradley Lepper notes,

> All that remains of the wondrously labyrinthine geometry are the Great Circle, the Octagon attached to its somewhat less great circle, a small fragment of the principal square, and a few additional shreds and patches of earthworks in front yards and scattered wood lots (Lepper 2004:73).

Nevertheless, the surviving Great Circle has a diameter of 1200 feet. The earthen walls forming the geometric figures rise from five to fourteen feet and "reflect a remarkably sophisticated understanding of geometry" (Lepper 2004: 75). Not only are there clear geometric relationships between the sizes of the various earthworks, but also with their distances to other features. Furthermore, some of the earthworks demonstrate celestial alignments.

Although lacking the huge temple mounds constructed at Cahokia and Etowah, the Hopewell earthworks are remarkable for their areal extent. Richard Townsend (2004:18) explains that Hopewell sites included "an impressive series of open circles, hexagons, and squares, defined by high earthen berms, some covering up to 50 acres (20 ha) and linked by long processional ways." Furthermore, "Hopewell people also elaborated upon funerary architecture for the priestly or highborn, with 70-foot earth mounds built over the incinerated remains of elaborate log tombs" (Nabokov and Easton 1989:101). Besides those described in Ohio, Hopewell period mounds are also found in Iowa at Effigy Mounds National Monument, although the mounds for which that site is most famous were erected over a millennia later by another people.

2.2.4 Poverty Point Culture

One of the oldest assemblages of elaborate earthworks in what is now the United States is found at the Poverty Point National Monument in northeastern Louisiana. These earthworks, whose construction occurred between 1730 and 1350 BC, include six concentric ridges that are separated by five outward radiating aisles (Gibson 1999). These ridges, which have now eroded to very gentle swales rising only several feet, enclose a plaza that is 1800 feet across. Behind the ridges and opposite the plaza is a large earthen mound, with its center being over 70 feet high. "Thought to represent a flying bird" (Gibson 1999:11), this mound had a wingspan stretching 640 feet, while its tail, which was pointed towards an aisle bisecting the ridges, extended 710 feet from the head. Several additional mounds are found nearby. While archaeologists remain uncertain about the exact role of the Poverty Point culture in the development of other mound building cultures, its residents received trade goods from throughout the Mississippi River Valley.

Even earlier mounds were erected about 100 miles farther west in Louisiana, overlooking the Ouachita River.

These earthworks, dated to about 5400 years ago, consist of "a ring of eleven irregularly sized mounds…[with] the biggest…as tall as a two-story house" (Mann 2006:286).

2.2.5 Burial Mounds and Middens

2.2.5.1 Burial Mounds

Burial mounds are typically smaller than the ceremonial and effigy mounds, although a times burials are also associated with the more elaborate mounds. As Knight (2004:739) explains, "Dome-shaped burial mounds of earth constitute a ritual tradition of long standing in the Southeast." Built during a period of over two millennia and extending into the historic period, "[b]urial mounds are found individually or in small groups, either at the margins of small settlements or at a short remove from them" (Knight 2004:740). Typically, such mounds are fifty to one hundred feet across, rising five to twenty feet above the countryside. Yet there is considerable regional variation in their construction—some represent a single ceremonial burial or reburial of existing graves, while others grew in size by continuous use for new burials. Such burial mounds are far less geometrical in design than the rectangular platform mounds so typical of Mississippian villages.

Burial mounds are found at literally thousands of sites throughout the eastern United States, and some have been so eroded in the centuries since their construction that their existence is only recognized today when excavations uncover skeletal remains. Unlike the cavalier treatment that such bones often received in the past, under the Native American Graves Protection and Repatriation Act, enacted in 1990, such remains are now to be carefully and respectfully reburied.

2.2.5.2 Shell Middens

Numerous mounds of garbage were created by prehistoric American Indians in addition to the thousands of temple mounds that are scattered throughout the Mississippi Valley and areas east to the Atlantic Ocean, extending from Canada to Florida. Although coastal middens are usually composed of oyster shells (Fig. 2.13), examples formed from conch shells exist along the west coast of the Florida Everglades.

These waste shell mounds surrounded the settlement sites of coastal settlers. Such middens once lined the Gulf and Atlantic Coasts, and many remain as the highest land near the coast. They may have served as hurricane refuges. A good example is Turtle Mound along the Atlantic Ocean north of Cape Canaveral. This twin peaked mound, rising to a height of 49 feet, is preserved in Florida's Canaveral National Seashore. Another conspicuous mound is located on Marco Island, Florida.

Oyster shell mounds were once far more common along southern coasts. However, many were destroyed for use as road gravel, given the lack of easily accessible local quarries

Fig. 2.13 Exposure of oyster shells along road cut through shell midden in Shell Mound County Park, located along the Gulf of Mexico north of Cedar Key, Florida

for stone when highways were first built in the early Twentieth Century. Earlier mining of shells from the mounds provided lime and tabby used in the construction of colonial buildings.

2.3 Agricultural Legacy of Native Americans

The agricultural practices of the Native Americans left a profound and lasting mark on the American landscape. At the time the Pilgrims first settled Massachusetts, they cultivated fields that had been recently abandoned by Native Americans and sought both advice and seeds from their indigenous neighbors. Because Native Americans lacked the plow, they used a digging stick, producing a field pattern that was distinctly different from that which would have been seen at that time in Europe, where draught animals were used to pull plows.

Although the passage of time and subsequent use of the land has long erased most of the evidence, archaeologists can still identify Indian field patterns and abandoned fields at certain locations. They have noted sites of raised and sunken fields, field walls, and water diversion and irrigation ditches. In Wisconsin 175 sites of Indian garden beds are known (Gartner 1997). The raised fields for corn are dotted with small conical mounds (rising about one foot) that were densely scattered across fields. In the Southwest, where Indian cultivation has continued to the present and the climate has aided in the preservation of prehistoric sites, field patterns are much more distinct, as we will see in Chap. 4.

One of the most conspicuous and important impacts of America's indigenous population upon the contemporary cultural landscape can be seen in the nation's farm fields. Beginning very early, European colonists adopted many Indian foods, resulting in a diet distinct from their practices in the Old World. The survival of the Pilgrims was dependent upon their quickly learning Indian practices of growing corn or maize. Corn was first domesticated in central Mexico, but it was growing throughout North America by the time Europeans arrived. Today, corn (maize) covers a greater acreage than any other crop grown in the United States, with the 2012 Census of Agriculture showing that over 87 million acres of corn are harvested for grain annually (Fig. 2.14). Nearly eight million additional acres are devoted to growing corn silage, sweet corn, and popcorn. By comparison, wheat, a staple crop brought to America by the Europeans, was harvested from 49 million acres in 2012 (NASS 2014a).

Numerous foodstuffs were native to the Americas, with prominent types varying regionally. Beans were grown in the southwestern and southeastern portion of what became the United States. Squash and pumpkins were grown in the same area, and thus to this day corn, beans, and squash are considered the traditional mainstay of the diet of Mexican-Americans and those Native American groups of the Southwest. Melons and sunflower seeds were grown in the Southeast, a region where sweet potatoes grew in the wild and where they were gathered. Wild rice was the staple crop for the Ojibwa, and it is still very important to Native American groups near the Upper Great Lakes. Today's American diet includes many Indian dishes: succotash, a combination of corn and lima beans; cornbread; hominy, which is corn bleached with lye; and tamales and tortillas, among others (Spencer and Jennings 1965).

Tree crops were also important. Pecans were gathered by Native Americans in the forests of Texas and the Southeast, and persimmons grew wild in the Southeast. Because of the existence of groves of pecan trees, "there is some suspicion that aboriginals in the Midwest and the South cultivated pecan" (Doolittle 2000:66). Likewise, Native Americans may have planted the Kentucky coffee tree and the Indian currant, among other trees and shrubs (Doolittle 2000). In the Northeast Native Americans tapped the sugar maple, a practice then learned by European settlers. Throughout much of the region east of the Mississippi River sassafras was gathered, providing a type of native tea. Acorns were collected and processed into a type of gruel in coastal California.

Tobacco was used by all Native Americans except the Inuit (or Eskimo) at the time of European contact. During many of its early years of colonial settlement, Virginia's primary export was tobacco. Cotton was first domesticated in the American Southwest, but it was grown for its fibers throughout much of the South. Indigo, a dye crop that was later produced on plantations in South Carolina, was first used by Native Americans. Turkeys were native to much of the United States, but they had been first domesticated by Indians in the Southwest.

Other crops were grown in the Americas south of the current United States border, yet they have subsequently become significant features in the American diet. These include the potato, first domesticated by the indigenous population in the Andes Highlands, and chili peppers, which were Mexican in their origin, but brought north into the American Southwest following Spanish settlement.

Several tropical crops first domesticated by Native Americans are excluded from most fields of the United States for climatic reasons, but they are consumed in great quantities. These include manioc (cassava); the pineapple which originated in Mexico; cacao, which was first used to make chocolate in Mexico; coca leaves, grown in the Andes and used to make cocaine as well as to provide flavoring in certain soft drinks; curare which came from the Amazon and is used in anesthetics; and cinchona tree bark, grown in Peru and from which quinine is now obtained.

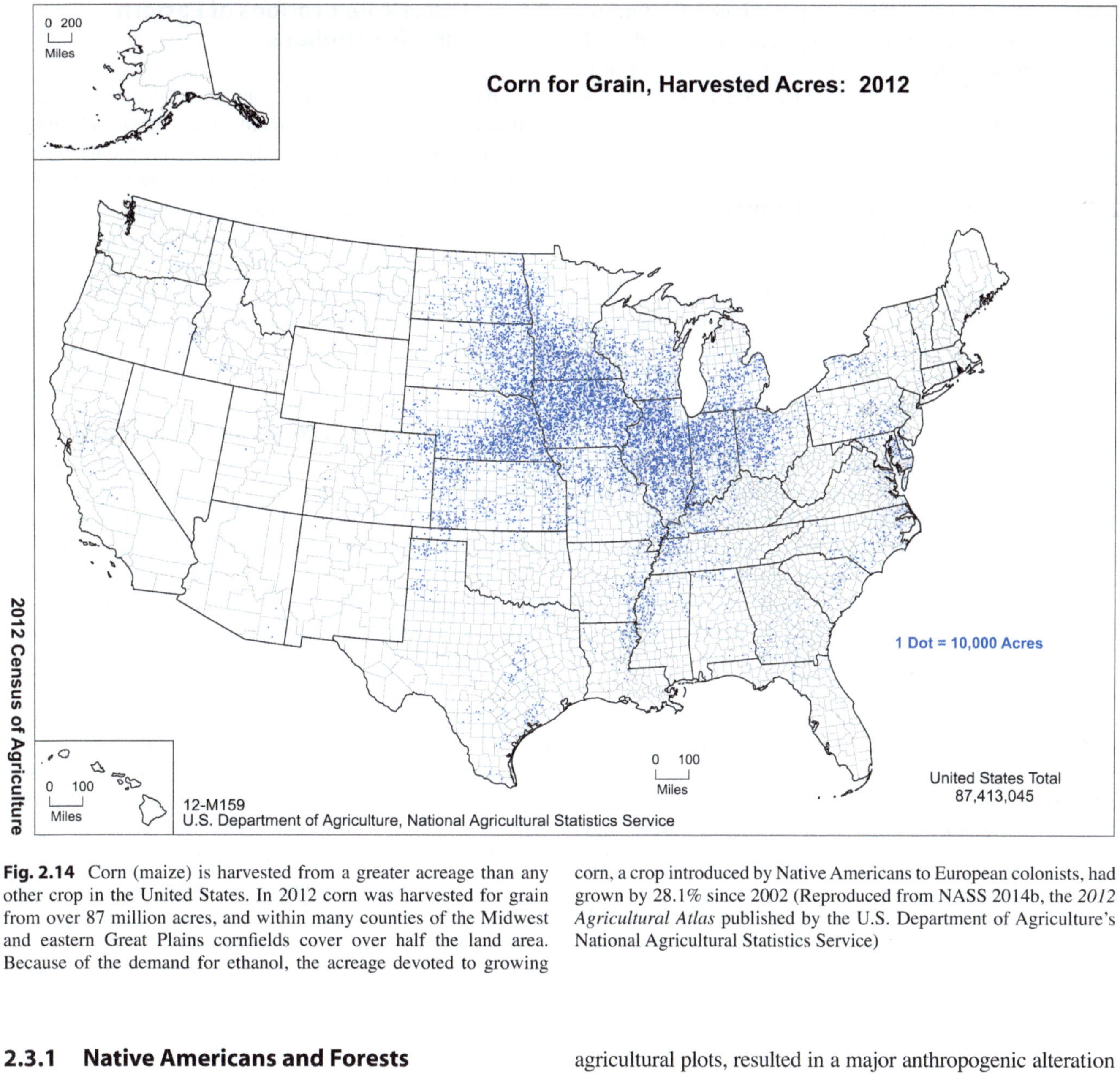

Fig. 2.14 Corn (maize) is harvested from a greater acreage than any other crop in the United States. In 2012 corn was harvested for grain from over 87 million acres, and within many counties of the Midwest and eastern Great Plains cornfields cover over half the land area. Because of the demand for ethanol, the acreage devoted to growing corn, a crop introduced by Native Americans to European colonists, had grown by 28.1% since 2002 (Reproduced from NASS 2014b, the *2012 Agricultural Atlas* published by the U.S. Department of Agriculture's National Agricultural Statistics Service)

2.3.1 Native Americans and Forests

Burning the surrounding forest twice annually, as occurred around Native American villages in southern New England, "promoted the mosaic quality of New England ecosystems, creating forests in many different states of ecological succession" (Cronon 2003:51). This use of fire not only enhanced the habitat for game, but permitted certain tree species to become more abundant, and provided favorable conditions for growing berries. Because of the Indian practice of favoring certain plants that provided sustenance, their seeds were often disproportionately dispersed near the locations of their settlements. In some cases, trees such as plums, the crabapple, and papaw and grape vines were deliberately planted by prehistoric Indians. These actions and the clearing of fresh agricultural plots, resulted in a major anthropogenic alteration of forest composition.

America's forest composition clearly reflects burning by Indians. William Denevan writes,

> In North America, burning not only maintained open forest and small meadows but also encouraged fire-tolerant and sun-loving species.... The longleaf, slash pine, and scrub oak forests of the Southeast are almost certainly an anthropogenic subclimax created originally by Indian burning, and maintained in part by fires set by subsequent farmers and woodlot owners. Lightning fires can account for some fire-climax vegetation, but Indian burning would have extended and maintained such vegetation (Denevan 1992:372, Reprinted with permission of American Association of Geographers).

The Southern Pine Forest represents a fire climax forest. Without the frequent burning that began with Native

Americans, and which has been continued by forestry officials today, pines would be replaced by assorted hardwoods. Similarly, the Kentucky bluegrass prairie began to revert to forest with the arrival of the region's first white settlers and a reduction of burning.

2.4 Native American Transportation Routes and Toponyms

Native Americans left a strong imprint that can be seen on maps of the United States today. They influenced the location of roads, the siting of many communities, and the naming of numerous geographic features. Toponyms or place names were often determined by the earliest settlers of a region.

Trails of Indians often became roads and highways during later times. For example, the Chickasaw Trail became the Natchez Trace, running from Natchez, Mississippi to Nashville, Tennessee. Indians were the first to locate many important passes over mountain ranges. Much of the early transportation in the United States utilized waterways. Not only had Native Americans located the best portages, but European colonists adopted Indian canoes and kayaks for their own usage.

Many cities as well as smaller communities are located on the sites of earlier Indian settlements. In fact, many European settlements and trading posts were deliberately established at the site of, or near, Indian settlements. Numerous examples can be cited, including Green Bay, Wisconsin; Detroit, Michigan; and Pittsburgh, Pennsylvania, just to name a very few.

Place names throughout much of the United States clearly demonstrate a lasting legacy of our Native American forebears, even in locations where no Indians are to be found today. Indian names have long out-lasted Indian settlement in many locations. Almost half of the U.S. states have Indian names: Alabama, Alaska, Arkansas, Connecticut, Idaho, Illinois, Indiana, Iowa, Kansas, Massachusetts, Minnesota, Mississippi, Missouri, Nebraska, North and South Dakota, Ohio, Oklahoma, Oregon, Tennessee, Texas, Utah, and Wisconsin (Spencer and Jennings 1965). Thirteen percent of all U.S. counties have Indian names, with such diverse names as Alachua, Arapahoe, Cayuga, Kewaunee, Multnomah, Outagamie, Seneca, Tallahatchie, and Winnebago. Numerous rivers, both large and small, bear indigenous names. Prominent examples include the Mississippi, Missouri, Ohio, Potomac, Tennessee, and the Tombigbee. Thousands of communities have Indian names, including Apalachicola, Chicago, Miami, Milwaukee, Natchez, Oshkosh, Spokane, Tallahassee, and Wichita. Although the conquering English renamed Dutch New Amsterdam, they retained the Indian name for the island upon which the center of the city is located, Manhattan.

2.5 Historic Relocations of Eastern American Indians

The spatial pattern of Indian population within the Eastern United States today varies greatly from that which existed before European contact. Not only were many groups literally exterminated or so seriously reduced in numbers that their remnants only survived by merging with other tribes; but many groups were forcibly relocated from those lands coveted by European settlers. Furthermore, conflict among Native American groups prior to the arrival of the Europeans, and continuing conflict after their arrival, resulted in many groups occupying lands different from their ancestral homelands.

2.5.1 Five Civilized Tribes and Trail of Tears

Relocation of the Five Civilized Tribes to Oklahoma removed a major portion of the Cherokee, Creek, Choctaw, Chickasaw, and Seminole from the southeastern United States (Thornton 1987). These five tribes were among the most acculturated of Indians, living as sedentary agriculturalists in housing similar to that occupied by European-Americans. Indeed, a visit today to New Echota State Historic Site in northern Georgia reveals log and wood frame buildings in the former Creek capitol similar to what would have been found in a prosperous white community of the time. These Indians had a well-developed political system, but unfortunately they occupied farmlands desired by whites, a problem exacerbated during Georgia's gold rush. The Indian Removal Act of 1830 provided for moving the Eastern tribes to west of the Mississippi River and for the establishment of Indian Territory, which eventually became Oklahoma.

The journey west, undertaken at the height of winter, has been termed the Trail of Tears. President Andrew Jackson ordered the relocation in violation of a U.S. Supreme Court ruling in the Indians' favor, but with strong support by Georgia state officials. The Indians moved between 1836 and 1838, suffering a high mortality en route and in the resettlements. Indeed, the Choctaws suffered 6000 dead, 15% of those who started the trip, and the Creeks lost 3500, or one third who began. Among the Seminoles, half died (Thornton 1987). In Indian Territory (now Oklahoma), lands were distributed among the Indian Nations. Later demand for lands by white settlers prompted the Oklahoma land rush of 1890, with much of the land that had been promised to the Indians being made available to others.

2.5.2 The Oneida

Farther to the north, the Oneida, an Iroquoian group, who had traditionally lived in a stockaded village near Oneida Lake in central New York, were scattered during both the

French and Indian War and the Revolutionary War (Campisi 1978). The Oneida, who favored the U.S. in its conflict with Britain, saw their League split and their villages burned. Several decades later they experienced pressures from white Americans to acquire their lands, with the Treaty of Buffalo Creek in 1838 providing for the removal of all Iroquois from New York state. Most relocated to either Wisconsin or Ontario, with fewer than a thousand remaining in New York.

With the arrival of the Oneida in Wisconsin, in 1838 they had a land grant of 65,426 acres. In 1892 these lands were allotted to individual Indians in fee simple under the General Allotment Act. Subsequently, many of the Oneida lost their lands through tax sales and foreclosure. During the mid-1920s only a few hundred acres remained in Oneida hands (Janke 1987). By 1972 the Oneida owned 2200 acres, spread in a checker-board pattern intermixed with white ownership as land was repurchased for tribal use by the federal government under the Indian Reorganization Act (Campisi 1978). Later, the tribe began purchasing additional lands on its own, using revenues from gaming, and it is enrolling private Indian owned lands. By 2004, the Oneida had control of 16,689 acres, a figure that had grown to over 25,000 a decade later. The Oneida eventually want to control three-quarters of its original reservation lands (Rodewald 2015).

The nearby Stockbridge Indian Reservation in Wisconsin also illustrates the rupturing of the links of a Native American group to its lands. Their initial reservation was sold, with the tribe being relocated five times. Within Calumet County the legacy of the departed Indians includes the delineation of the defunct reservation boundaries on the U.S. Geological Survey's topographic maps, with the land surveyed in grants, not the normal township and range system used elsewhere in the county. Place names, including Stockbridge, Brotherton, and Quinney Road (named after their chief), reflect the prior Indian occupation of the lands, yet only a few parcels remain in Indian ownership. Sacred ties continue to the area, as Indian graves remain in local cemeteries, even though the original home of the Stockbridge was in western Massachusetts.

2.6 Contemporary Indian Reservations

Lands set aside by the federal or state governments for Native Americans form the Indian reservations of today (Fig. 2.15), yet there is a huge diversity in the arrangements that were made with the various indigenous groups. Some reservations have been abolished or altered since their establishment. Some tribes never received reservations, while some tribes have extinguished their special government recognition. The reservation status is governed by treaty, with not all tribes having the same arrangements. In 2015 there were 566 Federally recognized tribes or communities in the continental United States who have special legal status as Indians or Alaskan Natives (Federal Register 2015), yet this listing does not include those Indians "who were never formally recognized by the federal government or their descendents" (Bailey 2008:7).

Land tenure within reservations varies greatly, reflecting the degree to which European-Americans imposed their concepts of land ownership upon the Indians. Many Indian groups historically held land by clan or family and had different concepts of ownership than their European neighbors. In the reservations in the eastern United States, most tribal lands were allotted. Thus Indians could sell their lands and paid real estate taxes, as they had received title (fee simple ownership) to the land. Overall, very little tribally owned land remains, with the Indians owning a small minority of their reservations. For example, in Wisconsin less than 5% of the lands on the Oneida reservation were owned by the tribe or Indian owners at its lowest point. Yet, on the nearby Menominee Reservation, the lands are largely Indian owned, as the Menominee refused to cooperate in the allotment process (Spindler 1978).

In contrast, most western tribal lands remain intact. In general, Indian tribal lands are held in trust by the federal government; thus individuals cannot sell the lands they occupy, nor are the lands subject to taxes. Lands are allocated to specific Indians for their use by tribal authorities. Yet without clear title to the individual lands they occupy, Indians cannot easily obtain mortgages to build houses or establish businesses.

2.6.1 Sovereignty of Indian Nations

Sovereignty of the Indian Nations over their territory has become increasingly recognized over the past several decades. Nevertheless, legal jurisdiction within Indian reservations varies, controlled by treaty. In general Federal laws apply on a reservation, with major crimes being tried in Federal court. State laws are usually invalid, as are local (white community) laws. The tribe has authority over local rules. Indians on a reservation are subject to tribal police and federal authority. For many years whites on reservations were not subject to the tribal laws, but this is now changing (National Congress of American Indians 2015). Indeed, on highways entering certain reservations signs have been posted reminding visitors of the tribe's sovereignty (Fig. 2.16). Because Indian tribes have sovereignty over their lands, one can sometimes see striking differences in land use controls and zoning between reservation lands and adjacent lands subject to local municipal or county control. Likewise, business activities that are legal on one side of the reservation boundary may be illegal on the other side.

Treaty rights are held by certain Indian Nations for lands outside their current reservation boundaries. These rights arose when Indian tribes sold their lands, but retained certain

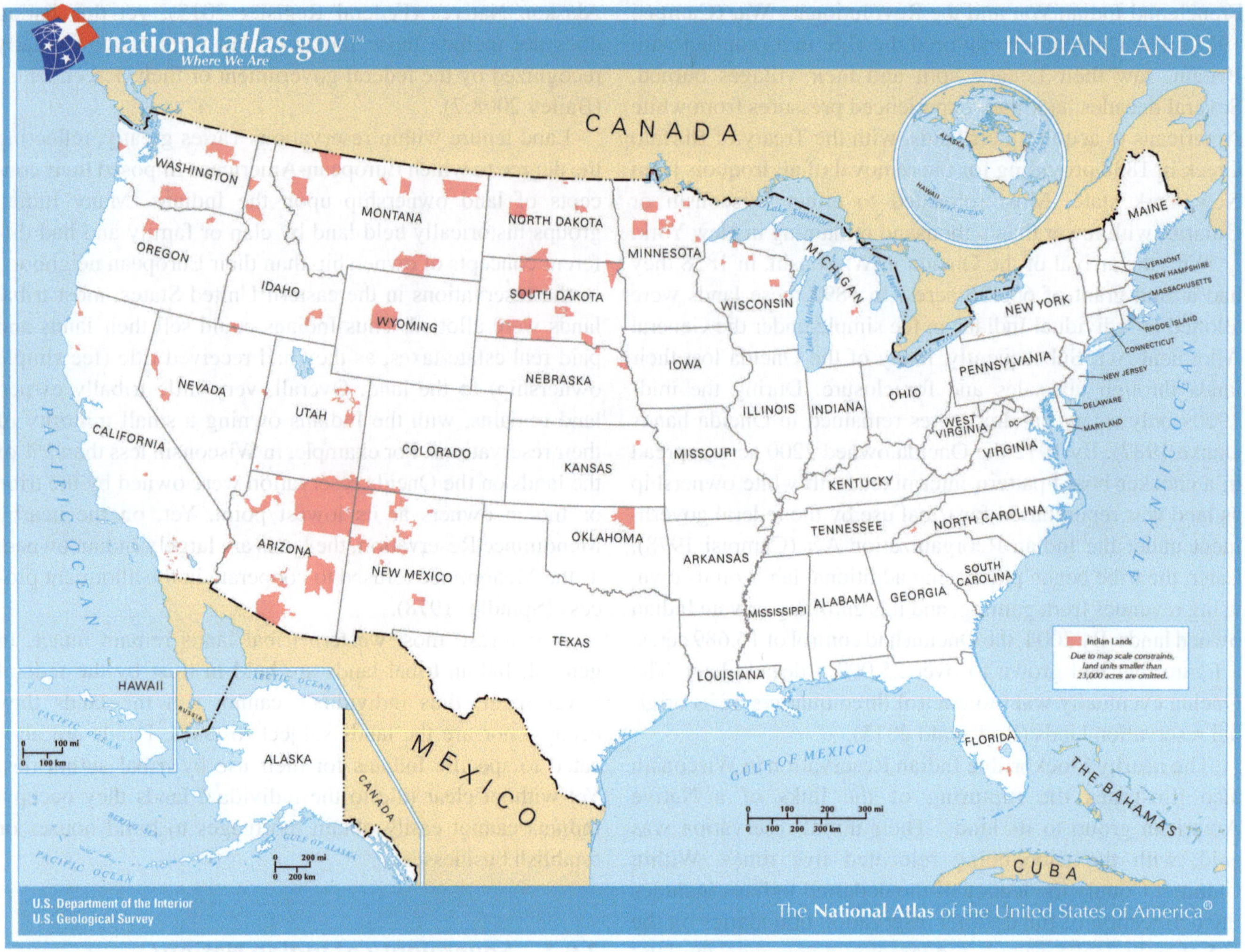

Fig. 2.15 Map showing the larger areas of Indian lands held as Federal Indian Reservations in the United States. Most Indian reservations in the United States are located west of the Mississippi River, with those in the eastern portion of the country far smaller than those located farther west. Because certain groups have altered their relationship with the Federal government and have transitioned from reservation status, the area covered by reservations in Oklahoma today is far less than it was over a century ago. Likewise, many reservations today are smaller than what they were at the time Indians were first moved onto reservations (Reproduced from map issued for the National Atlas by the U.S. Geological Survey (2004). Readers wanting to see all of the Indian Reservations shown on a larger-scale map should view the map published by U.S. Bureau of Indian Affairs in 1996 (https://www.nps.gov/nagpra/documents/RESERV.PDF))

off-reservation privileges, especially rights to hunt or fish in specific areas or rights to water. The ability to exercise these rights for many decades was often prevented by state or local authorities (Fig. 2.17). The courts are now increasingly recognizing the validity of treaty rights. A good example can be seen in the Chippewa Treaty Rights in Wisconsin. In the 1837 Treaty with the Chippewa, it was written:

> Article 5. The privilege of hunting, fishing, and gathering the wild rice, upon the lands, the rivers and the lakes included in the territory ceded, is guaranteed to the Indians, during the pleasure of the President of the United States (quoted in Satz 1991:156).

Such privileges are termed usufructuary rights, and can be considered as part of the payment for the lands. These rights were upheld by a Federal Court ruling in 1991, granting Indians the rights to fishing or spearing of fish and hunting on ceded lands that are publicly owned; but the Indians do not have these rights on private lands. Although controversial, the Indian harvest at the time of the court ruling was very small compared to the sport fishing harvest by the white majority. For example, in 1990 the Indians harvested 3842 deer, compared to 399,311 by non-Indians, and the Indians took 16,399 walleye, compared to the 623,525 taken by whites (Satz 1991:113,115). Because of the treaty rights, only Native Americans can legally spear fish—which is typically done at night, and the Lac du Flambeau promote this heritage not only via their name, but in the Lake of the Torches Resort and Casino that they operate.

Fig. 2.16 Sign along highway entering Santa Clara Indian Reservation in New Mexico informing visitors about tribal sovereignty

Fig. 2.17 Sign protesting Indian land claims near Powers Corner in Cayuga County, New York. Efforts to reclaim lands and regain rights has met with opposition in many parts of the United State

2.6.2 Indian Gaming and Gambling Casinos

Casino gambling and Indian gaming has become a very lucrative economic activity for many—but not all—Indian Nations. Such gaming enterprise has been termed "the new buffalo economy" (Johnson 1994). Total profits to all the Indian nations exceeded $1 billion in 1993, yet 5 years later the Foxwoods and Mohegan Sun casinos in Connecticut alone had $1.5 billion in gaming revenues. By 2005 Indian casinos earned $22.7 billion nationwide (Archibold 2006), an amount that peaked in 2007 before the Great Recession and slightly declined to $28 billion by 2013 (USGAO 2015). In 1999 104 of the nation's 557 recognized tribes had casinos, while 15 years later that number had grown to 240. Nationwide, Indian casinos and associated hotel or resort complexes provide employment, either directly or indirectly, to over 600,000 persons. Over two-fifths of the nation's casino gambling now takes place in Indian casinos.

Because Federal courts have ruled that states cannot outlaw casino gambling or gaming if the state permits any type of lottery, construction of huge casinos has proceeded on many reservations. Some Indian casinos are located on tribal lands off reservation, which is legal if the land was formerly tribal territory, although by 2006 permissions for new operations came under more severe restriction. Thus, Indian casinos can be found within Milwaukee, Wisconsin and Niagara Falls, New York, among several other major cities. However, while all but four of the states west of the Mississippi River (Missouri, Arkansas, Utah, and Hawaii) have Indian gaming, 17 eastern states lack them, partly in response to several states prohibiting all forms of gambling, while others lack viable Native American nations with historic links to the states that would enable them to establish casinos.

The Seminole Tribe in Florida led the way in the establishment of Indian gaming, beginning with its high stakes bingo hall in 1979. Although opposed by local authorities, Federal courts found that tribal sovereignty exempted the tribe from state regulation of their gaming. Operating five casinos by the beginning of the twenty-first century, the profits from gambling have transformed the tribe's economy. Not only did the casino profits provide over 95% of the tribe's budget, but it enabled new services to be provided to tribal members.

> Seminole casinos funded comprehensive tribal social services, from health clinics to tribal law enforcement, from the K-12 Ahfachkee School to new housing. Seminoles devoted large sums to cultural projects, such as the Ah-Tah-Thi-Ki Museum, language preservation programs, and cultural festivals (Sturtevant and Cattelino 2004:446).

By 2006, the Seminole's gambling revenues had risen to "approximately $600 million, and the tribal budget was over 95% funded by gaming" (Cattelino 2008:151). Similar accomplishments can be cited for many other Indian nations. Yet not all Indian nations have embraced gaming, nor have all who tried been successful.

Casinos have had revolutionary impacts upon the economies of many Indian communities, which historically have been far poorer than their white neighbors (Fig. 2.18). With the operation of the casinos, the tribes can provide significant employment to a population that has historically had high levels of unemployment, and the profits can be used for investment, to improve housing, to build new homes, to construct new tribal fire and police stations, to establish new tribal schools and colleges, and to develop museums and new

Fig. 2.18 Potowatomi Casino, just south of downtown Milwaukee, Wisconsin. The glass enclosed conical tower in the center is designed to evoke the Potowatomi's identification as the "Keepers of the Fire." Flags of the Potowatomi Nation, the United States, and Wisconsin fly outside the entrance to the casino, reminding the visitor of the tribe's sovereignty

powwows that showcase Native American culture. Many tribes have invested to improve other employment prospects for their members. For example, the Lakota Sioux are raising buffalo, while the Oneida in Wisconsin have been buying farms. Some tribes have provided cash grants to tribal members. While in a few rare cases these payments were huge, such as on the small Shakopee Mdewekantan Reservation in Minnesota where each tribal member received $1 million annually (Williams 2012), in most cases members receive far more modest payments, if any at all. Recent research indicates that those tribes that have been most successful in reducing poverty have not provided any payments to tribal members, but instead have utilized the profits to enhance social services, to promote economic development, or to provide employment (Economist 2015). Indeed, "in North Dakota most tribes considered their casinos successful because they created jobs for tribal members" (Cattelino 2008:151). Furthermore, some Indian tribes have rejected the notion of hosting gambling on their lands, while others are poorly located with respect to population centers that could provide customers. A few Indian casinos have closed because they were unprofitable. Nevertheless, with casino gambling there is hope that many reservations can rise above being landscapes of poverty, which was the norm during most of the twentieth century. Yet at the same time, questions have been raised about whether the profits from gambling that advance tribal self-determination might also result in cultural loss (Cattelino 2008).

2.6.3 Economic Development and Native American Culture

Profits from gaming have been used not only to enhance various social services that are provided to Native American groups, but to also finance investments in a variety of businesses. While these activities range from expanding resource development, including the forest products industry, to establishing manufacturing plants that employ tribal members, Native American cultural values often guides such development.

> Four values have been identified as enveloping the concept of Indian-centered economic values: community, connectedness, the seventh generation, and humility. Community encompasses the tribal idea of balance in the world and the idea that reciprocity must exist in all exchanges. Connectedness describes the interrelationship of all actions that must be accounted for to maintain balance. The…seventh generation involves a concept of stewardship to provide resources for the children yet to be born. The value of humility embraces the natural place of humans in the world as subject to the consequences of the development undertaken (Keen and Eaglewoman 2008:131, Reprinted with permission of Dept. of Anthropology, National Museum of Natural History).

Given these values, the routes to economic development within the Native American community may diverge from that taken by the majority of European-Americans. For example, the purchase in 2003 of the proposed Crandon mine site by the Sokaogon Chippewa of Wisconsin to protect the site from mine development is fully in keeping with several of these key Native American values.

Casino profits have also underwritten efforts to preserve and showcase Native American culture. With financing from their Foxwoods Resort and Casino, the largest casino in the Western Hemisphere (Sokolove 2012), the Mashantucket Pequot Museum and Research Center was erected in Connecticut. This museum, larger than the National Museum of the American Indian in Washington, D.C., included a huge diorama of a seventeenth century Pequot village, plus cultural archives (Schlossberg 2014). Many other tribes have also showcased their heritage, but far less extravagantly. Preservation of indigenous languages helps in the preservation of Native American culture, and efforts to record, transcribe, and teach tribal languages have been financed with casino profits throughout the nation. Such preservation is particularly critical with languages for which only a few elderly speakers remain, for with the loss of their language, a key instrument for identifying their culture and sovereignty is also lost (Onishi 2012).

Tribal corporations and Indian Nations are exempt from all Federal and state income taxes as well as property taxes on reservation lands. State and local regulations and labor regulations also do not apply to these entities because of tribal sovereignty, thus providing additional economic incentives for Native American manufacturing, retail, and tourism development. Thus, the array of businesses and their spatial arrangements (or lack thereof, depending upon whether tribal zoning regulations are in effect) within tribal jurisdictions may differ from those outside the reservation. This is readily apparent east of Massena, New York, where service stations selling gasoline exempt from New York state's fuel tax line the highway in the St. Regis Indian Reservation, just to name one example.

2.7 Native Americans in Eastern United States Today

Acculturation of those Native Americans remaining in the eastern United States has progressed to the point that culturally identifiable Native Americans are found in only scattered ethnic islands. Most persons of Indian ancestry became so highly assimilated that they "became virtually indistinguishable from their counterparts of Euro-American ancestry" (Porter 1992:120). Furthermore, as Finger and Perdue (2004:161) note, "Small, rural groups with no native language except English tended—in outward behavior, at least—to be much like their White and African-American neighbors." Acculturation took place both by adoption of cultural traits of the dominant population as well as by intermarriage.

Early intermixing was particularly strong in the Midlands (the Delaware River Valley), especially among persons of Finnish and Swedish ancestry. Mixing continued across the Appalachians, where Terry Jordan and Matti Kaups (1989) perceptively explain that the success of the European pioneers often depended upon their assimilation of Native American culture. Noting that interracial cohabitation frequently occurred and that half-breeds were numerous on the frontier, they summarized the consequences of frontier miscegenation:

> Considerable white blood entered the Indian population, to the extent that there are probably no pure Native Americans left today among the remnants of the eastern woodland tribes…. But the most significant outcome of miscegenation was the introduction of partial Indian ancestry into a substantial part of the Midland 'white' population…and left a permanent mark on the genetic and cultural makeup of Anglo-Americans (Jordan and Kaups 1989:87–8).

Thus, many Americans have Indian ancestry, yet it is so diluted by time and other ethnic contributions as to be virtually unidentifiable, and it is often unknown to the individual.

2.7.1 Native American Ethnic Islands

Native American ethnic islands exist in locations that are often considered marginal environments. For example, consider the Cherokee in North Carolina's Smoky Mountains and the Seminoles in the Florida Everglades. Because of their remoteness, isolation permitted the groups to maintain a distinctive culture, as isolation limited social interaction and intermarriage. Several types of Indian communities remain in the East Coast states (Porter 1992). These include reservation communities, which are mostly state, rather than federal. By the early 1990s only five federal Indian reservations existed along the east coast, two in Florida and one each in North Carolina, Maryland, and Maine, groups that survived by maintaining land possession. Most groups held only very small fragments of what lands they once held. An extreme example is in "Massachusetts [which] had only one reservation remaining in 1974, the Hassanamisco reservation in Grafton. It was, at that point, 11.9 acres, with only one resident" (Conkey et al. 1978:186).

Many Indian groups reside largely off-reservation. Folk communities developed as Indians adopted the Euro-American land tenure system, yet maintained a separate identity because of their isolation and strong kinship ties. Racial discrimination and segregation were important in maintaining the groups, yet within the eastern United States there was widespread mixing of the races, white, black, and Indians.

2.7.2 Mixed Ancestry

Racial intermixing resulted in several distinct groups. Some groups have significant non-Indian blood, but culturally identify themselves as being Indians. Certain groups primarily mixed with Europeans, others with Africans, while some intermarried with both. For example, the Seminoles evolved as Creeks moved among Timucuans in Florida and also mixed with escaped African-American slaves. The Lumbees are a grouping of remnant tribes in North Carolina including Croatans, who intermarried with black slaves and some whites (including those possibly from the lost Roanoke Island colony).

Other distinct groups maintain their identity, emphasizing neither their black nor Indian heritage (Berry 1978), and while their proportion of Indian blood varies, it may be small. Examples include the Sand Hill Indians (or "Pineys") in New Jersey, accepted as whites, but thought to be triracial, and the Melungeons in Tennessee who were described even in the 1700s as a mixture of Indians and Europeans. Writing about the Melungeons, Brewton Berry notes, that they are largely "undistinguishable economically, culturally, and physically from their White neighbors, although some do have brown complexions and straight black hair" (Berry 1978:292). Yet, they remain a distinct cultural group. The heritage of the Melungeons, as with many of these mixed groups, is obscure, yet remains sufficiently intriguing to warrant review in a major news magazine (Economist 2016). Some have made unproven claims of ancestry to a lost colony of Portuguese dating to the 1500s. Their origin is diverse, families coming from several roots, which may be a "product of a society diffusing westward from an eastern source" (Price 1951:267). Being seen as different, they suffered discrimination at the hands of others, yet their cultural landscape is not different from their neighbors.

Regardless of how the acculturation or mixing occurred, the vast majority of the Native American population in the eastern United States is of mixed racial ancestry. Indeed, pure-blooded Indians are extraordinarily rare in many of the eastern tribes. Membership in many tribes requires individuals to be one-fourth Indian, yet within the Cherokee, there is no blood quantum, merely the requirement that individuals prove that they are descended from a Cherokee who was listed over a century ago. Although the Cherokee blood is highly diluted, they are by far the largest group of Native Americans, with 819,105 Cherokees enumerated by the 2010 Census (Norris et al. 2012). They have promoted their heritage through cultural tourism. For example, the Eastern Cherokee in North Carolina have established a tribal museum, a historical recreation of a Cherokee village, and schedule musical performances in a gathering that "also reinforces Cherokee crafts, dances, and games in an annual event that economically helps the community" (Finger and Perdue 2004:160).

2.8 Contemporary American Indian Landscapes in the Eastern U.S.

The eastern United States provides few examples of contemporary landscapes that clearly illustrate Native American cultural traditions. Yet some elements of indigenous ethnicity do remain, and sovereignty has given many tribes the ability to create thriving casino developments where it otherwise could not be a part of the landscape.

2.8.1 Native American Casinos

The modern casinos have not only revolutionized the economy of many Native American nations, but their presence has altered the cultural landscape, both figuratively and literally. The casino generates revenue, permitting tribes to improve public housing—which at times reflects ethnic values in its design or spatial arrangements; to construct new community centers, schools, and tribal museums; and to build ever larger gaming halls and associated hotels. While some of the earlier casinos were not much more than warehouses filled with bingo tables or slot machines, other facilities clearly demonstrate Native American architectural influences, either in their overall design or in their decorative motifs. Although Carol Krinsky (1996:225) notes that "the nature-referential embellishments may reflect the Native American connection to Mother Earth, few claims are made for the 'Indianness' of most casinos," ethnic values are often associated with the selection of appropriate building materials, such as the erection of log structures, and the avoidance of construction on culturally sensitive ground. The Foxwoods Casino, operated by the Mashantucket Pequot Indians in Connecticut even before the construction of their huge museum, incorporated several features to illustrate the tribe's heritage. These included a room describing Pequot history, a replica wigwam, arrowhead and circle ornamentations, and cornices featuring "rectangular designs evoking wampum belts" (Krinsky 1996:227). Similarly, the Shakopee Mdewakanton Dakota Casino in Minnesota utilizes a dozen high-powered spotlights directed skyward to replicate the poles of a tepee, whose image rises several hundred feet over the casino at night. These, and other examples, display how Native Americans have used modern technologies and architecture to display elements of their ethnic heritage.

2.8.2 The Longhouse

The longhouse was intimately tied to the Iroquois religion (Tooker 1978). Longhouses, which archaeological excavations indicate were huge structures, were the "dominant symbol of social solidarity for Iroquois" (Nabokov and Easton 1989:76). The longhouse evolved out of dwelling of an extended family. It traditionally was a pole framed structure, with a vaulted roof of poles bent like a bow, and it was walled with bark. While Algonquins used birch bark, the Iroquois used elm bark in sheets typically six feet high by one foot wide. As the family grew (including in-laws and children) a new portion, 20–25 feet long was added, centered around another fire-pit, thus elongating the structure. These could run 100–400 feet long, housing about 150 persons. While nothing remains of these early structures, the design of the longhouse has been replicated in modern structures, some built for religious purposes, and others for community use (Fig. 2.19).

Longhouses today continue to have religious significance with the Code of Handsome Lake, but they are much smaller. They are typically a long, narrow wood frame buildings, with a fireplace or stove at or near each end, and having two rows of benches along both long walls. The buildings are otherwise modern in appearance and are the site of religious ceremonies and council meetings. Nearly four decades ago eleven were counted in New York state, and one in Oklahoma (Tooker 1978). Subsequently, new longhouses have been built in New York and elsewhere (Krinsky 1996).

2.8.3 The Wigwam

The wigwam was traditionally a one or two family house having a round or oblong floor plan built by woodland Indians. The word is from Algonquian, simply meaning "dwelling." It had a sapling frame of bent poles, lashed together, with horizontal connections covered with reed or grass mats, or sheets of birch, elm, or chestnut bark. The wigwam was typically seven to twenty feet in diameter and either elongate or round. Because the Indians were migratory, the frames were often left standing, with only the outer coverings being moved when the Indians migrated.

Variations in wigwam design existed throughout the Woodlands, the result of varying environments and preferences of different tribes. It was once the most common dwelling type except among the Iroquois. Nearly a century ago a geographer described this variation in wigwam design.

> The material used to cover the house was of considerable variety. In the Atlantic region the most popular covering was bark, which was readily available, both in large quantities and in large pieces, from a variety of trees–among them birch, elm, hickory, and ash. In areas where large sheets of bark could not readily be obtained, other materials were substituted. Thus, beyond the westward margin of the forested area, on the plains of Arkansas, houses of a similar framework of poles were thatched over with grass, as among the Wichita. Among the Ojibwa and Menominee, in the lake region of Minnesota, matting made of rushes was employed as covering material, being there more readily available than bark. It is noteworthy in a general way that as we go farther north the houses become simpler and simpler, until in the far north, among the Cree or the Salteaux, the structure is just a simple framework of light poles in the shape of a flat cone,

Fig. 2.19 Traditional Huron longhouse preserved at historical museum in St. Ignace, Michigan

> covered over in any fashion with sheets of bark, matting, or what not (Waterman 1924:4, Reprinted with permission of the American Geographical Society).

The author noted that there were still Sac and Fox Indians living in wigwams in the early 1920s.

Wigwams today no longer provide housing, and most Indians live in modern American-style houses. However, it is quite common for wigwam-type structures to be used as sweat lodges. A sweat lodge is basically a wigwam of small bent sapling frame covered with quilts or tarps around a fire pit. Modern sweat lodges among the Winnebago Indians in Wisconsin have been noted consisting of pole frames covered with plywood and tarpaper. Anthropologists have noticed a resurgence in the use of sweat lodges for both ritual cleansing and bathing, and one of this author's students described how her father maintained his traditional sweat lodge in their back yard.

2.8.4 Native American Cemeteries

Burial practices of some Indian tribes in the northeastern and north central United States remain distinctive. For example, within a cemetery of Chippewa graves on Wisconsin's Madeline Island, small wooden grave houses can be seen erected over older burials (Figs. 2.20 and 2.21), and ribbons and offerings have been placed on the cemetery fence. The grave houses have shingle roofs and windows facing east. In cemeteries scattered across the Upland South, extending from North Carolina into eastern Texas and Oklahoma, can be found gravesheds, described as "a small, low, roofed structure with its sides either open or enclosed by pickets, lattices, boards, or wire" (Jordan-Bychkov 2003:76). While most of these gravesheds are found in Anglo-American cemeteries of many denominations, they are most common within regions where surviving Native American communities exist or where the white population is more likely to have at least a little Indian blood, and they are attributed to Native American burial traditions. In a like manner, Jordan-Bychkov notes the diffusion of the camp meeting ground. Nevertheless, the grave houses and gravesheds are often the exception, with many Native Americans buried in graves indistinguishable from those of other Americans.

2.8.5 Seminole Indian Landscapes

The cultural landscape of the Seminoles differs significantly between modern communities in Oklahoma and some more traditional outposts in southern Florida, and it demonstrates both continuity and change. Although many of the Florida Seminoles are highly integrated into their surrounding communities and several groups operate huge modern casinos, there remained several very isolated communities of Seminoles in the Everglades in the 1980s. Not accessible by road, and only reachable by boat, were small camps of four

Figs. 2.20 Miniature wooden gravehouses placed above burials in Chippewa cemetery on Madeline Island, Wisconsin, along the southern shore of Lake Superior

Fig. 2.21 Various sacred offerings placed on fence surrounding Chippewa cemetery on Madeline Island, Wisconsin. Note the offerings of coins, corn, flowers, stones, and cloth bundles containing various items

or five buildings, in which live four to twelve persons related on the female line. These camps were located on small hammocks, which are slightly elevated locations in the Everglades (Nabokov and Easton 1989).

These villages, which are hidden among tropical vegetation, consist of chickees, which are small dwellings constructed of cypress, pine, or palm logs. These typically consist of a low platform raised on stilts, with palm or palmetto frond thatched roofs and walls. Members of dispersed hamlets gathered for religious festivals at isolated ceremonial grounds. An anthropologist and architect who visited an isolated Seminole camp noted that it was located on a low hammock surrounded by shallow marshes, and that it consisted of a "central cook house…surrounded by larger chickees for sleeping and working" (Nabokov and Easton 1989:119–20). Although the U.S. Bureau of Indian Affairs built more modern cinder block houses near the highway for the Seminoles in 1960s, many of the Misccosukee Seminole group abandoned them in favor of chickees. As recently as 1966, 58% of those Seminoles still occupied chickees (Sturtevant and Cattelino 2004). Even as the Seminoles transitioned to the newer single-family cinder block houses, many lamented what they lost. Besides complaints about poor ventilation and similarities among the tract housing, the Seminoles "disliked the density of their new developments: whereas clan camps had been built far apart, in housing developments families from different clans lived side-by-side in close proximity" (Cattelino 2006:711). Thus, chickees have remained attractive to many Seminoles, and with greater tribal wealth and sovereignty, the Seminole "Tribe now financially and administratively supports housing construction that promote 'traditional' and hybrid dwelling structures and practices" (Cattelino 2006:718). In addition, there has been a movement towards a more scattered and matrilocal settlement pattern (Cattelino 2006).

Chickees remain an important landscape feature into the twenty-first century, as Sturtevant and Cattelino explain.

> More than any other element of the built environment, chickees have marked reservation spaces as distinctly Indian. Most Seminoles in 2004 lived in government housing, generally concrete block structure homes, but chickees dotted reservation landscapes and were used in a variety of ways, including for relaxation, decoration, ceremonial activity, storage, and picnic and vehicle cover (Sturtevant and Cattelino 2004:447).

Hybrid homes, incorporating traditional elements into modern buildings, include "frame houses with thatched roofs" (Sturtevant and Cattelino 2004:447), "chickee-like outbuildings… with metal roofs (instead of thatch) to match the roof" (Cattelino 2006:718) of the house. Others erected separate free-standing chickees by their houses to function as a detached garage, and chickees have been erected at businesses and a school. Some non-Seminoles in South Florida have also recently erected chickees.

Thus the Seminole or the Miccosukee, as part of the group is now called, are more conspicuous in their visual adherence to many of the traditional building ways. In contrast, most Indians within the eastern United States live in houses and communities that differ little, if at all, from that of their non-Indian neighbors.

2.9 Précis

The Native American contribution to the landscape within the eastern United States largely reflects the past, seen in tens of thousands of mounds and earthworks, thousands of toponyms, and on the millions of acres devoted to growing crops that they first domesticated. After four centuries of interactions with European and African immigrants, the material culture of many Native Americans—or those individuals who have some Native American lineage—is hardly distinguishable from that of their neighbors. Yet with the renewed recognition of tribal sovereignty and a resurgence of ethnic pride—as well as greater prosperity—Native Americans in the eastern United States have constructed huge gaming facilities which sometimes display Indian motifs and have attempted to regain control of certain lands which have major historical or religious significance to them.

References

Archibold, Randal C. 2006. Governor's Push to Expand Indian Casinos Fails in California. *New York Times,* September 2.

Bailey, Garrick. 2004. Continuity and Change in Mississippian Civilization. In *Hero, Hawk, and Open Hand: American Indian Art of the Ancient Midwest and South*, ed. Richard F. Townsend and Robert V. Sharp, 82–91. Chicago: The Art Institute of Chicago.

———. 2008. Introduction. In *Handbook of North American Indians. Volume 2. Indians in Contemporary Society*, ed. Garrick A. Bailey, 1–9. Washington: Smithsonian Institution.

Berry, Brewton. 1978. Marginal Groups. In *Handbook of North American Indians. Volume 15. Northeast*, ed. Bruce G. Trigger, 290–295. Washington: Smithsonian Institution.

Birmingham, Robert A. 2010. *Spirits of Earth: The Effigy Mound Landscape of Madison and the Four Lakes*. Madison: University of Wisconsin Press.

Birmingham, Robert A., and Leslie E. Eisenberg. 2000. *Indian Mounds of Wisconsin*. Madison: University of Wisconsin Press.

Brown, Ian W. 2004. Prehistory of the Gulf Coastal Plain After 500 B.C. In *Handbook of North American Indians. Volume 14. Southeast*, ed. Raymond D. Fogelson, 574–585. Washington: Smithsonian Institution.

Butzer, Karl W. 1992. The Americas Before and After 1492: An Introduction to Current Geographical Research. *Annals of the Association of American Geographers.* 82 (3): 345–368.

Campisi, Jack. 1978. Oneida. In *Handbook of North American Indians. Volume 15. Northeast*, ed. Bruce G. Trigger, 481–490. Washington: Smithsonian Institution.

Cattelino, Jessica R. 2006. Florida Seminole Housing and the Social Meanings of Sovereignty. *Comparative Studies in Society and History.* 48 (3): 699–726.

———. 2008. Gaming. In *Handbook of North American Indians. Volume 2. Indians in Contemporary Society*, ed. Garrick A. Bailey, 148–156. Washington: Smithsonian Institution.

Conkey, Laura E., Ethel Boissevain, and Ives Goddard. 1978. Indians of Southern New England and Long Island: Late Period. In *Handbook of North American Indians. Volume 15. Northeast*, ed. Bruce G. Trigger, 177–189. Washington: Smithsonian Institution.

Cronon, William. 2003. *Changes in the Land: Indians, Colonists, and the Ecology of New England.* first revised ed. New York: Hill and Wang.

de las Casas, Bartolomé. 1992. *The Devastation of the Indies: A Brief Account.* Trans. Herma Briffault. Baltimore: Johns Hopkins University Press.

Denevan, William M. 1976. *The Native Population of the Americas in 1492.* Madison: University of Wisconsin Press.

———. 1992. The Pristine Myth: The Landscape of the Americas in 1492. *Annals of the Association of American Geographers* 82 (3): 369–385.

———. 2011. The 'Pristine Myth' Revisited. *The Geographical Review* 101 (4): 576–591.

Doolittle, William E. 2000. *Cultivated Landscapes of Native North America.* Oxford: Oxford University Press.

Economist. 2015. Gambling and Poverty: Of slots and sloth. *The Economist* 414 (8921, January 17): 32.

———. 2016. An American mystery: Down in the valley, up on the ridge. *The Economist* 420 (9004, August 27): 19–21.

Federal Register. 2015. Indian Entities Recognized and Eligible to Receive Services from the United States Bureau of Indian Affairs. *Federal Register* 80 (9): 1942–1948.

Finger, John R., and Theda Perdue. 2004. History of the Old South Since removal. In *Handbook of North American Indians. Volume 14. Southeast*, ed. Raymond D. Fogelson, 152–161. Washington: Smithsonian Institution.

Fowler, Melvin L., and Robert L. Hall. 1978. Late Prehistory of the Illinois Area. In *Handbook of North American Indians. Volume 15. Northeast*, ed. Bruce G. Trigger, 560–568. Washington: Smithsonian Institution.

Gartner, William Gustav. 1997. Four Worlds Without an Eden: Pre-Columbian Peoples and the Wisconsin Landscape. In *Wisconsin Land and Life*, ed. Robert C. Ostergren and Thomas R. Vale, 331–350. Madison: University of Wisconsin Press.

Gibson, Jon L. 1999. *Poverty Point: A Terminal Archaic Culture of the Lower Mississippi Valley.* Anthropological Study Series Number 7. Baton Rouge: Louisiana Archaeological Survey and Antiquities Division.

Griffin, James B. 1978. Late Prehistory in the Ohio Valley. In *Handbook of North American Indians. Volume 15. Northeast*, ed. Bruce G. Trigger, 547–559. Washington: Smithsonian Institution.

Hally, David J. and Robert C. Mainfort, Jr. 2004. Prehistory of the Eastern Interior After 500 B.C. In *Handbook of North American Indians. Volume 14. Southeast*, ed. Raymond D. Fogelson, 265–285. Washington: Smithsonian Institution.

Janke, Ronald A. 1987. The Loss of Indian Lands in Wisconsin, Montana and Arizona. In *A Cultural Geography of North American Indians*, ed. Thomas E. Ross and Tyrel G. Moore, 126–148. Boulder: Westview Press.

Jennings, Jesse D. 1968. *Prehistory of North America.* New York: McGraw-Hill.

Johnson, Dirk. 1994. Economies Come to Life on Indian Reservations. *New York Times* 143 (49,746 July 3, 1994): 1, 10–11.

Jordan, Terry G., and Matti Kaups. 1989. *The American Backwoods Frontier: An Ethnic and Ecologial Interpretation.* Baltimore: Johns Hopkins University Press.

Jordan-Bychkov, Terry G. 2003. *The Upland South: The Making of an American Folk Region and Landscape.* Santa Fe: Center for American Places.

Keen, Taylor, and Angelique Eaglewoman. 2008. Tribal Sovereignty and Economic Development. In *Handbook of North American Indians. Volume 2. Indians in Contemporary Society*, ed. Garrick A. Bailey, 129–139. Washington: Smithsonian Institution.

Knight, Vernon James. 2004. Ceremonialism Until 1500. In *Handbook of North American Indians. Volume 14. Southeast*, ed. Raymond D. Fogelson, 734–741. Washington: Smithsonian Institution.

Krinsky, Carol Herselle. 1996. *Contemporary Native American Architecture: Cultural Regeneration and Creativity.* New York: Oxford University Press.

Lenzendorf, Dennis. 2000. *Effigy Mounds: A Guide to Effigy Mounds National Monument.* Fort Washington: Eastern National.

Lepper, Bradley T. 2004. The Newark Earthworks. In *Hero, Hawk, and Open Hand: American Indian Art of the Ancient Midwest and South*, ed. Richard F. Townsend and Robert V. Sharp, 72–81. Chicago: The Art Institute of Chicago.

Mann, Charles C. 2006. *1491: New Revelations of the Americas Before Columbus.* New York: Vintage Books.

Nabokov, Peter, and Robert Easton. 1989. *Native American Architecture.* New York: Oxford University Press.

National Agricultural Statistics Service (NASS). 2014a. *2012 Census of Agriculture: United States Summary and State Data, Volume 1, Geographic Area Series, Part 51.* Washington: U.S. Department of Agriculture.AC-12-A-51.http://agcensus.usda.gov/Publications/2012/Full_Report/Volume_1,_Chapter_1_US/usv1.pdf. Accessed 28 Nov 2016.

———. 2014b. Corn for Grain, Harvested Acres: 2012. In *2012 Census of Agriculture: Agricultural Atlas.* Washington: U.S. Department of Agriculture. https://www.agcensus.usda.gov/Publications/2012/Online_Resources/Ag_Atlas_Maps/Crops_and_Plants/Field_Crops_Harvested/12-M159-RGBDot1-largetext.pdf. Accessed 28 Nov 2016.

National Congress of American Indians. 2015. *Tribal Nations and the United States: An introduction.* Washington: National Congress of American Indians.

Norris, Tina, Paula L. Vines, and Elizabeth M. Hoeffel. 2012. *The American Indian and Alaska Native Population: 2010.* Census 2010 Brief C2010BR-10. Washington: U.S. Department of Commerce.

Onishi, Norimitsu. 2012. With Casino Revenues, Tribes Push to Preserve Languages, and Cultures. *New York Times*, June 17, A14.

Porter, Frank W. 1992. American Indians in the Eastern United States. In *To Build in a New Land: Ethnic Landscapes in North America*, ed. Allen G. Noble, 119–135. Baltimore: Johns Hopkins University Press.

Price, Edward T. 1951. The Melungeons: A Mixed-Blood Strain of the Southern Appalachians. *The Geographical Review* 41 (2): 256–271.

Rodewald, Adam. 2015. Power Struggle over Tribal Land Hits Green Bay. *Green Bay Press-Gazette*, April 25.

Satz, Ronald N. 1991. Chippewa Treaty Rights: The Reserved Rights of Wisconsin's Chippewa Indians in Historical Perspective. *Transactions of the Wisconsin Academy of Sciences, Arts and Letters* 79 (1): i–251.

Saunt, Claudio. 2004. History Until 1776. In *Handbook of North American Indians. Volume 14. Southeast*, ed. Raymond D. Fogelson, 128–138. Washington: Smithsonian Institution.

Schlossberg, Tatiana. 2014. An Indian Tribe Faces Its Eroding Fortunes. *New York Times*, December 1, A22.

Smith, Marvin T. 1989. Indian Response to European Contact: The Coosa Example. In *First Encounters: Spanish Explorations in the Caribbean and the United States, 1492–1570*, ed. Jerald T. Milanich and Susan Milbrath, 135–149. Gainesville: University of Florida Press.

Sokolove, Michael. 2012. A Big Bet Gone Bad. *New York Times Sunday Magazine*, March 18, MM36.

Spencer, Robert F., and Jesse D. Jennings. 1965. *The Native Americans: Prehistory and Ethnology of the North American Indians.* New York: Harper and Row.

Spindler, Louise S. 1978. Menominee. In *Handbook of North American Indians. Volume 15. Northeast*, ed. Bruce G. Trigger, 708–724. Washington: Smithsonian Institution.

Sturtevant, William C., and Jessica R. Cattelino. 2004. Florida Seminole and Miccosukee. In *Handbook of North American Indians. Volume 14. Southeast*, ed. Raymond D. Fogelson, 429–449. Washington: Smithsonian Institution.

Thornton, Russell. 1987. *American Indian Holocaust and Survival: A Population History Since 1492*. Norman: University of Oklahoma Press.

Tooker, Elisabeth. 1978. Iroquois Since 1820. In *Handbook of North American Indians. Volume 15. Northeast*, ed. Bruce G. Trigger, 249–265. Washington: Smithsonian Institution.

Townsend, Richard F. 2004. American Landscapes, Seen and Unseen. In *Hero, Hawk, and Open Hand: American Indian Art of the Ancient Midwest and South*, ed. Richard F. Townsend and Robert V. Sharp, 15–35. Chicago: The Art Institute of Chicago.

U.S. Geological Survey. 2004. Indian Lands. In *The National Atlas of the United States of America*. Reston: U.S. Department of the Interior. http://nationalmap.gov/small_scale/printable/images/pdf/fedlands/BIA_2.pdf. Accessed 28 Nov 2016.

U.S. Government Accountability Office. 2015. *Indian Gaming: Regulation and Oversight by the Federal Government, States, and Tribes*. GAO-15-355.

Washburn, Wilcomb E. 1978. Seventeenth-Century Indian Wars. In *Handbook of North American Indians. Volume 15. Northeast*, ed. Bruce G. Trigger, 89–100. Washington: Smithsonian Institution.

Waterman, T.T. 1924. North American Indian Dwellings. *The Geographical Review* 14 (1): 1–25.

Williams, Timothy. 2012. $1 Million Each Year for All, as Long as Tribe's Luck Holds. *New York Times*, August 9, A1.

3 Native American Landscapes in the Plains and Northwest Coast

The cultural landscape of Native Americans living on the Great Plains and along the Northwest Coast has undergone tremendous change during the past two centuries, but these changes have been very well documented. The Lewis and Clark Expedition between 1804 and 1806 brought back a large volume of written descriptions and drawings illustrating the life of the Plains Indians before their society was devastated by disease, warfare, and the destruction of the vast buffalo herds upon which they were so dependent. Only a few decades later artists, such as George Catlin and Karl Bodmer, painted many scenes of village life and the buffalo hunts. Later in the nineteenth century and early twentieth century pioneer photographers, including William Henry Jackson and Edward S. Curtis on the Great Plains and others along the Pacific Coast, provided a wonderful documentation of a way of life that was shortly to disappear (Fig. 3.1). Thus, in viewing the ethnic landscape of these American Indians one has recourse to a well-illustrated historic record that helps in the interpretation of the contemporary landscape.

Even at the time of the journey of Meriwether Lewis and William Clark, Native Americans living within the Great Plains had already experienced over two centuries of change initiated by European explorers and colonists. With the arrival of European settlers along the east coast, many existing tribes began shifting towards the west, displacing groups already there. Indeed, the Sioux shifted west from the Lake Superior Region into the north central plains, while some groups shifted from a more sedentary village life to a migratory existence. Spanish settlements in the provinces of New Mexico and Texas changed the dynamics of relationships between the Pueblo and Plains peoples (Calloway 2003). The introduction of the horse and firearms dramatically altered not only political relationships among Native American peoples, but they fundamentally changed their economy. The horse permitted Indian hunters, particularly coupled with firearms, to range over a far greater area in pursuit of the buffalo and additionally enabled these Indians to transport much larger tepees. Thus, the Plains Indians' material culture that was encountered and described by the Lewis and Clark Expedition differed significantly from that seen in 1541 by Francisco Vázquez de Coronado on his quest for Quievera in modern-day Kansas.

Similarly, contact between Europeans and Northwest Coast Indians around the beginning of the nineteenth century provided iron axes, which permitted a proliferation in the number and size of elaborately carved totem poles that were seen in the Panhandle of Alaska during the early twentieth century. Farther south in California efforts by the Spanish to acculturate the Indians by converting them to Christianity began at the time of the American Revolution. Their relocation into farming communities around the missions resulted in coastal groups that had been radically changed and amalgamated by the time that the United States acquired California from Mexico.

The contemporary Native American landscape displays elements of the prehistoric past, modified and reshaped by contact with European ethnics and the authority of the Federal government. Indian lands today reflect a combination of the indigenous Native American culture and the influence of Christian churches and the federal government, in particular its Bureau of Indian Affairs and Public Health Service (Vogeler and Simmons 1975). It has been noted that "life on reservations is a struggle between Indian rights and values and those of the dominant western culture" (Vogeler and Simmons 1975:98). While the largest concentrations of Native Americans remain on Indian reservations, many Indians reside in off-reservation urban areas, where acculturation is strongest.

3.1 Contemporary American Indian Populations

Outside of the American Southwest (which is discussed in detail in Chap. 4), the largest regions of the United States that continue to be predominately inhabited by Native Americans are found within the Great Plains and in Alaska. Not only do many Indians reside on the lands of their ancestors, but nearly

J.A. Cross, *Ethnic Landscapes of America*, DOI 10.1007/978-3-319-54009-2_3

Fig. 3.1 Tepees in Atsina Indian village in Montana photographed by Edward S. Curtis in 1908 (Photograph from Library of Congress, Prints & Photographs Division, Edward S. Curtis Collection, [reproduction number: LC-USZ62–66669 (b&w film copy neg.)])

three hundred thousand Indians live upon the prairies and in the Flint Hills of Oklahoma, many who are descendents of various tribes that were relocated to Indian Territory during the nineteenth century. Furthermore, tremendous ethnic and linguistic diversity exists among the various Native American nations—even within a particular area of the United States, although regional similarities do exist with respect to certain aspects of their material culture.

3.1.1 Indians of the Great Plains

The Sioux were the third largest tribe in the United States in 2000, having a tribal population of 108,272, and their numbers had grown to 112,176 in 2010. When considering persons who also claimed other ancestry, the Sioux numbered 170,010 in 2010, the largest group in the Great Plains, but the fifth largest tribe with U.S. cultural roots (Norris et al. 2012). (If we include Mexican American Indians, the Sioux ranked sixth.) The Sioux comprise the vast majority of the residents in several counties in South Dakota (Fig. 3.2). For example, Shannon County, South Dakota is 9% Indian, having the lowest percentage of white population of any county in the U.S. Nearby Todd County, South Dakota is 89.8% Native American, while Buffalo County, South Dakota is 84.8% Indian. Sioux County, North Dakota is 86.9% Native American. In total, five counties within the Dakotas are at least three-quarters American Indian in ancestry, either in whole or in combination with another race, while an additional five counties in the two states are at least 50% Native American. Three counties within Montana are also over half Native American.

Many western states have substantial Indian populations, yet relatively small total populations, resulting in Indians accounting for sizeable proportions of their populations. In South Dakota the 2010 Census reported 82,073 Indians or part-Indians, who constituted 12.5% of the state's population. In North Dakota, its 42,996 Indians comprised 6.4% of the population, while in Montana, its 66,320 Native Americans represented 7.9% of the state's residents. The Plains state with the most Indians is Oklahoma, where 391,949 persons reported Indian ancestry in 2000, and 482,760 were counted in 2010, with 321,687 reporting only Indian ancestry, representing nearly one in eight state residents (Ogunwole 2002; Norris et al. 2012). The large Oklahoma Indian population includes some indigenous Plains

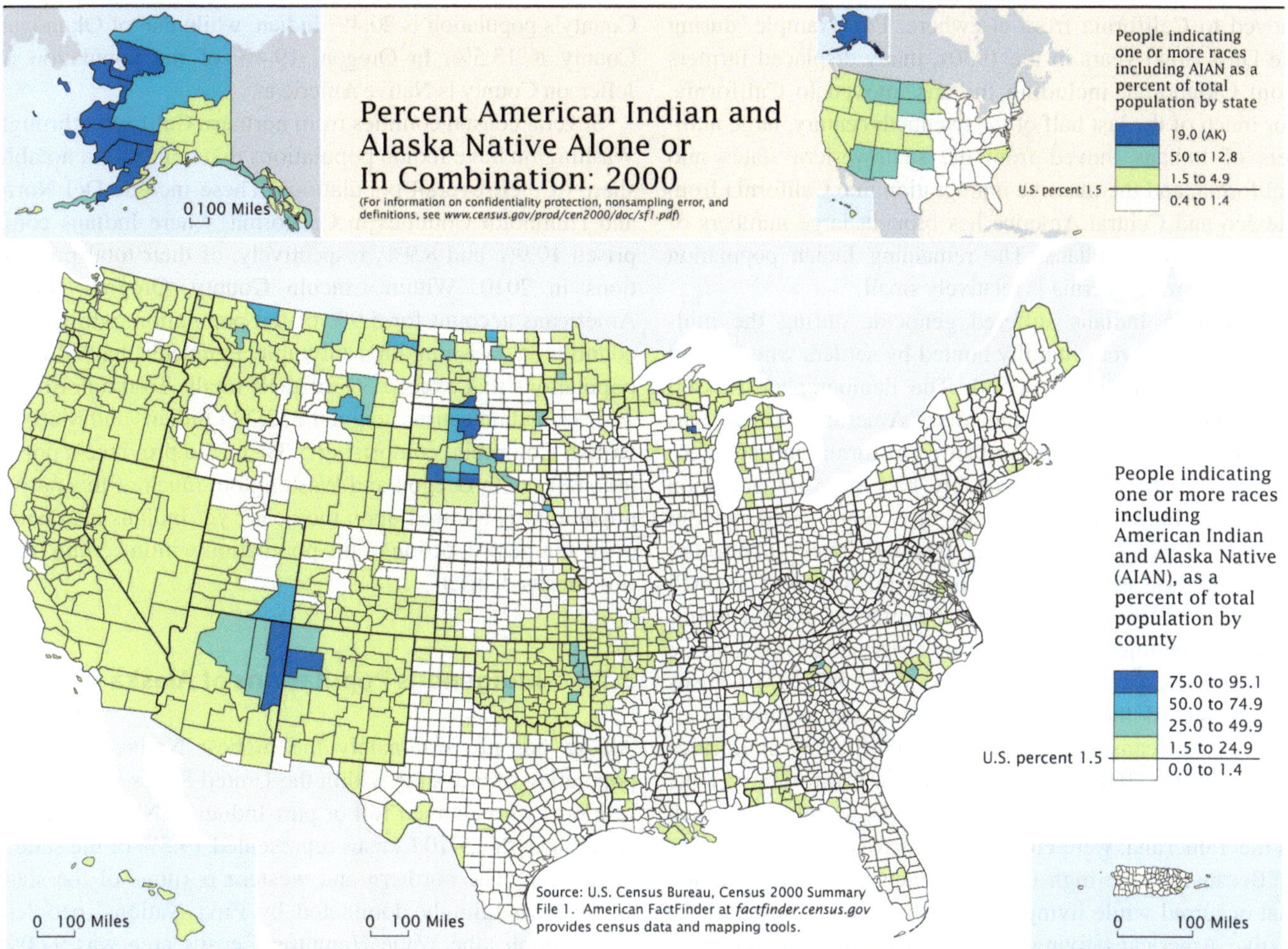

Fig. 3.2 Proportion of the population reporting Native American ancestry in each county at the time of the 2000 Census. Note that those counties with the greatest percentages of their populations being Native Americans are largely located within the Great Plains and in portions of the Southwest (Reproduced from U.S. Census Bureau map within Ogunwole (2002))

Indian groups, but it is largely the legacy of the removal of thousands of Indians from the eastern United States who were forcibly relocated to what was then referred to as Indian Territory.

Although most of the Indian Nations within Oklahoma have long since lost their reservation status, extensive Indian reservations are found within many states of the Great Plains and Great Basin. Indeed, four reservations, each of which covers at least 1 million acres, are found within Montana. These include the Crow Indian Reservation with 2.3 million acres, the Fort Peck Reservation of 2.1 million acres; the Blackfeet Reservation of 1.5 million acres, and the Flathead Indian Reservation, with 1.2 million acres. All of the million acre reservations, with the exception of the last, are located within the Great Plains, while the latter is located within the valleys along the Rocky Mountains. The Sioux, or Lakota, occupy the adjacent Pine Ridge and Rosebud reservations that together cover 3.5 million acres of southern South Dakota. Farther north, the Sioux and Dakota jointly occupy the 2.3 million acre Standing Rock Indian Reservation, astride the border between the two Dakotas. Immediately to the south is the Cheyenne River Indian Reservation, in which the Sioux occupy an additional 2.7 million acres. Within Wyoming, the Arapaho and Shoshone share the 2.2 million acre Wind River Reservation. In addition, there are smaller reservations within the region.

3.1.2 Indians of California

California now has the largest number of persons claiming full or part Indian ancestry of any state, with 723,225 counted in the 2010 Census. However, because California is also the nation's most populace state, the proportion of its population who are Indian is small, only 1.9% (Norris, et al. 2012). A significant share of California's Indian population is not native to the state, as it includes Native Americans who moved to California from elsewhere. For example, during the Dust Bowl years in the 1930s, many displaced farmers from Oklahoma, including Indians, moved to California. For much of the last half of the twentieth century, large numbers of Indians moved from the southwestern states into California, and the massive immigration into California from Mexico and Central America has brought large numbers of Latin American Indians. The remaining Indian population indigenous to California is relatively small.

California Indians suffered genocide during the mid-1800s. Indians were literally hunted by settlers who arrived during the Gold Rush period. The demographer Russell Thornton explains that "genocide of American Indians was probably most blatant in northern California and southern Oregon Territory…[and] probably had more impact upon American Indian peoples there than other incidents did among tribes elsewhere in the United States" (Thornton 1987:107). He explains, "It was not uncommon for small groups or villages of Indians to be attacked by the immigrants, sometimes in the name of a particular war, and be virtually wiped out overnight" (Thornton 1987:107). Census reports record the precipitous drop in the population of Native Americans in California. In 1800 Spanish authorities estimated that 260,000 Indians lived in California, yet the 1890 U.S. Census counted about 18,000. Many tribes, such as the Yahi Yana, were completely destroyed.

Because of the high degree of their ethnic assimilation that occurred while living at Spanish missions, some of the Native Americans living in southern California today are simply known as "Mission Indians," having lost their previous tribal affiliations and displaying a mixed cultural heritage (Shipek 1978). While these groups represent very small proportions of the local population, Indians within several relatively sparsely populated counties within California's Sierra Nevada comprise a much larger share. For example, within Alpine County Indians accounted for 22.3% of the population in 2010 and within Inyo County they were 13.4%. Nearby Mineral County, Nevada, within the western Great Basin, was 17.8% Indian.

3.1.3 Indians of the Pacific Northwest

The Northwest Pacific Coast Indian cultural area extends from northern California through Oregon, Washington, and British Columbia into the Panhandle of Alaska. In Oregon and Washington, the 2010 Census reported Indian populations of 109,223 and 198,998, representing 2.9% and 3.0% of the state populations, respectively. In both of those states many small Indian tribes occupy coastal sites, while other groups reside east of the Cascade Range. It is in several counties immediately east of the Cascade Range that Indians comprise the greatest proportions of the total population. In Washington, Ferry County's population is 20.4% Indian, while that of Okanogan County is 13.5%. In Oregon, 19.4% of the population of Jefferson County is Native American.

Several coastal counties from northern California through Washington have Indian populations that comprise a notable share of their overall populations. These include Del Norte and Humboldt Counties in California, where Indians comprised 10.9% and 8.9%, respectively, of their total populations in 2010. Within Lincoln County, Oregon, Native Americans account for 5.9% of the population, while three counties in Washington's Olympic Peninsula have Indian populations that exceed 5% of their totals. To the north the 2011 Canadian census counted 232,295 Indians and Métis in British Columbia, comprising 5.3% of the province's population. As with Oregon and Washington, much of this native population is focused upon the coast, yet Indians also comprise a large proportion of the population within certain interior census districts.

3.1.4 Indigenous Populations of Alaska

Alaska has proportionately the highest Native American population of any state within the United States. Its 138,312 residents who claimed full or part-Indian or Native Alaskan ancestry in the 2010 Census represented 19.5% of the state's population. The northern and western portions of the state are overwhelmingly dominated by First Nations' peoples. For example, the Wade Hampton Census area was 97.0% Native American in 2010, the Northwest Arctic Borough (Alaska's equivalent of a county, although huge in area) is 87.1%, and the Lake and Peninsula Borough, covering much of the Aleutian Peninsula, is 74.6%. Alaska also has significant numbers of Native Americans in its southeastern Panhandle, with Yakutat City and Borough being 49.8% Native American and Sitka City and Borough being 24.6%, for example.

The indigenous population of Alaska includes Indians, Aleuts, and Inuits (who are also called the Iñupiat in Alaska, and who were once called Eskimos). Indians primarily reside within the coastal areas near the Gulf of Alaska and in the southeastern Panhandle, as well as portions of interior Alaska. The Inuit occupy the far northern regions, particularly near the Bering Strait and the Arctic Ocean, while the Aleut live primarily on the Aleutian Islands and the adjacent Aleutian Peninsula, including Kodiak Island.

As natives of the Aleutian Islands, which extend towards Siberia, the Aleuts display many Russian influences, a legacy of the time when Alaska was Russian territory. Many Aleuts have Russian surnames, and because of intermarriage, many are multicultural. The Russian Orthodox Church is prominent within many Aleut settlements. (See Chap. 13)

Inuits are much more recent immigrants to North America than the Indians. The ancestors of the Inuit first began came to North America from Siberia about 4500 years ago (Raff et al. 2015), and large scale movements by foot over great distances of land and frozen ocean during the winter continued well into the early twentieth century. Today the Inuit live from Greenland and Canada across North America west to the Siberian East Cape. Today the Inuit are far less migratory, having been settled into villages by government officials, although hunting of large sea mammals, including seals, walrus, and whales, remains culturally important. The village landscape of the Inuit today contrasts sharply from their nomadic existence of a century ago.

3.1.4.1 Alaska Native Claims Settlement Act

The Alaska Native Claims Settlement Act of 1971 resulted in a very different type of relationship between the Federal government and the Native Americans of Alaska in comparison with those in the other states. There is only one Indian Reservation in Alaska, the 86,000 acre Metlakatla Reservation of the Tsimshian Indians on Annette Island, and it predates the act. The act of 1971 has not given the Native Americans in Alaska the sovereignty that was provided by treaty to Indians elsewhere. To extinguish the land claims of the indigenous population of Alaska, the Federal Government provided a payment of $1 billion and distributed 44 million acres of land to the native groups. This land was not provided directly to individuals, but to community and tribal corporations, of which 12 regional corporations were established, plus one for Natives living outside the state. Although these corporations were originally prohibited from selling the lands for 20 years, much of the prohibition was subsequently extended. The 1991 Amendments to the Alaska Native Claims Settlement Act protected undeveloped lands held by the regional corporations from external acquisition, however developed lands became subject to taxation or seizure should a bankruptcy occur. Many of the regional corporations use the products of their lands to support their local economies and have developed sawmills, fish packing facilities, and mines that employ their residents. Yet in some villages, unemployment remains high, as the only jobs available are either for the regional corporation or the government.

The goal of the Alaska Native Claims Settlement Act was to accelerate the assimilation of Alaska's indigenous peoples into the broader society. Yet, as Rosita Worl (2008:147) explains,

> Alaska Natives initially saw corporations as a vehicle to maintain control and ownership of their land. Less than 10 years after the passage of ANCSA, they discovered the dangers that the corporate structure posed to their cultural survival.

Amendments to the initial legislation enhanced the ability of the Alaska Natives to safeguard their cultural heritage. While the various regional corporations are designed to provide an economic underpinning of the Native Alaskan economy, they are unlike other businesses.

> They pursue both business and non-business goals, seeking to be profitable while at the same time protecting their Native way of life, including their subsistence hunting and fishing traditions (Worl 2008:147).

Such subsistence hunting and fishing remain fundamental economic activities for many rural Alaskans and are intimately tied to many cultural values (Fig. 3.3).

3.2 Historical Landscapes of the Plains Indians

Many excellent written accounts and photographs show the lifestyle and material culture of various Plains Indian groups well before they were settled onto reservations. Traditionally, Indians from the Great Plains included both village and migratory people. Those from the High Plains were migratory, living in tepees year-round. The Indians of the Prairies, an area roughly east of the Missouri Valley and also including the eastern portions of Nebraska, Kansas, and Oklahoma into Texas, lived in large, multifamily houses that were often dug partly into the ground, particularly within the northern half of the Prairies (DeMallie 2001).

3.2.1 Migratory Plains Indians and the Tepee

The migratory Indians on the High Plains were dependent upon the buffalo (bison) for food, clothing and shelter. They made use of all parts of the animal, and many tribes did not include any farmers. They followed buffalo herds during their migrations and typically camped along streams near firewood and water during the winters. In the spring, summer, and fall they followed the buffalo, relocating their camps every few days (Fig. 3.4). Thus, their material culture differs from that of groups living in agricultural settlements. Their possessions had to be easily transportable and housing easily moved. Thus, horses played a crucial role in their lives after they were obtained from the Spanish. Indeed, before the 1600s Plains Indians had no large beasts of burden, and their loads were pulled by dogs or humans.

The historical landscape of the Plains Indians greatly differed from that of prehistoric times. With horses there was a considerable expansion of tribes into areas previously unoccupied, and prairie and woodland tribes expanded into the Great Plains. For example, from the east came the Cheyenne, Arapaho, and Crow. These tribes were much more nomadic during the 1600s to the 1800s than they had been previously.

Fig. 3.3 Elevated log food storage cache in Klukshu Indian village, Yukon, near the Alaska border. Note how the poles supporting the log structure are covered with metal to prevent animals from climbing into the structure

Fig. 3.4 Tepees in Shoshoni village, circa 1880 (One source indicates the scene was photographed by William Henry Jackson in 1868.) Note that the central tipi is covered with the decorated skins (Photograph from Library of Congress, Prints & Photographs Division, National Photo Company Collection, [reproduction number: LC-USZ62–115466 (b&w film copy neg.)])

Among the nomadic hunters tepees (or tipis) provided their primary shelter. Basically, the tepee was a portable conical tent covered with buffalo skins. These skins were tanned, sewn together, and often painted, especially by the southern Kiowa and northern Blackfeet. The painted designs depicted battles or supernatural guardians, with designs specific to the various families. The doorway of a tepee usually faced east. The flaps at the tip of the cone could be adjusted to deflect the wind and release smoke from a centrally located fire.

The character and size of tepees changed with the introduction of horses. Pre-horse tepees had short poles and were covered with six to eight buffalo robes. Post-horse tepees doubled in size. They were covered by nearly 20 buffalo robes, with longer, stronger, and heavier poles of

Fig. 3.5 Tepee rings are formed from stones used to secure the bottom of those skins covering the lodge poles. This tepee ring is located near the boundary between Montana and Wyoming in the Bighorn Canyon National Recreation Area

lodgepole pine. Tepees were widely utilized, both on the Great Plains and by certain Indian groups in the northern part of the Great Basin (Nabokov and Easton 1989). For example, the Shoshone used either tepees or small conical lodges of sagebrush grass or willow branches.

Regional differences in the design and style of tepees existed. In the southern plains a three pole tripod frame dominated, while in the northern plains a four pole frame was prevalent. Upon the basic frame of three or four poles, typically 20–25 feet long, about 20 lighter poles were added by leaning against the "V" formed by the framing poles.

Tepee rings marked the perimeter of the tepee (Fig. 3.5). The tepee ring was a circle of stones used to anchor the skins of the tepees. In many locations tepee rings remain visible a century or more after their last use and comprise the "oldest traces of Plains Indian architecture" (Nabokov and Easton 1989:123). Frison (2001:141) notes, "Tepee rings or stone circles are probably the most visible feature on the Northwestern Plains in Late Prehistoric times but they began to appear in considerable numbers during the Late Archaic." Although Frison (2001) indicates that the ethnographic record clearly demonstrates the use of the stones to hold tepees skins during the more recent period, not all were used for that purpose, and the significance and usage of the oldest stone rings are less clear. Nevertheless, the stone rings are very abundant and are often found in groups in a wide variety of topographic locations, but particularly on ridges, atop buttes, and along stream terraces.

Tepees were often arranged in circles, and the circle has great cultural significance. Its importance was described by an Oglala Sioux to the anthropologist James Walker a century ago.

> The sun and the sky, the earth and the moon are round like a shield.... Since the Great Spirit has caused everything to be round mankind should look upon the circle as sacred for it is the symbol of all things in nature except stone.... The day, the night, and the moon go in a circle above the sky. Therefore the circle is a symbol of these divisions of time and hence of all time.
>
> For these reasons the Oglala make their tipis circular, their camp circle circular, and sit in a circle at all ceremonies. The circle is also the symbol of the tipi and of shelter (Walker 1917:160).

The circle continues to have cultural significance to many contemporary Indians.

3.2.2 Plains Indian Villages

Villages on the Great Plains were typically located along rivers. These sites were occupied before the arrival of the horse, being settlements of farmers cultivating fields on floodplains. Unlike nomadic groups, those Indians who relied on agriculture needed to stay near their crops.

Fig. 3.6 Mandan earthern lodge in North Dakota photographed in 1908 by Edward S. Curtis (Photograph from Library of Congress, Prints & Photographs Division, Edward S. Curtis Collection, [reproduction number: LC-USZ62–114582 (b&w film copy neg.)])

Prairie Indian villages typically consisted of twenty to over a hundred earth lodges clustered close together (Fig. 3.6). An earth lodge was basically a large, partly subterranean dwelling built around four central posts. The central posts were joined at their top by cross beams and circled by shorter posts and beams. A slanting sidewall of planks with sticks or brush supported an earthen or sod cover and roof. A central smoke hole was located between central the posts. Earth lodges were large, typically 40–60 feet in diameter, while some up to 90 feet across are known (Nabokov and Easton 1989).

Earth lodges are associated with the Mandan, Hidatsa, and Arikara in the Dakotas and the Pawnee in Kansas and Nebraska. The Lewis and Clark Expedition spent the winter of 1804–05 with the Mandan, providing the explorers with an excellent opportunity to become thoroughly familiar with their cultural landscape, as indicated in their account.

> There are…dwellings [all] about me, and they are purely unique—they are all covered with dirt….
>
> The groups of lodges around me present a very curious and pleasing appearance, resembling in shape…so many potash-kettles inverted. On the tops of these are to be seen groups standing and reclining (quoted in Ahler et al. 1991:14).

The Mandan earth lodges were arranged around a plaza or square. Key religious officials occupied the innermost buildings. The dome-shaped earth houses were closely spaced, being arranged roughly in rows. The village itself was surrounded by a low ditch and post palisade.

An excellent location to see the ruins of such Plains Indian villages is the Knife River Indian Villages National Historic Site in central North Dakota, very near the Missouri River (Fig. 3.7). This village was already in a steep decline by the time Lewis and Clark visited it, yet it was to figure prominently in the success of the expedition. The park preserves the village in which Sacagawea lived at the time she joined the expedition. That village had formed by a combination of Hidatsa, Mandan, and Arikaras, after a smallpox epidemic had decimated their populations in 1780. All of the Knife River villages were abandoned by 1845, following another devastating smallpox epidemic in 1837. Today the landscape is covered with pock-marked indentations showing the site of the ruined village. The evidence is visible both at ground-level and from the air, consisting of numerous pits (forming hummocky lands) showing the locations of the earth lodges. Archaeological work at the Big Hidatsa Village at Knife River "revealed 113 recognizable depressions within the main part of the village, each marking the former location of an earthlodge" (Ahler et al. 1991:85). Characteristically, "these depressions are distinctly circular in form, about 35–45 feet in diameter, and from less than a foot to more than two feet in depth" (Ahler et al. 1991:85). The outline of the village palisades is still visible, as are linear mounds in

Fig. 3.7 Undulating ground created by the collapse of the roofs of abandoned earth lodges results in intricate arrangements of shallow pits on the landscape. This former village site is preserved in Knife River Indian Villages National Historic Site in central North Dakota

the villages. The purposes of the mounds are unknown, but they were not for burials. In addition, trails remain visible, as travois eroded paths in the sod as horses dragged loads. Unfortunately, many additional nearby sites were flooded by the construction of the Garrison Dam in the 1940s.

While the earth lodges of the northern Prairie Indians left the most conspicuous marks on the landscape, some of the Prairie Indians, such as the Santee Sioux and Kansa, occupied "rectangular-bark covered lodges. The Kansa also built oval or rectangular lodges covered with bark, hides, or mats" (DeMallie 2001:10). The Wichita Indians covered their "beehive-shaped houses with bundles of grass" (DeMallie 2001:10). Regardless of house type, all of the Prairie Indians occupied their dwellings during times of planting and harvesting, yet some of them lived the rest of the year in tepees, following the buffalo herds.

3.2.3 Sacred Sites

The Plains Indians recognized both specific physical landmarks and constructed works as sacred sites. Medicine wheels, which are large circles of boulders, are found on hills or ridges and are aligned to rock cairns many miles away. The best known is Medicine Wheel atop the crest of the Bighorn Mountains in Wyoming (Fig. 3.8). At 9640 feet in elevation and having a diameter of seventy feet, it both overlooked the Big Horn Basin and was conspicuous. Lines and stone cairns point to key solar and stellar positions.

The Black Hills are considered a sacred site by the Plains Indians Sioux. As Christafferson (2001:836) notes, "Spiritually, the Black Hills were the center of the universe, believed by many to be the site of creation." The Black Hills provide an excellent example of how conflicts over the definition of what is sacred and the use of such sacred space can exist between different ethnic groups. In 1868 a treaty gave the Sioux Indians all of South Dakota west of the Missouri River as their reservation, including the Black Hills. Yet by 1874 the presence of gold in the Black Hills was confirmed by General George Custer. Black Hills gold seekers defied both the treaty and the U.S. Army and invaded the goldfields. By 1877 the U.S. government forced the Indians to cede the Black Hills. One may still find contemporary religious offerings, such as pebbles, feathers, and certain types of leaves, at sites in the Black Hills. Yet the sacred mountains were long the site of the largest gold mine in the United States, an operation that has permanently scarred the landscape.

The Sioux sued the Federal government. After nearly 60 years of litigation, in 1980 the U.S. Supreme Court affirmed the authority of the federal government to take the Black Hills, but it found that the government had failed to provide compensation. While the Sioux were awarded $17,553,484, the value that it was estimated to have had at the time it was seized in 1877, excluding the value of the mineral

Fig. 3.8 Medicine Wheel, formed by a careful arrangement of rocks, on a foggy stormy day atop Wyoming's Bighorn Mountains. Note the cloth prayer offerings tied to the fence surrounding the site. The arrangement of stones delineates solar positions

wealth subsequently extracted, plus interest compounded at 5% per annum from that time, the Sioux have refused payment of the sum that had grown to about $500 million by the beginning of the twenty-first century (Christafferson 2001). To the Sioux acceptance of the payment would preclude the possibility that the sacred lands would be returned to them.

More recently, the construction of a bar and campground catering to motorcyclists two and a half miles from Bear Butte, "which dozens of tribes hold as one of the most sacred sites in North America" (Robbins 2006), led to protests that attracted national attention. Although the 1100 foot high volcanic plug itself is preserved within a South Dakota state park to the north of the Black Hills, opponents wanted to keep development that was viewed as incompatible with the sacred site to a location at least five miles distant.

The standoff that started in 2016 between the Standing Rock Sioux, who have been joined by many supporters from other Native American tribes, and the diggers of the Dakota Access pipeline is simply the latest in a long stream of disagreements over sacred burial grounds, the desecration of ancestral lands, and access to natural resources (Harris and Gonchar 2016). The construction of the Oahe Dam along the Missouri River in 1958 flooded most of the Standing Rock Sioux's riverfront forests, orchards, and fertile valley farmland, necessitating the relocation of Indian villages and inundating burial grounds. Fluctuating water levels in the reservoir have caused erosion, exposing burials. It is with this context that the proposal to extend the oil pipeline across lands just north of the current reservation boundary met strong resistance from both the Sioux and environmentalists, who criticized the environmental assessment as inadequate and who feared the consequences of the pipeline construction and operation upon sacred places and the water quality downriver in the Missouri Valley (Archambault 2016). The encampment of thousands of Indian protestors on the banks of the Cannonball River, which forms the northern boundary of the Standing Rock Sioux Reservation, evoked a modern version of the traditional Plains Indian camp, complete with numerous tepees—just as one might see at a summer ceremonial encampment, as described in the next section.

Other sacred sites have been preserved in less contentious manners. Thermopolis Hot Springs is now a Wyoming State Park. It was known by the Indians for the medicinal powers of its waters and it was given by the Indian chief for public use. Likewise, Platt National Park, consisting of springs noted by the Indians for their healing powers, was promoted by the Chickasaw Indians for all citizens to enjoy. It is now part of the Chickasaw National Recreation Area in Oklahoma, yet its demotion "reflects a process of cultural marginalization" (Parker 2010:170). In Wyoming, towering 1267 feet above a nearby river is Devil's Tower, a giant plug of intrusive igneous basalt (Fig. 3.9). It is considered a sacred site by many Plains Indian tribes. Several Indian nations call it the "Bear's tipi" or "Bear Lodge," names which also highlight

Fig. 3.9 Devil's Tower, also know as Bear Lodge or Bear's Tipi, located in northeastern Wyoming. This geologic feature, formed by the exposure of intrusive igneous rock by the erosion of surrounding sedimentary rock, is considered a sacred feature by many Plains Indian groups. Preserved in Devil's Tower National Monument, park officials have long needed to deal with conflicting values of those who view the site as sacred and those who seek its challenge as rock climbers

the role of the sacred bear in its formation. Not only is the site regularly visited by many Plains Indians, but sacred bundles and cloth offerings have been left around the stone tower and ceremonials are held in its vicinity. This site is now preserved as Devil's Tower National Monument. Yet even here, the desire of Native Americans to preserve the sacred space conflicts with the longing of rock climbers who wish to scale the nearly vertical rock face.

3.3 Contemporary Plains Indian Landscapes

The contemporary Plains Indian region is a 'Landscape of Poverty.' Their reservations are centers of high unemployment and little industry. The elimination of the vast buffalo herds destroyed both their lifestyle and economy. It was a deliberate effort by the U.S. Army to subdue the Plains Indians that led to the slaughtering of the buffalo herds. The herds had included between 35 and 75 million buffalo in 1820, yet their numbers fell to around 1000 by 1890. There has been little integration of the Plains Indians into the overall American economy, and on many reservations the tribe—using federal subsidies—is the largest employer. Given the low populations of the regions where many of the Plains Indian groups live, casino gaming offers little promise to generate the type of profits that many tribes nearer major population centers have received.

Few of the Sioux or Crow Indians who actually own land either farm or graze their properties today. Instead, the grazing rights are often leased to nearby white ranchers. For example, on the Crow reservation trust land at the beginning of the twenty-first century, only 74 of 765 farmers were Indians and only 60 of 176 ranchers were Indian (Voget and Bernardis 2001). Similar findings were reported earlier by Vogeler and Simmons (1975), who noted that very few Native Americans worked as farmers or ranchers, partly a consequence of difficulties in borrowing funds to finance purchases of farm or irrigation equipment. In an effort to stimulate their economy, as well as to enhance pride in their cultural heritage, the Rosebud Sioux are trying to rebuild their buffalo herd, which now numbers well over one thousand head. While one can now see the buffalo, the tepee is no longer a symbol of these Indians' nomadic lifestyle, but it is erected during summer powwows and religious ceremonies.

The most prosperous communities on the Plains Indian reservations, other than those which are positioned to operate profitable casino businesses, typically include the Bureau of Indian Affairs administrative centers for the reservation. At these centers jobs in tribal administration and education are concentrated. There and in some of the smaller hamlets, public housing developments have been focused. Yet, outside these planned communities, following urban styles of white Americans, other Indians live in small dispersed reservation settlements. Over a quarter century ago, housing within these South Dakota settlements was described as "the adjustment Indians have made to material impoverishment. Log cabins, old trailers, tents, transported dwellings, clapboard shacks, and various combinations of these structures appear" (Vogeler and Simmons 1975:105). Unfortunately, such impoverishment remains. The 2000 Census reported that the median per capita income on South Dakota's Pine Ridge Indian Reservation was $6143, less than a third of the $21,587 overall national median. Although the 2010 Census did not ask questions regarding income, the American Community Survey conducted from 2010 through 2014 indicated per capita income on the Pine Ridge Indian Reservation had risen to $10,976, far below South Dakota's overall per capita income of $26,311, which was over $2000 less than the national figure. The nearby Rosebud Sioux Reservation was nearly as poor as the Pine Ridge Reservation.

The tepee has influenced modern architecture, both on and off Indian reservations and by Indians and non-Indians alike. A variety of modern buildings have been designed in the shape of tepees or have features that were inspired by tepees. For example, the architectural historian Carol Krinsky, citing examples of churches from Montana and South Dakota, comments that a "Native American Church congregation may…prefer to worship in a building reminiscent of the tipi because this Plains building type is commonly used for their prayer services" (Krinsky 1996:111). Other modern tepee shaped buildings have been constructed by Plains Indians to house community centers and a hotel. Concrete and steel shelters shaped like tepees shade picnic tables at rest stops along the highways in one western state.

Circular buildings, reminiscent of both earthlodges and roundhouses, have been constructed on many reservations. Such structures now house chapter houses, such as one on the Pine Ridge Indian Reservation of South Dakota, museums, community centers, and schools. The Native American architects who designed the cylindrical Four Winds School on North Dakota's Fort Totten Reservation linked their building "to the wind diagram or medicine wheel…[as they] wanted a design that would represent the tribe" (Krinsky 1996: 118). Thus, contemporary Indian communities continue to respect the traditional spirituality of the circle.

Housing in Plains Indian communities is typical of what is often found within lower-income white communities, even though the internal usage of that space might vary from European-American norms. Yet elements of the traditional lifestyle may remain visible on the outside of the houses. Tepee poles are sometimes seen resting against the sides of houses or stored nearby, and they are used during the Summer Ceremonial Encampment. At that time the tepees are erected, and large numbers may be occupied during summer powwows. While the tepees are of traditional designs, they are usually constructed of modern canvas material.

Wigwam-framed sweat baths, basically a small dome covered with tarps or quilts over a firepit to create steam, are as Nabokov and Easton (1986:25) explain, "[o]ne Plains structure that is now found throughout Native America." They are even more common than tepees. These may sometimes be seen near modern housing.

Among the Kiowa of western Oklahoma one material feature that ties them to their homeland is the brush arbor or "Indian shade" that they often erect behind their houses (Schnell 2001). While at one time the Kiowa cooked and slept under such an arbor during the summer, now their construction is more symbolic. Ranging from brush-covered shades to permanent outbuildings, "the building of an arbor today is a statement of identity. It is a way for Kiowas to feel that they are living, at least in part, in the way their ancestors did" (Schnell 2001:143).

3.4 Historical Landscapes of Northwest Coast Indians

Indians of the Northwest Coast, from northern California through southeastern Alaska, lived in villages placed in coastal and riverine locations. These villages of permanent winter houses were "usually sited along the curving beachfronts of sheltering coves" (Nabokov and Easton 1989:229). In some locations, such as around the Puget Sound area of Washington, "most settlements held from three to five multiple-family houses, usually laid out in rows with the long side facing the shoreline" (Nabokov and Easton 1989:233). Some of the villages were surrounded by a wooden palisade. During the summer, the Indians camped near salmon spawning grounds or berry patches.

3.4.1 Plank Houses

Houses were typically built of cedar planks, initially split with stone adzes and later by iron axes. Cedar splits very easily and evenly, as the wood is straight-grained with very few knots, resulting in massive boards that were used to construct substantial wooden buildings. Walls were cedar planks, and roofs were split cedar boards. The style and size of plank houses varied across the Northwest Coast.

In the region of Washington's Olympic Peninsula and Puget Sound, as well as British Columbia's Vancouver

Island area, shed houses that were basically a very long single-pitch shed roof structures were once common. British Naval Captain George Vancouver in 1792 described one single shed house near present-day Seattle that was 380 yards long. That structure housed over 600 Dwamish residents, and it was divided into eight sections, each housing a clan chieftain and his followers, with a totem pole erected for each clan.

In the central portion of the Northwest Coast, largely coinciding with the coast of British Columbia, two architectural traits were common to family houses. These were massive scale construction and iconography of supernatural beings, which displayed the heritage of the dwelling's occupants (Nabokov and Easton 1989). The iconography included mural-like paintings on front walls, carved figures on front posts, and decorative entrances to houses that were mouth-shaped or mimicked the open beak of a bird. The wooden houses had pole-framed gable roofs resting on four great posts. Such houses were arranged along the beach in two or three rows.

In the northernmost section of the Northwest Coast, including British Columbia's Queen Charlotte Islands and the Alaskan panhandle, the wooden houses were smaller, but the display of totem poles was most conspicuous (Fig. 3.10). The houses of the Haida, Tlingit, and some Tsimshian typically measured forty by forty feet and were arranged shoulder to shoulder, with their gable-ends forward. The houses were traditionally built of four foot-wide planks, tightly joined by mortise and tenons. The Haida had six beamed houses, with poles running from gable to gable.

3.4.2 Totem Poles

Totem poles, consisting of a carved tree or wooden pole displaying a family lineage by crests and figures of various animals, were placed at various locations throughout the village. The crests on the totem poles represented important events in a family history. Several varieties of totem poles existed (Drucker 1955). The memorial pole was erected by the heir of a deceased chief and it showed the heir's assumption of title and rights. It was typically located on the beach fronting a village, and among the southern Tlingit and Tsimshian, it was the main type of totem pole. The mortuary pole either constituted the grave of a deceased chief or it was erected beside the grave. If the pole formed the grave, the burial was in a niche at the back of the pole or the pole supported a burial aloft. The house-portal totem pole was "built onto the front of the house and rose high above it, with a large opening, forming the doorway, at or near the base" (Drucker 1955:191). Finally, there were special privilege poles that indicated the special right of owner. For example, among the southern Kwakiutl and Nootka "tall slender poles surmounted by a bird-like figure marked the house of the Beach-Owner—the chief who had inherited the

Fig. 3.10 Totem poles erected by tribal building near Hazelton, British Columbia

prerogative of being the first to invite important visitors to the village to a feast" (Drucker 1955:191).

Totem poles were much more common during the historic period than the pre-contact period, as anthropologists Douglas Cole and David Darling explain.

> European metal tools, especially chisels, the most highly prized of early trade goods, made wood carving much easier, especially on large sculptures. The early European explorers and traders found various kinds of carved poles among different Indian groups at contact. While reports of interior house posts are common in the eighteenth century, the large, freestanding 'totem poles' were unusual…. Poles soon appeared, partly as a result of the greater availability of metal tools, which made carving easier, and partly as a result of the closer relations that the fur trade fostered and that helped promote intertribal cultural exchange (Cole and Darling 1990:132, Reprinted with permission of Dept. of Anthropology, National Museum of Natural History).

Among all of the groups erecting totem poles, those erected by the Haida were the most complex. True totem poles were not erected south of Vancouver Island, yet less elaborate poles were erected to the south in the Columbia Basin and among the western Inuit in Alaska. In an effort to preserve the legacy of the monumental totem poles and large community houses at a site that had been recently abandoned, Old Kasaan National Monument was established in southern Alaska. Unfortunately, high humidity and the passage of time caused the poles as well as adjacent plank houses to rot. The park no longer exists, although some totem poles were relocated.

The potlatch was a redistribution of wealth for which many Northwest Coast Indians were famous. Special buildings were constructed as potlach houses. These potlatches, or festivals in which lavish gifts were bestowed upon the attendees, became more common and more widely distributed with increased trade and interaction among tribes after contact with Europeans. Clans often threw a potlach for tribesmen who helped build dwellings and clan houses.

3.5 Contemporary Landscapes of Northwest Coast Indians

Village settlements remain, but some have declined, and certain reservations are no longer viable. Contemporary house architecture is often virtually indistinguishable from that used by white American neighbors. Yet occasionally a house will have decorative motifs, and some of native designs are incorporated in the advertising seen on commercial buildings. None of the original plank shed houses survived in the southern part of the Northwest Coast region, victims of rot and changes in architectural style. On the campus of Evergreen State College in Olympia, Washington a native longhouse was constructed two decades ago utilizing massive log beams and timbers, resembling traditional Northwest Coast Indian structures, with a large carved and painted thunderbird mounted above its entrance (Fig. 3.11). Used for a variety of

Fig. 3.11 Longhouse constructed in mid-1990s on campus of Evergreen State College in Olympia, Washington, utilizing massive log beams and timbers. It resembles traditional Northwest Coast Indian structures, with a large carved and painted thunderbird mounted above its entrance

educational programs supporting Native American students and local indigenous communities, it was designed to provide a culturally sensitive meeting place that articulates with traditional architecture. In the northern part of the region some plank buildings have survived, and there has been a revival of architecture, especially for tribal council buildings, community houses, and museum structures.

Totem poles are much rarer today, remaining in front of the homes in only a few villages, such as near Hazelton, British Columbia. Several of the surviving poles from Old Kasaam were relocated to Alaska's Sitka National Monument. In addition many new totem poles have been commissioned for display by various public buildings and in parks throughout the Northwest Coast region. To the south, the tribal center building of the Quileute Indian Reservation in La Push, Washington, which displays a shed roof that resembles that on traditional plank houses, is flanked by a totem pole (Fig. 3.12). Another totem pole featuring a bird above a whale stands outside a restaurant in the center of La Push. On the opposite side of the Olympic Peninsula, the Jamestown S'Kallam Indian Reservation displays several totem poles flanking its administration building and lodge-casino, which also displays carved animal motifs on the support posts incorporated in its walls. Near the northern edge of the Northwest Coast cultural zone in Juneau, Alaska, a 35 foot tall totem pole, carved in 1940 and depicting four Haida stories, stands just across the street from the Alaska Capitol Building. Several blocks away, on the site of the century-old Juneau Indian Village, the Auke Tribe Building was erected as a community center. Now renovated and renamed Gajaa Hit, or Safe House in the Tlingit language, this building operated by the Tlingit-Haida Regional Housing Authority displays a mural painting of Raven and Eagle, the two halves of the Auke tribe, covering a two-story wall on its gabled end. Flanking this building that mimics a traditional clan house are two 26 foot totem poles (Fig. 3.13). Extensive displays of Northwest Coast Indian totem pole art and architecture are displayed at many other sites, including Sitka National Monument in Alaska and Thunderbird Park, adjacent to the Royal British Columbia Museum in Victoria, British Columbia. Other totem poles, although not necessarily as true to traditional motifs, have been erected for advertising purposes throughout the Pacific Northwest.

Fishing practices and fishing rights have been points of contention with non-Indians. Conflict with white commercial fishermen over the dwindling catch on the Columbia River arose because certain tribes were guaranteed salmon by treaty. The reduced catch is largely caused by hydroelectric dams that hinder the salmon's migration to their spawning grounds. As a result of court action, Indians are now guaranteed half of the catch. One can still see Indians continuing their traditional fishing practices, such as using fish traps both along the ocean and on rivers and streams (Fig. 3.14), and netting salmon along waterfalls and rapids. Many wooden

Fig. 3.12 Quileute Indian Reservation Tribal Center building in La Push, Washington that displays a shed roof resembling that on traditional plank houses. It is flanked by a totem pole

Fig. 3.13 Gajaa Hit building in Juneau, Alaska, with a mural painting of Raven and Eagle covering its gabled end. Flanking this building, operated by the Tlingit-Haida Regional Housing Authority and which resembles a traditional clan house, are two 26 foot totem poles, also displaying the raven and eagle

Fig. 3.14 Fish trap constructed across stream by Klukshu Indian village, Yukon, near the Alaska border. This trap was designed to catch salmon

Fig. 3.15 Platforms, from which Indians net salmon, erected on the rocky bank of the Columbia River near The Dalles, Oregon

fishing shacks and platforms from which Indians still net salmon are located along the Columbia River near The Dalles, Oregon (Fig. 3.15).

3.6 Contemporary Landscapes of the Alaskan Inuit

Within the lifetime of many of the Inuit elders, their society has transitioned from a nomadic hunting society, marked by many temporary encampments, to year-round settlements occupying weatherized European-American wood-frame buildings. The dog sled has been replaced by the snowmobile, and grocery stores have replaced the outdoors as the source of many foods. Nevertheless, many of the rural villages remain at least partly subsistence based, with hunting and fishing providing much of their diet. As Behnke (2002:161) notes, "the landscape is invested with meaning by village residents—imprinted with cultural and social significance.... From the village, the surrounding landscape is seen as a homeland." However, the way it is seen is different for those participating in subsistence activities, leading to conflict with non-Native sports fishermen and hunters.

The vast expanses of northern and western Alaska occupied by the Inuit and central Alaska inhabited by Athabaskan Indians are marked by very low population densities.

> From the air, much of rural Alaska appears to be wilderness, unused and unaffected by people. Village clearings, ribbons of trail, scattered campsites, and the overgrown remnants of old camps are the only signs of human use, and these are often difficult to see even from the ground (Behnke 2002:159).

These lands remain best known to the Natives, and they support their livelihoods. Yet the incredibly low density of population within much of Alaska results in few items of visible material culture upon the landscape. Indeed, in 2010 the vast Yukon-Koyukuk Census area, encompassing 147,842 square miles, had a population of only 5588—three-quarters whom are Native Americans, giving it a population density of one person per 26.5 square miles. Nevertheless, the Natives have given their names to many of the area's geographic features and have evolved an intimate relationship with and understanding of the wildlife.

Within the Seward Peninsula approximately 12,000 reindeer are herded upon tundra vegetation by Native herdsmen, a continuation of an activity that was introduced to the area over a century ago. Although far fewer than the hundreds of thousands of reindeer that grazed western Alaska over a half century ago, reindeer herding remains important to certain Native Alaskans. By federal law, non-Natives are prohibited from owning reindeer.

3.7 Many Western Native American Landscapes

This chapter has briefly discussed several of the numerous indigenous landscapes that are scattered across the western part of the United States. Yet, there are many additional tribes whose cultural landscapes can be seen. For numerous groups the geographic area that is occupied is small and often isolated in easily overlooked locations. In contrast, the Inuit are the dominant population across the vast region of northern and western Alaska—as well as much of northern Canada and Greenland, yet most of that land remains wilderness, only dotted with widely scattered settlements—most which bear little resemblance to the Inuit's traditional nomadic culture. Still, given their great distance from the national mainstream, the ethnic landscape of the Inuit is peripheral to the national core at best. Geographers have done little to study the ethnic landscapes of many of these smaller or more remote groups, yet this should not be construed as evidence that the group has not shaped all or part of the cultural landscape that one sees today.

References

Ahler, Stanley A., Thomas D. Thiessen, and Michael K. Trimble. 1991. *People of the Willows: The Prehistory and Early History of the Hidatsa Indians*. Grand Forks: University of North Dakota Press.

Archambault, David, II. 2016. Taking a Stand at Standing Rock. *New York Times*, August 25, A19.

Behnke, Steven. 2002. Alaska's Contested Rural Landscapes and the Subsistence Claims of Alaska Natives. In *Geographical Identities of Ethnic America: Race, Space, and Place*, ed. Kate A. Berry and Martha L. Henderson, 149–173. Reno: University of Nevada Press.

Calloway, Colin G. 2003. *The Vast Winter Count: The Native American West Before Lewis and Clark*. Lincoln: University of Nebraska Press.

Christafferson, Dennis M. 2001. Sioux, 1930–2000. In *Handbook of North American Indians, Volume 13, Plains*, ed. Raymond J. DeMallie, 821–839. Washington: Smithsonian Institution.

Cole, Douglas, and David Darling. 1990. History of the Early Period. In *Handbook of North American Indians, Volume 7, Northwest Coast*, ed. Wayne Suttles, 119–134. Washington: Smithsonian Institution.

DeMallie, Raymond. 2001. Introduction. In *Handbook of North American Indians, Volume 13, Plains*, ed. Raymond J. DeMallie, 1–13. Washington: Smithsonian Institution.

Drucker, Philip. 1955. *Indians of the Northwest Coast*. Garden City: Natural History Press (printing 1963).

Frison, George C. 2001. Hunting and Gathering Tradition: Northwestern and Central Plains. In *Handbook of North American Indians, Volume 13, Plains*, ed. Raymond J. DeMallie, 131–145. Washington: Smithsonian Institution.

Harris, Kate Michael Gonchar. 2016. Battle Over an Oil Pipeline: Teaching About the Standing Rock Sioux Protests. *New York Times*. November 30. http://www.nytimes.com/2016/11/30/learning/lesson-plans/battle-over-an-oil-pipeline-teaching-about-the-standing-rock-sioux-protests.html.

Krinsky, Carol Herselle. 1996. *Contemporary Native American Architecture: Cultural Regeneration and Creativity*. New York: Oxford University Press.

Nabokov, Peter, and Robert Easton. 1986. Native Americans. In *America's Architectural Roots: Ethnic Groups that Built America*, ed. Dell Upton, 16–35. Washington: Preservation Press.

——. 1989. *Native American Architecture*. New York: Oxford University Press.

Norris, Tina, Paula L. Vines, and Elizabeth M. Hoeffel. 2012. *The American Indian and Alaska Native Population: 2010*, Census 2010 Brief C2010BR-10. Washington: U.S. Department of Commerce.

Ogunwole, Stella U. 2002. *The American Indian and Alaska Native Population: 2000*, Census 2000 Brief C2KBR/01-15. Washington: U.S. Department of Commerce.

Parker, Albert J. 2010. A Park of the People: the Demotion of Platt National Park, Oklahoma. *Journal of Cultural Geography* 27 (2): 151–175.

Raff, Jennifer A., Margarita Rzhetskaya, Justin Tackney, and M. Geoffrey Hayes. 2015. Mitochondrial Diversity of Iñupiat People from the Alaskan North Slope Provides Evidence for the Origins of the Paleo- and Neo-Eskimo Peoples. *American Journal of Physical Anthropology* 157: 603–614.

Robbins, Jim. 2006. For Sacred Site, New Neighbors Are Far From Welcome. *New York Times*, August 4, 2006.

Schnell, Steven M. 2001. The Kiowa Homeland in Oklahoma. In *Homelands: A Geography of Culture and Place across America*, ed. Richard L. Nostrand and Lawrence E. Estaville, 139–154. Baltimore: Johns Hopkins University Press.

Shipek, Florence C. 1978. History of Southern California Mission Indians. In *Handbook of North American Indians, Volume 8, California*, ed. Robert F. Heizer, 610–618. Washington: Smithsonian Institution.

Thornton, Russell. 1987. *American Indian Holocaust and Survival: A Population History Since 1492*. Norman: University of Oklahoma Press.

Vogeler, Ingolf, and Terry Simmons. 1975. Settlement Morphology of South Dakota Indian Settlements. *Yearbook of the Association of Pacific Coast Geographers* 37: 91–108.

Voget, Fred W., and Timothy A. Bernardis. 2001. Crow. In *Handbook of North American Indians, Volume 13, Plains*, ed. Raymond J. DeMallie, 695–717. Washington: Smithsonian Institution.

Walker, James R. 1917. *The Sun Dance and Other Ceremonies of the Oglala Division of the Teton Dakota*, Anthropological Papers of the American Museum of Natural History, Vol. 16, Part 2. New York: American Museum of Natural History.

Worl, Rosita. 2008. Alaska Native Corporations. In *Handbook of North American Indians. Volume 2. Indians in Contemporary Society*, ed. Garrick A. Bailey, 140–147. Washington: Smithsonian Institution.

4 Native American Landscapes in the American Southwest

The southwestern United States contains numerous wonderful examples of cultural landscapes shaped by Native Americans. Unlike the eastern United States where so much of the indigenous legacy can only be seen in scattered earthen mounds and other archaeological sites, the Southwest has numerous robust archaeological and contemporary Native American landscapes. Here are the oldest continuously occupied settlements in North America. Acoma Pueblo in New Mexico has been constantly occupied for over 1000 years, while a portion of the Oraibi Pueblo of the Hopi in Arizona is similarly ancient, and Indians have occupied the area of that pueblo for the past 1500–1600 years.

The Southwest is also the homeland of the nation's largest concentration of Indians (See Fig. 3.2). Within Arizona the 2010 Census counted 353,386 Native Americans, who comprised 5.5% of the state's total population. In neighboring New Mexico 219,512 Native Americans were enumerated, accounting for 10.7% of that state's population. In Colorado another 107,832 Indians were counted (Norris et al. 2012). Unlike in Oklahoma and California, the two states with the largest Indian populations, with 482,760 and 723,225, respectively, the Native American populations of Arizona and New Mexico are occupying their ancestral homelands. In contrast, the preponderance of the Native Americans living in Oklahoma were resettled there from those states east of the Mississippi. Likewise, the majority of Indians living in California are also newcomers. They have come from both the core Native American homelands in the Southwest and the transplanted Indian Territory that became Oklahoma, as well as other parts of the country, Mexico, and Central America. Furthermore, nearly half of the Native Americans in California claim multiple ancestry, indicative of mixing with other races. In contrast, only one-seventh of New Mexico's Indians have such mixed ancestry. Likewise, only one-fifth of Arizona's Indians report multiple ancestry (Norris et al. 2012).

In many parts of the Southwest contemporary Native American societies exist in the vicinity of abundant archaeological remains of prehistoric civilizations. In some cases, the current residents are clearly the descendents of those who built the ruined pueblos, while in other cases, such as with the Navajo, the contemporary population is highly aware of the surrounding ruins, but avoid them, to keep from interacting with the spirits of individuals who were not their ancestors.

4.1 Prehistoric Cultures in the Southwest

Four major prehistoric culture groups have been identified in the Southwest: the Mogollon, the Hohokam, the Hakataya, and the Anasazi. These groups not only left impressive ruins of large buildings, irrigation systems, and roads, but modern Pueblo Indians are descendents of the Anasazi. As Alfonso Ortiz notes, "Pueblo peoples are the only one of the cultural groups identifiable as long ago as two millennia that have survived with clearly unbroken cultural continuity" (Ortiz 1979:3). The Pueblo Indians continue occupying their centuries-old villages into the Twenty-first Century, villages that clearly display connections in their architecture and material culture with their prehistoric ancestors.

4.1.1 Mogollon

The earliest of the Southwestern groups, the Mogollon, lived in southeastern Arizona and southwestern New Mexico beginning around 500 BC, and may have been the ancestors of the Hohokam and Anasazi. By the time the Mogollon had abandoned many of their village sites around 1400 A.D., their culture had become closely connected with that of the Anasazi (Martin 1979). While agriculture was important to the Mogollon, it was not nearly as dominant as with the other groups. The Mogollon's primary crops were corn (maize) and beans. The ruins of one of their small settlements, protected under a rock shelter, are preserved at Gila Cliff Dwellings National Monument in southern New Mexico.

J.A. Cross, *Ethnic Landscapes of America*, DOI 10.1007/978-3-319-54009-2_4

4.1.2 Hohokam

The Hohokam inhabited the arid Salt River Valley in the Phoenix area of central Arizona. George Gumerman, an archaeologist who spent his career studying this group, notes that Hohokam ruins are far less impressive than the cliff dwellings of the Anasazi. Other than "the ruins of ball courts and platform mounds, the most one sees even on the largest Hohokam sites is low, unimpressive heaps of hearth and a scattering of potsherds" (Gumerman 2007: 142). In contrast with their largely ruined village sites, the Hohokam created a highly sophisticated network of irrigation canals, whose evidence was conspicuous into the twentieth century. William Doolittle (1992, 2000), a geographer who has written extensively about Native American agriculture, describes how their irrigated agricultural successes set apart the Hohokam:

> The prehistoric Hohokam people who lived in the Salt and Gila river valleys distinguished themselves by developing some very large canal irrigation systems. Main canals ranged up to 10 m wide and 4 m deep. Some extended in length up to 30 km, had hundreds of smaller branch canals, and irrigated thousands of hectares (Doolittle 1992:389).

The extensive network of their abandoned canals clearly marked the ground upon which the city of Phoenix grew rapidly during the twentieth century. While modern urban development has covered most of the former Hohokam fields, portions of their canal system are still visible. For example, in Phoenix's Park of Four Waters near the Salt River, "spoil banks on each side of the canal are the largest prehistoric agricultural earthworks remaining in North America" (Doolittle 1992:390). So extensive were the irrigation works of the Hohokam that William Denevan notes that the "[c]omplex canal systems…in the Salt River Valley irrigated more land in prehistory than is cultivated today" (Denevan 1992:375). While modern farming activities and urban development have destroyed most of the prehistoric canal systems, it is known that some of the region's contemporary irrigation canals simply represent deepened or renovated Hohokam canals, or paralleled them (Doolittle 2000).

Although their canal system is the most widespread archaeological legacy of the Hohokam, they also left many house and village ruins. While most houses were pits that were surrounded by brush and mud walls (wattle and daub) that have long since eroded away, the Hohokam also built multi-storied rectangular structures, using puddled adobe. In these structures of puddle adobe or puddled mud, layer upon layer is added to the wall and dried, building them ever higher. The best remaining example is preserved at Casa Grande Ruins National Monument in Arizona (Fig. 4.1). This fourteenth century great house of four stories, constructed of caliche

Fig. 4.1 Hohokam buildings were typically constructed using puddled mud. Although this structure was built of less erodible caliche, a large protective structure was erected to help preserve it at Casa Grande Ruins National Monument in Arizona

Fig. 4.2 Portion of earthen ball court at Casa Grande Ruins National Monument in south central Arizona

(a type of sedimentary rock formed from soil naturally cemented by carbonate deposition), was in ruins even when first seen by Spanish explorers. Few other dwelling ruins are as readily identifiable, as most building sites are now low mud mounds, with the mud walls having been eroded over time, even in the arid environment. Hohokam ball courts (Fig. 4.2), which "appear as large, oval depressions with raised embankments" are scattered among 160 locations, with larger village sites sometimes having two, such that over 200 ball courts have been identified (Elson 2007:50). Ranging in size "from 60 to 250 feet long and from 30 to 90 feet wide" (Elson 2007:51), these ball courts apparently played a crucial role in the social and ritual system of the Hohokam, likely influenced by cultural elements from Mesoamerica. The ball courts became less culturally conspicuous beginning early in the twelfth century, and the Hohokam culture had disappeared before the first Spanish explorers came into Arizona. However, the O'odham Indians, who were once called Pima, are descendants of the Hohokam (Gumerman 2007). Besides Casa Grande, with its spectacular great house and large ball court, the archaeological site known as Snaketown in central Arizona is the best known of the other Hohokam ruins.

4.1.3 Hakataya

While the Hohokam were a dirt-oriented people who constructed canals, mounds, and structures of mud, the contemporaneous Hakataya "were 'rock-oriented' people—note their rock-outlined jacales, gravel or boulder alignments, rock-filed roasting pits, rock-pile trail shrines, thick dry-laid, low-walled rock or boulder structures" (Schroeder 1979:100), among other implements and ornaments of rock. Both the Hohokam and Hakataya constructed ball courts, with excavated mud embankments or stone walls, respectively (Fig. 4.3). Occupying that area of central Arizona north of the Hohokam and southwest of the Anasazi, the Hakataya were a more migratory people who only built pueblos for year round occupancy in parts of their eastern territory. To the west, along the Colorado River and across it into the Mojave Desert of southeastern California, trails and campsites are their more common archaeological legacy. Thus, considerable regional variation exists in the material culture of the Hakataya.

The Sinagua lived within the northeastern part of the Hakataya cultural area, an area broadly centered around the present-day city of Flagstaff, Arizona. While these people

Fig. 4.3 Stone walls surround a ball court at Wupatki ruin in Wupatki National Monument, northwest of Flagstaff, Arizona

were influenced by both the Hohokam and Anasazi, and it appears that colonies of these groups may have lived in their midst, they built smaller pueblos and generally lacked the ceremonial kivas of the Anasazi. Sinagua ruins are preserved in a several national monuments, including Montezuma's Castle, Tuzigoot (Fig. 4.4), Walnut Canyon, and Wupatki, all located in northern or central Arizona.

4.1.4 Anasazi

The Anasazi lived in the Four Corners Area, especially within southwestern Colorado, northeastern Arizona, and northwestern New Mexico. Of the Southwest's four major prehistoric cultures, they left by far the most extensive and conspicuous ruins. Numerous examples of ruins and abandoned buildings can be found within Mesa Verde National Park, Colorado; Canyon de Chelly National Monument, Arizona; Navajo National Monument, Arizona; Chaco Culture National Historic Park, New Mexico; Bandelier National Monument, New Mexico; Salinas Pueblo Missions National Monument, New Mexico; and Hovenweep National Monument along the Utah-Colorado border, plus a variety of other national, state, and tribal parks and monuments.

All these civilizations were substantially in retreat before the arrival of the first Spanish explorers in 1540. A major abandonment of many cliff dwellings and pueblos, including those at Mesa Verde and Chaco Canyon, had occurred about 1300, probably initiated by an extreme drought that occurred throughout the region between 1273 and 1297 (Benson and Berry 2009), with droughts also leading to abandonment of Mogollon and Hohokam settlements. Nevertheless, not all of the pueblos were abandoned, and many pueblos within central New Mexico remained occupied when the first Spanish explorers arrived. Some are still inhabited.

4.2 Prehistoric Anasazi Landscapes

Ruins of cliff dwellings and pueblos, together with remnants of ancient irrigation systems and roads, constitute some of the most widespread and conspicuous Anasazi landscapes (Figs. 4.5 and 4.6). The numbers of such ruins are extremely impressive. Within Mesa Verde National Park over 4000 ruins and 600 cliff dwellings have been discovered. Chaco Culture National Historical Park in northwestern New Mexico likewise includes nearly 4000 archaeological sites, including the ruins of the largest pueblos in the nation. While hundreds of additional pueblo ruins and cliff dwellings are found in other national parks and monuments scattered across the Four Corners region, numerous additional Anasazi ruins exist on the Navajo, Hopi, and Ute Indian Reservations as well as other public and private lands. Overall, the number of Anasazi ruins in the region runs into the many thousands.

Fig. 4.4 Southern Sinagua culture pueblo ruin atop ridge in Tuzigoot National Monument, Arizona

Fig. 4.5 This cliff dwelling, named White House Ruin, is located in an alcove in the sandstone cliff well above the canyon floor in Canyon de Chelly National Monument in northeastern Arizona. Note the large petroglyph located on the canyon wall below the cliff dwelling

Fig. 4.6 Multiple pueblo ruins are scattered across the valley floor in Chaco Culture National Historical Park in northwestern New Mexico. The ruins of Pueblo Bonito are shown in the foreground, while the ruins of another pueblo are visible in the distance

4.2.1 Pueblos and Cliff Dwellings

Both cliff dwellings and pueblos shared many similar features. They were multi-room structures with square rooms (Fig. 4.7). They were of multi-story construction, with masonry walls, log roofs or ceilings, with many of the rooms built for specific special uses. Yet pueblos and cliff dwellings differed in their protection from the environment and attack. Pueblos were constructed on the open ground. Today, most are largely in ruins, with their foundations and lower walls being best preserved. Cliff dwellings were built under rock overhangs, providing a rock shelter, with rock above, behind, and to their sides, shielding the structure from erosion and providing protection from enemies (Fig. 4.8). The approach to a cliff dwelling was by ladders, ropes, and toe or hand holds cut into rock. Thus, cliff dwellings and their contents had far superior preservation than pueblos, being protected from raindrop erosion and kept dry, with centuries-old wood poles and supports remaining (Fig. 4.9). Archaeologists discovered cotton cloth and stores of corn remaining when they first visited some cliff dwellings. In some areas, such as Mesa Verde National Park, both types of dwellings are found and they were occupied simultaneously.

The density of settlement was fairly high in some Anasazi communities. Villages could contain many buildings. Within many areas, such as in Mesa Verde and Chaco Culture National Parks, several pueblo ruins and cliff buildings are found in close proximity. Pueblos ranged in size from several rooms to hundreds (Fig. 4.10). Pueblo Bonito in New Mexico's Chaco Canyon is considered the largest, rising five stories high and containing approximately 800 rooms (Cordell 1979). It has been estimated to have had a population possibly as high as 1000. So massive was this structure that "it was the largest 'apartment house' in the world until a larger one was erected in New York in 1882" (Wormington 1947:86). Cetro Ketl Pueblo, also in Chaco Canyon, included about 500 rooms, used both for habitation and storage (Cordell 1979).

Multi-story construction was typical for both cliff dwellings and pueblos. Both today's still occupied Taos Pueblo and the long-abandoned Pueblo Bonito had five floors. The buildings had various watchtowers, which could be square or circular. Doorways and windows for both cliff dwellings and pueblos utilized stone or pole lintels. Because of the necessity for defense, pueblos did not have exterior windows on their lower floors. Their lower rooms were entered by a

Fig. 4.7 Square stone masonry rooms comprise much of the pueblo ruins preserved at Aztec Ruins National Monument in northwestern New Mexico. This Anasazi site, notwithstanding its name, had no connection with the Aztecs of Mexico

Fig. 4.8 Well preserved stone rooms and towers of large cliff dwelling protected by rock ledge. Cliff Palace is one of the over 600 cliff dwellings constructed by the Anasazi in Mesa Verde National Park, located in southwestern Colorado

Fig. 4.9 Masonry rooms with their log roof beams have been preserved for over seven centuries since their abandonment at the Betatakin cliff dwelling in Navajo National Monument in northern Arizona. While presently surrounded by the Navajo Nation, the cliff dwelling was constructed by the Anasazi, who were not ancestral to the Navajo

Fig. 4.10 Tyuonyi Pueblo ruins within Frijoles Canyon is estimated to have once had around 400 rooms, spread among three floors, of which only the ground floor remains visible. This pueblo is located in Bandelier National Monument in north central New Mexico

Fig. 4.11 Stone masonry walls of Pueblo del Arroyo in Chaco Culture National Historical Park in New Mexico. Note the careful layered arrangement of the stones and the formation of the windows

ladder through their roof. Nevertheless, not all pueblos or cliff dwellings were multi-floor structures, and Chaco Culture National Historic Park contains many smaller pueblos, averaging about ten rooms each. In addition, pueblos in some regions, such as south of Santa Fe, New Mexico "consisted of a large number of buildings, some connected, some separate" (Fliedner 1975:364).

4.2.1.1 Types of Walls

Stone masonry walls were standard among the Anasazi, and archaeologists have identified three distinct styles. Chacoan walls, found in today's Chaco Culture National Historical Park, are the most sophisticated (Fig. 4.11). Chacoan walls are typified by the use of thin tabular sandstone slabs to create a banded masonry, with layers of large stones being interspersed with layers of closely fitted tablet sized stones. The stones were faced by the Anasazi masons, so they fit together very closely. Lower walls averaged about four feet in thickness. The Mesa Verde style walls used well-shaped, relatively uniform stone blocks in a single wall, with larger stones outside and smaller inside. Mesa Verde style walls typically range from 1 to 3 feet thick. The Kayenta masonry style used undressed, irregular stones, with larger proportions of mud mortar. Kayenta style walls are one stone thick.

Painting covered many of the Anasazi walls. Interior walls were plastered with mud, with the lower walls painted reddish, while the upper walls were white or cream colored. Visitors to ruins to Bandelier National Monument can still see the colors within some of the cliff dwelling rooms.

Pueblos, and even cliff dwellings within large rock shelters where space was available, were often constructed around a plaza or courtyard, with the open areas used for sacred and festive activities. This space had a great ritual significance to the community, as an anthropologist and architect explain.

> The 'heart' of a pueblo was sometimes architecturally represented in a modest rock shrine or by an offering beneath the central plaza.... [T]he plaza itself seemed to incarnate this idea of a sacred, shared center space, with domestic rooms where people actually lived piled protectively around it (Nabokov and Easton 1989:368).

Several distinct pueblo layouts have been identified. In the plaza type the arrangement of houses defines a center space, with kivas and later churches located off-plaza, such as at today's San Felipe Pueblo. In the street type of pueblo arrangement the organization is linear, with the widest middle street serving as a dance ground, such as at Santo Domingo Pueblo. In the combination type, there is an elongated wide street, with no parallel streets, as seen at Zia Pueblo.

Pueblos and cliff dwellings had three main types of rooms: living rooms, storage rooms, and rooms for religious activities. The habitation or living rooms were used primarily for sleeping, because most activities took place outside

Fig. 4.12 Kiva exposed by collapse of its roof in the plaza of Pueblo Bonito in Chaco Culture National Historical Park, New Mexico

when possible. Storage rooms, consisting of many very small rooms, were used as granaries to store maize or other items. These lower rooms were primarily for food storage and food preparation. Near the storage rooms may be seen manos and metates, the grinding stones for preparing cornmeal by hand. The metates are represented by small hollows in the rock upon which the grain was ground using the hand-held mano. The most distinctive rooms for religious activities are the kivas.

4.2.1.2 Kivas

Kivas are circular underground rooms used by religious societies (males only), typically linked by kinship groups. The kiva belonged to the men of a clan and also served as men's club rooms, and are viewed as "one of the clearest hallmarks of Puebloan culture" (Ferguson and Rohn 1987:28). Every kin group had its own kiva, typically located south or southeast of the houses or dwelling rooms. Thus, pueblos had numerous kivas, with Pueblo Bonito having 37 (Fig. 4.12). Therefore, at many pueblo ruins, the walls of these circular rooms are a conspicuous part of the landscape. The kiva was entered by a ladder through a roof hole, which also served as a smoke hole. Within the kiva, there was a North-South orientation of the sipapu, a tube-like pit in the floor providing an opening into the underworld, and the hearth, with its firepit, a deflector to keep air circulating around the kiva, and a ventilating tunnel for air entry. Along the sides of the kiva were benches, as well as pilasters to support its roof.

Great Kivas were large, often 45–75 feet in diameter, much larger than the more typical kiva (Fig. 4.13). Four square columns supported the roof of a great kiva, and entry was by staircases from both the north and south sides. They had masonry fireboxes and floor vaults, and rooms were attached to the main chamber. The great kiva, such as the magnificent example preserved at the deceptively named Aztec Ruins National Monument in New Mexico, served as a site for community ceremonies.

4.2.2 Agricultural Legacy

Cliff dwellings and pueblo ruins are just part of the landscape shaped by the Anasazi. As Dietrich Fliedner (1975:371) explains, "Former fields can be recognized by the relief, by the condition of the soil, and by the relics of agricultural use." Among these relics are small terraces, marked by a riser of small stones following the contour of the field. To irrigate their fields of maize, beans, cotton, and squash, among other crops, the Anasazi constructed water diversion structures. With irrigation being critical, drainage ditches on Mesa Verde one mile long are visible. Other water diversion structures in Mesa Verde National Park include a reservoir with masonry walls that measures 100 feet across, terracing of fields with rock walls, small check dams along streams, and field creation with water diversion schemes into rock-bordered gardens. In many locations within the Southwest

Fig. 4.13 Great Kiva in ruin of Chetro Ketl Pueblo, located in Chaco Culture National Historical Park, New Mexico

archaeologists and geographers have identified earthen or rock weirs upstream of fields to divert runoff waters into irrigation channels (Doolittle 2000).

4.2.3 Roads of the Anasazi

The Anasazi created an extensive system of roads to provide transportation, as trade was important to them, and some of these roads or trails remain. The best examples are near Chaco Canyon, from which they seem to radiate. These roads are described by Linda Cordell.

> They are distinctive features in being virtually straight lines, not contoured to topographic relief. When they change direction, this is accomplished by making a sharp, angular turn, rather than by curving. Some of the roads are lined with masonry curbs; others are visible as slight swales on the landscape. Although some of the roads are narrow, others are nearly 30 feet wide. Well marked, planned, straight, wide roadways represent a great labor investment for people who did not have wheeled vehicles. It has been suggested that shrines along these roads may have served as signaling stations (Cordell 1979:140, Reprinted with permission of Dept. of Anthropology, National Museum of Natural History).

While not all of these roads are readily visible on the ground, "segments stand out distinctly at dawn or dusk and a few appear as a line of changed vegetation" (Obenauf 1991:34). From the air they are far more distinctive, and overall 640 kilometers (398 miles) of prehistoric roads have been mapped in the San Juan Basin, which includes Chaco Canyon (Mathien 1991).

Roads have been discerned at other prehistoric sites in the Southwest, but they are much narrower than those found near Chaco Canyon. For example, those near Gran Quivira are only half as wide, and those near Jemez are much narrower (Mathien 1991). Elsewhere Native American paths have been mapped that crossed the extensive volcanic flows within El Malpais National Monument in New Mexico.

Roads were more than just cleared paths. While some are clearly constructed with curbs, others trails have a braided appearance, such as those in Bandelier National Monument where "multiple stairs and trail segments form twisting and intertwined routes worn or constructed up the sides of canyons" (Snead 2002:760) have been observed. Conspicuous cut stairways within the cliffs at Chaco Canyon as well as masonry steps have also been described (Windes 1991). Some of the routes, whether a formal road or the more vernacular path, there and in Bandelier National Monument, are marked by stone cairns, "typically piles of tuff cobbles marking routes in places where no worn trail segments are visible or identifying locations along canyon rims where trails descend" (Snead 2002:761). While stacked rocks also indicate trails, occasionally petroglyphs marked trails and others led to rock art sites.

Fig. 4.14 Large assemblage of petroglyphs carved into stone wall at Newspaper Rock State Historical Monument, near Canyonlands National Park, Utah

4.2.4 Petroglyphs and Petrographs

Petroglyphs and petrographs were both created by the Anasazi and they are very common (Fig. 4.14). These are rock carvings and rock paintings, respectively, and one of the best places to see them is in Petroglyph National Monument, where thousands of carvings were made on lava rocks across the Rio Grande from downtown Albuquerque, New Mexico (Fig. 4.15). Many other locations throughout the Four Corners Region contain such petroglyphs. While some are found by cliff dwellings or pueblos on nearby cliffs, others exist where such ruins are not as evident (Fig. 4.16). The petrogylphs depict both real and mythical animals as well as many geometric shapes.

4.3 Contemporary Pueblo Indian Landscapes

Today's Pueblo Indians represent a cultural continuation from Anasazi times, although the peoples have moved considerably since 1300 and there was a great drop in number of settlements during the Pueblo Abandonment of the fourteenth century. Since the earliest Spanish exploration of the Southwest, an additional 61 pueblos were abandoned. Thirty-three pueblos remain occupied, with most being located in the Rio Grande Valley, ranging from Isleta, south of present-day Albuquerque, New Mexico to Taos in the north (Schroeder 1979). Besides Taos (Fig. 4.17), some of the better know pueblos along the greater Rio Grande Valley include the Santo Domingo, Santa Clara, and the Jemez Pueblos. To the west of Albuquerque lie the Laguna and Acoma Pueblos, while farther west, near the border with Arizona is the Zuni Pueblo. In northeastern Arizona is the Hopi Pueblo, which includes ten villages, of which the oldest is Orabi. Although the occupants of these communities are all called Pueblo Indians, in reality they consist of a variety of peoples, many with their own languages.

Among today's Pueblo Indians, "their sense of place and identity places the land itself, their place-in-the-world in a preeminent role both as homeland and sanctuary, and also as a place with religious associations that govern people's understanding of themselves" (Grugel 2012:798). Their religion ties them to their pueblo and their sacred places, and their rituals take place at specific locations at the pueblo, such as at the plaza or kiva. Thus, "the close interrelationships between space, sacred traditions and the system of values and standards predestine the reservation, in principle, to be the sole possible settlement location for any Pueblo Indian wishing to remain part of his or her society and religious system" (Grugel 2012:801).

Fig. 4.15 Petroglyph pecked onto basalt rocks in Petroglyph National Monument near Albuquerque, New Mexico. Park officials estimate that 20,000 such images are found in the monument

Fig. 4.16 These petroglyphs of human-like figures were carved into the sandstone wall of Fremont River Canyon, Capitol Reef National Park, Utah. These figures were carved by the Fremont People, a prehistoric group that was contemporaneous with and strongly influenced by the Anasazi

Fig. 4.17 The northern portion of Taos Pueblo rises five floors, and it faces a plaza and stream, across which the lower southern portion of the pueblo is constructed

4.3.1 Pueblo Housing's Historic Legacy

Pueblo housing is clearly a legacy from the past. Acoma and the Hopi Pueblos were established over 500 years before the arrival of the first Spanish explorers. This linkage to the past is eloquently noted by John Brinckerhoff Jackson, who describes the importance of the room in Pueblo construction.

> It is generally agreed that the basic unit in Pueblo architecture is the room. Since the width of a room is determined by the length of the roof beams, we rarely find a room much wider than fourteen feet, and a room of twelve by twenty feet would be accounted large. Many rooms, possibly used for storage, are no more than five feet square….
>
> Every Pueblo building, no matter what its size, is actually a cluster of such cells. Even the largest community house, like the one at Taos, six stories high and a quarter of a mile long, is a honeycomb of innumerable small rectangular rooms. It may be objected that this is also true of every modern European American building: are they not all essentially clusters of cells of varying dimensions? Yes, but they are designed and constructed as a unit subsequently divided. With us the frame…comes first; with the prehistoric Pueblo builder the cell or room came first and was then duplicated (Jackson, *A Sense of Place, A Sense of Time*, 1994:30–32, Copyright © 1994 by Yale University Press, Reprinted by permission of Yale University Press).

While both cliff dwellings and pueblos were configured from rooms, no contemporary Indian group occupies cliff dwellings.

The arrival of the Spanish in the late 1500s and early 1600s resulted in some changes in the appearance of many pueblos. The most conspicuous modification was the construction of a Roman Catholic church in a prominent location within the pueblo (Fig. 4.18), an institution whose presence for over three centuries now illustrates "the centrality of the churches in Pueblo identity" (Blake and Smith 2000: 378). An example of cultural mixing, including the integration of traditional religious beliefs with Catholic practices, the presence of these churches has been viewed as symbolizing the permanence and identity of their Pueblo residents. The massive fortress-like structures were built by Pueblo labor under the guidance of Spanish priests, have been maintained by continuing labor of the Pueblo members, are located adjacent to a cemetery in which centuries of Indians have been buried in layers of burials, and have come to both strengthen Pueblo identity and "inspire the Pueblo Mission Revival and Santa Fe architectural styles" (Blake and Smith 2000:370). These mission churches (see also the Hispanic Chaps. 5 and 6) can be found today within virtually all of the pueblos with the exception of those of the Hopi, although not all remain active sites of Catholic worship. Even in several pueblos abandoned after Spanish contact, such as Gran Quivera and Pecos, whose ruins are preserved in Salinas Pueblo Missions National Monument and Pecos National Historical Park, respectively, the massive masonry walls of the churches remain as conspicuous parts of the pueblo ruins.

Fig. 4.18 Ruins of Mission of San Buenaventura from Las Humanas Pueblo, part of the Gran Quivira group in Salinas Pueblo Missions National Monument in central New Mexico. This site was abandoned in the 1670s

4.3.1.1 Hopi Pueblo

Because of geographic isolation and minimal contact with the Spanish, the Hopi are among the Pueblo Indians least acculturated by contacts with either other Pueblo Indians or Europeans. With a reservation population of 7185 in 2010, the Hopi continue living in their millennia-old village while tilling their cornfields in an area without even intermittent streams. Their maize is often planted by hand in floodplains or on dry farmed sand dunes near the base of Black Mesa, in rows that are distinctively less regular than those that are plowed by a tractor (Fig. 4.19). In addition, complex rock-faced terraces are used by the Hopi at First Mesa to create level irrigated fields on steep slopes, although the area cultivated today is far less extensive than in the past (Doolittle 2000).

Hopi housing is typical of pueblo buildings in the Southwest and is "clearly a continuation of the building tradition of the Anaszai culture which occupied much of the Colorado Plateau during the eleventh to fourteenth centuries" (McIntire 1971:511). Because the Hopi pueblos consist of multi-dwelling units, often of several stories, with each room built separately, the age of the structure varies, just as the age of buildings varies along the main street of most American communities. In parts of the pueblo are new rooms, some parts are under construction, while other portions are in ruins. Likewise, because of the continual construction of new rooms and demolition or decay of abandoned rooms, the overall shape and configuration of pueblos is constantly evolving over time.

Each building or house within the pueblos of the Hopi is built by a specific clan, adjacent to other houses of the same clan. This resulted, as explained by Eliot McIntire, in the creation of

> [l]ong rows of terraced two- and three- story structures developed by the gradual addition of rooms. The lower story seldom had any openings leading directly outside. Entry was usually through a roof opening. The roof was reached by means of a similar ladder.... Doors and windows were much more common on the upper floors, and the roof entry was not usually found above the ground floor (McIntire 1971:512).

The adjacent buildings that constituted the pueblo were themselves arranged around enclosed courtyards.

Many modern changes have occurred in the Hopi pueblos, which are a conglomeration of the ancient and modern styles. For example, even four decades ago, new houses were often separate from the compact village, doors are placed on the ground level, larger windows are used, and the structures have gabled rather than flat roofs. At the same time, many aspects remained little changed, such as the continuing use of masonry walls and a single room is still a complete living unit (McIntire 1971). Yet new houses are being built at locations away from the mesas, and more in the valleys by the Hopi today.

Fig. 4.19 Cornfields planted in irregular rows outside Indian village of Moenkopi, Arizona

4.3.2 Blending Ancient and Modern Construction

Overall changes are occurring in the pueblos of many different tribes. During the last half of the twentieth century the architectural historian Bainbridge Bunting (1976:30) lamented that, "pueblos are rapidly losing their architectural integrity as foreign materials …disrupt the old harmony of adobe and timber that was so simple and so beautiful." He listed a variety of new types of construction materials, including concrete block and hard stucco, that were being utilized in walls, and he noted that windows have been added to many walls, and that the front walls of many structures in the pueblos are now more aligned than in the past.

Similar changes have been noted at Acoma Pueblo (Figs. 4.20 and 4.21), whose reservation had as population of 3011 in 2010. With the construction of tract housing by the federal government 50 years ago and stronger nuclear families less tied to kinship groups, many of the Acoma moved off the butte top pueblo to Acomita and McCarty's, which are more modern communities located near the expressway (Nabokov and Easton 1989). Nevertheless, these Indians "retain strong emotional and proprietary ties" to the historic pueblo, visiting it during traditional feast days and providing a small number of members to "guard the deserted community year-round and guide tourists" (Nabokov and Easton 1989:379). Yet not all modern accoutrements have entered the historic pueblos. For example, neither Taos nor Acoma have permitted electrification in their old pueblos, nor has modern plumbing entered the historic core of the pueblos.

Even with their changes, many of the remaining pueblos provide a wonderful window into the past, and some, such as those at Taos and Acoma, welcome visitors while at the same time retaining a strong ethnic pride in preserving their heritage. Within the Taos Pueblo visitors can roam the plaza separating the two distinct clusters of pueblo buildings, can see the ramadas or wooden drying racks, carefully inspect the adobe walls of the buildings and enter a variety of shops that occupy some of the ground floor rooms facing the plaza, some which are warmed with adobe fireplaces in their corners. Traditional crafts, as well as fry bread and bread baked in outdoor ovens called hornos (and described in Chap. 5), are sold to visitors (Fig. 4.22). While visitors can wander much of the exterior area of the pueblo, they are excluded from the private dwelling spaces of the residents and cannot approach the area of the kivas. While the Taos and Acoma Pueblos welcome visitors, charging them an entry fee, requiring photographers to pay an additional fee, and providing a steady stream of customers to shop owners, some pueblos remain private communities, not open to outsiders. Even the Acoma and Taos Pueblos are closed to outsiders at the time of certain ceremonials.

While a few Pueblo Indians still maintain year-round homes within the traditional pueblos, far more only occupy

Fig. 4.20 Street of contiguous houses within Acoma Pueblo in central New Mexico. Ladders are used to reach rooms on upper floors, and the different types and age of masonry construction of the buildings are clearly visible. While doorways open to the street today, at one time pueblo buildings lacked doors and windows on their lowest levels

Fig. 4.21 Flat roofed buildings in Acoma Pueblo, New Mexico. Note that the building to the left has been recently plastered with adobe and that several cars are parked in the pueblo. At one time the pueblo was only accessible by a narrow footpath climbing the cliff to reach the mesa top, upon which the pueblo was constructed for defensive purposes

Fig. 4.22 Hornos are outdoor ovens in which bread can be baked and pottery fired. This horno against the wall is one of many at Taos Pueblo, in north central New Mexico. Often hornos are freestanding structures located away from the walls

their ancestral home at times of ceremonials, or have no house at all at the pueblo. By the mid-1990s, "[s]ome Pueblo Indian landscapes ha[d] already been significantly altered by the construction of new houses that resemble tract bungalows found in housing developments in many American suburbs" (Carlson 1996:58). Yet even with all the new development, the ancient ethnic landscape of the Pueblo Indians is well preserved in a region that also encompasses the Hispano homeland and is not far from the Navajo homeland.

4.4 Navajo Landscapes

The Navajo, also called Dine, occupy the Four Corners Region, particularly that part within Arizona and New Mexico, and their "homeland is among the most remote from Euro-American population centers in the lower 48 states" (Arthur and Diamond 2011:911). The Navajo Nation is the largest Indian group in United States living in a contiguous parcel. Unlike most tribes whose numbers fell precipitously following contact with whites, the Navajo population today is much larger than it was when Europeans first encountered the group. Their remoteness from early Spanish and American settlers, the region's environment, compatible with "the Navajos' life-style of dispersed families practicing a mixed economy of hunting, herding, and farming, but unattractive to Euro-Americans," and "their cultural flexibility and selective borrowing…may have contributed to the Navajo population explosion" (Arthur and Diamond 2011:912). With a 2010 Census population of 332,129, up from 298,197 in 2000, the Navajo population is second only to the Cherokee. However, more Navajo report just Native American ancestry than among the Cherokee, who counted 819,105 in total, yet over two-thirds of these individuals reported combinations of ancestry (Ogunwole 2002; Norris et al. 2012). The Navajo have more tribal members on their reservation, with 173,667 living on either the reservation proper or on off-reservation trust lands in Arizona, New Mexico, and Utah in 2010, down from 181,269 a decade earlier. The Navajo have stricter requirements for tribal membership (one-fourth Navajo) than the Cherokee, who only require that members document a single ancestor with Cherokee heritage. The Navajo Nation Reservation covers 27,000 square miles, including parts of three states, and it is by far the largest Indian reservation in the United States.

Because of their growing population, the Navajo have expanded considerably to the east, south, and west of their reservation onto U.S. Bureau of Land Management lands since the middle of the twentieth century. The U.S. Congress legislated a settlement of a territorial dispute between the Navajo and the Hopi, with whom the Navajo formerly shared a joint use area. Because of the Navajo's expanding population, they had increasingly pushed the Hopi out of the shared area, and the arrangement gave the Navajo more lands elsewhere.

Fig. 4.23 Scattered rural homestead clusters of the Navajo in the Monument Valley near Goulding, in southeastern Utah. Note how several of these clusters contain modern mobile homes

Nevertheless, many Navajo have resisted relocation from the Hopi lands that they and their ancestors have occupied for over a century.

The Navajo have lived in the Southwest for a shorter time than the Pueblo peoples. They are an Athabaskan people who arrived between around 1000 and 1400 AD, yet there was some intermixing with Pueblo Indians, particularly during the period of the Pueblo Rebellion between 1680 and 1692 (See Chap. 5). The intermixing influenced their agricultural practices and resulted in a blend of Apache and Pueblo culture (Jett 1992a). Contact with the Spanish, particularly via Pueblo groups, brought livestock raising (sheep and goats) to the Navajo.

4.4.1 Navajo Settlement Pattern

The settlement pattern of the Navajo is almost the exact opposite of what is found among the Pueblo nations. The population of the Navajo is widely dispersed over their large reservation, which ranges from desert plains to forested mountains and mesas (Fig. 4.23). Stephen Jett, a geographer who has extensively studied the Navajo, notes that "[t]here are practically no urban areas and very few villages in the Navajo Country, and those few that do exist are sites of commercial, administrative, health, and educational services with significant non-Navajo staffs" (Jett 1978:359). The Navajo are only found in small agglomerations, which are "economically cooperating extended families" (Jett 1980:101). While the Navajo own rights to use the lands occupied by their homestead and the lands they farm (if they grow crops), traditionally lands were not owned by individuals (Jett 1998). Similarly, grazing land are communal, although families have rights to grazing permits. Because individuals and families do not hold formal ownership of the lands, fencing "is still absent over large areas" (Jett 1998:117).

Their homestead groups or camps are shaped by the Navajo's matrilineal society that determines residence groups.

> The residence group is the fundamental unit of Navajo social organization. It is organized around a head mother, a sheep herd, a customary land-use area, and sometimes agricultural fields.... The residence group is both a social and economic unit. It is a cooperative unit organized, structured, and integrated by the symbols of motherhood.
>
> The personnel of the residence group are organized around a head mother. All rights of residence within the group are ultimately derived from the head mother or the unit.... A Navajo may live wherever his mother has the right to live. In addition, a Navajo may live wherever his or her spouse has the right to live (Witherspoon 1983:525, Reprinted by permission of Dept. of Anthropology, National Museum of Natural History).

Because of their population growth, many Navajo now live outside their traditional residence groups. While the area occupied by the Navajo is rapidly expanding, they still maintain close ties with their birth clan, even though they have dispersed considerable distances. Ties to the land are highly spiritual. As Stephen Jett (1998:119) explains, "More traditional Navajo believe that they are literally connected with Mother Earth at birth, through local burial of their umbilical cord and placenta." Such strong ties to place enhanced the sense of dislocation by those Navajo forced from the lands of their birth by the Navajo-Hopi land dispute.

The family homestead of the Navajo is characterized as having multiple sites, with each homestead group maintaining permanent buildings at two or more camps. These sites are occupied seasonally, depending on pasturage, water, firewood, farming, and other work opportunities. Thus, many scholars argue that many Navajo are either nomadic or seminomadic, and this is particularly true of those who range over wide areas with their herds of sheep and goats.

4.4.2 Navajo Homestead Buildings

The Navajo family homestead usually contains several buildings (Fig. 4.24). As described by Stephen Jett,

> a Navajo homestead may consist of anything from a simple log and-brush summer shade (ramada) or tent, to a grouping of 20 or more hogans (traditional circular or polygonal dwellings), houses, and shades, one or more corrals, and often other sundry structures such as privies, beehive ovens, animal shelters, root cellars, sudatories [sweat baths], and the like. A camp may be a permanent homesite or temporary camp of an individual, a nuclear family, or a man and his plural wives and their children, or it may comprise the several dwellings and other structures belonging to members of a homestead group which…may include grandparents, siblings and their spouses and children, and occasionally other persons. Sometimes, too, loose clusters of dwellings of adjacent homestead groups may be fairly close to each other, forming what appears to be a single large camp or small hamlet. There is, therefore, no 'typical' Navajo homestead, although a couple of hogans (or, today, a house or two and a hogan), plus a shade and a corral would be a common configuration (Jett 1980, "The Navajo Homestead: Situation and Site," *Yearbook of the Association of Pacific Coast Geographers,* 102–3; Reprinted by permission of the American Association of Geographers).

Typically, two or more related families often live nearby. Each family will have its own home along with several outbuildings, such as a brush corral for animals, dugouts for storage, and a ramada (a four posted shade). Today's homesite typically contains more buildings than in the past, and by the beginning of the twenty-first century many of the rural homesites included a mobile home.

Houses found on Navajo homesteads today show a mix of modern American and indigenous styles. While modern-style houses are increasingly dominant, especially frame houses and mobile homes, many traditional style dwellings remain and continue to be built. Hogans are the vernacular Navajo dwelling, of which several types were built (Kluckhohn and Leighton 1962).

Fig. 4.24 Navajo eight-sided cribbed log hogan and ramada on the floor of Canyon de Chelly in northeastern Arizona. The Navajo have retained the right to farm, graze, and live upon the lands within Canyon de Chelly National Monument, which is surrounded by the Navajo Indian Reservation

4.4.3 Hogans

The hogan has been described as "the most distinctive landscape manifestation of Navajoness" (Jett 1998:121). While the traditional hogan had three forked poles for support, which then held up a large number of other poles, it is very rare today. Also termed a male hogan, it was "essentially an adobe pyramid with an extended doorway" (Nabokov and Easton 1989:28). A six sided structure, with four forked poles for support, is the more common form. It was basically six walls around a square. Other hogans are eight sided while some are round. These could have walls built of small logs, slightly notched at their ends, or might have walls of rock, stone, mud, or even rarely concrete block. The eight walled or rounded structure has been termed a female hogan (Nabokov and Easton 1989) or a corbelled-log hogan (Jett 1992a). The roof logs of the hogan were gradually built towards center, forming a crib-work roof shaped like the top of a beehive. This corbelled-log roof was covered with earth, with an opening in the roof center for smoke hole. Contemporary hogans often have a woodburning stove with a central chimney. In addition, more recently built hogans have utilized more modern building materials, such as tarpaper roofing and plywood siding (Noble 1984). Doors of hogans face east.

The traditional hogan has been characterized as demonstrating great environmental adaption, given that the thick walls help keep the occupants warm in the winter, yet cooler in the summer (Kluckhohn and Leighton 1962). With its centrally placed fire, those surrounding it were kept warm. While many contemporary hogans have been built in traditional ways, others are actually sold as prefabricated structures to be assembled on site. Hogans not only provided shelter, but they also have great religious significance to the Navajo. Stephen Jett (1998:121) notes that hogans are "perceived as a microcosm of the Navajo cosmos." As Kluckhohn and Leighton (1962:89) explain, "curing chants can be carried on only in a hogan." They explain that to the Navajo, the hogan was more that just "places to eat and sleep," as it "occupies a central place in the sacred world" (Kluckhohn and Leighton 1962:89). The door into the hogan faces the morning sun. The hogan should not be defiled by any death, and bodies should not be removed through the doorway, but through an opening made in the northern wall of the hogan. If a death occurs in a dwelling, particularly a violent or unexpected death of someone other than an elderly person, the structure may be destroyed or abandoned. Likewise, the Navajo may destroy a hogan if bad luck is associated with it or if it is hit by lightning (Jett 1980).

Traditionally hogans were located in sheltered secluded locations. As Stephen Jett explains, "the preferred dwelling site generally has been a sheltered cliff base or slope, a ravine head, a canyon alcove, or a valley, generally with southern or eastern exposure and preferably near or on the edge of a grove of trees and not far from water" (Jett 1980:113). The Navajo avoided construction of homes in box canyons and near caves. Likewise, the Navajo avoided prehistoric ruins, as they were thought to be ghost infected (Jett 1980). Today, many modern dwellings are more conspicuous and few Navajo live in hogans. Yet, hogans can be found in many residence camps. Indeed, as Jett (1992a:344) explains, "By the late twentieth century, frame houses had become by far the predominant Navajo owner-built dwelling, with mobile and modular homes becoming increasingly common."

4.4.4 Ramadas, Sheep Herding, and Trading Posts

Ramadas provide a simple shelter during the summer. They are typically of light construction, are an easily erected collection of poles roofed with branches, brush, or grasses, and typically open on three sides for ventilation. Only the side facing the prevailing wind is blocked by poles or brush to provide shelter. Ramadas are frequently found in summer grazing camps (Fig. 4.25).

Herding of sheep and goats has been an important economic activity of the Navajo for several centuries. Herds are organized around a residence group. The sheep are tended in a common herd, with herding tasks shared. Poverty is widespread, because far fewer persons can be supported at an adequate income level by such grazing activities, given the aridity of the region (Aberle 1983). While Navajo weavers have woven traditional rugs using the wool from their sheep and goats, it is a very laborious process. When one considers the effort to dye the wool, to spin the thread and to weave the rugs, it was estimated to provide an average wage of $1 per hour for rugs in the late 1980s. A decade earlier it was reckoned that it took 345 hours to make a three by five foot rug (approximately one by one and a half meter), including the time required for shearing of the sheep; dying, carding, and spinning the wool into thread; stringing the loom; and weaving the rug, bringing the weaver a payment of about thirty cents an hour (Roessel 1983). Although the price of rugs in the twenty-first century is now higher, the sale of the rugs, whether through a traditional trading post or from one of the numerous small stands lining the main highways, often brings the weaver an income that is far below the federal minimum wage.

Through the first three-quarters of the twentieth century trading posts played a major role in economic life of the Navajo. Operators of these trading posts first encouraged the weaving of rugs instead of traditional blankets for market. The trading posts bought silver jewelry and marketed crafts. Navajo silversmiths craft a wide variety of belt buckles, bolo ties, and jewelry for both their own use and sale to tourists.

Fig. 4.25 Herd of sheep gathered around hut, ramada, and corral on Navajo Indian Reservation north of Rough Rock, Arizona

In its role as pawnbroker, especially of silver, the trading post served as a primary source of consumer credit in many areas of the reservation. Today the role of the trading post has been greatly diminished by improved roads and the establishment of shopping centers in various cities, yet a few remain, including Hubbell Trading Post in Ganado, Arizona, which is now operated by the National Park Service.

4.4.5 Sacred Places in the Navajo Landscape

The Navajo recognize many sacred places in their landscape. Mountains are the most sacred of Navajo places. Four mountains serve as the "symbolic and preordained frontier markers of Navajo territory" (Jett 1992b:31). These are Blanca Peak in south central Colorado to the east, Mt. Taylor (near Grants, New Mexico) to the south, the San Francisco Peaks (near Flagstaff, Arizona) to the west, and Hesperus Peak (in the La Plata Mountains of southwestern Colorado) to the north. To the Navajo, these mountains are sacred, and Blessingway Singers, or spiritual healers, keep bags of soils from each of these peaks, which are to be revisited every 12 years. The multicolored soils are utilized for the creation of sand-paintings that figure so prominently in sacred healing rituals. Various other buttes, lakes, mountains are also sacred, including Canyon de Chelly and Rainbow Bridge.

> The power of these sacred places sanctifies the people, and it can be tapped for specific ritual purposes, mostly curing through restoration of harmony, by Singers (medicine men). But to the noninitiate Navajo, the power is dangerous, and sacred places are, for them, not to be trespassed upon (Jett 1992b:30).

The administrative headquarters building of the Navajo Nation faces Window Rock (Fig. 4.26), a two hundred foot high natural arch that figures prominently within the Water Way Ceremony, with a nearby spring providing sacred water for hallowed rites that seek abundant rainfall.

Conflicting values placed upon the landscape have resulted in what the Navajo view as desecration of their sacred places. For example, roads have been built, mines dug, power lines strung, and communication towers erected on or across sacred spots. Volcanic features, which the Navajo view as sacred, are excavated and used for road gravel. To the Navajo these desecrations lead to anxiety, and "some attribute…social disintegration to these desecrations" (Jett 1992b:37).

The Navajo have also created numerous works of rock art, both by carving and painting. Although far less studied than the works of the Anasazi, hundreds of examples of Navajo rock art have been noted, including human forms, animals, and linear designs (Kolber 1998).

The Navajo homeland, as explained by Stephen Jett (2001:182), "is the most notable of America's ethnic homelands" because of its size, extent, and homogeneity.

Fig. 4.26 Navajo Nation Veterans Memorial is located by the sacred Window Rock. The capital of the Navajo Nation in Window Rock, Arizona faces this geological formation

He argues that the Navajo sense of homeland is particularly strong, given their "perception of the sanctity of the foreordained Dinétah, laid out specifically for the People," which now constitutes the Navajo reservation, where they can practice "many aspects of a traditional culture that exists in clear contrast to that of mainstream America" (Jett 2001:183). Yet as we have seen, the Navajo homeland overlaps the even more ancient homeland of the Hopi, and there are many Indian groups besides the Navajo and Pueblo peoples who view the region as their ancestral home.

4.5 Other Southwestern Indian Landscapes

While the cultural landscapes of the Navajo and the various Pueblo groups are the most conspicuous of the Southwestern Indians, many other groups have also made a distinct ethnic imprint on the landscape. Although geographers have not studied all of these cultural landscapes in detail, several nice overviews have been published and provide the examples cited here.

4.5.1 Mescallero Apache Landscapes

The Mescallero Apache are another Athabaskan people whose traditional housing aids their cultural survival. Thus, "the vernacular built-form environment on the Mescalero Apache Reservation in New Mexico is a blend of traditional and modern architecture" (Henderson 1992:16). Traditionally the Mescallero Apache favored temporary locations and rapidly built structures, such as an oak arbor, the brush wickiup, and the tipi, adopted from Plains culture. The federal government promoted dispersed, agricultural settlements. Today we see the resettlement of these Indians into matrilocal groups. Because of their matrilocal society, it is important to them to maintain a close proximity to female kin. Sisters may live together in the same household. As Henderson (1990:237) concludes, "Traditional preferences for matrilocality and proximity of residences, along with a desire to maintain large expanses of land for extensive resource use, are exhibited in the modern cultural landscape of the Mescalero reservation."

Modern reservation housing is American-style, yet reminders of the ethnic traditions are visible. Henderson explains,

> Modern Mescalero residences include both contemporary house forms and evidence of arbors, wickiups, and tipis. Tipi poles are often leaned vertically against modern houses. Occasionally, arbors are constructed in the summer and continue to provide outdoor space for food-processing, child-rearing, and social interaction. Tipis create outdoor space for sleeping during the summer.
>
> More importantly, arbors and tipis are constructed today for ceremonial purposes (Henderson 1992:25).

Thus, the Native American imprint on the landscape of the Mescallero Apache is not as distinct as that of the Navajo, yet material culture is not all that defines a homeland.

In contrast with the Mescalero Apache, the Chiricahua Apache were forcibly relocated from their ancestral homeland in southeastern Arizona and southwestern New Mexico in 1886, some being settled among the Comanche and Kiowa on their reservation near Fort Sill, Oklahoma, while the remainder joined the Mescalero Apache reservation in New Mexico. Yet, some of the Chiricahua Apache seek to regain part of their homeland, which at present is best defined by historical heritage sites and toponyms. Mountains were central to marking the homeland, and as Daniel Arreola (2012:117) explains, "mountains were central to the spiritual and belief systems of Chiricahua Apache, as they were for other Apache groups, a centrality that reinforces attachment to place." Canyons, springs, and isolated refuges also figured prominently within their homeland, and the Chiricahua Apache have acquired small landholdings to reestablish its presence within the traditional homeland—strong evidence of heritage ties to homeland when all that remains are sites of historic events that may be visited on pilgrimages. Yet as Arreola (2012:128) concludes, "Places, therefore, become part of the memory of a past that shapes a consciousness in the present and thereby validates that past for people."

4.5.2 O'odham (Pima and Papago) Landscapes

The O'odham Indians, until recently termed the Pima and Papago Indians, occupy an area along the Gila and Salt Rivers in south central Arizona. They are distributed similarly to the earlier Hohokam, and they are irrigation farmers much like their predecessors. Indian water rights are increasingly important in maintaining their traditional agrarian economy, and these water rights for many years were not exercised fully. Extensive irrigated fields of various grain and other crops are now seen on their reservation. The Indians view agriculture as a way of life and vital economically, while non-Indian farmers have been rapidly selling their water rights for urban usage. Some geographers have argued that if such behavior continues, within a few decades the Pima and Papago Indians may be only remaining farmers in southern Arizona.

The traditional housing of these Indians was the ki, a brush and mud-covered structure, typically excavated a foot and half deep into the soil, and arranged around four support posts in a hemispherical fashion. Later kis were rectangular, built with saguaro cactus rafters, walls consisting of a lattice work of twigs and other materials. Both the roof and walls were covered with mud and stray mud plaster. In the summer these Indians used ramadas like the Navajo.

Contemporary housing of the O'odham evolved to adobe buildings, or a "sandwich house" with adobe placed between boards vertically stacked. Today, many use modern American homes, however traditional ceremonial structures remain. These include the Rain House, which displays the traditional ki construction methods, built to protect saguaro cactus wine used in a summer festival. The Pima and Papago always have a watto (ramada) by the Rain House.

4.5.3 Many Ethnic Landscapes in the Southwest

Other Native American tribes also live within the Southwest. The Apache themselves are broken into several additional groups, with the San Carlos Indian Reservation covering a large area about a hundred miles east of Phoenix, Arizona. A variety of smaller Indian groups are scattered along the Colorado River, including the Havasupai and Hualapai within the Grand Canyon, several groups of Mojave Indians farther downriver, and Yuman groups near the Mexican border. South central Arizona is home to both Tohono O'Odham and Yaqui Indians. The Coachella and Imperial Valleys of California include a variety of small Indian territories, some arranged in a checkerboard pattern because of federal land grants. Both Nevada and Utah are home to many sizeable Indian groups, including the Pyramid Lake Reservation occupied by Paiutes and the Ute Mountain Reservation, respectively. Even though this chapter described Pueblo Indians in detail, it concentrated upon only a few of these disparate indigenous nations.

The visitor to the Southwestern region is rewarded with a richly diverse variety of ethnic landscapes, where homelands of several groups intersect. For example, the Navajo Nation surrounds the homeland of the Hopi, a Puebloan people, yet its eastern boundary abuts the homelands of other pueblo groups as well as the Hispano Homeland. Thus, the ethnic landscapes of the Southwest were not only shaped by indigenous peoples, but also by the Spanish, as discussed in the next two chapters (Chaps. 5 and 6).

References

Aberle, David F. 1983. Navajo Economic Development. In *Handbook of North American Indians, Volume 10, Southwest*, ed. Alfonso Ortiz, 641–658. Washington: Smithsonian Institution Press.

Arreola, Daniel D. 2012. Chiracahua Apache Homeland in the Borderland Southwest. *The Geographical Review.* 102 (1): 111–131.

Arthur, Roman, and Jared Diamond. 2011. Understanding Tribal Fates. *Science* 334: 911–912.

Benson, Larry V., and Michael S. Berry. 2009. Climate Change and Cultural Response in the Prehistoric American Southwest. *KIVA*

The Journal of Southwestern Anthropology and History 75 (1): 89–119.

Blake, Kevin S., and Jeffrey S. Smith. 2000. Pueblo Mission Churches as Symbols of Permanence and Identity. *The Geographical Review*. 90 (3): 359–380.

Bunting, Bainbridge. 1976. *Early Architecture in New Mexico*. Albuquerque: University of New Mexico Press.

Carlson, Alvar. 1996. Cultural Persistence in New Mexico's Rio Arriba: The Case for the Designation of a Rural Historic Culture Region. In *Ethnic Persistence and Change in Europe and America: Traces in Landscape and Society. Veröffentlichungen der Universität Innsbruck 213*, ed. Klaus Frantz and Robert A. Sauder, 47–66. Innsbruck: University of Innsbruck.

Cordell, Linda S. 1979. Prehistory: Eastern Anasazi. In *Handbook of North American Indians, Volume 9, Southwest Pueblo*, ed. Alfonso Ortiz, 130–152. Washington: Smithsonian Institution Press.

Denevan, William M. 1992. The Pristine Myth: The Landscape of the Americas in 1492. *Annals of the Association of American Geographers* 82 (3): 369–385.

Doolittle, William E. 1992. Agriculture in North America on the Eve of Contact. *Annals of the Association of American Geographers* 82 (3): 386–401.

——. 2000. *Cultivated Landscapes of Native North America*. Oxford: Oxford University Press.

Elson, Mark D. 2007. Into the Earth and Up to the Sky: Hohokam Ritual Architecture. In *The Hohokam Millennium*, ed. Suzanne K. Fish and Paul R. Fish, 48–56. Santa Fe: School for Advanced Research Press.

Ferguson, William M., and Arthur H. Rohn. 1987. *Anasazi Ruins of the Southwest in Color*. Albuquerque: University of New Mexico Press.

Fliedner, Dietrich. 1975. Pre-Spanish Pueblos in New Mexico. *Annals of the Association of American Geographers* 65 (3): 363–377.

Grugel, Andrea. 2012. Culture, Religion and Economy in the American Southwest: Zuni Pueblo and Laguna Pueblo. *GeoJournal* 77: 791–803.

Gumerman, George J. 2007. The Hohokam: The Who and the Why. In *The Hohokam Millennium*, ed. Suzanne K. Fish and Paul R. Fish, 141–146. Santa Fe, NM: School for Advanced Research Press.

Henderson, Martha L. 1990. Settlement Patterns on the Mescalero Apache Reservation Since 1883. *Geographical Review* 80 (3): 226–238.

——. 1992. Maintaining Vernacular Architecture on the Mescalero Apache Reservation. *Journal of Cultural Geography* 13 (1): 15–28.

Jackson, John Brinckerhoff. 1994. *A Sense of Place, a Sense of Time*. New Haven: Yale University Press.

Jett, Stephen C. 1978. The Origins of Navajo Settlement Patterns. *Annals of the Association of American Geographers* 68 (3): 351–362.

——. 1980. The Navajo Homestead: Situation and Site. *Yearbook of the Association of Pacific Coast Geographers*. 42: 101–117.

——. 1992a. The Navajo in the American Southwest. In *To Build in a New Land: Ethnic Landscapes in North America*, ed. Allen G. Noble, 331–344. Baltimore: Johns Hopkins University Press.

——. 1992b. An Introduction to Navajo Sacred Places. *Journal of Cultural Geography* 13 (1): 29–39.

——. 1998. Territory and Hogan: Local Homelands of the Navajo. In *Diné Bíkéyah: Papers in Honor of David M. Brugge*, ed. Meliha S. Duran and David T. Kirkpatrick, 117–128. Albuquerque: Archaeological Society of New Mexico.

——. 2001. The Navajo Homeland. In *Homelands: A Geography of Culture and Place across America*, ed. Richard L. Nostrand and Lawrence E. Estaville, 168–183. Baltimore: Johns Hopkins University Press.

Kluckhohn, Clyde, and Dorothy Leighton. 1962. *The Navaho*. Garden City: Doubleday and Co.

Kolber, Jane. 1998. The Human Figure in Navajo Rock Art, with an Emphasis on Mockingbird Canyon, Chaco. In *Diné Bíkéyah: Papers in Honor of David M. Brugge*, ed. Meliha S. Duran and David T. Kirkpatrick, 157–168. Albuquerque: Archaeological Society of New Mexico.

Martin, Paul S. 1979. Prehistory: Mogollon. In *Handbook of North American Indians, Volume 9, Southwest Pueblo*, ed. Alfonso Ortiz, 61–74. Washington: Smithsonian Institution Press.

Mathien, Frances Joan. 1991. Political, Economic, and Demographic Implications of the Chaco Road Network. In *Ancient Road Networks and Settlement Hierarchies in the New World*, ed. Charles D. Trombold, 99–110. Cambridge: Cambridge University Press.

McIntire, Eliot G. 1971. Changing Patterns of Hopi Indian Settlement. *Annals of the Association of American Geographers* 61 (3): 511–521.

Nabokov, Peter, and Robert Easton. 1989. *Native American Architecture*. New York: Oxford University Press.

Noble, Allen G. 1984. *Wood, Brick, and Stone: The North American Settlement Landscape, Volume 1: Houses*. Amherst: University of Massachusetts Press.

Norris, Tina, Paula L. Vines, and Elizabeth M. Hoeffel. 2012. *The American Indian and Alaska Native Population: 2010*, Census 2010 Brief C2010BR-10. Washington: U.S. Department of Commerce.

Obenauf, Margaret Senter. 1991. Photointerpretation of Chacoan Roads. In *Ancient Road Networks and Settlement Hierarchies in the New World*, ed. Charles D. Trombold, 34–41. Cambridge: Cambridge University Press.

Ogunwole, Stella U. 2002. *The American Indian and Alaska Native Population: 2000*, Census 2000 Brief C2KBR/01–15. Washington: U.S. Department of Commerce, U.S. Census Bureau.

Ortiz, Alfonzo. 1979. Introduction. In *Handbook of North American Indians, Volume 9, Southwest Pueblo*, ed. Alfonso Ortiz, 1–4. Washington: Smithsonian Institution Press.

Roessel, Ruth. 1983. Navajo Arts and Crafts. In *Handbook of North American Indians, Volume 10, Southwest*, ed. Alfonso Ortiz, 592–604. Washington: Smithsonian Institution Press.

Schroeder, Albert H. 1979. Pueblos Abandoned in Historic Times. In *Handbook of North American Indians, Volume 9, Southwest Pueblo*, ed. Alfonso Ortiz, 236–254. Washington: Smithsonian Institution Press.

Snead, James E. 2002. Ancestral Pueblo Trails and the Cultural Landscape of the Pajarito Plateau, New Mexico. *Antiquity* 76 (293): 756–765.

Windes, Thomas C. 1991. The Prehistoric Road Network at Pueblo Alto. In *Ancient Road Networks and Settlement Hierarchies in the New World*, ed. Charles D. Trombold, 111–131. Cambridge: Cambridge University Press.

Witherspoon, Gary. 1983. Navajo Social Organization. In *Handbook of North American Indians, Volume 10, Southwest*, ed. Alfonso Ortiz, 524–535. Washington: Smithsonian Institution Press.

Wormington, H.M. 1947. *Prehistoric Indians of the Southwest. Popular Series No. 7*. Denver: Denver Museum of Natural History.

5 Hispanic Landscapes of the America Southwest

The term "Hispanic" includes all persons of Spanish descent. Thus it includes Mexicans; Central Americans; Caribbean Hispanics from Cuba, the Dominican Republic, and Puerto Rico; most South Americans, other than those from Brazil; plus Spaniards coming directly from the Iberian Peninsula. As Kevin McHugh aptly notes, "Other than having common ancestral ties to Latin America or Spain, peoples of Spanish origin in the US are highly diverse" (McHugh 1989:429). Indeed, because of this diversity and the large numbers of Hispanics within America, their ethnic landscapes will be reviewed in two chapters. This chapter provides an overview of Hispanic populations within the United States, discusses the ethnic landscapes of those who settled the northern provinces of New Spain, creating the vibrant Hispano and Tejano homelands of northern New Mexico and southern Texas, respectively, and explores the ethnic landscapes being shaped by recent immigrants from Mexico. Chapter 6 will consider the colonial legacy of the Spanish settlement of Florida that was focused upon Havana rather than Mexico City, and it will review the ethnic landscapes created by Cubans, Dominicans, and Puerto Ricans who immigrated from the former Spanish colonies in the Caribbean.

5.1 Subgroups of Hispanics

Many different ethnic terms are used to refer to specific subgroups of Hispanics: Chicano to refer to Mexicans (particularly those of Indian heritage); Latino as a generic reference to Latin Americans; Hispano to signify inhabitants of the centuries-old New Mexican settlement, although Nuevomexicano is a more inclusive term that includes both these long-time Hispanic residents and descendents of more recent arrivals (Smith 2015); and Tejano to describe Texas-Mexicans, particularly those coming from the Lower Rio Grande Valley. (Because Spanish is a gender specific language, the final "o" in each of these terms should be replaced by an "a" if the persons described are female.) Sometimes terms are indicative of national origin, such as Spanish-American, Mexican-American, or Cuban-American. Race is not a criteria in defining Hispanics, as Hispanics can be white, black, or have Native American heritage. Even one Asian group, the Filipinos, have Hispanic heritage.

In the United States, most Hispanics display a mixed Indian-European heritage. The word Mestizo is used to describe a person of mixed European and Indian ancestry. Indeed, Mestizos comprise about 60% of the Mexican population. Thus, Hispanics of specific national origins frequently differ with respect to their racial composition. For example, most Dominicans are of African or a mixed African-European heritage (often termed mulatto), while Cuban-Americans are typically of either European or mulatto heritage. Nevertheless, their ethnicity transcends racial boundaries.

The Hispanic population within the United States is rapidly growing, increasing by 57.9% between 1990 and 2000, over four times the nation's overall growth rate (Guzman 2001). The following decade it increased by 43%, "accounting for over half of the 27.3 million increase in the total population of the United States" (Ennis et al. 2011:2). This growth reflects not only high birth rates in areas of the country that have historically had large Hispanic populations, but the rapid immigration of Hispanics. Indeed, the U.S. Hispanic population was 9.1 million in 1970; by 2010 it was 50.5 million—accounting for 16.3% of the nation's population. In 2010 Mexicans comprised 63.0% of all U.S. Hispanics, with Puerto Ricans accounting for 9.2%, Cubans 3.5%, and Dominicans 2.8%. The biggest change from the 2000 Hispanic statistics was continued growth in the Mexican and Central American share of the Hispanic population (Guzman 2001; Ennis et al. 2011).

5.2 Spatial Distribution of Hispanic Americans

The contemporary Hispanic distribution reflects the historic legacy of the Spanish colonial period, continuing immigration of Hispanics to the United States, and internal movements of

J.A. Cross, *Ethnic Landscapes of America*, DOI 10.1007/978-3-319-54009-2_5

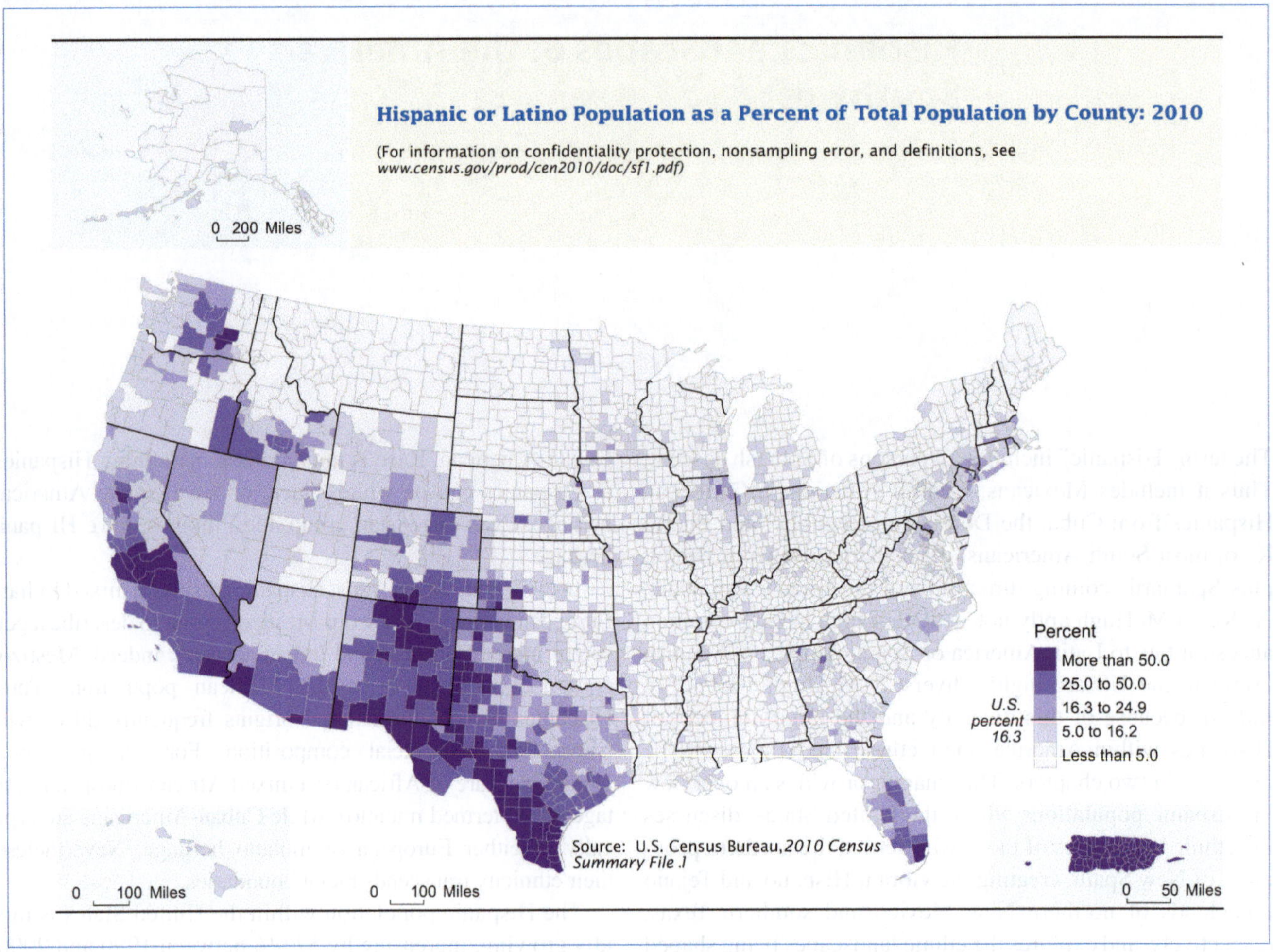

Fig. 5.1 Proportion of the population reporting Hispanic ancestry in each county at the time of the 2010 Census. Note that those counties with the greatest percentages of their populations being Hispanics are largely located from Texas to California, north of the Mexican border. Concentrations are also noted in southern Florida and in several major metropolitan areas (Reproduced from U.S. Census Bureau map within Ennis et al. (2011))

Hispanic peoples after their earlier arrivals (Fig. 5.1). Considerable diversity exists within the Hispanic population, not only based upon national origin, but also regional variations among ethnics whose roots are within large nations, such as Mexico.

5.2.1 Concentration of Hispanics in Border States

Hispanics are concentrated in states adjacent to Mexico, yet significant Hispanic populations are found in certain major cities in the Midwest and Mid Atlantic regions, as well as in agricultural areas of the Pacific Northwest. California, with 14,013,719 Hispanics in 2010, had more than any other state, and within California they comprised 37.6% of its population, with over four-fifths of its Hispanics indicating Mexican ancestry. Similarly, Texas, with the second largest number of Hispanics (9,460,921 in 2010), reported that 85% of these ethnics were of Mexican ancestry. In contrast, New Mexico, whose 953,403 Hispanics comprised 46.3% of the state's population—the highest share nationwide, was settled by Hispanics whose predominate ancestry was neither Mexican nor West Indian, reflecting the early influence of Spaniards in the state's settlement. Florida, with 4,223,806 Hispanics (22.5% of the state population), was populated by a diverse population with non-Mexican origins. There, Cubans account for over a quarter of the total, Puerto Ricans a fifth, with South Americans also prominently represented (Ennis et al. 2011).

All of these border states were Spanish territory at one time. New Spain included all of the states from California through Texas, plus Nevada, Utah, and Colorado, plus small parts of several other states. Florida, which at one time extended west to the Mississippi River, was a separate

Spanish colony, administered from Havana. Florida's Spanish population departed when it was transferred to English control and few of its residents can directly trace their heritage to the second period of Spanish colonization. In contrast, significant shares of the Hispanic population in what was once New Spain, particularly within Texas and New Mexico, are descendents of the populations who were living in those Mexican provinces at the time they were acquired by the United States. Thus within those areas, the historic landscape is largely a Hispanic landscape.

Within South Texas, the dominance of the Hispanic population is particularly pronounced. Two decades ago Daniel Arreola noted that 85% of the population within a 17 county area of South Texas were Mexican-Americans and that the "Mexican American homeland in South Texas is thus one of the largest ethnically pure concentrations in the United States" (Arreola 1993:63). In South Texas, as in much of northern New Mexico, Hispanics dominate both the rural and urban landscape. Furthermore, they have expanded far from the border in the past two decades.

5.2.2 Concentrations of Hispanic Populations in Cities

In contrast, those Hispanic populations outside the western states are largely concentrated in specific cities. For example, while New York State has the nation's fourth largest Hispanic population, two-thirds of that population lives within the city limits of New York City, where they comprise 28.6% of that city's population (Ennis et al. 2011). New York's Hispanic population is primarily of Puerto Rican and Dominican heritage. Chicago also has a large Hispanic population, comprising 28.9% of the city's residents, and it is split between those of Puerto Rican and Mexican ancestry.

The concentration of Hispanics in central cities of metropolitan area has been noted for several decades. Three decades ago Kevin McHugh wrote,

> In 1980, 81% of Mexican-Americans resided in metropolitan areas. Other Hispanic groups show greater levels of metropolitan concentration: 96% for Puerto Ricans, 94% for Cuban-Americans, and 96% for Central/South Americans (McHugh 1989:433).

Today, large Hispanic population clusters exist within certain American cities, with New York City and Los Angeles each having well over a million Hispanic residents and Chicago, Houston, and San Antonio each having over three-quarters of a million, yet rapid growth has also been occurring in many rural regions.

5.2.3 Growing Hispanic Populations in the South, Midwest, and Plains

During the past two to three decades significant numbers of Hispanics have been drawn to many smaller cities and rural communities across the Midwest and Great Plains, as well as the Pacific Northwest, where agricultural jobs and other employment opportunities exist. While the influx of Mexican workers into meat packing facilities in many rural counties of the Great Plains is well known, Hispanics have also contributed to the growing ethnic diversity of counties that are experiencing overall population loss (Barcus and Simmons 2013). Over half of the nation's rural counties experiencing high-growth rates of their Hispanic populations are in the South, and "without new Latino residents, large areas in the region would be experiencing population decline" (Furuseth and Smith 2008:14). Furthermore, there are now some rural communities in the South where Hispanic immigrants now outnumber African-Americans.

5.3 Spanish Colonial Settlement in the Western U.S.

Florida received the earliest settlement attention from the Spanish—indeed the first colonizing effort by any European nation in the lands that eventually became the United States (Fig. 5.2). Spanish Florida initially extended from the Mississippi River east to the Atlantic Ocean, running from the Straits of Florida north to Virginia. Spanish colonization within the American Southwest was tied to New Spain, as Mexico was called during the colonial era. It was within the Southwest that several Hispanic homelands developed, two of which remain quite identifiable on the contemporary landscape.

5.3.1 New Mexico

The Spanish colonization of New Mexico several decades following the establishment of St. Augustine, Florida led to the development of the Hispano Homeland, an area where the ethnic legacy is both widespread and pronounced. Of all the areas in what is now the United States, New Mexico was the area of greatest Spanish settlement, the focus of the earliest and largest Spanish colonization of the Hispanic-American Borderland (Nostrand 1970). Moving north from Chihuahua, Mexico and El Paso del Norte, presently Ciudad Juarez, Mexico across the Rio Grande from El Paso, Texas, the settlers followed the Rio Grande northwards into an area

Fig. 5.2 Much of the territory in the southern and western parts of the United States was acquired either directly from Spain or other nations that had obtained their territory from Spain. California and the other southwestern states were acquired directly from Mexico—which a quarter century earlier had acquired its independence from Spain—as a result of the Mexican-American War or the Gadsden Purchase. Texas was annexed nearly a decade after it had defeated the Mexican Army in its quest for independence. While Florida, which extended west to the Mississippi River, was acquired directly from Spain, the Louisiana Purchase was sold to the U.S. by France, yet it had been administered by the Spanish for 40 years before it was sold. Other territorial acquisitions are associated with the British, yet parts of coastal Georgia and South Carolina were once part of Spain's colonial realm (Reproduced from the *National Atlas* published by the U.S. Geological Survey (2005))

termed the Rio Arriba in the Upper Rio Grande Valley. Even though the Spanish moved north from Mexico, many of the settlers who accompanied Don Juan de Oñate had themselves been born in Spain, leading to this region's stronger identity today as being Hispano rather than Mexican.

The first Spanish settlement in New Mexico was established in 1598 at the confluence of the Rio Grande and Rio Chama, near a Native American pueblo (Carlson 1992). With the establishment of Santa Fe in 1609 and the movement of Oñate's followers there the following year, both San Juan (now known by its pre-Hispanic name, Ohkay Owingeh) and San Gabriel declined, with the latter being abandoned. Santa Fe became the northernmost Spanish colonial capital. For nearly a century, missions and encomiendas were centered on Indian villages (pueblos), with Santa Fe being the only authorized community. Estacias, which were Spanish farms or ranches on land grants, were also usually located near Indian settlements.

The Spanish settlements in New Mexico were established to provide a buffer zone between important mining settlements in Mexico and Indians considered as being hostile. The religious conversion of local Indians was part of the Spanish pacification effort, yet friction between the priests and the Indians led to the Pueblo Rebellion in 1680. Many Catholic priests in the missions were slain and the missions destroyed, while all the Spaniards in Santa Fe fled to El Paso del Norte, the vicinity of today's Ciudad Juarez and El Paso. Although a mission had been established there in 1657, its growth as a community of both Hispanics and sympathetic Pueblo Indians dates from the rebellion (Weber 1992). Today's small Indian reservation and its historic mission church just outside of El Paso, Texas are a legacy of that region's connection with New Mexico.

The Reconquest in 1692 gave renewed vigor to the Spanish colonial settlement in New Mexico. Santa Fe received a presidio (military garrison) when it was reoccupied. Additional Spanish settlements (termed villas) were established at sites of former Indian villages. Examples include Santa Cruz, founded in 1695 north of Santa Fe; Bernalillo, started before 1706 along the Rio Grande; and Albuquerque, established in 1706 farther to the south along the same river. To the north of Santa Fe, many sites in the Chama Valley and the Española Valley were settled by the 1730s, with the community of Las Trampas begun in 1751. Most of these new settlements represented a growing population moving out from existing settlements.

The isolation of the Rio Arriba settlement from the rest of New Spain played a key role in the history and cultural legacy of the region. To the north the Rio Grande was confined within a narrow canyon. To the east were the Sangre de Cristo Mountains, rising to over 10,000 feet, and the Plains Indians that lived to the east of those mountains were viewed as a major threat by both the Spaniards and the Pueblo Indians. To the west were the Jeméz Mountains and extensive tracts of desert, while to the south lay more desert, especially in the region south of Albuquerque. The distance to Chihuahua, Mexico, the nearest external trading center, was nearly 600 miles. An official trading expedition linked Santa Fe with Chihuahua once every 3 years.

Isolation forced the Rio Arriba area to evolve its own distinctive set of cultural traits (Nostrand 1992). The culture was different from Mexican, with the population's self identity being Spanish, not Mexican. The language was largely a Castilian dialect, including the preservation of many archaic words, expressions, and constructions. A distinctive folk art arose, including santos, carved and painted religious images. The Penitente Brotherhood, a lay society for penance and mutual aid based upon practices of sixteenth century Spain, evolved to meet many spiritual needs in the absence of regular visits by Catholic priests in many of the smaller Hispano settlements (Carlson 1990). Indeed, "considering the vital role (both religious and social) that the Penitentes have played in community life, and the pervasiveness of their existence throughout the upper Rio Grande region, numerous scholars contend that no other aspect embodies the essence of the Hispano culture more than the Penitentes" (Smith 2015:28). Because New Mexico's settlers arrived earlier than other Hispanics, were more likely to have come directly from Spain than elsewhere in the borderlands, and their settlements survived for two centuries in near isolation, this region evolved into the distinctive "Hispano Homeland."

The independence of Mexico in 1821, following over a decade of conflict, had little change upon the area's architecture and Hispano landscape. Indeed, the only consequence of Mexican independence from Spain was to perhaps "exaggerate the isolation and self sufficiency of the remote northern province" (Bunting 1992:1). With Mexican independence, the first Anglo-American traders arrived in New Mexico via the Santa Fe Trail from Missouri. Before independence, such contact was strictly prohibited by Spanish authorities. Two and a half decades later, as a culmination of events begun in Texas, New Mexico was under U.S. military control by 1846, yet the foundations of a vibrant and expanding Hispano Homeland of these Spanish ethnics had been firmly placed on the landscape. Over the next half century the Hispano Homeland was to expand towards the east in New Mexico and north into southern Colorado.

5.3.2 Texas

Texas, or Tejas as it was called at the time, was settled by the Spanish as a buffer against the French, yet by the time that France lost Louisiana to Spain in 1763 as one of the consequences of the French and Indian War (or Seven Years War), it was the least populated province of New Spain, with only

Fig. 5.3 Mission Concepción, built in the mid 1700s, is one of five mission churches constructed near San Antonio, Texas. Although religious services still occur in these structures, they are preserved as part of San Antonio Missions National Historical Park

several thousand Hispanics. The first Spanish settlement in Texas occurred at Nacogdoches, where five missions were established in 1690, yet they failed after 3 years. The Spaniards reestablished themselves in 1716 at Las Adaes near the French border (actually in what is now northwestern Louisiana), and the town of Nacogdoches was founded in 1779, rivaling San Antonio in the later colonial period.

Colonial Texas's most prominent community, San Antonio de Bejar (now spelled Bexar) was established as a presidio, a town, and the site of missions in 1718. Five stone walled missions were constructed in the area (Fig. 5.3), more to house Christianized Indians relocated from eastern Texas than to serve the needs of the local Indian population. In 1731 nearby San Fernando de Béxar was established by settlers from the Canary Islands, islands colonized by Spain off the coast of Africa.

In 1749 the Spanish established Mission Espiriti Santo and Presidio La Bahia about ninety miles southeast of San Antonio, near the present city of Goliad. Five years later another mission was established six miles distant. A settlement developed around Presidio La Bahia, and the slaughter of Texan captives there in the Goliad Massacre of 1836 took more lives than the fall of the Alamo.

Other earlier Spanish communities just across the Rio Grande spread settlers into what became far southern Texas, lands that during Spanish colonial times were in the provinces of Coahuila and Nuevo Santander, south of Tejas. For example, just across the Rio Grande from present day Eagle Pass, Texas, San Juan Bautista (now Piedras Negras), Mexico was founded in 1700. By the mid 1700s, a series of civil settlements were established farther downriver along the Lower Rio Grande by the Spanish, who awarded land grants called "prociones" to the settlers. Although most of the settlements established between 1749 and 1755 were on the southern side of the Rio Grande, thus on today's Mexican side, the city of Laredo, Texas was established on the opposite side in 1755.

As these ranching and farming communities grew, additional settlements were established on both sides of the river, with Roma, Texas being founded about 1770 and San Ygnacio, Texas receiving its first residents in 1830. In several cases, cities simply grew across the river which later became a border. For example, Matamoros, Mexico was established in 1774 as a ranching center, and its ranchers had expanded into what is now the Brownsville, Texas area before the end of the eighteenth century. The growth of Laredo, Texas led to the establishment of Nuevo Laredo, Mexico, following the Mexican-American War. The Lower Rio Grande, which remained navigable through the 1800s, was not a boundary to the Mexicans who settled the area, but a connection between settlements on both of its sides, and commercial, social, and family linkages across the border have historically been very strong (Thompson 1994). Indeed, these linkages remain today.

Fig. 5.4 Mission San Xavier del Bac, completed in 1797 on what is now the Tohono O'odham Indian Reservation, located southwest of Tucson, Arizona. Facing a large plaza, this baroque style church illustrates the ethnic influence of the Spanish upon the indigenous populations of New Spain's Sonoran entrada

The War for Texas Independence involved both Anglo-Americans who had settled near Mexican communities, such as San Antonio, Goliad, and Victoria, as well as many Mexicans who were unhappy with the dictatorship of President Antonio López de Santa Ana. Laredo itself served as the de facto capital of the short-lived Republic of the Rio Grande, which having lands on both sides of the Rio Grande, fought for its own autonomy between 1839 and 1840. Again, both Anglos and Mexicans fought together, although Mexicans comprised the majority of those opposing the Mexican army.

One of the biggest ethnic holidays celebrated today by Mexicans and Mexican-Americans commemorates the victory of the Mexican Army over the invading French at the Battle of Puebla on May 5, 1862. The leader of the victorious Mexicans was General Ignacio Zaragoza, who was born in Goliad, Texas. Thus, his reconstructed home in Goliad has been preserved in a park, and a large memorial statue has been erected to this Mexican hero in Texas. The celebration of "Cinco de Mayo" is just one example of the historical and cultural ties that southern Texas has to both nations.

5.3.3 Arizona

The Spanish settled a small portion of southern Arizona as part of the Sonoran entrada. While most of the Sonoran colonization effort occurred within what is now the Mexican state of Sonora, a portion occurred in the area between the present Mexican-American border and the city of Tucson. Within this region that the Spanish called Pimería Alta, or the upper region where the Pima Indians lived, missions were established very early. For example, Mission San Xavier del Bac was established in 1700 on what is now the Tohono O'odham Indian Reservation. In keeping with traditional Spanish urban practice, the ornate baroque mission church completed in 1797 (Fig. 5.4) faces a large open plaza. As James Griffith (1992:150) notes, the church is similar to most Mexican baroque churches, being "built on a simple cruciform plan, with a dome at the crossing. Two towers, which would be symmetrical if both were finished, flank the portal at the south end." The ornate portal itself contains statues of five saints, four of which occupy niches, while centered above the door is a carved coat of arms of the Franciscan order. Another mission church, La Purísima Concepción de Tumacacori, built in the 1820s, is preserved within Tumacacori National Historical Park. Less accessible ruins of several additional Spanish colonial churches are located nearby.

The Spanish ethnic influence remains visible within the Native American landscape in the over 60 folk-catholic chapels that are found on the Tohono O'odham Indian Reservation. Simple in form and almost universally facing east, these chapels were usually of adobe construction, requiring continuing renewal. These chapels, "which [are] usually the only non-domestic building in the village"

Fig. 5.5 Mission Dolores, constructed in 1776 in San Francisco, California. San Francisco was also the site of a Spanish military outpost, called the Presidio, which became an American military base that operated until near the end of the twentieth century. Such mission buildings are California's most visible architectural legacy from the Spanish colonial period

(Griffith 1975:21), function in a manner quite different from a Roman Catholic church. Yet they are symbolic of how the ethnic traditions of the Spanish have been both informed and modified by the heritage of another ethnic group, so commonly seen within the American Southwest.

Presidios were established against the Apache threat in Pimería Alta. Tubac, just north of the current border city of Nogales, was established in 1752, while Tucson was founded as a presidio in 1776. Being established as a presidio, rather than a civil community or pueblo, the Tucson street grid was more irregular than what would be found in a planned city. An urban renewal project that began in 1967 destroyed much of the city's Hispanic historic district.

> Bulldozers erased a uniquely irregular street network and obliterated a rich heritage of adobe structures, some dating back more than one hundred years. Three hundred nineteen homes were torn down and more than one thousand residents were forcibly relocated. A unique three-plaza settlement was lost (Gomez-Novy and Polyzoides 2003:94).

Nevertheless, a few over century old adobe buildings—many altered with Anglo-style hipped or gabled roofs—remain near the surviving Plaza de las Armas in Tucson. As a consequence of the urban renewal project, the long-term Hispanic residents were dispersed into other lower-class neighborhoods of the city. Although Phoenix lacked the longevity of Tucson's Hispanic settlement, it similarly lost its historic downtown Mexican core to urban renewal (Arreola 2012).

Between Tucson and the present city of Tombstone the Spanish also established a presidio at Terrenate in 1776, relocating a garrison that had previously been located at a presidio by the same name in the northern part of the Mexican state of Sonora. Abandoned for over two centuries, today all that remains are the ruins of adobe walls.

5.3.4 California

Spanish settlement in California began in the late 1760s to counter expansion by the British and Russians. Three settlement types were used from the beginning. Presidios, or military outposts, were established at San Diego in 1769; in Monterrey, which served as the colony's capitol, in 1770; at San Francisco in 1776; and at Santa Barbara in 1782. San Francisco's military base housed first Spanish, then Mexican, and finally American troops, only closing in 1994. Much of its land is now incorporated in an urban park, part of the Golden Gate National Recreation Area. Pueblos, or civil communities to provide food to the presidios, were founded at San Jose and Los Angeles in 1777 and at Santa Cruz in 1797. Missions to convert natives and "teach them to become loyal Spanish subjects" (Hornbeck 1990:55) were established along the coastal area of California, from San Diego to just north of San Francisco.

Twenty missions were established in California during the Spanish period (Fig. 5.5), plus one during the Mexican period. These missions operated very differently from those earlier established ones in New Mexico. A major goal was the intensive acculturation of the Indians, and today descendents of those Indians are referred to as Mission Indians, having lost their identity with prior tribal affiliations. Although the Mexican government had secularized the missions by the 1830s, the missions played a crucial role in California's development. Indeed, in explaining California's landscapes, their influence upon today's urban pattern is undeniable.

> Of all the Spanish settlement institutions brought to California, the missions became the most dominant, and their locations form the basis of California's present-day urban structure. Mission locations were used as building sites for American cities, which currently account for almost 60 percent of the state's population (Kenzer et al. 1992:16).

Monterey served as the Spanish colonial capitol of California, and from there came the Monterey house in the 1830s. Typically of two stories, it was built of adobe brick, with the exterior plastered with lime. Its distinctive feature was "the long balcony supported on the extended vigas of the first floor ceiling" (Noble 1984:86). The second floor rooms were entered through doors that faced the balcony, which served as a hallway.

California was formally acquired by the United States from Mexico in 1848, the same year gold was discovered along the American River. While tens of thousands of gold seekers rapidly transformed the landscape of the Sierra Nevada east of Sacramento and the San Francisco Bay area, little development occurred in southern California during most of the nineteenth century. The Hispanic landscape of California today reflects not only the historical legacy of the time that California was a Spanish colony and then Mexican territory, but the heritage of the millions of Mexican immigrants who arrived to harvest its crops and to settle within its cities during the last half of the twentieth century.

The Spanish legacy in California remains in the heritage of the missions, most which have been restored, after many had fallen into ruins. Plazas remain in some cities, such as in Los Angeles, San Juan Bautista, and in San Diego, where it is preserved in Old Town State Historic Park. Many California place names reflect the early Spanish settlement, and land grants by the Spanish or Mexican governments shaped the division of the land. Popularization of Hispanic architecture, for example the Mission Revival Architecture, provides additional reminders of the state's ethnic heritage. Mission Revival architecture incorporated Hispanic elements beginning in 1880s and its use continues. This style incorporated stucco walls, red tiles (especially in roofs), arched loggias, and bell towers. The historian David Weber explains this popularization:

> The californios of the Spanish era had lived in simple one-storey adobes, many with flat, tar-covered roofs, and few with wooden floors, glass windows, or tree-shaded landscaping. Anglo Americans reimagined those modest structures as elegant two-story, red-tile-roofed structures, with carved woodwork and cantilevered balconies that looked into tree-filled patios where water played in fountains (Weber 1992:343).

Thus, Mission Revival style houses and public buildings, incorporating elements from both the dwellings of the californios and their missions, proliferated in more upscale communities. Some of the most conspicuous mission style architecture can be found in railroad stations built by the Santa Fe Railroad.

5.3.5 Spanish Louisiana

From Texas through California numerous reminders of the Spanish colonial landscape, as well as the landscapes created by more recent Hispanic immigrants, are visible. Although most of the remainder of this chapter will describe in detail several of the most distinct and conspicuous Hispanic landscapes, the impact of Hispanics upon the built landscape can be seen in many other areas of the United States. For example, although most Americans associate New Orleans with its French settlement, the city was under Spanish administration from 1763 until 1803, nearly as long as it was under French rule. Indeed, most buildings that one sees in the French Quarter today were built following several devastating fires that occurred when the city was a Spanish city, and the city's famous Jackson Square provides a wonderful example of a Spanish plaza facing a cabildo and cathedral. Recent Hispanic immigration is creating new contemporary Hispanic landscapes in communities across America that were established, and until recently predominately inhabited, by members of other ethnic groups.

5.4 Hispanic Urban Landscapes

Settlements established by Spanish colonizers, regardless of whether they are located in the United States, Mexico, the Caribbean, or Central and South America, share many distinct patterns. The similarity among Hispanic urban landscapes is a direct consequence of the Law of the Indies issued in 1681 or "Recopilacion de leyes de los reynos de las Indias" (Veregge 1993). The first instructions were given by Spain's King Ferdinand in 1501 and the subsequent revisions "were formalized by Philip II in 1573 as the Law of the Indies" (Treib 1993:18). Within this Law were "The Royal Ordinances for Laying Out of New Towns." These guidelines applied to

Fig. 5.6 Church (Iglesia de Santa Cruz de la Cañada), business building, and community building facing the plaza within Santa Cruz, established in 1695 as a villa, or formal organized settlement. Today this community is overshadowed by the city of Espanola, New Mexico, whose center is located several miles west of Santa Cruz

civil settlements that the Spanish called "pueblos", which depending on size (large to small) were termed ciudades, villas, poblaciones, and plazas.

5.4.1 Requirements of the Law of the Indies

The Law of Indies consisted of 148 ordinances, of which nearly 100 dealt with the selection of a town site, and the financial and legal considerations of the city. For example, Ordinance 111 described the selection of a good townsite, with an elevated location, "good soils and sufficient land for crops and pasturage", "sufficient resources of timber, fuel, and water", "near an indigenous population," and "accessible for transporting goods and people" (Veregge 1993:380).

The Law also specified a grid pattern found in most Latin American cities today. Ordinance 112 provided the specifications for the establishment of plaza in the center of a city (Figs. 5.5 and 5.6), or at a landing should it be a port city. The plaza was to be a square or rectangle, which Ordinance 113 stated must be at least 200 by 300 feet in size up to a maximum of 535 by 800 feet, with a typical plaza recommended to cover an area of 400 by 600 feet. The plaza was sized to be proportional to the expected population.

Additional ordinances described the construction of streets and buildings along the edges of the plaza. Ordinance 114 stated that streets were to run along each side of plaza, with the main streets running perpendicular to the middle of each side of plaza. Thus, a grid was established with the plaza in its center. The next ordinance noted, "buildings around the edge of the plaza should have portales, as should those on the four principal streets" (Veregge 1993:381). The portales are arcades, or covered walkways, along the front of the buildings. Ordinance 117 stated, "The streets should run from the plaza in such a manner as to allow for substantial growth without inconvenience or adverse effects on appearance, defense, or comfort" (Veregge 1993:381).

The relationships between the plaza, the community's main church (or cathedral), its cabildo (government administration building or council house), and other important buildings were also stipulated in the Law of the Indies. For example, the cabildo and customhouse should be located between the church and the plaza (Veregge 1993). The church or cathedral, according to the 124th Ordinance, "shall not be placed on the square but at a distance and shall be separated from any other building or from adjoining buildings; and ought to be seen from all sides so it can be decorated better, thus acquiring more authority" (Treib 1993:20). Nevertheless, the church or cathedral was often

Fig. 5.7 Palace of the Governors facing plaza within Santa Fe, New Mexico. This building served as the governor's residence and seat of government from 1610 until 1909, and it has housed the state museum ever since. Native American vendors can be seen displaying their handicrafts under the portales of this classic example of a cabildo

located facing the plaza. While the buildings near the plaza were typically the businesses and homes of the more prosperous residents, other ordinances mandated that certain undesirable activities should be located away from the city center, including the hospital for contagious diseases, the cemetery, tanneries, and slaughterhouses (Veregge 1993).

The plaza was the center of the Hispanic community, typically with the church, government building (cabildo), and residences facing plaza. This was true within larger communities, such as Santa Fe, New Mexico; Albuquerque, New Mexico; San Antonio, Texas; and St. Augustine, Florida, and within much smaller settlements, such as Santa Cruz, New Mexico; and Las Trampas, New Mexico. Plazas are not limited to communities established following the period of Spanish colonial rule, but are central in La Mesilla, New Mexico and San Ygnacio, Texas, both which were established during the Mexican regime. Some cities created their plazas during American administration, as Arreola (1992) describes in Texas. While commercial buildings might surround the plaza in the larger communities, with their first floors or front room used for business, and the back rooms or upper floors used for residences, the portales along plaza provided space for a market. An excellent example is the daily market of Native American handicrafts that fills the space under the covered sidewalk in front of the Palace of the Governors in Santa Fe, New Mexico (Fig. 5.7).

5.5 Cultural Landscape of the Hispano Homeland

The ethnic landscape created by the early Spanish settlers within northern New Mexico and incorporating influences from their centuries of interaction with nearby Pueblo Indians and with Anglo-Americans for the past century and a half has been intensely and devotedly studied by some of the most prominent cultural geographers and landscape scholars of the twentieth century. These include the geographers Alvar Carlson (1990, 1992) and Richard Nostrand (1992), among others, as well as the landscape essayist John Brinckerhoff Jackson (1994, 1997), who spent part of his life living in a small Hispano village. More recently, Jeffrey Smith (2002a, b, 2015) has written extensively about the region. Much of the Hispano Homeland has now been designated as comprising the Northern Rio Grande National Heritage Area in New Mexico and the Sangre de Cristo National Heritage Area in Colorado, both under the sponsorship of the National Park Service. Such cultural recognition and "[h]eritage development makes cultural difference more recognizable. It brings culture, identity, and the past into consciousness and into view, lifting people, places and social practices out of everyday existence and holding them up for inspection" (Guthrie 2013:9). This region is best understood when considering that it is tri-cultural—Native American, Hispanic, and Anglo—and has a double colonial history, inasmuch as "Hispanos simultaneously occupy the positions as colonizer and colonized" (Guthrie 2013:98).

The Hispano population within rural northern New Mexico was concentrated in villages, located where irrigation waters could be channeled to water their fields and orchards. Similar to what would have been found in Medieval Europe, many residents lived within villages, which provided the community protection. Although civil communities are often called pueblos by the Spanish, in New Mexico they were often termed *villas*. Smaller, more dispersed settlements were called *ranchos*, which were loose groupings of small farmsteads. The word plaza was also used to describe a small farm village, and indeed, a central plaza faced by a Catholic church is still visible today within numerous Hispano communities. New Mexico did have its larger Spanish settlements, such as Santa Fe with its presidio and Albuquerque, but the heart of the Hispano Homeland is found within the smaller villas. The arrangement of many communities followed the guidelines of the Law of the Indies, yet within many villages the necessity of irrigating fields resulted in linear, rather than grid, forms.

5.5.1 Land Tenure

Land tenure practices and the need to irrigate crops shaped the form of many Hispano villages. Spanish land grants were given to individuals and to communities. Individual families received *solares*, or village house lots, as well as *suertes*, or floodplain farm plots. The number of suertes a man was granted depended on his family size and marital status. Agriculture was largely subsistence, meaning that the family grew most of what it consumed, with little external market for surpluses. In contrast to the individually owned cultivated fields and homesites, grazing and forest lands were held in common by all of the residents of the village. In a confiscation that remains controversial to this day, much of this common land was lost upon the U.S. annexation of New Mexico. While the Spanish land grants were recognized, the local population did not understand the Anglo property system or the language. When taxes were often not paid (as the concept of property taxes was new to the Hispanos), the lands were alienated. Much of the U.S. Forest Service lands in northern New Mexico were once such common land. In contrast, in the San Luis Valley of Colorado descendents of the original grant settlers have now received resource access rights "to graze, gather firewood, and cut timber" upon the grant that had become private property (Lindner 2012:372). In response to the Colorado Supreme Court ruling, a cultural mapping project was undertaken "to demonstrate historic and deep ties between the local Hispano community and La Sierra…, thereby shoring up the community's legitimacy to use the mountain and its resources" (Lindner 2012: 377).

The suertes were arranged along irrigation ditches (Figs. 5.8 and 5.9). Water acquisition was critical to the Hispano community, resulting in organized institutions for providing water. Digging an *acequina madre,* or major irrigation ditch, was typically one of the first activities in establishing a new community. Community members were responsible for volunteering labor for its construction, maintenance and regulation, and an elected *mayordomo* directed these activities at a local, community level (Carlson 1990). Nearly three centuries later, the "Alcalde Acequina Association remains active, and the annual cleaning of the ditches evokes nineteenth-century cultural rhythms" (Wright and Campbell 2008:559). While not all acequinas remain in use as in Alcalde, their development has had a lasting impact upon the settlement pattern in many valleys.

Field patterns displayed long lots laid out perpendicular to rivers, in an indigenous system unique to the Hispano Homeland (Fig. 5.10). As Alvar Carlson (1990:23) explains, "Long-lots ultimately became the basis for the development and retention of much of the Spanish-American subculture." The narrow ribbonlike fields extended from the acequina madre on the upland side of the field to the river's edge. Fields were typically 90–140 yards wide, with trees usually left growing along riverbanks. With this arrangement of the fields, irrigation water flowed by gravity from the ditch to the bottom of the field. Farmsteads were located on the upland side of ditch or acequina, with houses arranged in rows, usually between a road and ditch. Over time, inheritance divided the fields into narrower strips, so that all fields still retained access to irrigation water. Furthermore, with smaller fields and less reliance upon agricultural employment, houses are now found on both sides of the acequinas, and in places houses have been built along lanes running perpendicular to roads paralleling the acequinas. Thus, these field patterns are still visible today, even in the Los Corralles area suburbs of Albuquerque (Fig. 5.11), where roads and home sites follow the old field boundaries.

5.5.2 Linear Villages

The land tenure system also influenced the cohesiveness of the rural population. Alvar Carlson, when studying the area, noted that

> Spanish-American villages tended to be complete social and economic units. ... Religious and kinship ties and village loyalties, campanilismo, fitted in nicely with the long-lot survey system, which allowed people to live relatively close together. The people in their individual valleys were somewhat isolated, and isolation also strengthened their ties with each other (Carlson 1975:55).

Fig. 5.8 Acequina supplying water to fields near Truchas, New Mexico. Note the ruins of the log barn in the distance, an architectural style that has been attributed to the influence of seventeenth century German miners who worked in Chihuahua. Fields slope to the right, away from the acequina

Fig. 5.9 Acequina follows the contour of the valley edge just south of Vallecitos, New Mexico. Fields that extended down the slope from the irrigation ditch could be easily watered

Fig. 5.10 Long narrow fields and apple orchard running downslope in the Rio Grande Valley near Velarde, New Mexico

The practice of dividing land among children led to population stability, maintaining the concentration of population there, and helped maintain a distinct cultural homeland. Unfortunately, it also contributed to the region's poverty, as farm families had smaller and smaller plots to cultivate.

This land tenure system shaped the configuration of many of the area's rural villages.

> The long-lot field pattern, which led to the establishment of line villages, was used extensively in the Rio Arriba, even outside the Spanish and Mexican land grants. Spanish-Americans employed the long-lot when they purchased parcels or squatted on Pueblo Indian Grant lands in the late 1800s. The same pattern was used by squatters in remote valleys and by the Spanish-Americans who settled the small floodplains in the southern part of the adjacent San Luis Valley of Colorado. The long-lot became commonplace throughout the upper Rio Grande Basin" (Carlson 1975:53, Reprinted with permission of American Association of Geographers).

These field patterns resulted in linear villages or line villages that can be seen along the Rio Chama and Rio Santa Cruz between Santa Cruz and Chimayo today (Fig. 5.12). Furthermore, the field pattern resulted in a more concentrated rural population than is found in most other farming areas. Richard Nostrand explains that "New Mexico's 'plazas' survive. Sprawling affairs of remarkably low density, with houses dispersed in linear fashion along the high sides of irrigation ditches or at intervals along roads, these loose agglomerations of people dominate the landscape of the Homeland" (Nostrand 1992:220). In the mid-1800s Hispanos established about two-dozen linear settlements in what is now southern Colorado, although fortified plazas were briefly erected for security and then subsequently abandoned (Smith 2002a).

5.5.3 Fortified Building Square

The Hispano village of Chimayo, even though situated in an area of linear villages, is concentrated around a plaza (Fig. 5.13). At its center is a fortified building square or walled plaza, the legacy of those fortified communities established in the 1760s and 1770s because of raids by Navajo and Plains Indians. During this time, among the Hispanos "the term plaza connoted the idea of a fortified place rather than a central square" (Bunting 1992:3). Within the Plaza del Cerro houses are arranged in a contiguous

Fig. 5.11 Long narrow fields with Hispano homes located perpendicular to the Rio Grande. Because of growing population and the splitting of the suertes in a lengthwise manner upon inheritance, the rural population density is quite high within the community of Corrales, located north of Albuquerque, New Mexico

fashion around an open plaza. The outer walls of these houses are windowless, and at one time the plaza entrances were gated. Within the plaza animals could be protected, while closing of the gates kept attackers outside the community. While examples of walled plazas also existed at Taos, Ranchos de Taos, and other villas, the best preserved example is at Chimayo. The prominent landscape historian, John Stilgoe, noting the necessity for defense following the Pueblo Rebellion, explains,

> The houses of Chimayo abut one another and in the seventeenth century had no windows or doors on their outside walls; from a distance the dwelling cluster appeared as a solid wall of adobe broken only by two portals. Chimayo measured roughly 420 by 330 feet and consisted of some forty houses, one or two stores, and a church, all facing a small plaza planted with trees, partially fenced off as a livestock corral, and cross by an acequia. Beyond the rectangle of houses lay the suertes and common fields, all abutting the Rio Santa Cruz or an acequia (Stilgoe 1982:40).

Yet even this walled plaza no longer provides the continuous unbroken line of occupied buildings that it once did, but it is the "most intact Spanish colonial plaza in New Mexico" (Guthrie 2013:106).

The buildings within the villages, whether tightly arranged around a fortified building square such as at Chimayo; arranged around a more open plaza such as at Santa Cruz or Las Trampas (which was likely a closed plaza in the past), tightly clustered about a church and tiny square as in Cordova (Fig. 5.14), or spread out along the valley such as in Velarde and Corrales, share many similarities in their architectural features. While an hacienda wrapped around a placita is often conceptualized as a characteristic Spanish house, in reality Hispano homes are more varied and too small to surround a small courtyard. As Bunting (1992:5) explains, "Only a few dwellings in the Taos area…were big enough to extend around a placita. Much more common are houses of a single-axis or L-shaped plan" (Bunting 1992:4–5). Thus, while one can find Hispano buildings, ranging from urban houses to rural ranches, constructed around an interior courtyard called a placita—in essence a domestic version of the village plaza—they are too grandiose to represent the vernacular of the common households. While these large and small houses might differ in their arrangements, they utilize common building materials and traditionally had similar roof construction.

5.5.4 Adobe Buildings

Adobe is the material from which the vernacular buildings of the Hispano Homeland were historically constructed. While not as dominant today, many new buildings continue to

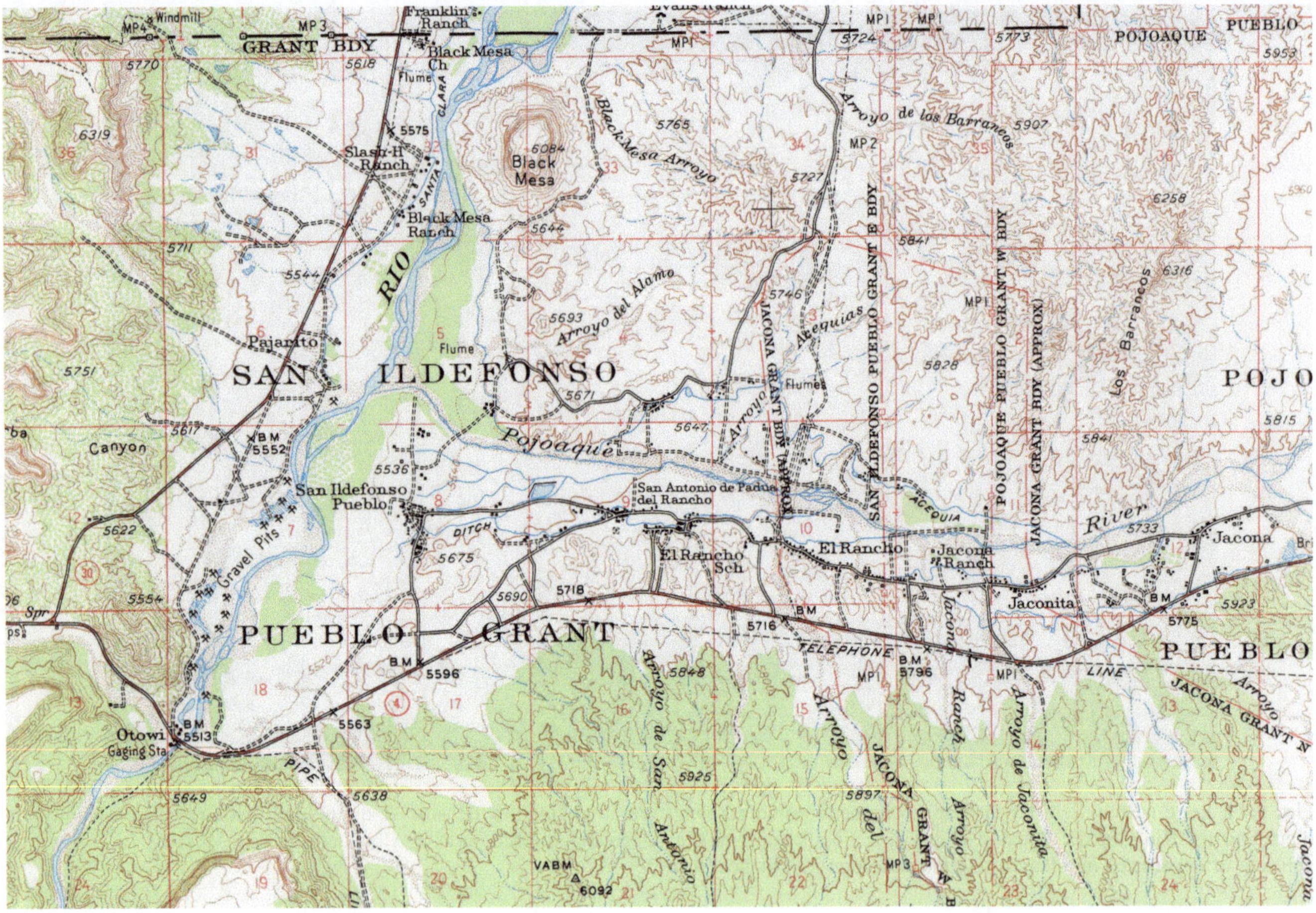

Fig. 5.12 Linear line village arrangement (see buildings along road in communities of El Rancho and Jaconita) in Pojoaque Valley to north of Santa Fe, New Mexico. This map also shows an acequina north of the river and several overlapping Spanish land grants, of which the San Ildefonso is most conspicuous, centered upon the San Ildefonso Indian Pueblo (Reproduction from U.S. Geological survey topographic map of Espanola 15-min quadrangle)

utilize adobe walls. Adobe is sun-dried brick in which a mixture of clay, soil, and straw are molded in a frame and then dried without firing (Fig. 5.15). Adobe buildings are often plastered with mud or a mud-straw mix, which provides protection against the erosion of the walls by rain. This process must be repeated every year or so. Adobe walls are often very thick, with a one-storey house having walls that are 18 inches to two feet thick. Within a larger building, such as a church, the lower walls may be nearly two meters or six feet thick (Fig. 5.16).

Adobe is a good example of environmental adaptation. The thick earthen walls heat slowly during the day, keeping the building interiors cooler than the outdoors, while at night they retain much of the heat they absorb during the day, staying warmer than the cool nighttime temperatures common in the dry climate (Treib 1993). Several differences can be seen between Spanish adobe and Indian adobe construction. Spanish adobe was shaped into bricks, while Indians simply used puddled adobe, by placing one layer of the adobe mud atop another, waiting for the puddled mud to dry before adding the next layer. In addition, Spanish adobe walls were much thicker and were placed on stone foundations, while Indians built directly on the ground (Bunting 1992).

The typical adobe house "was usually of simple construction: rooms of rectangular mud blocks with few window openings and a door" (Treib 1993: 22). The house was often long and narrow, with its width limited by the length of the vigas, or logs, spanning its walls (Treib 1993). The roofs of adobe buildings were typically flat, constructed of logs that extended from wall to wall, protruding from the exterior walls. Atop the log vigas smaller branches were placed, atop which was placed a layer of clay. Although most of the houses remained small, if they grew larger, "the string of rooms bent around an enclosed placita, thereby limiting the external openings to one or a few while increasing the structure's defensibility" (Treib 1993:22). Regardless of size, the traditional adobe houses had few if any exterior windows, and most were one story. If they had windows, these had wooden shutters, and most windows opened into central courtyard. Houses were often L-shaped, and additional

Fig. 5.13 Contiguous wall of adobe buildings within fortified plaza village. Plaza del Cerro in Chimayo, New Mexico, is the best preserved example of this eighteenth century Hispano village arrangement in which a continuous wall of buildings facing the plaza provided security

Fig. 5.14 San Antonio de Padua Church facing small plaza in center of village of Cordova, New Mexico. Note how both the church and the wall surrounding it are constructed of plastered adobe brick

Fig. 5.15 Drying and curing of stacks of adobe bricks in the sun near Alcalde, New Mexico. While fewer new buildings are now constructed of adobe brick than in the past, sufficient demand in the Hispano homeland remains to keep this adobe brick manufacturer in business

Fig. 5.16 Massive adobe walls buttress the San Francisco de Asis Church in Ranchos de Taos, New Mexico. This church was constructed around 1730

rooms with outside doors were added for married children. A wall often joined house with outbuildings or corrals. Farm outbuildings were typically small. However, houses could differ from the ideal.

Within several of the Hispano villages, such as in Las Truchas, in the Highlands at the foot of the Sangre de Cristo Mountains, pole and log houses were once common. Although a variant from the normal adobe buildings, the log

Fig. 5.17 Strings of dried chilies called ristras are often used for decoration in the Hispano homeland. These ristras are hanging from the portales facing the plaza in the Old Town section of Albuquerque, New Mexico

wall (or palisaded log walls, where the logs were placed upright in the ground) were often covered with adobe plaster (Gritzner 1974). Unusual within a Hispanic setting, geographers (Gritzner 1974; Winberry 1974) have speculated that German miners who worked in Mexico for the Spanish in the 1500s introduced log construction techniques that were brought north by the earliest Spanish settlers. Not only can one see old houses of log construction in several Highland New Mexico communities, but also ruins of small barns and gristmills.

Hanging from the porches or from vigas of many houses and shops within the Hispano Homeland are ristras. Ristras are strings of chilies that have been tied together and hung to dry (Fig. 5.17). While such one or two foot long ristras are often seen hanging on rural homes, and in season vendors sell them in roadside stands, they are also used as decorations in urban areas. Today, even chains of electric lights shaped and colored like chilis can be seen. Within placitas or near some houses can be seen *hornos* or outdoor bake ovens in which bread or other corn or wheat items are baked. These dome shaped ovens constructed of adobe were first introduced to the region by Oñate four centuries ago and are commonplace today in the Hispano Homeland (See also Fig. 4.20). Because they are rarely seen in Mexico or other Hispanic areas of the American West, the "hornos are an item of material culture that sets the Homeland apart" (Nostrand 1993:53).

5.5.5 Hispano Churches

The Roman Catholic churches were centers of their communities. Similarly, within nearby Indian pueblos the mission churches and convento were built on the most physically suitable and prestigious site possible. Both Spanish colonial parish churches and mission churches are architecturally very similar and have remained very consistent in their design since the 1600s. Indeed, the "same building techniques continue in use even today" (Treib 1993:29).

The general characteristics of these churches are described by Marc Treib (1993) in his monumental study of Hispano sanctuaries. He noted that "the thick-walled nave, the flat roof, the articulated apse, and the basic long and low profile wedded to the ground—combined to create the distinctive form of the Spanish sanctuaries of the colonial period" (Treib 1993:29). San Miguel Church in Santa Fe, with its angled apse and built in the early seventeenth century, is a classic example. The façade of Hispano churches is either planar or twin-towered. Those churches with a planar façade have a flat front, with the highest wall above front door. The highest point is marked by a cross, and sometimes a bell is hung below. Churches with a twin-towered façade usually have a balcony between their towers above the outside door.

The church walls, regardless of façade plan, are invariably of adobe or rough stone masonry, usually plastered with adobe (Figs. 5.18 and 5.19). The floor was initially earthen (with burials beneath), although later milled wood floors were installed. The roof is supported by vigas (logs) laid between the walls, with latillas (poles laid across logs), covered with grass, reeds or a mat topped with earth roof—simply a higher and more elaborate version of the roofs of many houses. Ceilings are not plastered, with the vigas and latillas being visible. Initially all the churches had such a flat roof, and many still do, but in some churches, such as the one in Santa Cruz, the original flat roof has been covered by a gabled roof built after the style was introduced by Anglo-Americans in the 1800s.

Fig. 5.18 San Jose de Gracia de las Trampas Church was built in the eighteenth century facing the plaza in Las Trampas, New Mexico. Note that this twin-towered adobe church has an exterior balcony, reached from an interior stairway, extending between the two towers. The adobe wall surrounds the church and its campo santo

Fig. 5.19 Interior of San Jose de Gracia de las Trampas Church. Besides the magnificent displays of Hispano religious art, one can clearly see the vigas supporting the church's roof

5.5.5.1 Church Grounds

The church grounds, termed the campo santo, are typically surrounded by an adobe wall about five feet high. Although one of the Law of the Indies ordinances suggested that cemeteries be located away from the center of the community, in reality "[b]urial within the campo santo was the normal practice" (Treib 1993:49). Indeed, it both northern New Mexico and Texas "most Spanish Americans [buried] their dead within the floors of their parish churches until the early nineteenth century" (Will de Chapparro 2014:1). Burial in the church occurred in both small Hispano villages as well as around the mission churches at Indian pueblos. Although in some of these grounds are individual grave markers, in others a single large cross commemorates the burials. In many of the Southwest's Mexican-American cemeteries, the oldest headstones are frequently from the late nineteenth century, as earlier grave markers have not survived. Fences or cerquitas, typically low concrete curbs arranged in rectangles, often separate graves in Mexican-American cemeteries, and in some less maintained cemeteries, they remain the primary indication of the grave after the loss of a wooden cross or grave marker (Will de Chapparro 2014).

5.5.5.2 Penitente Chapter House

In some of the smaller Hispano communities can be found a Penitente chapter house, termed a morada (Fig. 5.20). It has been argued that "No building distinguished the Spanish Americans' vernacular landscapes more than the Penitentes' (Los Hermanos de Jesús) unpretentious chapter meetinghouse" (Carlson 1990:142–3). Because of the scarcity of Roman Catholic priests in New Mexico during much of the eighteenth and nineteenth centuries, most communities lacked the presence of a priest, with the Penitentes providing religious services, conducting charitable activities, and maintaining community order (Smith 1999). The plainly colored moradas have been described as "a simple, private chapel and meeting place" (Treib 1993:61), located typically at the village edge in a secluded spot. They are often adjacent to a graveyard. They have the configuration of a small, single nave church. Alvar Carlson notes that typically "moradas have no belfries, cupolas, or crosses that would identify them as religious structures.... [yet sometimes] Latin crosses, without ornamentation, and occasionally Celtic pomee crosses, painted on the exterior walls and doors, help to identify a morada" (Carlson 1990:150). While few crosses appear affixed to the moradas, a "large wooden cross used during Holy Week" (Treib 1993:61) is often stored against the exterior walls, and a heavy wooden cross is sometimes erected near front entrance of the morada. Today, most moradas have fallen into disuse, many have disappeared, and some have been converted into other uses, yet several continue to hold Lenten services (Smith 2015).

The influence of religion throughout the Hispano Homeland is strong, and is seen by more than just the churches, moradas, and cemeteries (Fig. 5.21). As Carlson (1990:144) explains, "Some families had built small private

Fig. 5.20 Penitente morada located north of San Pedro, Colorado, within the San Luis Valley. This morada is no longer in active use, and its religious artifacts have been removed (Smith 2015). Many moradas have fallen into ruin or have disappeared

Fig. 5.21 Religion plays a major role in the life of Hispano homeland residents. This hand painted sign along the highway, commemorating the Virgin of Guadalupe, indicates the location of a Catholic Church in Velarde, New Mexico

chapels on their land to satisfy their immediate spiritual needs." Examples can still be seen in several northern New Mexico communities. Thousands of pilgrims are drawn to the Santuario de Chimayo to seek cures from its healing muds and "along roads and in the countryside are a variety of religious shrines" (Nostrand 2001:164). Together with numerous religious place names that often commemorate patron saints, Nostrand concludes that in "a religious sense what differentiates the homeland from other sections of the borderlands is the sheer quantity of these things" (Nostrand 2001:164).

Several miscellaneous rural landscape features distinguished the Hispano Homeland. These included the molino, which was a gristmill, unique in that the water wheel turned horizontally. In several communities a torreon or a circular tower built as a watchtower can be seen. These also functioned as defensive tower and provided a site of shelter during attack. Most of these have now been lost, yet a few still remain.

5.5.6 Perseverance and Change

Time and acculturation by Anglos have modified the ethnic landscape within the rural Hispano Homeland, yet it remains richly and distinctly different. Over time, elements contributed by Anglo-Americans have been added, while some communities have been reconfigured. For example, one can see the placement of a gabled roof on early churches or other buildings, such as the 1740's church at Santa Cruz, near Espanola, New Mexico. Yet the adobe plastered church with its twin towers remains, still displaying its ethnic heritage. Likewise, while the movement of many Anglos into parts of the Hispano Homeland led to the establishment of "nodal-shaped settlements with clearly defined business and commercial districts," (Smith 2002a:187), many elements of the traditional ribbon-shaped settlements remain upon the landscape.

5.5.6.1 Change and Persistence in Rural Areas

The change in the rural landscape is perceptively described by John Brinckerhoff Jackson:

> When I first came to New Mexico in the 1920s, I was attracted by the Spanish-American villages scattered throughout the ranch country.... They seemed very foreign, very un-American. In those times most families supported themselves by farming and raising cattle or sheep....
>
> The villages were half hidden in the immense open rangeland near a stream that watered the small fields of corn and chili and beans. The surrounding landscape was organized in an almost medieval manner. Easterners are not always aware that communal control of the land and its use, with a large common for the livestock, existed in the Spanish Southwest before New England had been heard of. Each household had the right to graze its cows or sheep on the very extensive community range, to use water from the community irrigation ditch, and to take wood from the community forest. Furthermore, the church

> belonged to the villagers, and so did the few roads and trails. Whoever used those facilities had to help keep them up, and this sharing of work and land and responsibility helps explain the strong sense of solidarity among the villagers. This kind of community, harking back to the late sixteenth century and brought from Spain by way of Mexico, had already begun to fall apart by the time I arrived in New Mexico (Jackson 1994:54).

He continued

> Sixty years, however, is a respectable length of time, and when I set out in 1990 to refresh my memory of the past, I found many changes in the New Mexico landscape, the most striking being what had happened to some of the villages I had once known. They had degenerated into rural slums of a very abject kind. Fields had reverted to second growth, houses were in decay, roads and irrigation ditches choked with rubbish and abandoned cars.... Farming no longer provided a living, most villages were too isolated, too lacking in resources, to try another way of life....
>
> Yet some villages have survived. The men now work for wages in the service sector or do odd jobs in town. Thanks to the automobile, the environment they depend on has expanded well beyond the village and ranch, and they think nothing of commuting thirty or forty miles to work....
>
> I was struck by the number of new dwellings and the decay and abandonment of the old. Adobe is no longer the most popular building material, cement block and frame houses are common, as well as houses trucked in from elsewhere.... The new freestanding houses seem to prefer the margins of the road leading out of the village to the traditional compact pattern of plazas (Jackson, *A Sense of Place, A Sense of Time*, 1994:57–58, Copyright © 1994 by Yale University Press, Reprinted by permission of Yale University Press).

Jackson has also lamented the frequency with which mobile homes are now seen.

> They are everywhere: tucked in between houses, attached to houses, even on top of houses; in alleys and gardens and out in the fields. In fact there are New Mexico villages where trailers outnumber conventional dwellings (Jackson 1990:361).

Nevertheless, a strong vibrant Hispanic imprint remains in rural New Mexico. Many of the rural churches have been restored, and the state has a special program for such architectural preservation. Better transportation permits Hispanos to obtain employment in cities, providing infusions of capital into communities whose fields are far too small to support even a subsistence lifestyle. In this respect, the Hispano Homeland is similar to virtually every agricultural region of the nation, in that most show strong change and decline in rural population centers and decreasing numbers of active farmsteads. Yet for all that change, many of the Hispano villages persist. Even when the rural Hispano population migrates to the cities, they retain visible links to their native villages, through murals they paint illustrating rural historical and religions themes, through continued internments in the village cemetery, and by commemoration of acequinas (Smith 2002b).

Alvar Carlson likewise laments changes in the Hispano Homeland. He titles a chapter "The Río Arriba: A Waning Cultural Region" in his *The Spanish-American Homeland*, published in 1990. He also writes, "Spanish-American rural landscapes have also been changed in their own ways, largely by the abandonment of old farmsteads and acequinas and the placement of nearby modern housetrailers or mobile homes" (Carlson 1996:60). However, he also notes that the mobile homes "are basically one-room deep like most adobe houses" (Carlson 1996:61). While many of the older traditional houses are in the poorer districts and have been abandoned, others have been acquired by wealthier Anglos who are sensitive to historical preservation. Carlson notes that many residents of the Hispano Homeland have taken steps "to raise their people's consciousness for cultural survival by restoring their traditional lifestyles and material culture" (Carlson 1996:61). This region remains one of the best examples of a clearly defined region-wide ethnic homeland in the United States, where both the rural and urban landscapes display the ethnic heritage of its residents.

5.5.6.2 Change and Persistence in New Mexico's Urban Centers

Change has also come to New Mexico's urban areas, but the Hispanic imprint remains. For example, within Santa Fe the Plaza remains, but it has been reduced in size. The Palace of the Governors, built about 1610, faces the plaza, having served as capitol and governor's residence until 1909. It is now the State Museum, with Indian merchants selling silver and pottery under the portales. The original cathedral, facing the side of the former plaza before it was reduced in size, was replaced by the current sanctuary built beginning in 1869 over and around the previous structure that dated to the 1600s, incorporating parts of the original. The streets near the plaza are narrow, portales or arcades line many of the sidewalks, and there are numerous adobe and flat roofed buildings. A few blocks from the plaza portions of the acequina madre remain, and adobe houses (Fig. 5.22) from the seventeenth century remain in use near the smaller San Miguel Church built between 1610 and 1628.

Overall, Santa Fe is generally well preserved, a circumstance that can largely be attributed to the fact that the AT&SF Railroad's main line bypassed the city by eight miles and to a popular revival of interest in Hispanic building style. This is eloquently explained by Nina Veregge:

> More subtle but in the long run of great importance to the preservation of Santa Fe's Hispanic environment was a shift in attitude: 'A psychological change toward the country and its Spanish-Mexican Indian culture'.... [A]rchaeologists... with others, promoted conservation of Pueblo and Spanish architecture and carried out a restoration of the Governor's Palace in 1916. Collectively, their influence on the future direction of planning in Santa Fe was strong (Veregge 1993:402, Reprinted by permission of *Journal of the Southwest*).

Fig. 5.22 El Zaguán hacienda in Santa Fe, New Mexico. This flat roofed adobe house was built in the early nineteenth century, not far from San Miguel Church

Much of the modern construction within Santa Fe incorporates many Hispano architectural elements, with walls around houses, use of adobe brick and tile in roofs, and various motifs.

Within Albuquerque, founded in 1706 and now New Mexico's largest city, the ethnic imprint clearly shows in the historic district. Its plaza remains, and many buildings with portales, both shops and residences, face the plaza. Likewise, along the plaza is the San Felipe de Neri Church (built in 1706), little changed other than the addition to its towers. The main business center of Albuquerque is located several miles distant, as the railroad bypassed Old Town by two miles. Thus, with the Anglo New Town being laid out to east, the attention of builders was diverted away from the Old Town area, contributing to its preservation, even though much of its activity is now devoted to tourism. Yet conflict in use of the Old Town space remains, not only involving non-Indians who would like to also sell their handicrafts from the sidewalks, but between Anglos and Hispanic ethnics over the types of business and religious activities occurring around the plaza. Furthermore, two-thirds of the property owners are non-Hispanic, even though the historic district was established to preserve Hispanic heritage (Dürr 2002).

A similar set of circumstances occurred in Las Vegas, New Mexico. Younger than either Santa Fe or Albuquerque, Las Vegas was formed from a Mexican land grant, and settlement began in 1835. Again, two distinct downtowns, about 1.5 miles apart, developed. The Plaza is the center of the old downtown, while a new downtown developed near the railroad station. Yet in Las Vegas preservation of the ethnic character of the plaza neighborhood was much less complete, as Nina Veregge notes.

> By 1895 the character of the plaza had changed completely. Brick and stone buildings, several of two stories, had replaced the adobe buildings around the square, now landscaped in the pattern of nineteenth century Anglo American parks…. Unlike Santa Fe and Albuquerque's Old Town, however, which retained their Hispanic character after the railroad shifted the focus of new development away from their plazas, the plaza of Las Vegas was transformed through demolition and replacement to reflect the growing influence of Anglo culture, and the 'old town' is a mix of flat-roofed adobes and pitched-roof frame houses (Veregge 1993:417–420, Reprinted by permission of *Journal of the Southwest*).

Nevertheless, while the architectural preservation of the buildings surrounding the plaza in Las Vegas is less complete, the character and community usage of the plaza is much more authentic. While most of the shops and restaurants abutting the plazas in both Albuquerque and Santa Fe now cater to tourists, those along the Las Vegas plaza serve many local residents, and the plaza continues to attract locals to its various community celebrations (Smith 2004).

5.6 Cultural Landscape of the Tejano Homeland

The ethnic landscape of Tejano South Texas, in particular that portion of the state that extends from San Antonio south to the Rio Grande and upriver to beyond Laredo, has been extensively studied and described by the geographer Daniel Arreola (2001, 2002a) as well as by architectural historians who have documented the area's heritage (Thompson 1994 and Freeman 1994). Indeed, Arreola notes that among "the Hispanic American borderland subareas only Hispano New Mexico demonstrates greater antiquity and cultural persistence than does the Tejano homeland of South Texas" (Arreola 2001:110). Largely dating from farming and ranching settlements made in the last half of the eighteenth century and the first decades of the nineteenth century, the Tejano homeland also incorporates communities that developed around Indian missions near San Antonio. The Tejano homeland reflects an ethnic landscape shaped by a Mexican-American population who represent over 90% of the population within many of the region's largest cities, such as Laredo and Brownsville, as well as the overwhelming majority of the rural residents. Thus, unlike the Hispano Homeland which has seen increasing acculturation by a growing Anglo population, the Tejano Homeland displays growing demographic and political dominance of its Hispanic population after Anglo domination during the first century of Anglo-American domination of Texas.

The porciónes that were surveyed along the Rio Grande during the Spanish, Mexican, and Republic of Texas periods were arranged perpendicular to the waterway, with their width fronting the river being one-fourth of their depth. While Jordan (1974) postulated that this surveying tradition reflects central European ways, the system is quite similar to what was described in parts of the Hispano Homeland. Even away from rivers, there was a long tradition in New Spain of surveying lots into rectangles, and such cadastre forms the basis of property boundaries in much of the Tejano Homeland.

The strong ties that the Tejano Homeland has with Mexico can be seen in both the well-preserved historic districts within small communities such as San Ygnacio and Roma and within the downtown area of Laredo, as well as in the contemporary business districts catering to a Mexican and Mexican-American clientele. Indeed, within the central business districts of the larger cities such as Laredo, Del Rio, and Brownsville, the majority of store signs and displays are Spanish-language, and large discount department stores are located near the border to cater to customers from both sides of the Rio Grande. Throughout the region numerous billboards along the highways advertise products and services in Spanish.

5.6.1 Plazas in Tejano Towns

The center of social life within many of the Tejano communities is the plaza, although with the exception of the oldest communities, these were platted under American laws and were not legally mandated. Daniel Areola notes, "Plaza towns are more numerous in south Texas than elsewhere in the Hispanic-American borderland states like California, Arizona, and New Mexico" (Arreola 1992:72), excluding informal defensive plazas. In total, 20 communities in South Texas have traditional plazas (Fig. 5.23), and several more have parks with playgrounds or picnic areas that partly fulfill the plaza function (Arreola 2002). Thus, most of the major communities have one or more plaza, several which were donated by prominent Anglos who promoted the development of their communities. In some communities, the plazas were established long after the settlements were founded, such as in Del Rio. These plazas "functioned as traditional social nodes during the nineteenth and early twentieth centuries" (Arreola 2001:111) and are still used for community celebrations.

Kioscos, or bandstands surrounded by gardenlike landscaping, are found in 80% of the plazas of South Texas (Arreola 2002). This open pavilion-like feature on a platform, which he relates to the period of French occupation of Mexico in the early 1860s, "became a standard element of Mexican townscapes during the prosperous era of the Porfirio Díaz regime" (Arreola 2002:78). Within most plazas the kiosco is centrally located, with walkways running to the plaza corners. Plazas sometimes contain monuments or statues. Sixty percent of the plazas in the Tejano homeland are fronted by a church (Arreola 2002).

5.6.2 Tejano Housing

Housing within South Texas displays the wide range between Spanish-Mexican ranch houses of the relatively wealthy and the "small, flat-roofed, cubelike" houses of the majority (Herzog 1999:101). The architecture of houses and other buildings within several of the older Tejano communities display strong connections to northeastern Mexican building traditions and the use of vernacular building materials. For example, within San Ygnasio, Texas three dozen blocky rectangular houses and former commercial buildings can be seen that were built of locally quarried sandstone and a caliche type of limestone, cemented together with a mortar mud and plastered with a mixture of sand and lime (Fig. 5.24). The thick stone walls were constructed in a style akin to the central Mexican *rejoneado* stonework. As Freeman (1994:184) explains, "This technique resulted in a pattern of larger stones and a secondary pattern of smaller stones coursed in

Fig. 5.23 Plaza in which a kiosco is located, with the church in background. This community of San Ygnacio, Texas is located along the Rio Grande

Fig. 5.24 Jesus Treviño House, built in 1830 in San Ygnacio, Texas. Originally constructed with a flat roof, gargolas can be seen along the right wall of this plastered stone masonry building. The oldest structures in this community were built without exterior windows for defensive reasons

the joints of the larger pattern." Although some of these buildings now sport hip or gabled roofs, many of the structures originally had flat roof covered with *chipichil*, defined as "a lime concrete with a pea-gravel aggregate" (Freeman 1994:185). This same material was also used for flooring.

These stone masonry buildings, most of which are one story in height, were built with few windows. Indeed, some have no windows, with opening the solid wooden doors providing the only ventilation, yet in some buildings both in the community and scattered on more isolated surrounding ranches, one can see where *troneras* or gunsights had been slotted into the walls (Simons and Hoyt 1996), a strong reminder that during much of the nineteenth century and continuing until the time of the First World War that political instability and banditry on both sides of the border necessitated homes having defensive elements.

The flat roofs of many of the single story houses display roof beams (*vigas*) similar to those noted in the Hispano Homeland. Some of the buildings display carved sandstone *canales* to drain the roofs, and one example in San Ygnacio has projecting rain spouts or *gargolas* (Simons and Hoyt 1996). Some of the older commercial buildings, such as in Roma, are built of adobe that is faced with brick, while others are a combination of stone and adobe. By the later part of the nineteenth century and continuing into the twentieth century, brick largely replaced stone as a building material in the Lower Rio Grande Valley.

Although the ranchos played a major role in the early development of the region, having a stone masonry house of the patron surrounded by wooden or jacal structures of the farm workers and their families, most Mexican-Americans of South Texas now live in the barrios of the various cities. Within these ethnic neighborhoods, separated from those of the Anglos and some of the wealthier Tejanos, "enclosed house properties, colorful building exteriors, small groceries, and neighborhood bars persist" (Arreola 2002:87).

5.6.3 Colonias in South Texas

Outside of the incorporated communities colonias are found, where Mexican-Americans were sold small lots on land contracts, upon which housing of a wide variety of design and quality has been constructed. Frequently lacking paved roads and utilities, including running water and sewers, the colonias nevertheless provide housing for a significant number of South Texans. Indeed, in 1995 an estimated 340,000 persons lived in Texas's 1400 colonias (Herzog 1999). By 2015 the number of colonias in Texas had risen to 2294, accommodating approximately a half million residents. Of the colonias in Texas, the greatest concentration is in counties adjacent to the Rio Grande. Indeed, within just six South Texas counties an estimated 369,500 persons live in colonias (Barton et al. 2015).

Over 20 years ago George Towers (1992:68) wrote, "Colonias are a new landscape unique to the United States' border region, a landscape developed from American land-use politics and peopled by Mexican immigrants." While recent study finds that "35 percent of colonia residents are foreign born" (Barton et al. 2015:3) and 96% are Hispanic, 73% are U.S. citizens. While initially established to house migrant farmworkers, colonias have grown to accommodate a large number of the working poor, with the quality of housing ranging from substandard to well-built. While most numerous in Texas, examples are found elsewhere in the Southwest. For example, it has also been shown that colonia developments in Arizona's Salt River Valley on the urban periphery of Phoenix played a major role in that metropolitan area's suburbanization process, with colonias developing on cheap lands subject to flooding or in proximity to the cotton fields or orchards upon which the residents labored (Lukinbeal et al. 2010).

Colonias, nevertheless, are not a conspicuous part of the Texas landscape, even though the largest colonia populations are located near Laredo, McAllen and Brownsville in the Lower Rio Grande Valley of South Texas. Colonias are often located away from the major highways or roads on floodplains or on lands unsuited for agricultural production (Herzog 1999). Within them the architecture varies widely, with virtually no control, from shacks to mobile homes, from wood-frame houses to masonry block. The one common element is poverty.

5.6.4 South Texas Border Cities

Many of the larger cities of the Tejano Homeland sit astride the U.S.-Mexican border, including cities such as Laredo-Nuevo Laredo, Roma-Ciudad Miguel Aleman, Rio Grande City-Camargo, Hidalgo-Reynosa and Brownsville-Matamoros. In all of these communities, the central business district on the United States side of the border contains numerous businesses that draw their customers from the Mexican side of the Rio Grande. Retail establishments in these Mexican-American shopping districts include clothing, jewelry, discount variety, and electronics stores (Arreola 2002). Store signage is in Spanish and many of the billboards along the highways advertise in Spanish. Within the cities Hispanic barrios, with their own sets of small neighborhood groceries and bars, are conspicuous. Indeed, racial discrimination initially separated these poorer Mexican-American neighborhoods from more upscale Anglo neighborhoods, yet rapid growth of population has resulted in a succession of newer barrios as one moves from the city center.

5.7 Contemporary Mexican-American Barrios

Throughout the American Southwest, and in many metropolitan areas in the Midwest, many Mexican-American neighborhoods or barrios developed during the twentieth century and continue expanding in the twenty-first century. Many of the best known barrios are found within the Los Angeles, California metropolitan area. Barrios developed in response to residential segregation that separated Hispanic populations from Euro-American neighborhoods, yet that separation facilitated the creation in barrios of "community-sustaining practices in segregated space that heighten cultural identity and reinforce an ethnic solidarity of place" (Arreola 2012:159). Although some Hispanic neighborhoods date back well over a century—and indeed Los Angeles was founded by the Spanish over two and a quarter centuries ago—other Hispanic barrios developed as growth of the region's Hispanic population by both immigration and natural increase resulted in Hispanics occupying neighborhoods that were originally built by other residents. While most of the best known barrios are in metropolitan areas, Mexican-Americans have a long history of employment within meat packing plants scattered across the Midwest and Great Plains. Thus, many of the ethic features described in larger urban settings are equally applicable to smaller communities with a long history of employment of Mexican-Americans, such as Garden City, Kansas (Stull and Broadway 2008).

The largest Los Angeles barrio, consisting of low- to moderately-priced housing, extends south from downtown Los Angeles (Fig. 5.25), yet growing Latino neighborhoods are found throughout the metropolitan area. As Allen and Turner (2002:22) explain, "Latino settlement in more distant suburbs symbolizes the upward mobility of the mostly U.S.-born Latinos." While many of the Hispanic barrios, such as the Boyle Heights district just east of downtown Los Angeles, were first occupied by a variety of residents, it has been almost entirely Mexican-American for over a half century. In contrast with this neighborhood which was urban oriented from its beginnings, other ethnic Mexican barrios in Los Angeles originated as colonias of Mexican farm workers. Over time these evolved into Mexican-American barrios, and during "the last half of the twentieth century most houses in the *colonias* were completely rebuilt or substantially improved" (Allen and Turner 2002:26). Examples are found near San Fernando, Canoga Park, Oxnard, and Santa Ana, just to name a few.

Fig. 5.25 Jewelry store on Broadway in downtown Los Angles, California. Spanish language signage on the store indicates that Hispanics now form a majority of customers in that retail district of the city, given the proximity of the East and South L.A. barrios

The development and expansion of urban Hispanic barrios has also occurred even in cities that long promoted their Anglo heritage, notwithstanding their Hispanic heritage, such as in Phoenix, Arizona. There it can be seen by upon the evolving ethnic landscape. Alex Oberle argues that,

> The proliferation of lower income apartment complexes and trailer parks … provided reasonably affordable accommodations and a level of relative invisibility for Hispanic immigrants. As a result, businesses rather than homes are the most visible manifestation for Hispanic heritage (Oberle 2006:161).

In contrast, in neighborhoods where many of the Hispanic residents are homeowners, such as in Phoenix's Garfield neighborhood studied by Arreola (2012:162), "homeownership…creates the opportunity to shape a Mexican housescape, a critical component to culturally sustainable placemaking." Thus, visible aspects of Hispanic ethnicity can be seen in both business and residential districts. In a similar manner the development of a barrio in the Pilsen neighborhood of Chicago, once known for its German and Bohemian population, is marked by its numerous Mexican restaurants and shops advertised in Spanish.

Culturally, and even linguistically, recent Mexican-American immigrants frequently differ from American Hispanics whose roots in the United States go back a century or more. Thus, while earlier arriving Mexican settlers in Phoenix came from Sonora and Chihuahua, and those two states along with Sinaloa continue to dominate immigration patterns, newer Mexican residents in Phoenix are also coming from the states of Michoacan, Guerro, and Oaxaca in southern Mexico. Thus, "regional identity is a critical variable in reading the landscape of Mexicanidad in Phoenix" (Oberle and Arreola 2008:190). Similar regional connections exist between many sending communities in Mexico and receiving communities in the United States, where clear regional connections exist, but where changes have occurred as the regional diversity of Mexican immigrants has increased. The barrios are dominated by more recent immigrants, and their origins accentuate their diversity.

5.7.1 Carnicerías and Hispanic-Oriented Businesses

Carnicerías are viewed as "the most widespread and visible examples of Latino expansion into previously non-Hispanic areas" (Oberle 2006:148). While the name translates as a meat market or butcher shop, carnicerías in actuality are far more, providing a variety of goods that are often associated with a mini-market, as well as international telephone cards and a place where funds can be wired home (Fig. 5.26). Functioning like a general store—but with an extensive array of fresh meats and sausages—the carnicería is typically the first Hispanic business to open within many neighborhoods, often opening when Hispanics are still only a small fraction of the population.

Fig. 5.26 Ornate mural on exterior wall of carnicería in barrio of East Los Angeles, California

Carnicerías not only stock a wide variety of ethnic foodstuffs, but they conspicuously display their ethnicity by their business names and exterior decoration. Not only are images of animals, often beef cattle, frequently displayed in logos among the colorful signage at the stores, but murals painted across their exterior walls display "rural, bucolic imagery" or "ranching scenes like cattle being roped and branded or livestock grazing in verdant fields" (Oberle 2006:158). As the Hispanic population grows, additional ethnic stores, including panaderías, joyerías, discotecas, yerberías, or dulcerías, among others, are added to the ethnic business district. In some cities, such as Los Angeles, much of the downtown central business district has come to be dominated by Hispanic shops and signage.

Cross-border shopping has grown tremendously, with many cities on the American side of the international boundary developing shopping districts and malls that increasingly cater to Mexican customers, those both living in the Mexican counterparts of the border cities as well as those living in cities farther from the border. As Daniel Arreola has noted in McAllen, Texas, not only have Mexican visitors become a major part of the customer base for its large shopping mall, but "this shopping bonanza has spilled beyond the mall to the traditional Anglo shopping district now converted to a Mexican downtown as upscale retail has been lured to suburban malls" (Arreola 2010:339). All along the border business districts have evolved catering to the special needs of customers from across the border, with discount clothing, general merchandise, jewelry, and electronics shops dominating the districts. Adding another ethnic element to the mixture, many of the shops have Asian proprietors (Arreola 2010).

5.7.2 Mexican-American Housescapes

The importance of the Mexican-American housescape, not only visually, but symbolically is highlighted by Daniel Arreola.

> The housescape represents a domestic cultural landscape, rooted in Spanish architectural heritage yet melded to Native American building traditions of Mexico and the greater Southwest. Today, the housescape remains a symbolic landscape in Mexican American communities, created and recreated, often unconsciously, as a representation of a cultural ideal (Arreola 2006:113).

Housing occupied by Mexican-Americans varies widely, from some whose architecture clearly speaks of its Mexican cultural legacy, to others that are typical of dwellings occupied by other Americans of a similar economic status. Sometimes it is more how the exterior of the building is decorated, how items are stored outside—as many Mexican immigrants came from rural areas where more everyday living occurred outdoors (Arreola 2006), and how yards are fenced (Fig. 5.27) that marks the Mexican-American ethnicity upon the barrio landscape (Arreola 1981).

Fig. 5.27 House within predominately Mexican ethnic barrio in South Phoenix, Arizona. Note that the concrete block masonry fence, topped by wrought iron, completely encloses the front yard. A shrine displaying the Madonna faces the yard, while the house is painted purple

The ethnic distinctiveness of the Mexican-American housescapes has been attributed to "the almost continuous extent of front-property enclosure through a variety of fence types, a sometimes riotous use of brilliant colors on house exteriors, and an occasional religious shrine in the front yard" (Arreola 1988:299). Daniel Arreola explains that "enclosure, exterior color, and religious statuary are not unique to the Mexican American barrio and can be found as individual landscape signatures in several other cultures and subcultures" yet, "the presence of the signatures together… suggests a specific form of cultural communication" (Arreola 1988:299). Other markers of Mexican-American neighborhoods, such as Jones (2008:144) describes in San Antonio, Texas, include "frontyard gardens, and maguey and prickly pear cactus" ornamenting their yards.

Mexican-Americans often have brightly painted houses, selecting bright colors of blue, green, orange, or pink. In some neighborhoods Arreola (1988, 2012) notes approximately half to two-thirds of houses are so painted, and some entirely painted, yet most with only brightly painted trims. This appears to come from their tradition of painted churches or churches with colored tiles. While most houses had entire walls of a color or only added another color on the trim, a few houses within Mexican-American barrios are painted in two or three colors. It has been suggested that the "use of multicolors may be a variation of the stereochromatic color scheme employed in Mexico in which two horizontal strips are painted across the façade with the darker hue along the base" (Manger 2000:21).

Fencing or enclosure of the yard is far more common among Mexican-Americans than among Americans of most European ethnicities. This is particularly true for the backyards, where Manger (2000) noted at least 99% compliance within the Phoenix barrio he studied, yet within many Mexican-American working-class neighborhoods approximately two-thirds of the houses have fencing or walls around their front yards. While the construction materials vary, with "wrought iron, chain link, concrete block, or stone" seen within Phoenix's Garfield neighborhood, Arreola (2012:163) noted "a distinctive wrought iron style, and characteristically…a sun fan." The use of the front yard as a living space, including furniture in the yard or on a front porch, is commonly associated with many working-class Mexican-American communities, in sharp contrast to neighborhoods that are non-Hispanic white, where residents, if outdoors, are more likely to seek the privacy of a fenced backyard. While the boundaries of the Mexican-American front-yard are delineated, they still permit the occupants to interact with their neighbors. This can be contrasted with a desire for the interior enclosed patio, especially common in middle to upper class Hispanic homes. The fencing is important in defining property boundary, meeting the cultural need for defining private space. In addition, ornamental wrought iron grates and bars over windows further enhance the privacy of the enclosed spaces.

5.7.2.1 Catholic Shrines

Religious shrines (Fig. 5.28), statuary, and year-round Christmas decorations adorn some Mexican-American yards. Daniel Arreola (1988:308) suggests the yard shrine may be "the most distinctive element of the housescape in Mexican American neighborhoods." He describes them as a "Roman Catholic religious feature that resembles a little house or church, the shrine is usually an open fronted enclosure that is permanently placed in a yard. Called nichos (niches) in Tucson, capillas (little chapels) in San Antonio, and grutas in Los Angeles, all yard shrines are similar in basic design and intent" (Arreola 1988:308). While the shrine is a clearly recognized diagnostic characteristic of the Mexican-American neighborhood, the vast majority of yards lack them. Although yard shrines are not typically found in Mexico, it is suggested that "yard shrines result from the adaptation of traditional Mexican symbols to a new spatial setting" (Manger 2000:22). Arreola (1988) indicates that home altars reflect pre-Catholic Mexican heritage of having indoor shrines. The shrines contain images of various saints, with the Virgin of San Juan being most common in San Antonio and Saint Jude dominating in Tucson. Nevertheless, such house shrines are the rarest of the distinctive characteristics in Mexican-American neighborhoods, being found with fewer than 1% of the homes.

5.7.2.2 Chicano Mural Art

Chicano mural art is especially common in cities (Fig. 5.29), and may cover entire walls of buildings or enclosures, while others have been painted on bridge supports. It has been noted that "murals tend to be found on prominent public and private buildings that ensure exposure to the greatest possible audience" (Arreola 1985:16). Such locations can be found on the sides of stores, community centers, schools, and housing projects. California alone was estimated to have 1500 painted murals as the twentieth century ended (Herzog 1999), and their numbers have increased. Murals typically display pre-Columbian and Mexican history, patriotic personalities, religious messages, and local images (Arreola 1985). In many communities, the murals depict the heritage or scenery of the region in Mexico from which the population has migrated (Smith 2002b).

Muralism has been connected to the rise of Mexican political consciousness, beginning following the Mexican Revolution of 1910–17. As Lawrence Herzog explains, "It became a way to inform and educate Mexico's basically illiterate rural population of the 1920s about the ideals of a new society and the evils of the past" (Herzog 1999:135).

Fig. 5.28 Brick masonry domed structure housing Catholic shrine in Mexican ethnic barrio in South Phoenix, Arizona. Religious statues are housed behind the glass windows of the shrine, which is festooned with Christmas lights and topped with a cross. The yard is enclosed by the iron fence, and the window frames and the gable are painted pink

Fig. 5.29 Mural covering side of building within the Pilsen neighborhood of Chicago, Illinois. This mural, one of several covering sides of commercial buildings within the one of Chicago's best known Hispanic neighborhoods, highlights the heritage of the residents' Mexican homeland and celebrates the accomplishments of many individuals

With the development of the Chicano movement in the United States during the 1960s, "[m]uralism, as it had in Mexico some four decades earlier, became a vehicle for protesting and airing demands on the daily urban landscape. It was also a way for Latinos to reclaim their place on that landscape" (Herzog 1999:135). Even the widespread painting of Latino gang-related graffiti can be viewed as a display of heritage and territoriality by members of this ethnic population.

5.8 Multiple Hispanic Landscapes

This chapter has touched upon only some of the multitude of ethnic Hispanic landscapes to be found in the United States. Some, such as the Hispano Homeland in New Mexico and the Tejano Homeland along the Rio Grande, span extensive areas that incorporate many ethnic farming villages. Others, such as those lands that attracted the Basques, who largely worked as solitary sheepherders, created far fewer visual landscape legacies (Douglas 1992). While the Hispano and Tejano homelands are several centuries old, others such as the Mexican-American neighborhoods within many Midwestern and Southern farming communities are only a decade or two old. While the discussion of the Hispanic ethnic landscapes in this chapter has focused upon the long-established Hispanic communities of the Southwest and large urban clusters elsewhere, the arrival of Hispanic minority populations has begun to reshape recipient neighborhoods elsewhere, even where Hispanic newcomers remain distinct minorities. For example, an "influx of Latinos initiated a profound change in the cultural landscape of the Texas Panhandle" (Montalvo and Estaville 2016:340), an area long dominated by Anglo Texans. These changes reflect the same themes of house architecture and adornment, murals, religious shrines and cemeteries, and Spanish language advertising and political signage that have been noted elsewhere. Recently arrived Hispanics in Charlotte, North Carolina, while meeting resistance from some anti-immigrant politicians, have nevertheless begun

> to remake social space and fostered Latino place-making processes.... Small scale entrepreneurship revitalized declining strip malls and filled empty spaces along suburban arterial roadways. Spanish-language signage created bilingual streetscapes. New types of street-oriented businesses appeared. In particular, *loncheras* (Mexican foodtrucks) and street vendors added a new layer of activity to sidewalks and parking lots (Furuseth et al. 2015:11, Reprinted with permission of the American Geographical Society).

The overwhelming majority of the operators of the loncheras or taquerias are Hispanic and they serve Mexican foods, yet few permanent structures display their ethnic landscape.

As Chap. 6 will discuss, the Cuban settlement in Miami-Dade County, Florida, dates from the arrival of anti-Castro refugees beginning in 1959, while the Dominican community in New York City was largely established following the assassination of the Dominican dictator in 1960. While some settlements, such as the Ybor City settlement of Cubans in Tampa and the Puerto Rican community in New York City, are older, others are even more recent, such as the Honduran and Salvadorian settlements within southern California. Elsewhere, the arrival of newcomers has altered pre-existing Hispanic settlements.

Hispanic architectural influences extend far beyond areas of early Spanish settlement. For example, Santa Fe style architecture included the use of adobe building materials, log roofs (use of vigas), and interior patios. This prompted the development of Spanish Revival Architecture, which became popular following World War I.

> Spanish Revival, which ranged from Renaissance to Baroque and contained a generous admixture of influences from throughout the Mediterranean and Mexico, had few genuine antecedents in America north of Mexico. Unable to find, much less revive, significant numbers of elaborate and decorative models in the nation's Hispanic past, American architects manufactured their own indigenous Hispanic architecture (Weber 1992:352–3).

Thus, the style was partly ethnic, partly contrived, yet it enhanced the perception by both residents and tourists of the area as being Hispanic. Furthermore, the style was copied and modified throughout much of the nation's Sunbelt, spreading far beyond the area of its origin. While many ethnic groups have assimilated into the American cultural mainstream, it is also true that portions of American culture have become increasingly Hispanicized.

References

Allen, James P., and Eugene Turner. 2002. *Changing Faces, Changing Places: Mapping Southern Californians*. Northridge: California State University Center for Geographical Studies.

Arreola, Daniel D. 1981. Fences as Landscape Taste: Tucson's Barrios. *Journal of Cultural Geography* 2 (1): 96–105.

———. 1985. Chicano Mural Art. *American Geographical Society's Focus* 35 (3): 14–19.

———. 1988. Mexican American Housescapes. *The Geographical Review* 78 (3): 299–315.

———. 1992. Plaza Towns of South Texas. *The Geographical Review* 82 (1): 56–73.

———. 1993. The Texas-Mexican Homeland. *Journal of Cultural Geography* 13 (2): 61–74.

———. 2001. La Tierra Tejana: A South Texas Homeland. In *Homelands: A Geography of Culture and Place across America*, ed. Richard L. Nostrand and Lawrence E. Estaville, 101–124. Baltimore: Johns Hopkins University Press.

———. 2002. *Tejano South Texas: A Mexican American Cultural Province*. Austin: University of Texas Press.

———. 2006. The Picture Postcard Mexican Housescape: Visual Culture and Domestic Identity. In *Landscape and Race in the United States*, ed. Richard H. Schein, 113–126. New York: Routledge.

———. 2010. The Mexican-US Borderlands through Two Decades. *Journal of Cultural Geography* 27 (3): 331–351.

———. 2012. Placemaking and Latino Urbanism in a Phoenix Mexican Immigrant Community. *Journal of Urbanism* 5 (2-3): 157–170.

Barcus, Holly R., and Laura Simmons. 2013. Ethnic Restructuring in Rural America: Migration and the Changing Faces of Rural Communities in the Great Plains. *The Professional Geographer* 65 (1): 130–152.

Barton, Jordana, Emily Ryder Perlmeter, and Raquel R. Marquez. 2015. *Las Colonias in the 21st Century: Progress Along the Texas-Mexico Border*. Dallas: Federal Reserve Bank of Dallas.

Bunting, Bainbridge. 1992. *Taos Adobes: Spanish Colonial and Territorial Architecture of the Taos Valley*. Albuquerque: University of New Mexico Press. (republication of 1964 edition).

Carlson, Alvar W. 1975. Long-Lots in the Rio Arriba. *Annals of the Association of American Geographers* 65 (1): 48–57.

———. 1990. *The Spanish-American Homeland: Four Centuries in New Mexico's Rio Arriba*. Baltimore: Johns Hopkins University Press.

———. 1992. Spanish Americans in New Mexico's Rio Arriba. In *To Build in a New Land: Ethnic Landscapes in North America*, ed. Allen G. Noble, 345–361. Baltimore: Johns Hopkins University Press.

———. 1996. Cultural Persistence in New Mexico's Rio Arriba: The Case for the Designation of a Rural Historic Culture Region. In *Ethnic Persistence and Change in Europe and America: Traces in Landscape and Society*. Veröffentlichungen der Universität Innsbruck 213, ed. Klaus Frantz and Robert A. Sauder, 47–66. Innsbruck: University of Innsbruck.

Douglas, William A. 1992. Basques in the American West. In *To Build in a New Land: Ethnic Landscapes in North America*, ed. Allen G. Noble, 379–398. Baltimore: Johns Hopkins University Press.

Dürr, Eveline. 2002. Contested Urban Space: Symbolizing Power and Identity in the City of Albuquerque, USA. In *Urban Ethnic Encounters: The Spatial Consequences*, ed. Aygen Erdentug and Freek Colombijn, 209–225. London: Routledge.

Ennis, Sharon R., Merarys Rios-Vargas, and Nora G. Albert. 2011. *The Hispanic population*: 2010. Census 2010 Brief C2010BR/4. Washington: U.S. Census Bureau.

Freeman, Joe C. 1994. Regional Architecture and Associated Crafts. In *A Shared Experience: The History, Architecture and Historic Designations of the Lower Rio Grande Heritage Corridor*, ed. Mario L. Sanchez, 2nd ed., 137–197. Austin: Texas Historical Commission.

Furuseth, Owen J., and Heather A. Smith. 2008. From Winn-Dixie to Tiendas: The Remaking of the New South. In *Latinos in the New South: Transformations of Place*, ed. Heather A. Smith and Owen J. Furuseth, 1–17. Burlington: Ashgate.

Furuseth, Owen, Heather Smith, and Paul McDaniel. 2015. Belonging in Charlotte: Multiscalar Differences in Local Immigration Politics and Policies. *The Geographical Review* 105 (1): 1–19.

Gomez-Novy, Juan, and Stefanos Polyzoides. 2003. A Tale of Two Cities: The Failed Urban Renewal of Downtown Tucson in the Twentieth Century. *Journal of the Southwest* 45 (1-2): 87–119.

Griffith, James S. 1975. The Folk-Catholic Chapels of the Papagueria. *Pioneer America* 7 (2): 21–36.

———. 1992. *Beliefs and Holy Places: A Spiritual Geography of the Pimeria Alta*. Tucson: University of Arizona Press.

Gritzner, Charles. 1974. Construction Materials in a Folk Housing Tradition: Considerations Governing Their Selection in New Mexico. *Pioneer America* 6 (1): 25–39.

Guthrie, Thomas H. 2013. *Recognizing Heritage: The Politics of Multiculturalism in New Mexico*. Lincoln: University of Nebraska Press.

Guzman, Betsy. 2001. *The Hispanic Population*, Census 2000 Brief C2KBR/01-3. Washington, DC: U.S. Census Bureau.

Herzog, Lawrence A. 1999. *From Aztec to High Tech: Architecture and Landscape Across the Mexico-United States Border*. Baltimore: Johns Hopkins University Press.

Hornbeck, David. 1990. Spanish Legacy in the Borderlands. In *The Making of the American Landscape*, ed. Michael P. Conzen, 51–62. Boston: Unwin Hyman.

Jackson, John Brinckerhoff. 1990. The House in the Vernacular Landscape. In *The Making of the American Landscape*, ed. Michael P. Conzen, 355–369. Boston: Unwin Hyman.

———. 1994. *A Sense of Place, a Sense of Time*. New Haven: Yale University Press.

———. 1997. In *Landscape in Sight: Looking at America*, ed. Helen Lefkowitz Horowitz. New Haven: Yale University Press.

Jones, Richard C. 2008. Cultural Retrenchment and Economic Marginality: Mexican Immigrants in San Antonio. In *Immigrants outside Megalopolis: Ethnic Transformation in the Heartland*, ed. Richard C. Jones, 135–160. Lanham (MD): Lexington Books.

Jordan, Terry G. 1974. Antecedents of the Long-Lot in Texas. *Annals of the Association of American Geographers* 64 (1): 70–86.

Kenzer, Martin S., Douglas J. Sherman, Robert A. Rundstrom, and Bernard O. Bauer. 1992. *California Landscapes*. New Brunswick: Rutgers University Press.

Lindner, Keith W. 2012. Geographies of Struggle in the San Luis Valley. *The Geographical Review* 102 (3): 372–381.

Lukinbeal, Chris, Daniel D. Arreola, and D. Drew Lucio. 2010. Mexican Urban Colonias in the Salt River Valley of Arizona. *The Geographical Review* 100 (1): 12–34.

Manger, William F. 2000. The 'Idealized' Mexican American Housescape. *Material Culture* 32 (1): 1–36.

McHugh, Kevin E. 1989. Hispanic Migration and Population Redistribution in the United States. *The Professional Geographer* 41 (4): 429–439.

Montalvo, Edris J., and Lawrence E. Estaville. 2016. Latino Political Participation in the Texas Panhandle, 1980–2010. *Journal of Cultural Geography* 33 (3): 339–355.

Noble, Allen G. 1984. *Wood, Brick, and Stone: The North American Settlement Landscape, Volume 1, Houses*. Amherst: University of Massachusetts Press.

Nostrand, Richard L. 1970. The Hispanic-American Borderland: Delimitation of an American Culture Region. *Annals of the Association of American Geographers* 60 (4): 638–661.

———. 1992. *The Hispano Homeland*. Norman: University of Oklahoma Press.

———. 1993. The New Mexico-Centered Hispano Homeland. *Journal of Cultural Geography* 13 (2): 47–59.

———. 2001. The Highland-Hispano Homeland. In *Homelands: A Geography of Culture and Place Across America*, ed. Richard L. Nostrand and Lawrence E. Estaville, 155–167. Baltimore: Johns Hopkins University Press.

Oberle, Alex. 2006. Latino Business Landscapes and the Hispanic Ethnic Economy. In *Landscapes of the Ethnic Economy*, ed. David H. Kaplan and Wei Li, 149–163. Lanham: Rowman and Littlefield Publishers.

Oberle, Alex P., and Daniel D. Arreola. 2008. Resurgent Mexican Phoenix. *The Geographical Review* 98 (2): 171–196.

Simons, Helen, and Cathryn A. Hoyt. 1996. *A Guide to Hispanic Texas*. Austin: University of Texas Press.

Smith, Jeffrey S. 1999. Anglo Intrusion on the Old Sangre de Cristo Land Grant. *The Professional Geographer* 51 (2): 170–183.

———. 2002a. Cultural Landscape Change in a Hispanic Region. In *Geographical Identities of Ethnic America*, ed. Kate A. Berry and Martha L. Henderson, 174–200. Reno: University of Nevada Press.

———. 2002b. Rural Place Attachment in Hispano Urban Centers. *The Geographical Review* 92 (3): 432–451.

——. 2004. The Plaza in Las Vegas, New Mexico: A Community Gathering Place. In *Hispanic Places, Latino places: Community and Cultural Diversity in Contemporary America*, ed. Daniel D. Arreola, 39–53. Austin: University of Texas Press.

——. 2015. Penitente Moradas: A Vestige of the Nuevomexicano Cultural Landscape. *Material Culture* 47 (2): 23–40.

Stilgoe, John R. 1982. *Common Landscape of America, 1580 to 1845*. New Haven: Yale University Press.

Stull, Donald, and Michael Broadway. 2008. Meatpacking and Mexicans on the High Plains: From Minority to Majority in Garden City, Kansas. In *Immigrants outside Megalopolis: Ethnic Transformation in the Heartland*, ed. Richard C. Jones, 115–133. Lanham: Lexington Books.

Thompson, Jerry D. 1994. Historical Survey. In *A Shared Experience: The History, Architecture and Historic Designations of the Lower Rio Grande Heritage Corridor*, ed. Mario L. Sanchez, 2nd ed., 17–77. Austin: Texas Historical Commission.

Towers, George. 1992. The Urban Colonias of El Paso County, Texas. In *Geographical Snapshots of North America: Commemorating the 27th Congress of the International Geographical Union and Assembly*, ed. Donald G. Janelle, 65–68. New York: The Guilford Press.

Treib, Marc. 1993. *Sanctuaries of Spanish New Mexico*. Berkeley: University of California Press.

U.S. Geological Survey. 2005. *Territorial Acquisitions. The National Atlas of the United States of America*. Reston: U.S. Department of the Interior. http://nationalmap.gov/small_scale/printable/images/pdf/territory/pagetacq3.pdf. Accessed 30 November 2016.

Veregge, Nina. 1993. Transformations of Spanish Urban Landscapes in the American Southwest, 1821–1990. *Journal of the Southwest*. 35 (4): 371–459.

Weber, David J. 1992. *The Spanish Frontier in North America*. New Haven: Yale University Press.

Will de Chapparo, Martina. 2014. Relics of Time. In *En Recuerdo de: The Dying Art of Mexican Cemeteries in the Southwest*, ed. Bruce F. Jordan, 1–12. Lincoln: University of Nebraska Press.

Winberry, John J. 1974. The Log House in Mexico. *Annals of the Association of American Geographers* 64 (1): 54–69.

Wright, John B., and Carol L. Campbell. 2008. Landscape Change in Hispano and Chicano Villages of New Mexico. *The Geographical Review* 98 (4): 551–565.

6 Hispanic Landscapes of the Eastern United States

The number of Hispanics in the United States is rapidly increasing. Hispanics, who comprised 12.5% of the nation's population in 2000, were second only to German ethnics, who then comprised 15.2% of the population. Hispanics comprised 16.4% of the nation's population when enumerated by the 2010 Census, while the American Community Survey for that year estimated that Germans were outnumbered by Hispanics by 2.6 million (Ennis et al. 2011, American Community Survey). Unlike the German ethnics, most of whom have lost their facility with the German language in the century or more since most of their ancestors immigrated, most Hispanics maintain their fluency in Spanish. While Mexicans comprise the majority (63.0% in 2010, up from 58.5% in 2000) of the Hispanic population in the United States, individuals of Mexican heritage, or those who can trace their ancestry to the various ethnic homelands that evolved within that portion of New Spain that now comprises the American Southwest, are not the only Hispanics to have imprinted their ethnic landscapes on the United States. While Chap. 5 examined the Hispanic ethnic landscapes that are best displayed in the American Southwest, including the Hispano and Tejano homelands and the Mexican-American barrios, this chapter explores the Hispanic landscapes that exist primarily within the eastern portion of the United States.

6.1 Hispanic Populations in the Eastern United States

The Hispanic heritage displayed in much of the eastern United States illustrates linkages between that area and the Spanish-speaking islands within the Caribbean. Not only was Spanish Florida governed via the Council of the Indies in Havana, but centuries later most Hispanic immigrants arriving in Florida and the rest of the eastern United States came from such former Spanish colonies as Cuba, the Dominican Republic, and Puerto Rico (Fig. 6.1). Puerto Ricans are U.S. citizens; thus they can move freely between their island home and the U.S. mainland. In 2010, Puerto Ricans comprised 9.2% of all Hispanics in the United States (down from 9.6% in 2000), Cubans accounted for 3.5%, and Dominicans comprised 2.8% (Ennis et al. 2011; Guzman 2001). Comparisons of the 2000 and 2010 Censuses indicated that the number of Puerto Ricans living in the United States increased by 1.22 million, Cubans grew by 544,000, while Dominicans gained 650,000, the largest proportional gain among the three West Indian groups.

Caribbean Hispanics are highly concentrated within several East Coast locations, although with population growth, the degree of that concentration has decreased over the past decade. For example, 68.0% of all persons in the United States claiming Cuban ancestry lived within Florida in 2010, and Miami-Dade County alone contained 43.2% of the nation's Cuban-Americans, down from 52.4% in 2000, as the group has begun dispersing to other parts of the state. Similarly, 60.6% of all Dominican ethnics living in the United States reside within the greater New York-Northern New Jersey consolidated metropolitan statistical area, down from 72.1% a decade earlier. Only Santo Domingo, the capital of the Dominican Republic, is home to a greater number of Dominicans. New York City alone was home to 41.8% of the nation's Dominicans, down from 53.2% in 2000, even though the city's Dominican population had increased by 214,000 over the decade. While nationally Puerto Ricans are both more numerous and somewhat less concentrated in a single location, nevertheless 15.4% of the nation's Puerto Ricans reside in New York City. Indeed, there are twice as many individuals of Puerto Rican ancestry living in New York City than in Puerto Rico's San Juan municipio, although more Puerto Ricans live within the San Juan metropolitan area than within New York City. As with the Dominicans, the percent of the nation's Puerto Ricans living in New York City has declined since 2000, when the figure was 23.2% (Ennis et al. 2011; Guzman 2001). As Puerto

J.A. Cross, *Ethnic Landscapes of America*, DOI 10.1007/978-3-319-54009-2_6

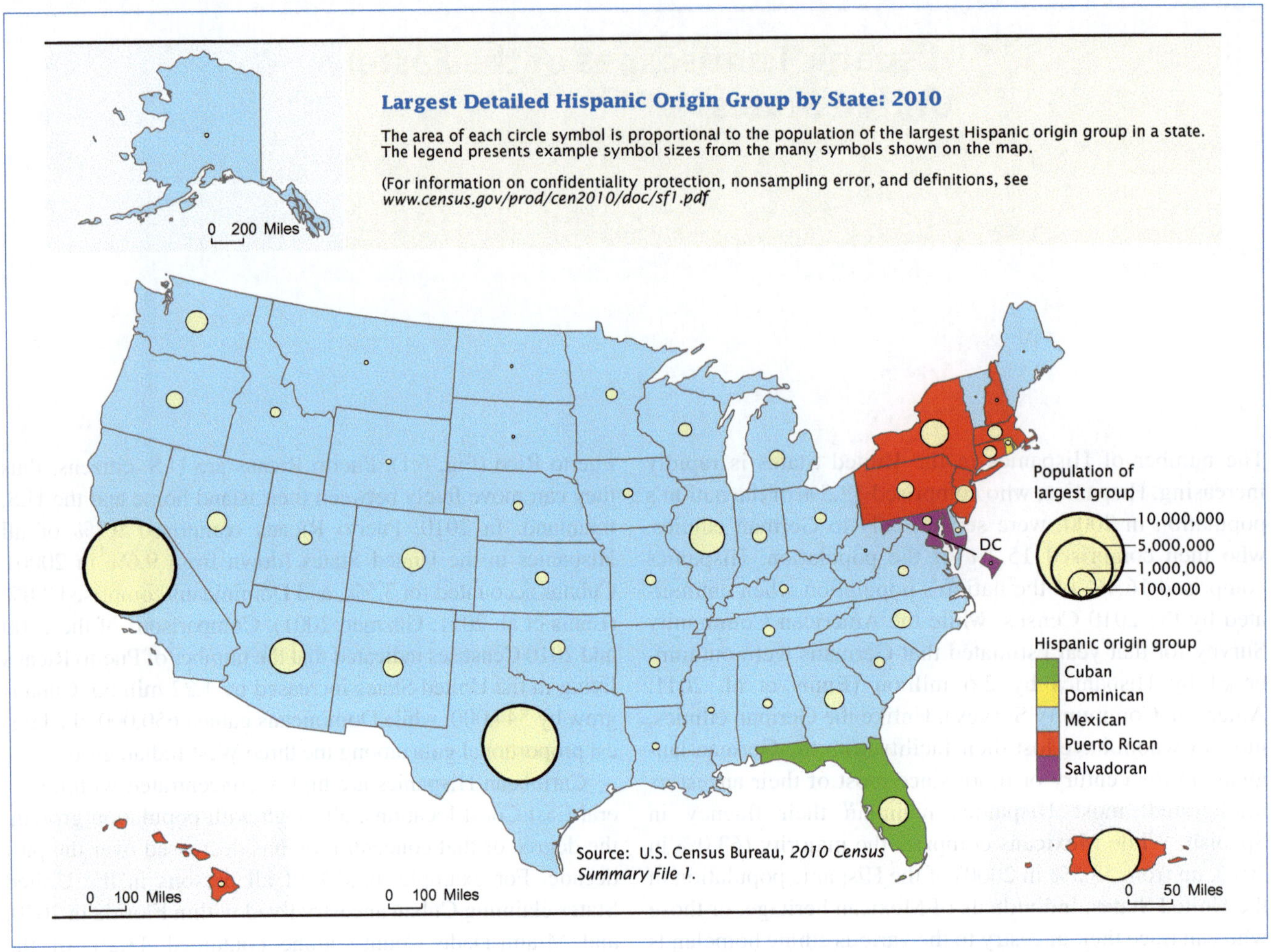

Fig. 6.1 Largest Hispanic groups by geographic origin in the various states in 2010. While Mexicans are the dominant group in 40 states, Cubans are the largest Hispanic group in Florida, Dominicans are the largest Hispanic group in Rhode Island, and Puerto Ricans are the largest Hispanic population in six Mid-Atlantic and southern New England states. Salvadorians from El Salvador in Central America are the largest Hispanic group in Maryland and the District of Columbia (Reproduced from U.S. Census Bureau map within Ennis et al. 2011)

Rican ethnics from New York have relocated to other parts of the nation, they may be considered as Nuyoricans.

While Caribbean Hispanics are found in large numbers in only a few communities, in these cities they comprise both large numbers and significant proportions of the population. Within New York City's Bronx borough, for example, Puerto Ricans comprised 20.6% of the population in 2010, while Dominicans accounted for 18.5%. Just across the Harlem River, Dominicans comprised 10.4% of Manhattan Island's population—and they are concentrated in the Washington Heights neighborhood at the north end of Manhattan Island. Yet most of the growth of the New York City's Dominican population over the past decade has been within the Bronx, where it nearly doubled. Likewise, 33.1% of Florida's Miami-Dade County's population was of Cuban ancestry in 2010. The city of Miami's population is 70.7% Hispanic, of which nearly half were Cuban, according to the American Community Surveys 5-year estimate in 2014, while nearby Hialeah was 73.9% Cuban and over 95% Hispanic. Ten percent of Florida's Monroe County is also of Cuban ancestry, representing waves of refugees who fled Spanish ruled Cuba in the late 1800s as well as those escaping the Fidel Castro government. No other county nationally has as many as 5% of its population reporting Cuban ancestry. In contrast, in three counties besides Bronx county Puerto Ricans comprise over 10% of the population: Osceola County, Florida (25.0%); Hampden County, Massachusetts (16.6%); and Cumberland County, New Jersey (13.8%). Nationally, 17 additional counties counted Puerto Ricans as comprising between 5% and 10% of their populations in 2000. New York and New Jersey each had four, and Connecticut, Pennsylvania and Florida each had three.

6.2 Spanish Colonial Settlement and Landscape Legacy in Florida

Florida received the earliest settlement attention from the Spanish—indeed the first colonizing effort by any European nation in the lands that eventually became the United States. Spanish Florida initially extended from the Mississippi River east to the Atlantic Ocean, running from the Straits of Florida north to Virginia, and it was in both the present states of Florida and South Carolina where settlements were attempted in the mid sixteenth century. Yet only in northern Florida are to be found any clearly visible landscape legacies from the Spanish colonial period.

6.2.1 Spanish Colonial Settlements

The first effort by the Spanish to establish a settlement in Florida occurred in 1559 at Pensacola, yet that community was abandoned in 1561. Pensacola was planned as a classic Spanish city, "Santa Maria de Filipino, the first in North America" (Weber 1992:68). It was re-established in 1596, and it relocated across the bay in 1698, where Fort San Carlos de Austria was built. Pensacola has been continuously occupied since then. The Spanish built nearby Fort San Carlos de Barrancas in 1797 (Fig. 6.2), portions which survive in Gulf Islands National Seashore.

Four years after the Spanish abandoned their initial Florida settlement, St. Augustine was established by Pedro Menendez de Aviles in 1565 (Fig. 6.3). It has been continuously occupied since that date, making St. Augustine, Florida the oldest European settlement in the United States. The year following the founding of St. Augustine, two hundred miles to the north the Spanish established Santa Elena, located at the site of today's Parris Island Marine Base in South Carolina. Santa Elena, which included both a fort (with 76 soldiers) and community of 171 settlers, was briefly *adelantamiento* of Spanish Florida before this administrative function was moved to St. Augustine (Weber 1992). Santa Elena was abandoned in 1587. Between 1566 and 1567 six other short-lived military garrisons were established in Florida, including ones at Tampa Bay, Charlotte Harbor, Tequesta (Miami), and San Mateo (along the St. Johns River).

The Spanish established settlements in Florida to protect shipping along its coast following the Gulf Stream from Havana. Silver fleets gathered annually at Havana for the trip to Spain. Because fleets were often attacked by pirates and privateers, locating military forces along the Florida coast reduced the threat. The Spanish also sought to establish effective Spanish claim to the area to counter French efforts

Fig. 6.2 Water battery (Baterie de San Antonio) built as part of Fort San Carlos de Barrancas, completed by the Spanish in 1797. Earlier Spanish fortifications in the Pensacola area were destroyed by the French in 1719. Parts of this fort, not shown, were significantly altered in the years leading up to the Civil War. This fort is surrounded by the Pensacola Naval Air Station, but the fort is operated as part of the Gulf Islands National Seashore

Fig. 6.3 Castillo de San Marcos, constructed in the mid seventeenth century as the largest stone masonry Spanish fort north of Havana, anchored the northeastern corner of the defensive walls that surrounded St. Augustine, Florida. The fortification is now preserved in Castillo de San Marcos National Monument

to colonize the region. Only St. Augustine survived of the initial settlements. The street grid within its historic core dates from 1598 when the Plaza de Armas was platted (Gordon 2002). Within St. Augustine today one can visit the massive stone-masonry Castillo de San Marcos (now a National Monument), built in 1672 to replace a series of earlier wooden forts, plus numerous structures within the city's historic district, either dating from the Spanish colonial period or having been reconstructed "to emphasize the period of the first Spanish occupation" (Boeschenstein 1999:272). Fifteen miles to the south, much smaller Fort Matanzas was built in 1743 to protect the southern approach to St. Augustine (Fig. 6.4). It remains a conspicuous feature along Matanzas Inlet, so named because over 200 French Huguenots (French Protestants) were executed by the Spanish near the site in 1565.

St. Augustine served as the colonial capitol of Florida for over two centuries, yet its population was always small, never much more than 3000. It was continually reliant upon Spanish subsidies and resupply from Havana. Crop production near Tallahassee and cattle raising near Alachua Prairie in the center of the Florida Peninsula also fed the settlement. South of Tallahassee, the Spanish established a presidio (military garrison) in the late 1670s at St. Marks, which they named San Marcos de Apalachee. Portions of that earthen and stone fort were incorporated into a Civil War era fort, the ruins of which still remain, and the stone foundations of several Spanish buildings are visible in a state park.

6.2.2 Spanish Missions

A string of 31 Spanish Catholic missions were built between St. Augustine and the Apalachee River between 1573 and the beginning of the 1700s in Florida (Gannon 1993). An additional five missions were located on coastal islands in what is now Georgia. These missions, while primarily established to convert Indians, also provided a buffer against outsiders and were important in providing foodstuffs to the Spanish in St. Augustine. The mission system was destroyed by conflict with English and American settlers from Georgia and South Carolina, chiefly between 1702 and 1704, at the time of Queen Anne's War. The missions were attacked and burned, and the Indians were sold into slavery. Today, nothing remains of the missions, as they were built of wood or wattle and daub. A replica of the massive timber-framed Mission San Luis complex has been reconstructed on the original site near Tallahassee. On the other hand, the exact locations of nearly half of the Florida missions remain unknown, although in the soils near some missions can be found fragments of Spanish olive jars and other archaeological evidence. Nearly half of the mission foundations have been found by archaeologists. For much of northern Florida, the continued use of Spanish place names is the most conspicuous legacy of the Spanish colonial landscape.

Two mass Spanish exoduses from Florida occurred. In 1763 Florida was transferred to the British as result of the treaty

Fig. 6.4 Fort Matanzas was built in the early 1740s to protect the inlet providing access to St. Augustine from the south. This coquina stone fortification is preserved in Fort Matanzas National Monument, located fifteen miles south of St. Augustine, Florida

ending the French and Indian War (or Seven Years War as Europeans saw it). The entire Spanish population of 3064 left for Havana, Cuba, and many Christian Indians went with them. In 1783 Florida was transferred back to Spain. Between 1812 and 1816 U.S. citizens invaded Florida on several occasions, destroying many small settlements. These incursions were prompted by U.S. concerns about Creek Indians and about free black settlements that were being established in Florida, where the Spaniards prohibited slavery. In 1818 Florida was invaded by the U.S. Army, led by Andrew Jackson, who for a short period occupied the old Spanish fort in St. Marks. Florida's annexation to the United States was made official by treaty in 1821.

The Spanish colonial legacy in Florida today can best be seen in the historic core of St. Augustine, the Spanish town plan for old Pensacola, scattered forts and ruins elsewhere, and in the numerous Spanish place names throughout the state. Furthermore, Spanish land grants set the pattern for land division in parts of Florida.

6.2.3 Example of Spanish Colonial City: St. Augustine

St. Augustine, Florida has long been recognized as being of both great antiquity and ethnically distinct. Over a century and a half ago Rufus King Sewall wrote,

> The city of St. Augustine is built in the style of an ancient Spanish military town. The plan of the city is a parallelogram.... Though not larger than many of our New England villages, the city is nevertheless regularly laid out (Sewall 1848:9).

Even in the earliest years of Florida's statehood, the Yankee author recognized that the ethnic landscape was different, even if he did not appreciate the city's narrow streets lined with buildings having projecting piazzas.

St. Augustine provides a good example of an early Spanish colonial city. Its plaza was located along the harbor, as the Law of the Indies indicated was appropriate for a city (Fig. 6.5). The Cathedral faces the plaza, which in keeping with requirements of the Law of the Indies, is aligned with the harbor on one side. At the edge of the old historic district, just beyond the stone city gate, is the Castillo de San Marcos that was built in 1672 to replace a series of earlier forts. This fortress, with bastions at each corner, was built of coquina, a native shellstone. It is the most massive Spanish fort in the United States, but it is quite small compared to the huge fortifications in San Juan, Puerto Rico and Havana, Cuba. The city itself was surrounded by a wall, and portions of that earthen and stone wall, along with one of the city gates, still remain (Fig. 6.6). Within the historic core of St. Augustine are a variety of structures dating from the first period of Spanish occupation. These include the Governor's House, parts of whose lower stone walls date from a structure built in the 1680s

Fig. 6.5 Cathedral facing the plaza in St. Augustine, Florida. Although the city with its plaza was laid out at the time of its founding in 1565, the current Cathedral Basilica of St. Augustine was constructed in the 1790s

Fig. 6.6 Gate through defensive wall in St. Augustine, Florida. Although a small portion of the walls have been reconstructed, most of the walls that once surrounded St. Augustine have been destroyed

upon which a new stone structure had been rebuilt by 1713, the St. Francis Barracks remodeled in 1769 by the British from a 1730s Franciscan monastery and now the headquarters building for the Florida National Guard, and the Old Spanish Treasurer's house (Fig. 6.7) constructed in 1741, plus houses and shops from the early eighteenth century remain. The Cathedral dates from the second Spanish occupation of Florida, having been built in 1793.

Fig. 6.7 The Old Spanish Treasurer's House, constructed around 1715, was the home of both the Spanish Royal Treasurer and later the governor of British Florida. Although remodeled, it retains it galleries and is in keeping with the idea that the cabildo faces the plaza

6.2.3.1 St. Augustine Style Architecture

Given its two periods of Spanish occupation, separated by two decades of British colonial occupation during which many buildings were made-over to suit English notions of appropriate architecture for the Caribbean and followed by American rule, St. Augustine style architecture is not entirely Spanish, but reflected a melding of several cultures. An architectural historian describes St. Augustine style houses.

> Leitmotifs of the colonial St. Augustine style are thick and cool masonry walls, second stories of stone or wood, arcaded loggias and galerías overlooking rear gardens and patios, pitched roofs, glazed window sashes, roofed balconies suspended on double beams, and houses built up to the edges of narrow streets and entered through side gardens or courtyard walls and exterior staircases (Gordon 2002:139, Reprinted with permission of the University Press of Florida).

The colonial period houses, or combinations of a shop on the first floor and a residence above, were located along narrow streets (Fig. 6.8). As stipulated in the Law of the Indies, these streets were arranged in a grid, paralleling the sides of the plaza. While many of the oldest buildings were initially constructed by the Spanish, they contain numerous modifications displaying other ethnic influences. Nevertheless, many structures have been restored to highlight their Hispanic roots.

The St. Augustine style house (Fig. 6.9) had a floor plan similar to contemporary houses commonly found along the East coast. The St. Augustine house was distinctive because of the presence of "a sheltered porch or a loggia (a room open on at least one side, but built within the body of the building) that opens onto the backyard of the dwelling" (Noble 1984:64). Overhanging the street at the front of a house was a balcony. As noted by Elsbeth Gordon,

> The St. Augustine style's most endearing characteristics are the suspended wood balconies, the arcaded stone loggias with wood gallery above.... The covered wood balconies were a centuries-old Spanish tradition, descended from Islamic Spain, where they had relieved the flat, unadorned Christian Romanesque walls. In St. Augustine they were constructed as they were in Spain and in its American colonies, cantilevered over the public streets on hewn double beams that were first-floor ceiling joists rammed through the walls (Gordon 2002:150, Reprinted with permission of the University Press of Florida).

Most of these houses were two stories high, and their doors opened to their side, thus not directly onto the street. These houses were typically constructed from wood, stone, and tabby. Tabby is a type of poured masonry in which shells (often from Indian mounds) or gravel are included in a mortar of lime and sand. Vertical wooden posts were often embedded in the tabby for support. New residences incorporating the St. Augustine style continue to be built—but incorporating modern concrete block building materials, not only in its namesake city, but elsewhere in Florida within some planned communities, such as Village of Windsor in Vero Beach (Gordon 2002).

Fig. 6.8 Restored and reconstructed buildings along St. George Street in the heart of St. Augustine's historic district. Shops, museums, and restaurants catering to tourists now dominate the narrow streets of St. Augustine

Fig. 6.9 Llambias House, constructed in the early 1700s, is a good example the St. Augustine style house. Note how the wood balcony projects from the second floor of this plastered stone masonry house. Arched stone loggias supporting a gallery are found on the opposite side of the house

Fig. 6.10 Buildings with galleries covering the sidewalk, viewed from Plaza Ferdinand VII in Pensacola, Florida. While these buildings postdate the Spanish colonial period, they display the mixing of French and Spanish architectural heritage upon the landscape within West Florida, which was occupied by the French from 1719 until the end of the French and Indian War

6.2.4 Pensacola

In contrast with St. Augustine, the Spanish imprint upon Pensacola is less noticeable, yet the city displays two distinct plazas and has a grid arrangement with streets running parallel and perpendicular to the plazas. Seville Square is actually a plaza faced by a church (although it is Protestant and was constructed in 1832, following the American annexation of Florida). Nearby houses were constructed during the second period of Spanish occupation of Florida, yet they more closely resemble French creole cottages, not surprising given French settlement in the area from Pensacola to New Orleans (See Chap. 7). In downtown Pensacola Ferdinand VII Plaza is faced by the former city hall, now a museum. While its location is in keeping with the dictates of the Law of the Indies regarding the cabildo, it was constructed 80 years after Spanish colonial rule of Florida ended. However, covered arcades shade the sidewalks of several buildings facing the plaza (Fig. 6.10), yet in design these more closely resemble French galleries than Spanish portales.

6.3 Cuban-American Landscapes in Florida

Cuban refugees have established several vibrant communities in Florida, the largest of which is centered in the Miami-Dade County area. Although those Cubans who fled their homeland during the dictatorship of Fidel Castro were by far the most numerous group of refugees, earlier waves of refugees fled Cuba during the final half century of Spanish colonial rule, which abruptly ended when the United States defeated Spain during the Spanish-American War. The refugees who arrived during the later 1800s primarily settled in Key West, Florida, only ninety miles across the Straits of Florida from Havana, Cuba, and in Tampa.

6.3.1 19th Century Cuban-American Settlements

The connections between Havana and both Key West and Tampa, Florida were strong, and both cities received large numbers of refugees from the Cuban revolutions in 1868 and during the 1890s. At the same time, their Cuban residents helped foment unrest in Cuba. The San Carlos Institute in Key West, a combination school, theatre, and cultural center, was the focus of the Cuban revolutionaries, included the famed José Martí. At the same time Cubans in Tampa were also supporting the independence movement, and a historical marker in Tampa's Ybor City neighborhood marks the "Cradle of Cuban Liberty," where José Martí delivered two speeches in 1891. José Martí Park is nearby (Fig. 6.11). It was in Key West in 1892 that the Partido Revolucionario Cubano was proclaimed. Indeed the San Carlos Theatre in Key West, once financially supported by the government of Cuba,

Fig. 6.11 José Martí Park in Ybor City district of Tampa, Florida. The statue of José Martí and the display of the Cuban and American flags commemorate the significance of Florida's Cuban immigrant community in the gaining of Cuba's independence from Spain. A nearby historical marker notes, "Cradle of Cuban Liberty. On this corner was located El Liceo Cubano, a tobacco stripping house converted into a Cuban social center in 1886. This is the cradle of Cuban independence. Here, on Nov. 26-27, 1891, Jose Marti delivered the two speeches, 'Con Todos y Para Todos' and 'Los Pinos Nuevos,' and drafted 'Las Resoluciones' which became the program of the united Cuban Revolutionary Party and eventually secured the independence of Cuba from Spain in 1898"

has long been considered a Cuban historic site, yet one that has significance to Florida, which recently helped fund its restoration. While only 9.3% of Key West's residents reported Cuban ancestry in 2010, many Cubans lived there in the late nineteenth century, working in cigar factories.

Tampa also received many Cuban refugees fleeing the Spanish colonial government. One of the most conspicuous landscape legacies of these early Cuban-American communities are the large cigar factories (Fig. 6.12). This is particularly visible in Tampa's Ybor City district where the largest ones were built and where the structures still remain, although only one still manufactures cigars, while another was recently remodeled for other uses. Ybor City displays the architecture of many Cuban-American homes and shops (Fig. 6.13). It contains brick commercial buildings and cigar factories dating to the late nineteenth and early twentieth century, along with residential neighborhoods where both walls of ornately colored tiled stucco houses can be seen along with West Indian houses that display a variety of ethnic influences.

6.3.2 20th Century Cuban-American Communities

The 2010 Census enumerated 1,213,438 persons of Cuban descent in Florida (Ennis et al. 2011). The two cities with the greatest Cuban population, Hialeah and Miami, are both located in Miami-Dade County, where over half the population is Hispanic. Miami itself is over two-thirds Hispanic, while Hialeah is over 95% Hispanic, nearly three-quarters of whom are ethnic Cubans. Unlike so many ethnic communities in the United States, the development of the Cuban-American homeland in South Florida is relatively recent. Indeed, Hispanics comprised only 4% of Dade County's population in 1950, yet Miami had received refugees fleeing Cuba during Gerardo Machado's rule in the late 1920s, to be followed by President Machado and his followers later when his government was overthrown in 1933 (Nijman 2011). The huge influx of Cuban refugees who began arriving in 1959, following the seizure of power by Fidel Castro, changed the ethnic character of the area.

Fig. 6.12 Cigar shop next to restaurant in converted cigar store along the main business street in Ybor City district of Tampa, Florida. The manufacture of cigars was long centered in Ybor City, and several multistory brick manufacturing facilities remain several blocks away, although remodeled for other uses. Cigar manufacturing and use today is still often associated with Hispanics from the Caribbean nations of Cuba and the Dominican Republic

Fig. 6.13 Columbia Restaurant in Ybor City district of Tampa, Florida. Note the tile roof and extensive colorful tile decorations under the awning of the arcade that surrounds the building housing the Cuban restaurant

Yet because Miami is such a young city, only founded in 1896, Cubans have directed nearly half of its existence.

Hispanics, especially Cubans, dominate the population and economy of Miami metropolitan area (Boswell 1993). In Miami-Dade County, Florida, 1,565,410 Hispanics were enumerated by the 2010 Census, with Hispanics accounting for 64.0% of the total population. Although Cubans account for half of Miami-Dade County's Hispanic population, there are cultural and racial differences among this group. For example, as Ira Sheskin (1992:43) notes, "Black Cubans are residentially segregated from White Cubans," with the more recent arriving Marielitos being far more likely to be non-white than the main wave of white middle class Cuban refugees during the 1960s. Furthermore, while the majority of the region's Cuban population is Roman Catholic, there were approximately 10,000 Jews from Cuba, who resided within the Miami Beach area (Sheskin 1992).

6.3.2.1 Little Havana

While Cuban ethnics are widely dispersed across Miami-Dade County, with the exception of the African-American neighborhoods—particularly those which are non-Hispanic, Tom Boswell (1990:59) found "that Cubans are most concentrated in the central cities of Miami and Hialeah." While many have now dispersed from "Little Havana," Cuban-American residential patterns remain different from those of Puerto Ricans as well as Nicaraguans, who have created a "Little Managua" in western Dade County. Unlike most other Hispanic groups in the United States, Cuban-Americans are a refugee population, with many viewing Miami as a surrogate homeland. For decades lacking the freedom for easy visits with Cuba, many Cuban-Americans returned to Miami for family visits, even when they live elsewhere in the United States. Maintaining linkages with their former communities, Cuban-Americans formed "municipios en el exilio," which are organizations of Cubans from specific home areas of Cuba (Boswell and Curtis 1991). Furthermore, Cubans have filled Little Havana "with symbols and meaning, and as such Little Havana is deeply 'localized'—but the symbols and meanings always related to the real Havana, a hundred miles away" (Nijman 2011:151). Thus, as Heike Alberts (2016:184) explains, "Cubans are not only among the most concentrated immigrant groups in the United States, but no other immigrant group dominates a city as Cubans do in Miami."

The arrival of the Cubans in Miami, a relatively young city that had not experienced any prior influx of immigrants, transformed the city's physical landscape, so that "Little Havana itself looked more and more like a Cuban city with its cafeterias serving Cuban coffee, old men playing dominos in domino park, and people placing flowers at Cuban monuments" (Alberts 2006:140). Little Havana not only is home to a large Cuban-American population, but it is the site of "anchoring institutions: Domino Park; the Bay of Pigs Monument; the old headquarters of Alpha 66 (the anti-Castro paramilitary organization); …specialty shops for cigars and *guayaberas*; and well-known *cafererías* and restaurants like El Pub and Versailles" (Nijman 2011:150).

The Cuban-American landscapes in Miami are both similar and quite different from other Hispanic landscapes in the United States. While the major shopping areas are overwhelmingly Cuban with the Spanish language used extensively in signage and outdoor advertisements, Cuban-Americans are far more likely to own the businesses. Indeed, they have the highest rate of Hispanic owned businesses in U.S. Cuban-Americans, as a group, are typically middle-class and strongly Republican, with an older median age than both other Hispanics and non-Hispanics. They are generally well-educated, with low fertility rates.

Even though portions of their community predated their arrival, ethnic influences can be seen in their landscapes. In neighborhoods there are small parks where tables are set out for residents to play checkers or chess and smoke cigars. Tom Boswell and James Curtis describe aspects of the Cuban-American landscape in Little Havana:

> The commercial landscape of Little Havana…does not have a unifying architectural identity because its buildings were built before the Cubans arrived.… Some of the small shopping plazas that have been built in the district recently have incorporated Spanish elements in their facades and decorative motifs (Boswell and Curtis 1991:157).

As with so many ethnic communities, the commercial area of Little Havana displays a wide variety of advertising signage, posters, and billboards in Spanish. Before stores marked with Spanish language names, Boswell and Curtis 1991:157) note that "chairs are usually lined up…in keeping with an old Spanish custom" (Boswell and Curtis 1991:157). Reminiscent of scenes from a century earlier in Key West and Tampa, over 30 years ago Curtis described the "acrid smell of cured tobacco wafts from the thirty or so small cigar factories where old men (tabaqueros) patiently roll cigars… by hand" (Curtis 1980:3). While today there are fewer cigar factories in business, one can still find a half dozen along Calle Ocho in Little Havana, and even more shops that sell cigars made elsewhere.

6.3.2.2 Housescapes of Little Havana

The residential area of Little Havana "is not nearly as 'Latinized' as the commercial district" (Boswell and Curtis 1991:158). Nevertheless, many Hispanic features that marked Mexican-American yards as described in Chap. 5 are similar to those seen in Cuban-American neighborhoods.

> Fences, for example, now enclose many front yards. Wrought iron grill work over windows and decorative Spanish tiles also embellish a lot of the homes. Certainly one of the most distinctive features of the residential district is the presence of yard shrines, with literally hundreds of them gracing the area (Boswell and Curtis 1991:158).

Many of these features have been added to housing that was built before the Cubans arrived (Boswell and Rivero 1985). While the presence of the fences does not distinguish the Cuban-American ethnic landscape from that within the Mexican-American barrios, their yard shrines are more unique.

In Miami Cuban-American yard shrines are most often located in the front yard near the sidewalk, yet they can be found anywhere in yard. The shrines, built of brick, cement, stone and glass with their exterior walls often covered with stucco or tile, always face the street, and they are usually rectangular in shape. Glass windows provide visibility of their interior, which typically contains statues of saints on altar. At the base of the statue can be seen offerings of flowers, candles, a crucifix, beads, jars of leaves or water, among other items placed by followers of Santeria (worshipers of saints). The saints that are most often honored include Santa Barbara, Our Lady of Charity (a Cuban patron saint), and Saint Lazarus (Curtis 1980). Thus in both design and in the selection of revered saints, the Cuban-American yard shrines are ethnically distinct.

The presence of the large Cuban population in Miami-Dade County has also influenced local agricultural practices. As Garcia (1996:88) explains, "Outside city limits, small farmers harvested vegetables important to the emigrés' diet: *malanga, boniato, carambola, yucca*, and *calabaza*." Fifty-seven percent of Miami-Dade County's farmers were Hispanic at the time of the 2012 Census of Agriculture, and many tropical fruits popular with Cubans and other West Indians are grown locally (NASS 2014).

6.4 Puerto Rican Landscapes in Northern Cities

Puerto Ricans played a major role in shaping the cultural landscape of the Spanish Harlem section of New York City as well as neighborhoods in several other cities within the Mid-Atlantic and Great Lakes regions. As with all Hispanic groups, Puerto Rican businesses typically display signage and advertising in Spanish and Roman Catholic churches traditionally dominate the religious landscape, even though evangelical Protestant groups are becoming increasingly common. While Puerto Rican residential landscapes display many features that are similar to other Hispanic groups, such as the brightly colored walls of Mexican ethnics and the yard shrines of both that group and Cuban-Americans, there are also regional differences. As with other Hispanic groups, not every house within the ethnic neighborhood displays any diagnostic landscape feature, and some items may be exhibited by only a small minority, but these items collectively are far more frequently displayed within the Hispanic neighborhood than within neighborhoods of other residents. Thus, together they clearly distinguish the ethnicity of the neighborhood.

6.4.1 Religious Shrines in Puerto Rican Neighborhoods

Religious shrines were observed in 13% of the yards within the Puerto Rican neighborhood of Lorain, Ohio, four times the frequency elsewhere in the city (Kent and Gandia-Ojeda 1999). Within the Puerto Rican community the "greatest numbers of yard shrines are of the 'Milagrosa'… [where] the Virgin Mary has open arms and wears a white mantle with a light blue canopy" (Kent and Gandia-Ojeda 1999:55). While Mexican-Americans favored shrines honoring the Virgin of Guadalupe and Cubans honored a variety of other saints, Puerto Ricans displayed the Virgin of the Milagrosa (Benedict and Kent 2004). Thus, the landscape elements are both similar but different.

The selection of colors and the display of Puerto Rican flags distinguish Puerto Rican communities from landscapes created by other Hispanic communities. Likewise, the display "of the *coquí*, a frog native to Puerto Rico that is often used as the unofficial symbol of the island" (Benedict and Kent (2004:198), marks Puerto Rican enclaves. Although most Puerto Ricans occupied structures constructed for previous occupants, "semi-fixed feature elements," as defined by Benedict and Kent (2004:188), distinguish their ethnic landscape. These include "a religious shrine placed in the yard, a house painted in bright pastel colors, perimeter fencing, or a well-kept front yard flower garden" (Benedict and Kent 2004:188).

Within their study of the Puerto Rican neighborhood of Lorain, Ohio, Kent and Gandia-Ojeda (1999) noted that gardens, often of marigolds, within the front yards of residences were far more common than in the yards of non-Hispanics. While only a minority of all residents in the city displayed potted plants in their front yard, these were twice as often present within the Puerto Rican neighborhood.

Although Puerto Ricans, like other Hispanic-Americans, may only display some of the ethnic landscape features, Benedict and Kent discovered that those who painted their houses in bright colors often displayed a wider component. They explained, based upon their observations within a Puerto Rican neighborhood in Cleveland, Ohio, that "the handful of houses that were brilliantly painted…almost always displayed multiple landscape elements" (Benedict and Kent 2004:202). Thus, one example they described is "painted light blue, has a crucifix hanging from the gable, a shrine to the Virgin of Milagrosa, a front yard garden with potted plants, and a decorative, yet secure, wrought-iron frame over the door" (Benedict and Kent 2004:202). Yet, the Cleveland survey determined that only about 5% of the Puerto Ricans distinguished their homes by these semi-fixed landscape elements. However, their presence clearly identified the ethnicity of their neighborhood.

6.4.2 Puerto Ricans in New York City

The largest concentration of Puerto Ricans—indeed 723,069 persons in 2010, nearly one-fifth of the nation's total—are found within New York City. There the traditional enclave of Spanish East Harlem, which developed in the decades following World War II as Italians departed for other neighborhoods, has begun to experience a resurgence of Puerto Rican "professionals, artists and intellectuals drawn by an authenticity they miss" (Berger 2007:34). During the 1970s and 1980s many Puerto Ricans left to escape crime and neighborhood decay within Spanish Harlem, yet its gentrification is attracting more middle-class Puerto Ricans. Strengthening this identification of Spanish Harlem with Puerto Ricans, even though many Dominicans and Mexicans had begun settling in the community, the landscape reflects their ethnicity. New York journalist Joseph Berger (2007: 41) describes "the carnival street scene—the colorful semi-illegal community gardens, the Mexican flower sellers, and the flamboyant murals that in some places strike a visitor as folk art." These scenes, as with the bodegas and botanicas selling ethnically familiar merchandise and foods—with a wide variety of shops bearing Spanish language signage and Puerto Rican symbolism in their decorations—comprise part of the visible ethnic landscape. Indeed, one of the leaders of the Puerto Rican independence movement has called "the neighborhood the 'symbol of the Puerto Rican diaspora'" (Berger 2007: 41). Because of the number of Puerto Ricans in New York City area, many reside in other neighborhoods of both the city and surrounding suburbs.

Within the Puerto Rican barrios of New York City may be found many community gardens that occupy the vacant lots of buildings that have been razed. These began to appear in New York City in the 1970s, and by the beginning of the twenty-first century there were approximately 750 community gardens scattered over 1274 plots of land (Smith and Kurtz 2003). Some of these are now formally named and labeled as "Community Garden Parks." While a variety of populations have been associated with these gardens, strong linkages exist between their cultivation and Puerto Rican ethnic populations (Fig. 6.14). These garden plots, often decorated with flowering plants, wind chimes, statues, flags of Puerto Rico, and murals painted on adjacent walls, function as small parks or children's play spaces, yet they may also house chickens or rabbits as well as have vegetable gardens.

Casitas, or small structures or "little shacks" that may function as a clubhouse, are found within many of these gardens (Fig. 6.15). Based upon her study in the Loisaida neighborhood of the Lower East Side of Manhattan Island, Karen Schmelzkopf (1996:369) notes that "primarily these offshoots of the casitas in the mountains of Puerto Rico are local gathering places, male domains where men drink beer, play dominos, socialize, and sometimes even sleep." Besides often displaying flags of Puerto Rico, she notes that statues of the Virgin Mary are often present in the community gardens, whether family- or casita-based, along with benches.

Fig. 6.14 Puerto Rican flag is prominently displayed in community garden along Madison Avenue in Spanish Harlem section of New York City. Note the variety of plants, decorations, and benches along the walk in this garden which occupies a narrow lot

Fig. 6.15 Casita in El Gallo Community Garden in Spanish Harlem neighborhood of New York City

The character of the plantings in the gardens varies widely, from decorative flowers, shrubs, and trees to vegetables, and some are much more orderly in appearance than others. Such spaces provide "a place [for children] to play and the only site in Loisaida where they can be around nature, where they can watch and make things grow" (Schmelzkopf 1996:374).

6.4.3 Puerto Ricans in Chicago

The second largest stateside urban population of Puerto Ricans is found in Chicago, Illinois (which had 102,703 in 2010), also the result of a migration that started following World War II. However, with subsequent population movements to the suburbs, now the city has 20,000 fewer Puerto Ricans than it did in 1990, with over 30,000 Puerto Ricans living in Cook County outside the Chicago city limits. Their concentration into the Chicago neighborhoods of West Town and Humboldt Park resulted in the formation of the "la Division barrio," including a wide variety of Puerto Rican groceries, restaurants, as well as other ethnically oriented shops. These neighborhoods had been established by European immigrants in the late nineteenth century, and thus the Puerto Rican ethnicity is largely displayed by the presence of both Puerto Rican shops and flag (Casuso and Camacho 1995). Recent growth has resulted in the development of new Puerto Rican enclaves in the Chicago suburbs of Naperville and Schaumburg. However, the rapid growth of Mexicans within Chicago (where they outnumbered Puerto Ricans by better than five to one in 2010) has resulted in most of the Hispanic communities in Chicago displaying ethnic landscapes that are more Mexican than Puerto Rican.

Puerto Ricans in Chicago have become far more assimilated than the city's Mexicans. Even two decades ago it was observed that "young Puerto Ricans and Mexicans come from different backgrounds and move in different worlds" (Casuso and Camacho 1995: 352). Puerto Ricans have lived in Chicago for a longer period and typically are less fluent in Spanish, while Mexican-Americans were more likely to be immigrants, and tied to Mexican culture and music (Casuso and Camacho 1995). While this remains true of recently arrived Mexicans, many Mexican ethnics in Chicago are now American born and more assimilated. Today, Puerto Ricans today are less concentrated in their West Town, Humboldt Park, and Logan Square neighborhoods, while significant growth of Mexicans has occurred in the city's other Hispanic neighborhoods.

6.5 Dominican Ethnic Landscapes in New York City

The traditional heart of the Dominican community in New York City is the Washington Heights neighborhood at the northern end of Manhattan Island. At one time home to many Irish immigrants, Eastern European Jews, and later

Fig. 6.16 Casa de las maletas, or luggage shop, with clothing hanging from the building and suitcases stacked on the sidewalk in Washington Heights neighborhood of New York City. Note the profuse use of Spanish language signage

Greeks, Dominican settlement grew rapidly beginning in the 1960s. Dominicans have dominated the neighborhood's population for the past three decades. Over the past decade spillover growth across the Harlem River has resulted with the Bronx having an even greater Dominican population, yet it is in Washington Heights that the Dominican presence is best displayed in the business district, through a wide variety of retail shops, institutions, and offices advertising their products and services in Spanish and restaurants, bakeries (panderias), and butcher shops (carnecerias) promoting their Dominican delicacies.

Washington Heights' landscape, as Miyares (2004:158) explains, "is clearly a Dominican one, strongly connected transnationally to its home country." Besides signage in Spanish, one sees symbols of national pride, such as the "use the red, white, and blue of the Dominican flag…[and] the use of Quisqueya, the Taino word for the island of Hispanola" (Miyares 2004:158–159). One sees many reminders of the close connection between the Dominican neighborhood and the homeland, including Spanish language billboards advertising frequent air flights between New York City and Santo Domingo, Dominican Republic and placards promoting candidates for an upcoming Dominican presidential election. Indeed, it is noted that "many Dominicans are members of hometown clubs aimed specifically at maintaining links with and supporting their communities of origin, and they actively participate in Dominican politics even from abroad" (Alberts 2016:193). This connection is so strong that a Washington Heights resident was elected president of the Dominican Republic in 1996, leaving his family in the city during his 4-year presidency (Miyares 2004).

The sidewalks in the Dominican business district of Washington Heights spill over with vendors displaying an incredible variety of wares. While some of these displays are simply local merchants using the sidewalk in front of their shops as additional retail space (Figs. 6.16 and 6.17), other vendors stack tables with their wares that are unloaded from trucks parked along the street (Fig. 6.18). The range of wares displayed by the sidewalk vendors is substantial. Food items are most common, particularly a wide array of tropical fruits and vegetables.

> Food, a key part of every culture, is central to the Dominicanness of this neighborhood. Street vendors sell tropical fruit, peeled and sliced for immediate consumption, as well as *empanadas de yuca*, *pastelitos*, and *kipes*—various types of fried pastries. Fresh tropical juices and *batidos* (milkshakes) are available (Miyares 2004:160).

In addition, many vendors' tables lining the street are piled with clothing, cosmetics, novelties, and publications, or next to stacks of suitcases.

6.6 New and Changing Hispanic Ethnic Landscapes

This chapter has touched upon only some of the multitude of ethnic Hispanic landscapes to be found in the eastern United States. Some, such as the ancient buildings in St. Augustine,

Fig. 6.17 Grocery store along St. Nicholas Avenue in Washington Heights neighborhood of New York City. Note the conspicuous display of tropical fruits, including papayas and plantains, as well as sugar cane and yucca facing the sidewalk near the entrance of this Dominican super market. Dominicans were the largest ancestry group within New York County in 2000

Fig. 6.18 Sidewalk vendors with display tables on sidewalk along St. Nicholas Avenue in Washington Heights neighborhood of New York City. While these tables are laden with purses and bottles of fragrances, other vendors a short distance away were selling tropical fruits and vegetables, as well as clothing

Florida, are several centuries old, but not as ethnically impressive or as expansive as the Hispano Homeland in New Mexico or the Tejano Homeland along the Rio Grande. Yet others such as the Cuban settlement in Miami-Dade County, Florida, date from the arrival of anti-Castro refugees beginning in 1959. Others are even more recent, such as the Honduran and Salvadorian settlements within southern California, or the Nicaraguan and Venezuelan settlements in Miami-Dade and Broward Counties, respectively, of South Florida.

6.6.1 Immigrants from Central America and South America

New York City has long attracted immigrants from Latin America, and while its Puerto Rican and Dominican neighborhoods are well known, over the past couple of decades it has attracted tens of thousands of Hispanics from Central and South America. The 2010 Census, which is likely to significantly underreport the numbers of recent immigrants, particularly those lacking appropriate documentation, counted 187,572 Ecuadorians, 99,419 Colombians, 55,505 Hondurans, 47,561 Peruvians, and 43,325 Salvadorians living within New York City, nearly double their numbers a decade earlier. While overall, 28.6% of New York City's population reported Hispanic or Latino heritage, the city was home to a disproportionate share of some of these smaller Hispanic ethnic groups (Price 2007). Indeed, within New York City's neighborhoods are to be found over 30% of all Ecuadorians living in the United States, and over a tenth of the Colombians, Panamanians, and Paraguayans residing within the United States. When the entire New York metropolitan area is considered, in 2010 the region was home to 56.0% of all Ecuadorians, 30.5% of total Uruguayans, 25.7% of all Colombians, and 21.1% of all Costa Ricans in the United States.

While many of the more recently arriving South and Central American Hispanics are widely dispersed among New York City's boroughs, certain neighborhoods have attained certain ethnic flavors. For example, Marie Price writes that "the streets of Jackson Heights pulse with the sound of cumbia (a Colombian musical form) while the smell of arepas (a round fat corn muffin) fills the air" (Price 2007:196). She notes that this neighborhood in Queens is the first place of settlement for many Colombian immigrants. The establishment of such new centers for immigrating ethnics is not just limited to groups that are relatively new to the United States. Indeed, New York City has recently begun attracting Mexicans. While Mexican-Americans comprised less than 9% of the city's Hispanic population in 2000, it was 12.8% in 2010. Their community is sizable, with 326,375 estimated by the 5-year American Community Survey for 2010–2014. Thus, New York City's Hispanic landscapes represent a large and growing variety of national groups, enhancing the city's diversity.

6.6.2 Changing Hispanic Landscapes

Elsewhere, the arrival of newcomers has altered pre-existing Hispanic settlements. New Orleans, Louisiana provides an excellent example of a city that reflects both well-established historical Hispanic influences as well as the very recent arrival of Central American workers (and a smaller number of Mexican-American laborers) who are helping rebuilt the city following Hurricane Katrina (Drever 2008). Historically, New Orleans strongly reflects its Spanish heritage, while French ethnicity was better displayed along the bayous. The city was under Spanish rule from 1763 to 1803, and it had been formally under French control for less than 3 weeks when the Louisiana Purchase was formally transferred to the United States. Because "fires obliterated the French-built city in 1788 and 1794," the city had been "rebuilt in Spanish style" (Weber 1994:61). Indeed, Jackson Square is an ideal plaza (Fig. 6.19), with both the St. Louis Cathedral and the Cabildo, built in the classical Spanish Baroque style, facing it. The buildings on the other two sides of the plaza display conspicuous cast iron balconies, an idealized form of portales shading the sidewalks.

While the historic French and Spanish districts survived Hurricane Katrina relatively unscathed, the city's population was one-third smaller a decade following the flooding. New Orleans ceased being a city that was overwhelmingly African-American, and the Hispanic population has grown. The 2010 Census showed a city that was over 5% Hispanic, while the metropolitan area was 7.3% Hispanic, yet more recent estimates show their share is growing. The 5-year American Community Survey for 2010–2014 shows that while Orleans Parish (which includes New Orleans) is 5.4% Hispanic, adjacent Jefferson and St. Bernard Parishes are 13.2% and 9.4%, respectively. Even more noteworthy is that it is "Other Hispanics," those other than Mexicans, Puerto Ricans, and Cubans, who comprise two-thirds of their Hispanic population. While it is too early to describe a new Latino landscape in a city with deep Hispanic roots, these new ethnic residents will undoubtedly add their ethnic imprint to the city's landscape.

Given that Hispanics continue to be America's most numerous immigrants, their ethnic imprint upon the nation's landscape can only be expected to expand and become more diversified. This will include the expansion of Mexican-American settlements into many farming communities—both in the Midwest and southeastern states where Hispanics until recently were largely absent. Yet, as Jamie Winders (2005:689) explains, "in many southern locales, Latinos have

Fig. 6.19 St. Louis Cathedral, flanked by the Cabildo (*left*) and Presbytere (*right*), faces the plaza, now called Jackson Square, in New Orleans, Louisiana. These eighteenth century buildings and their arrangement far better showcase Spanish baroque architecture and Spanish city planning than the fact that they are located in the French Quarter

been accepted as *workers* but not as community *members*, a distinction with important implications for long-term community change." Nevertheless, with thousands of Mexican-Americans working in the carpet mills of Dalton, Georgia, numerous chicken slaughterhouses and packing plants across the rural South, and in crab packing facilities of the Carolinas, along with thousands taking laboring jobs in Nashville, Tennessee as well as other urban centers, their presence raises questions about both the type of impact they will have upon the cultural landscape and the consequences of their employment upon the opportunities available for African-Americans.

In addition, the spread of Puerto Rican communities into more cities and suburban communities within the Mid-Atlantic states and the establishment of Hispanic ethnic neighborhoods composed largely of nationalities that are poorly represented at present in the U.S. will undoubtedly alter the cultural landscape of these regions. Just as the coming to power of Fidel Castro resulted in the rapid growth of a Cuban ethnic community within South Florida, the regime of Hugo Chavez in Venezuela prompted a similar exodus of middle to upper-class Venezuelans. Thus, the Venezuelan population in South Florida more than doubled within the first 6 years of the twenty-first century to an estimated 70,000 persons, and it is still increasing (Semple 2008). Within some areas—Miami-Dade County, Florida simply represents an early example—multiple Hispanic ethnic landscapes will form, as distinctly separate immigrant groups congregate in their own neighborhoods while better established Hispanic ethnics increasingly assimilate into the broader community, while still maintaining many aspects of their ethnic heritage, not unlike that of many European ethnics of a century earlier.

References

Alberts, Heike. 2006. The Multiple Transformations of Miami. In *Latinos in the New South: Transformations of Place*, ed. Heather A. Smith and Owen J. Furuseth, 135–151. Burlington: Ashgate.

———. 2016. Caribbean Hispanics. In *Contemporary Ethnic Geographies in America*, ed. Christopher A. Airriess, second ed., 171–200. Lanham: Rowman & Littlefield.

Benedict, Albert, and Robert B. Kent. 2004. The Cultural Landscape of a Puerto Rican Neighborhood in Cleveland, Ohio. In *Hispanic Places, Latino Places: Community and Cultural Diversity in Contemporary America*, ed. Daniel D. Arreola, 187–205. Austin: University of Texas Press.

Berger, Joseph. 2007. *The World in a City: Traveling the Globe through the Neighborhoods of the New New York*. New York: Ballantine Books.

Boeschenstein, Warren. 1999. *Historic American Towns along the Atlantic Coast*. Baltimore: Johns Hopkins University Press.

Boswell, Thomas D. 1990. Hispanic Segregation Patterns in Metropolitan Miami. *The Florida Geographer* 24: 34–66.

———. 1993. The Cuban-American Homeland in Miami. *Journal of Cultural Geography* 13 (2): 133–148.

Boswell, Thomas D., and James R. Curtis. 1991. The Hispanization of Metropolitan Miami. In *South Florida: The Winds of Change*, ed.

T.D. Boswell, 140–161. Washington: Association of American Geographers.

Boswell, Thomas D., and Manuel Rivero. 1985. Cubans in America: A Minority Group Comes of Age. *The American Geographical Society's Focus* 35 (2): 2–9.

Casuso, Jorge, and Eduardo Camacho. 1995. Latino Chicago. In *Ethnic Chicago: A Multicultural Portrait*, ed. Melvin G. Holli and Peter d'Alroy. Jones, 346–377. Grand Rapids: William B. Eerdmans Publishing Company.

Curtis, James R. 1980. Miami's Little Havana: Yard Shrines, Cult Religion and Landscape. *Journal of Cultural Geography* 1 (1): 1–15.

Drever, Anita I. 2008. New Orleans: a re-emerging Latino destination city. *Journal of Cultural Geography* 25 (3): 287–303.

Ennis, Sharon R., Merarys Rios-Vargas, and Nora G. Albert. 2011. *The Hispanic Population: 2010. Census 2010 Brief*. Washington, DC: Census Bureau (C2010BR/4).

Gannon, Michael. 1993. *Florida: A Short History*. Gainesville: University Press of Florida.

García, María Cristina. 1996. *Havana USA: Cuban Exiles and Cuban Americans in South Florida, 1959–1994*. Berkeley: University of California Press.

Gordon, Elsbeth K. 2002. *Florida's Colonial Architectural Heritage*. Gainesville: University Press of Florida.

Guzman, Betsy. 2001. *The Hispanic Population, Census 2000 Brief*. Washington, DC: Census Bureau (C2KBR/01–3).

Kent, Robert B., and Augusto F. Gandia-Ojeda. 1999. The Puerto Rican Yard-Complex of Lorain, Ohio. *Yearbook, Conference of Latin Americanist Geographers* 25: 45–60.

Miyares, Inés M. 2004. Changing Latinization of New York City. In *Hispanic Spaces, Latino Places: Community and Cultural Diversity in Contemporary America*, ed. Daniel D. Arreola, 145–166. Austin: University of Texas Press.

National Agricultural Statistics Service (NASS). 2014. *2012 Census of Agriculture: Florida State and County Data, Volume 1, Geographic Area Series, Part 9*. Washington: U.S. Department of Agriculture. AC-12-A-9 (https://www.agcensus.usda.gov/Publications/2012/Full_Report/Volume_1,_Chapter_1_State_Level/Florida/flv1.pdf).

Nijman, Jan. 2011. *Miami: Mistress of the Americas*. Philadelphia: University of Pennsylvania Press.

Noble, Allen G. 1984. *Wood, Brick, and Stone: The North American Settlement Landscape, Volume 1, Houses*. Amherst: University of Massachusetts Press.

Price, Marie. 2007. Andean South Americans and Cultural Networks. In *Contemporary Ethnic Geographies in America*, ed. Ines M. Miyares and Christopher A. Airriess, 191–211. Lanham: Rowman and Littlefield.

Sewall, Rufus King. 1848. *Sketches of St. Augustine with a View of Its History and Advantages*. New York: George P. Putnam. (Reissued by University Presses of Florida in 1976).

Schmelzkopf, Karen. 1996. Urban Community Gardens as Contested pace. *The Geographical Review* 85 (3): 364–381.

Semple, Kirk. 2008. Rise of Chávez Sends Venezuelans to Florida. *The New York Times*, January 23 (http://www.nytimes.com/2008/01/23/us/23florida.html?hp).

Sheskin, Ira M. 1992. The Miami Ethnic Archipelago. *The Florida Geographer.* 26: 40–57.

Smith, Christopher M., and Hilda E. Kurtz. 2003. Community Gardens and Politics of Scale in New York City. *The Geographical Review.* 93 (2): 193–212.

Weber, David J. 1992. *The Spanish Frontier in North America*. New Haven: Yale University Press.

Winders, Jamie. 2005. Changing Politics of Race and Region: Latino Migration to the US South. *Progress in Human Geography* 29 (6): 683–699.

7 French Ethnic Landscapes in America

Colonial settlement was largely responsible for shaping the contemporary distribution of the French across North America. Those immigrants from France during the past century were likely to disperse across the nation (Fig. 7.1), rather than just settling in areas where French ethnics derived from the colonial settlements lived (Allen and Turner 1988). In studying the impact of French colonial settlement, it is important to note that Quebec once included the entire Great Lakes region while Louisiana extended up the Mississippi River Valley from the Gulf of Mexico and included the Missouri River watershed. While the Great Lakes region and the Mississippi Valley were the only parts of the United States to receive effective colonization by the French, French colonization efforts in both Canada, which resulted in the formation of the distinctive Québécois homeland that is seen today, and in the West Indies, particularly Saint Domingue, led to subsequent movements of French-Canadian and French Creole populations into the United States, where they contributed to the unique French ethnic landscapes that remain distinctive to the regions where they settled.

7.1 Colonial Settlement by the French in North America

Early French colonial activities focused upon the shores of the St. Lawrence River and the Gulf of St. Lawrence, as well as along the Atlantic coast running from Cape Breton to south of the Bay of Fundy. Initially, French fishermen and traders had visited these areas during the summer, as several ill-planned colonization efforts had resulted in failure. Beginning in 1604 a renewed effort was made to establish a year round settlement, first in Acadia and several years later at Quebec (Fischer 2008). Although these lands are now part of Canada, Quebec quickly expanded to include the American Midwest and many Acadians had relocated to lands that became part of the United States before the American Revolution. Thus the colonial French ethnicity brought to the United States via Canada not only reflected the regional heritage brought from France to the two areas of French settlement, but also the cultural adjustments that were made in Acadia and Quebec. Farther south, the French established a sugar plantation-based colony within the West Indies. Beginning in the late seventeenth century French colonization occurred along the Gulf of Mexico, and during the eighteenth century French settlement spread north in the Mississippi Valley. Finally, refugees from Saint Domingue in the late eighteenth century brought a very different type of French ethnicity to the United States.

From its earliest founding, the colonization efforts of the French differed significantly from those of the English, the Dutch, and the Spanish. Not only was the number of French settlers quite small, but the French differed in their economic orientation and manner of establishing colonies. James Kornwolf (2002:341) explains, "The French approach to establishing permanent colonies was totally different from that of the English, the Dutch, or even the Spanish. It clearly reflected the strong centralized government of France, with officially designed architecture and towns and a colonial policy that placed greater emphasis on an effective military." The French also differed in their attitude towards the indigenous population. As David Hackett Fischer notes, the French interaction with the native population "was very different from the English in Massachusetts and Virginia, who settled apart from the Indians, kept them at a distance, annexed large tracts of land, and cultivated an attitude of distrust and contempt" (Fischer 2008:218). This interaction resulted in the formation of a hybrid French-Indian Métis population that spread French culture far into the interior of North America.

7.1.1 Quebec

The initial settlement of what became the city of Quebec occurred in 1608. With its Upper Town atop the bluff and its Lower Town along the St. Lawrence River at the foot of the cliff, the city retains a wealth of French colonial architecture (Fig. 7.2). As the French geographer Raoul Blanchard wrote,

J.A. Cross, *Ethnic Landscapes of America*, DOI 10.1007/978-3-319-54009-2_7

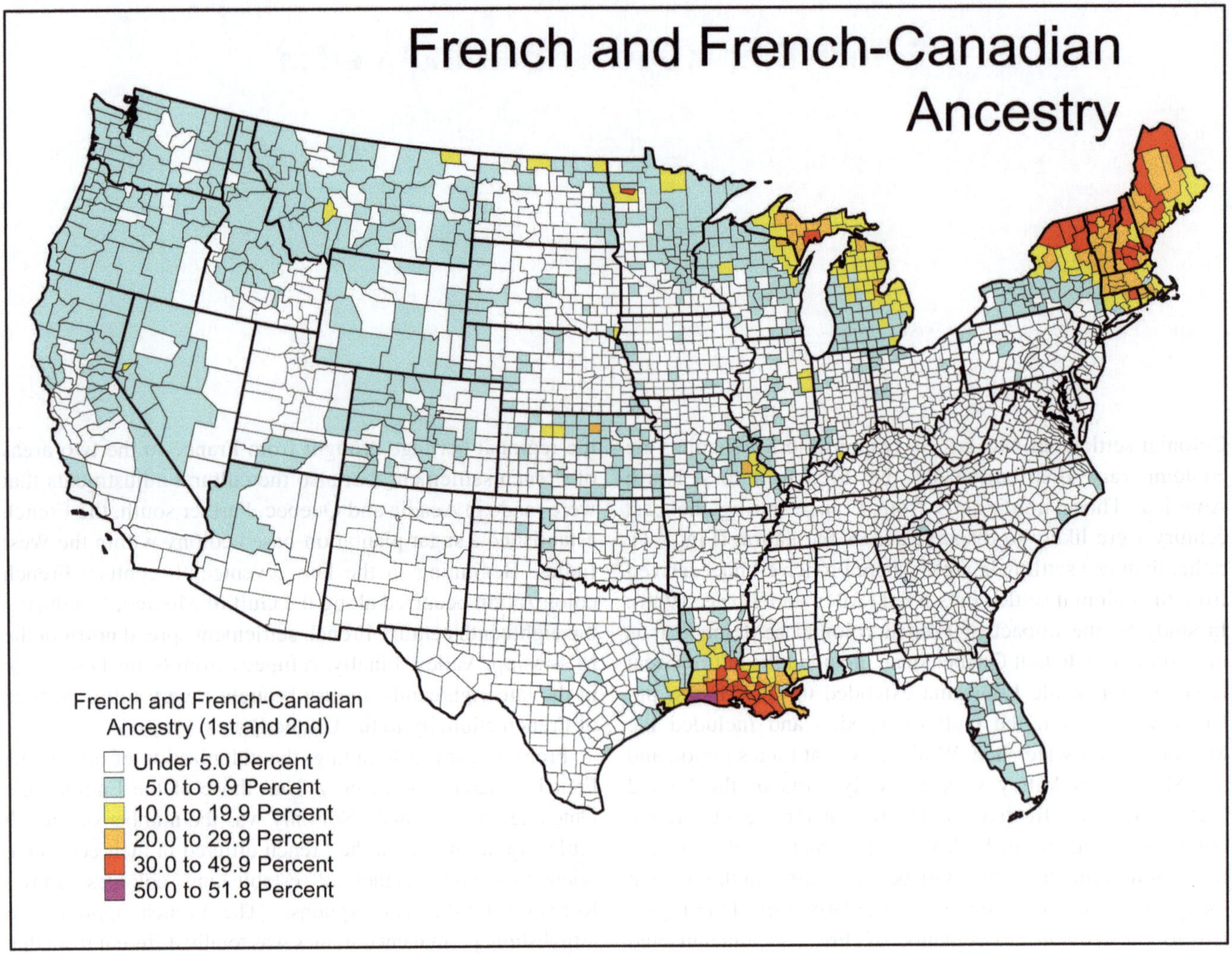

Fig. 7.1 Map showing the percentage of individuals who reported their ancestry on the 2000 Census indicating that French, French-Canadian, or Acadian was their primary or secondary ancestry. Note the largest concentrations of ethnic French are found in New England, upstate New York, and southern Louisiana, yet French ethnics are also prominent within the Great Lakes region. French ethnics also account for over 5% of the population reporting their ancestry throughout many of the counties of the West, from the Rocky Mountains to the Pacific Coast

"the façade of the *haute ville* which, enclosed in its fortifications, is a bit of a museum full of monuments where old French houses, tall and stately, squeeze together along the steep and sinuous streets, a spectacle quite rare in America" (translation quoted in Villeneuve 1992:114). The Lower Town, or the *basse terre*, was the site of Champlain's early settlement. Although Samuel Champlain's home was destroyed by fire, its site is now occupied by Notre-Dame-des-Victoires Church, built between 1687 and 1723, facing Place Royale in basse terre (Fig. 7.3). Other prominent French colonial buildings include the Monastèere des Ursulines and the Hôpital Général, which contain major portions constructed in the 1600s, along with other structures, such as the Cathedral, which only retain segments of their early walls. The seventeenth century walls surrounding the Upper Town and its fortifications remain, with Quebec being the only walled city remaining in North America, as the walls surrounding New Orleans, Mobile, and Detroit have long disappeared. As Kornwolf explains, "the numerous large-scale, institutional structures and the dramatic site and fortifications meant that Quebec was unrivaled in terms of both architectural and urban monumentality on the North American continent by 1700" (Kornwolf 2002:212). Furthermore, "the strong architectural and urban character of Quebec was due in no small way to the unrivaled position accorded the colony's architects, masons, and joiners" (Kornwolf 2002:213). Today most of the remaining structures in Quebec that were built two centuries ago—both major public institutions as well as family houses—are of stone masonry construction. Two or three story townhouses (Fig. 7.4), with their gables facing parapeted party walls from which chimneys rise, are conspicuous in parts of Lower Town Quebec (Kornwolf 2002).

Fig. 7.2 Maison Chevalier, constructed in 1752 in the *basse terre* area of Vieux Quebec. Note the enclosure of the large chimneys within the stone walls and the Quebec flag

Fig. 7.3 Eglise-Notre-Dame-des-Victoires, built in seventeenth century and reconstructed following severe damage incurred by bombardment during French and Indian War (Seven Years War) in 1763 in Place Royale the *basse terre* area of Vieux Quebec. Note how the chimneys are enclosed within the wall and gable in the stone building to the left of the church. The church was erected on the site of Samuel Champlain's house that was destroyed by fire

Fig. 7.4 Maison Milot, constructed in 1689 along Rue Notre-Dame in the *basse terre* area of Vieux Quebec

It has been argued that it is "hardly accurate to invoke Old Quebec, the site of the founding of New France, as a prime example of the 'French' landscape in North America" (Noppen and Morisset 1999:12), given the subsequent overlay of British influence and the consequent evolution of a new Quebec culture. Nevertheless, while the presence of the British substantially altered the Quebec landscape, just as the presence of the Spanish, Africans, and Anglo-Americans mutated New Orleans' landscape, the French set the stage with their survey of the land, their urban plan for a city behind fortified walls, and the way they viewed property. Many French architectural traits remain conspicuous upon Quebec's urban landscape. These include thick stone fire resistant walls that extended beyond the roofline, separating adjacent buildings, high roofs with "French capuchin-shaped dormer windows" (Noppen and Morisset 1999:21), and double windows.

Outside the city of Quebec many rural settlements developed along both the north and south shores of the Gulf of St. Lawrence and the St. Lawrence River during the French colonial period. While noting the "distinctive historical landscapes" of Quebec and Montreal, William Wyckoff (2014:340) claims that "the rural settlement landscape lining the St. Lawrence River contains the most enduring and recognizable landscape signatures of persisting French influence and regional identity." Land was originally granted to a seigneur, who held a feudal relationship with those farm families who settled long narrow parcels of land in the seigneuries. While a linear arrangement of farmsteads developed along roads, which will be discussed later in this chapter, more concentrated settlements developed around the Catholic church that was erected to serve the parish, whose boundaries typically corresponded with the seigneuries, and whose name was often applied to the settlement. Among "the oldest St. Lawrence seigneuries and the rural hearth of the French culture" (Villeneuve 1992:132) is Ile d' Orléans, located just east of Quebec, where four churches dating from the French colonial period remain (Fig. 7.5). Scattered among the many seigneurial centers along the St. Lawrence are numerous reminders of French Norman architecture (Fig. 7.6). Thus, the urban and rural legacy of Quebec in Canada today is far different than that of its far flung trading posts and settlements in the American Great Lakes states, yet we can see its connections to the landscape there and in the Mississippi River Valley.

7.1.2 The Great Lakes and Mississippi Valley

Over several decades following the founding of the city of Quebec, French explorers, trappers, and traders spread up the St. Lawrence River into the Great Lakes region, with colonial settlement following. Fur trapping and trading with Indians were the economic foundation of Quebec, and the quest for furs took French voyagers far into the interior of North America. Within these remote trading centers intermarriage

Fig. 7.5 Eglise Sainte-Famille in village of Sainte-Famille on Île d'Orléans, Quebec. This stone church was erected in 1734 and is the center of the community, and a stone processional chapel is located along the road about a quarter mile distance

Fig. 7.6 Norman style French stone cottage in Lotbiniere, Quebec

with Indian women was very common, resulting in the creation of the Métis, a people who were culturally a mixture of both races, and who have maintained a separate and distinct identity to this day (Ens 1996). The Métis themselves were very important in conducting the fur trade.

7.1.2.1 French Fur Trade in the Great Lakes Region

The French fur trading posts were established early. These included Fort Detroit in 1701; Saint-Francois Xavier (at present DePere, Wisconsin) in 1669; Fort de la baye des Puants (at

Green Bay, Wisconsin) in 1717; Fort Saint-Ignace (near Mackinac, Michigan) in 1627; Fort Michilimackinac (along the Straits of Mackinac connecting Lakes Michigan and Huron) in 1712; Fort Saint-Nicolas (at Prairie du Chien, Wisconsin) in 1686 and later the village of Prairie du Chien in 1754; and Fort Duquesne (Pittsburgh, Pennsylvania) in 1753. Although the major cities of Detroit and Pittsburgh were to develop at the sites of early French posts, most of the trading posts were small remote settlements during French colonial times. The two largest were Fort Detroit and Fort Michilimackinac. These were described by Cole Harris as "entrepots laid out in a grid of streets and defended by cannon mounted in small angled towers at the corners of curtain walls" (Harris 1990:65). By the mid 1700s "Detroit had a sizeable community of habitants, permanently settled agriculturalists" (Ekberg 1998:14), whose fields of wheat, corn, barley, oats, and peas extended back from the Detroit River and Lake St. Clair. All of these posts were connected by the voyagers using large canoes.

Portages permitted fur traders to journey from the Great Lakes system and the St. Lawrence into the Mississippi River watershed. The most prominent portages were located at Portage, Wisconsin, where 1.5 miles separated the Fox River and the Wisconsin River; south of Superior, Wisconsin, where the St. Croix River was reached from Lake Superior; just a few miles southwest of the current center of Chicago, where a low divide separated the Chicago River from the Illinois River; and near Fort Wayne, Indiana, where the French established a trading post in 1680 and a fort 8 years later, at the location where voyagers traveled from the Maumee River to the Wabash River.

French Canadians played a major role in the American fur trade after the end of the French colonial period well into the 1800s. They moved up the tributaries of the Mississippi River across the Great Plains into the Rocky Mountains. Métis communities remain as far west as Montana, and French toponyms dating from the fur trade can be found in the Rocky Mountains.

7.1.2.2 French Settlements in the Mississippi Valley

The Mississippi River provided the French with access into the center of the continent from the south, and the entire area focused upon the river became the colony of Louisiana. Biloxi (now in Mississippi) was the first successful French colony along the Gulf of Mexico, being established in 1699. Mobile (now in Alabama) was founded in 1702, while Natchitoches (on the Cane River, then the main channel of the Red River in present-day Louisiana) was established in 1714. New Orleans was founded in 1718 along the Mississippi River, and additional settlements were established up the Mississippi River and its tributaries. These included Baton Rouge in 1719; Natchez, Mississippi on the site of Indian village in about 1700; Ste. Genevieve, Missouri around 1735; and Kaskaskia, Illinois by an Indian village in 1703. Kaskaskia was the first capitol of the state of Illinois (between 1818 and 1820), but the site was ravaged by floods and the shifting of the channel of the Mississippi River, leaving little trace of the early French settlement other than its relocated cemetery. During the eighteenth century the French also established Fort de Chartres (Fig. 7.7) in 1720 near Kaskaskia; Vincennes,

Fig. 7.7 Reconstructed wall and original powder magazine of Fort de Chartres near Kaskaskia, Illinois. The powder magazine, built by the French in the 1750s, is the oldest structure erected by Europeans in Illinois. The fort is preserved in Fort de Chartres State Historic Site

Fig. 7.8 French creole cottages were constructed throughout the areas of the Mississippi River valley that were colonized by the French. This French creole cottage shows poteaux-sur-sole construction, galleries, and a roof with a changing slope. Located in Ste. Genevieve, Missouri, the Nicholas Janis house was erected in 1791–2

Indiana as a French mission in 1702 and as a fortified trading site about 1732; and St. Louis, Missouri in 1764.

The six communities that were established along the Mississippi River in what is now Illinois, plus Ste. Genevieve in Missouri and Prairie du Chien in Wisconsin, provided their settlers with both household lots in the village and long narrow farm lots nearby. "The house, outbuildings, and a garden were located on the village lot. Crops were planted on the farm lot, on which there might also stand a barn" (de Julio 1996:37). Not only do a few buildings remain from this early French settlement, particularly in Ste. Genevieve, Missouri, which contains, according to the U.S. National Park Service (2016:20), "the greatest concentration of [French vernacular vertical log buildings] in North America" (Fig. 7.8), but the legacy of the French settlers also remains in the plats of their fields. Similarly, a small number of French colonial style cottages and an eighteenth century church utilizing vertical log construction remain in the vicinity of Cahokia, Illinois.

7.1.3 Acadia and Cajuns

French settlement of Acadia was focused upon the Bay of Fundy, particularly that portion that is now within the Canadian province of Nova Scotia, as well as the area facing the Atlantic Ocean near La Hève. As part of the early settlement of coastal Acadia, a Jesuit mission was established on Mount Desert Island, in what is now the state of Maine, in 1611. Although it was destroyed by the English 2 years later, subsequently a French seigneury was granted covering a portion of the Maine coast, and "traces can be seen at Hulls Cove" (Fischer 2008:471) on the eastern part of Mount Desert Island, within today's Acadia National Park.

Acadian settlers in both Nova Scotia and New Brunswick developed extensive systems of ditches and dikes within the coastal marshlands by their agricultural settlements. Experts were recruited from the Saintonge region of western France to develop the dikes and aboiteaux (wooden sluices which controlled the drainage) in the marshlands. Thus, "French families of Acadia won this land from the sea by their unceasing labor, and they became strongly attached to it" (Fischer 2008:489). Although control of Acadia switched back and forth between the French, English, and Scots during the early seventeenth century, and traded hands four times between 1650 and 1740, the French settlement continued growing, but conflict was to disperse and fragment this group.

In 1755 the French Acadians were forced to leave Nova Scotia by British forces in an effort that the historian John Faragher (2005:xiii) has likened "to events of ethnic cleansing in our own time." Their expulsion was an episode that led up to the French and Indian War (or the Seven Years War, as it was called in Europe). The Acadians had already developed an ethnic distinctiveness in their relatively isolated and self-sufficient settlements along the Bay of Fundy, and they were different from other French and French Canadian communities (Dormon 1983). Furthermore, many French settlers of the Acadian community were drawn from the Poitou region of west central France, yet arrivals from the Basque region, other parts of France, as well as Ireland and Scotland also contributed to the bloodlines of a population that further intermixed with the native Mi'kmaw Indians (Griffiths 1973, Faragher 2005). In contrast, settlers of Quebec were largely drawn from Normandy and other parts of northwestern France and were far less likely to intermix with the Indian population than were either the Acadians or the French Canadian voyagers who penetrated the Great Lakes region.

As a result of "The Expulsion," or *le grand dérangement*, 10,000 Acadians were exiled in 1755, initially to the Eastern Seaboard of British North America, or returned to France. A small number of Acadians avoided deportation by the English, moving into the interior of the Canadian Maritimes from their coastal settlements. Descendents of this group occupy the Madawaska Valley along the border between Maine and New Brunswick, and comprise many of the residents of Madawaska, Caribou, Presque'isle, Saint-Francis, Saint-Luce, and Saint-Joseph, Maine (Fischer 2008). These people, who intermixed considerably with Quebecers, now identity with either Quebec or Acadia, depending upon whether they live on the Canadian or American sides of the border (Comeaux 1996). French Canadians from Quebec elsewhere in New England vastly outnumber those of Acadian descent living in Maine. A major migration of French Canadians into the New England states occurred during the 1800s and early 1900s.

7.1.3.1 Acadians to Louisiana

After the French and Indian War, 1200 Acadians relocated from France to Louisiana between 1765 and 1768. Another 1600 Acadian exiles traveled from France in 1785, after the American Revolution, in an effort by the Spanish to increase the number of settlers in their colony. By 1790 over 4000 Acadian refugees had relocated to Louisiana. The Acadians settled along the Mississippi River downstream from Baton Rouge and along Bayou Teche, west of the Atchafalaya River. James Dormon points out that

> the Acadians made a tacit decision to avoid the older, Creole-dominated river settlements, and to seek out their own people to the south and southwest. There they could work their small farms and live in close communion with little interferences from the outside world (Dormon 1983:237).

There were many similarities in the environment of both Acadia and Louisiana, the former having tidal marshes versus Louisiana's swamps. Levee construction and dikes were necessary in both locations. Within this environment the Acadians "maintained a system of community structure in much the same form as had prevailed in Acadie" (Dormon 1983:237). Indeed, their relative isolation in Louisiana's swamps, away from both the French Creoles and Anglo-Americans, was a major factor in the continued evolution and perpetuation of Cajun ethnicity during the nineteenth century.

7.1.3.2 Louisiana Cajuns

Over time the Acadians became Cajuns, as Lawrence Estaville eloquently explains:

> Once settled in their new homeland, the Acadians rapidly changed their culture to better fit South Louisiana's subtropical environment. They abandoned wheat, barley, oats, flax, and apples for corn, rice, okra, cotton, and figs.... The Canadian woolen shawl and poteaux-en-terre structure became the Louisiana cotton garde-soleil and wooden Acadian house. All the while, irony shrouded the Acadians: they were highly religious but strongly anticlerical; they despised their wealthier Creole brethren, the ancienne population, but sought economic upward mobility; they feuded with the Indians but adopted many of their ways; and they prided themselves on a tradition of egalitarianism but became Southern slaveholders. The Acadians were indeed becoming Cajuns (Estaville 1993:33–34, Reprinted by permission of *Journal of Cultural Geography*).

The Acadians thus continued making cultural adaptations that had begun when they first moved to the New World, and they adopted various Native American ways in insulating the walls of their buildings. In Louisiana, the Cajuns observed the architecture that the French Creoles had brought from the West Indies and borrowed heavily to create a new architecture better adapted to their environmental conditions (Estaville 2001).

Many outsiders today erroneously associate Cajuns with all of the French ethnics in Louisiana. Although the Cajuns are the largest of the French ethnic groups in Louisiana, their heritage differs from the Creoles. However, the Cajuns "have extensive blood and cultural ties to both the Creoles and their former slaves" (Rushton 1979:5). Long isolated, first as settlers in Acadia, far from the more prominent French colony of Quebec, and then by two centuries of living in the swampy bayou country of Louisiana, it is argued that "the Cajuns are the living repository of a unique set of pre-industrial agrarian values" (Rushton 1979:5).

Cajun culture also extends to those who lack French blood heritage. This includes the Black Cajuns, who are descendants of slaves owned by French settlers. In parts of Louisiana some blacks still speak French. Likewise, French-speaking Indians, particularly the Houmas, also contribute to French Louisiana's regional identity (Trépanier 1991). Many individuals from other groups were also absorbed into Cajun culture, including Anglos who lived among the Cajuns, as well as Italian, Spanish, and German immigrants. As Malcolm Comeaux (1996:33) explains, "Many people of Spanish and French descent scattered about south Louisiana also consider themselves to be Cajuns (with the notable exception of the descendents of the French and Spanish settlers of New Orleans who maintained their 'Creole' identity)." The route of the French Creoles into Louisiana was very different from that of the Cajuns.

7.1.4 Saint Domingue

Acadians were not the only French ethnics who settled in Louisiana as refugees. Louisiana received French ethnic settlers, who both developed sugar plantations and engaged in commerce, both directly from France and via several French colonies in the West Indies, with Saint Domingue, the western portion of the island of Hispaniola that is now Haiti, being particularly important. French settlement occurred along the Gulf of Mexico coast not only within what is now Louisiana, but also farther east, including major settlements at Biloxi, Mississippi; Mobile, Alabama; and Pensacola, Florida.

The word creole was originally utilized to indicate a person of European heritage born in the New World. Later the word came to indicate people of a mixed ancestry. French Creoles were very important in Louisiana, as some of its earliest settlers came from Saint Domingue, at the time the most important and wealthiest French colony. The historical geographer Donald Meinig notes that "Louisiana could be viewed as an outlier of St. Domingue in social and cultural as well as commercial terms" (Meinig 1986:200). Just as black slaves were brought to Louisiana both directly from Africa and via Saint Domingue, hundreds of Natchez Indians were sent as slaves to Saint Domingue (Meinig 1986). There were only 7000 white settlers in Louisiana at the end of French control in 1763, including Alsatians and their descendents who settled the "German Coast" of the Mississippi River between New Orleans and Baton Rouge in 1721 and who became assimilated early into the French culture. In addition, there were thousands of Gallicized African slaves, and even some of the slave owners were free blacks.

When the slave rebellions began in Haiti in the 1790s, many French whites as well as individuals of mixed African and French ancestry, who were also despised by the rebelling slaves, fled to Louisiana, significantly adding to its French Creole population. Within Louisiana the French Creoles concentrated in the New Orleans area.

French settlement in the United States thus represents several groups. Besides the Quebecois and the Métis along the Canadian border and in the Midwest and Acadians in northern Maine, within Louisiana both French Creoles and Cajuns contributed their own distinctly different French cultural heritage. Other migrants from France brought their own distinctly non-French ethnicity, including Alsatians who had a strong German ethnicity and Basques who were similar to their brethren on the Spanish side of the Pyrenees.

7.2 French Ethnic Population in the Contemporary U.S.

The number of persons claiming French ancestry fell by over two million between the 1990 and 2000 U.S. Censuses, falling from 10,320,935 to 8,309,908. Much of this drop can be attributed to changes in the way the questions were worded and the data coded (Bragg 2001), as some of the statistical results are clearly unreliable. Indeed, the 1990 Census reported an Acadian/Cajun population of 668,271, while only 85,414 Acadians and Cajuns were counted in 2000. Undoubtedly many of the Cajuns were counted as French in 2000, and the French were the second largest ancestry in Louisiana, second only to African-Americans. French Canadian numbers actually increased to 2,349,684 by the time of the 2000 Census, up from 2,167,127 in 1990. More recent results from the 5-year American Community Survey conducted between 2010 and 2014 indicate 8,431,568 Americans reporting French ancestry, plus an additional 2,052,979 claiming French-Canadian ancestry. Cajuns were estimated at 113,869, half of whom lived in Louisiana. Nevertheless, over time attitude changes have resulted in Acadiana, Cajun Country, and French Louisiana becoming more synonymous (Trépanier 1991).

The greatest concentrations of French ethnic population in the United States are located in both Louisiana, settled by both Acadians/Cajuns and the French Creoles during the colonial period, and within those counties of New England near the borders of Quebec and New Brunswick. It is in these two areas that the French language continues to be most widely spoken at home. The 2000 Census indicated that in several of the parishes within Louisiana's French Triangle, that area southwest of the Mississippi River, over 20% of the population still spoke French at home (Suchan et al. 2007). Beginning in the mid nineteenth century and continuing until the mid twentieth century, French Canadians could move freely across the international border, and most persons of French heritage in the northeastern United States are descended from those who moved south for industrial jobs or to work in the forests.

7.3 French Legacy in the United States

There are many lasting legacies from the French colonial and fur trading periods. Some of the most widespread are toponyms, the names of places; as French names abound on the landscape. Consider numerous French place names in the Fox River Valley of Wisconsin, such as DePere, Grand Chute, Fond du Lac, Butte des Mortes, and Omro, among many others. Upriver, the location of Portage was named after the French word for the process of unloading and carrying their canoes between the Fox River and the Wisconsin River, and downriver the French gave names to Prairie du Sac and Prairie du Chien. The French propensity to utilize waterways for transportation routes and the fact that the French were the earliest Europeans to journey in much of the central part of North America explains this legacy of French names. Most French toponyms "are applied to natural features such as the Missouri Coteau of the plains states, Mille

Lacs in Minnesota or Lac du Flambeau in Wisconsin, rivers like the Des Moines, Au Sable, St. Croix, or Eau Pleine, or scattered occurrences of descriptive words like 'slough' or 'prairie'" (Johnson 1974:2). Yet sometimes the French name only survives in translation, as "one would not recognize Green Bay as French without knowing its derivation from the earlier La Baye Verte" (Johnson 1974:2). One may question why French names are so numerous in the Great Lakes region and the Upper Mississippi River Valley when few of the area's contemporary residents are of French descent. As was noted in Chap. 1, the role of early arrival is very important in determining a group's influence on the landscape. The French arrived in the region before other Europeans and landscape habits are reshaped during the critical initial period of settlement and names are placed on the maps. Similarly, when one asks why there are so few other French legacies in the region when French settlers arrived first, it must be remembered that the French settlers were widely dispersed and never numerous. When later immigrants arrived into the sparsely settled area, they were able to establish their own imprint. Thus, within the middle section of the United States there are only scattered ethnic islands that display a clear French cultural legacy. Within Louisiana, that imprint is far more pervasive.

7.3.1 Long Lot Survey System

The long lot survey system is intimately associated with areas of early French settlement. The earliest long lot grant at Detroit was made in 1707, having a width to depth ratio of one to ten, in keeping with customary practice already well established along the St. Lawrence (Ekberg 1998). The way the French surveyed their lands influenced how they arranged their fields and positioned their farmsteads, and how they laid out their villages. As Allen Noble writes, "the most significant French settlement feature, and the easiest to identify, is the long-lot system of land division" (Noble 1984a:88). The land survey system adapted along the Mississippi River had its roots in the seigneurial system that evolved within French Canada (Fig. 7.9). Under that system large grants of land called seigneuries were given to a seigneur, typically members of the nobility or religious officials, who had certain feudal rights and was responsible for recruiting settlers. The seigneur allocated portions, termed rotures, of his grant to individual farm families (Héroux 1992). The settlers had the rights to sell their lands or bequeath them to their children, yet for many years they also paid an annual rent to the seigneur, among other obligations. At the same time, the seigneur was obligated to provide a grist mill for the settlers. The seigneurie

Fig. 7.9 Long, narrow fields extending perpendicular to the shore of the St. Lawrence River west of Berthier-sur-Mer, Quebec. Note that a line of farm houses and outbuildings are located along the coast road paralleling the shore, producing a type of string village lined with the long-lots

Fig. 7.10 French long lot field arrangement along Mississippi River upriver from New Orleans, Louisiana. Note that several of the narrow fields have been subdivided to create a narrow line of houses extending away from the river

boundaries typically coincided with the boundaries of the Roman Catholic parish, and the seigneur usually played a role in determining where the church was located. Just as seigneuries were rectangular in shape, extending back perpendicularly from the edge of the St. Lawrence, rotures typically were smaller versions of seigneuries, with the first rang (or row) of them extending along the shore or river.

The long lot system is viewed by Edward Price, a geographer who studied numerous survey systems, as well suited to the topography of southern Louisiana (Fig. 7.10).

> Longlots cut across the levees, giving each holder a proportional share of the lighter, better-drained soils near the river and the heavier, wetter lowland as well as access to fishing and stream transportation and to the roads that would run along the levees.
>
> With their houses at the ends of the narrow lots, farm families lived close to their neighbors in a continuous line village that might develop central nodes with churches, schools, stores, and other services. The long dimensions of the lots facilitated drainage by gravity and permitted plowing furrows as long as the farmer might desire. Marking off the longlots was relatively simple: only the corners at the stream end and the directions of the sides leading toward the backswamp had to be fixed at first (Price 1995:289–290, Reprinted courtesy of University of Chicago Press).

In Louisiana the surveying of land into long lots is sometimes termed the arpent system, named after the unit of measure that was used by the French. An arpent equals approximately 192 feet. Long lots were typically a few arpents wide along a river or bayou and 40 arpents (or about 1.5 miles) deep. Surveying a roture with a depth of 40 arpents was more standard in Louisiana than in Quebec. As Richard Campanella (2008:131) explains, “Straight portions of the river yielded neat rectangular long lots; where the river meandered, lots diverged on the convex side and converged on the concave side, forming a radiating pattern of elongated triangles and trapezoids.” Longlots were divided lengthwise upon inheritance, resulting in most lots narrowing, even to the point of almost ridiculousness, “sometimes tapering to a point a mile and a half down the levee slope” (Price 1995:298).

In Louisiana long lots were initially granted either in allodium, a fee simple outright ownership, or with feudal quitrents, requiring an annual payment of money and/or labor to a landlord. Over time the long lot property survey has been perpetuated. Upon the Spanish acquisition of Louisiana in 1763 the new government recognized existing land divisions and expanded the system into newly settled areas. Upon the U.S. annexation of Louisiana Territory in 1803, the earlier surveys were recognized, but American surveyors found that the standard American public land survey using townships and ranges was not acceptable to the French ethnic settlers, who wished to occupy newly surveyed lands. Thus, Congress authorized the surveying of lands along Louisiana’s rivers and bayous into 160 acre long lots (Price 1995). Therefore,

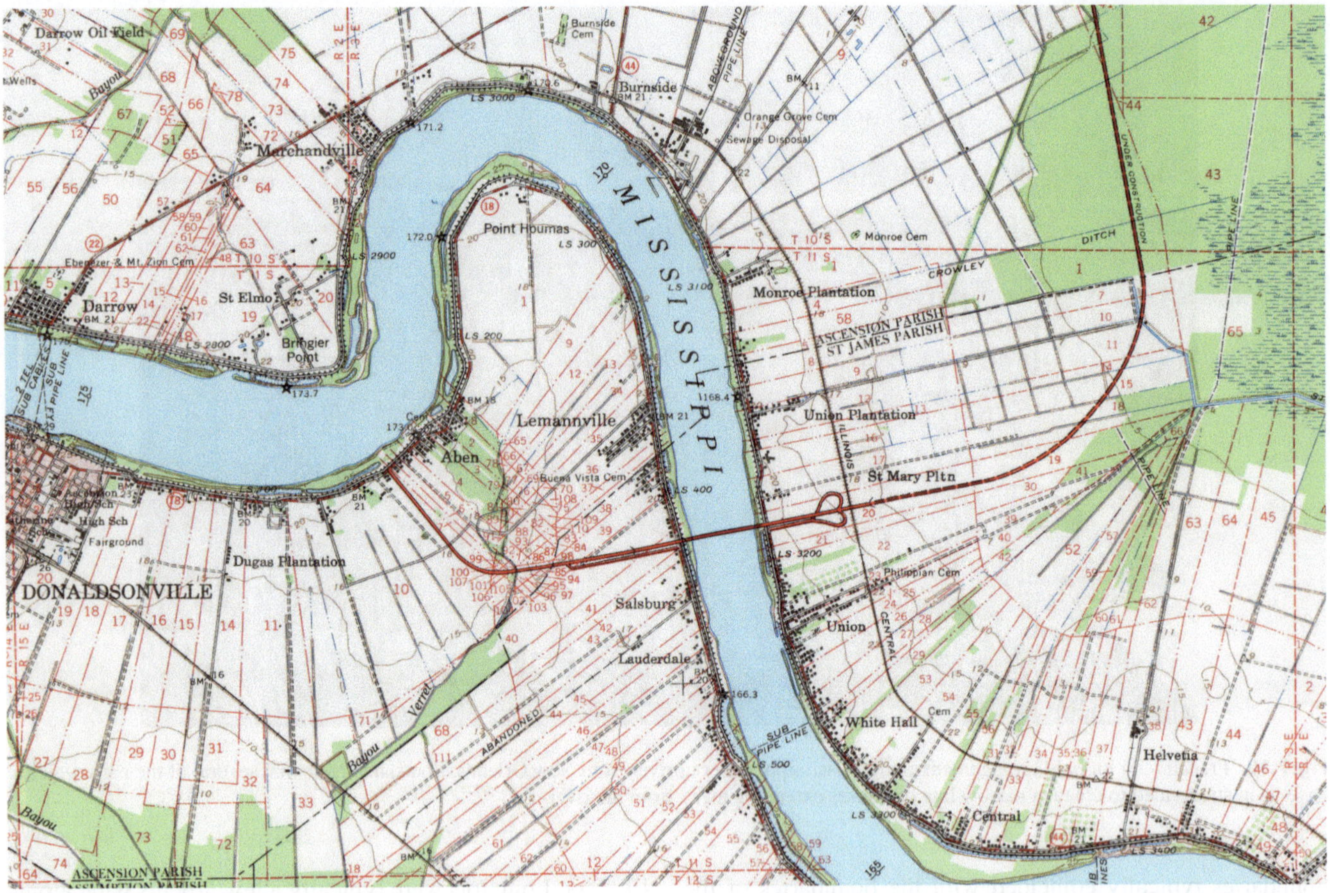

Fig. 7.11 Property lines radiating back from the meandering Mississippi River, just downriver from Donaldsonville, Louisiana, illustrating the French inspired long lot survey system. Note how the survey lines, delineated in red, are arranged perpendicular to the curve of the river and run to another line over a mile from the river, demarking the second rang of long linear lots. Reproduction from U.S. Geological Survey topographic map of Donaldsonville 15-min quadrangle, Louisiana

not all long lots one sees in Louisiana today date to the colonial period. The long lots mapped out by American surveyors were approximately five by forty arpents in size (Fig. 7.11). As in Quebec, a second range (rang) of long lots was surveyed in areas of French settlement where there was room for expansion.

The long lot influence on the modern New Orleans landscape is fairly evident. While the Vieux Carre, or today's French Quarter, was surveyed into square blocks, the then rural area upriver was surveyed into long lots. Given the sharp bend in the river, one sees the converging pattern of streets following the long lots' property lines in the area of Canal Street and in the Garden District. Thus, streets in downtown New Orleans and the Garden District run perpendicular from the curving river. In the rural countryside field lines, drainage ditches, and side roads all display the lasting legacy of the French long lots. Once survey lines are laid out, they tend to remain, as later land purchasers acquire parts or all of previously surveyed properties. When flying over the Mississippi River or one of the bayous the imprint is obvious, with the main road running along the levee, with side roads and fences running at right angles away from the waters. Many plantations in Louisiana, particularly ones established by French Creoles and alike ones in the French West Indies, had a linear plantation settlement pattern, where the houses of the workers and sometimes the plantation store were all laid out in a line (Rehder 1999), paralleling the direction of the property lines.

French long lot survey system remnants also remain outside of Louisiana, although it is most evident in either the French parishes of Louisiana or in Quebec. French ethnic islands with longlots can be found along Lake Champlain in New York; near Detroit and Monroe, Michigan; south of Green Bay, Wisconsin near DePere; along the Mississippi River near Prairie du Chien, Wisconsin; south of St. Louis, Missouri along the Mississippi River, particularly near Ste. Genevieve, Missouri and Prairie du Rocher, Illinois; and near Vincennes, Indiana. As Michael Roark (1988:111) explains, longlots were created along the Mississippi River from fifty miles north of St. Louis to about 200 miles south, and roughly one hundred miles up the Missouri River. In addition, square Spanish land grants that were made in the lead mining district near Potosi, Missouri, after Spain acquired Louisiana following the French and Indian War, were divided by the French into longlots.

Although the agricultural long lots have long disappeared, their legacy remains in the street arrangement of Detroit, Michigan. Johnson similarly notes, "Other such areas existed, but since the sites selected had advantages for water transportation, they have been covered by the urban sprawl of cities like St. Paul or St. Louis" (Johnson 1974:2).

7.3.2 Parishes in Louisiana

In Louisiana, as with Quebec, the local administrative area was typically associated with ecclesiastical boundaries. Thus, Louisiana is divided into parishes, rather than counties. While the parishes share many functions of counties, portions of Louisiana law are based upon French civil law, not English common law that provides the legal foundation within the other states. Furthermore, Louisiana parishes are governed by "police juries," which have both legislative and administrative authority, rather than by county commissions that are typical in many other states.

7.4 French Imprint upon New Orleans

Americans often think of Louisiana, and in particular, New Orleans, as the epitome of French architectural heritage (Fig. 7.12). Indeed, the naming of the city and its streets strongly reflects its French origins.

> [T]he French chose the name "Nouvelle Orléans" to project a specific image of European regality.... The original street names also indicate the spirit in which New Orleans was founded. ... In addition to "Royal Street," streets were named after prominent French aristocratic families, such as "Chartres" (for the duc de Chartres). "Bourbon," "Dauphin" (for the eldest son of the king of France), and "Bienville" (for the governor)" (Radding and Western 2010:404, Reprinted with permission of American Geographical Society).

These names remained during the subsequent four decades of Spanish rule and are seen in the French Quarter today. As Peirce Lewis so eloquently explains, "From the onset, New Orleans was a foreign city, and it has never completely lost its foreign flavor" (Lewis 2003:5). Not only was it a foreign city, but, as Karen Kingsley (2003:26) notes, "Almost all of the eighteenth-century architects, engineers, and surveyors in Louisiana were French born, and they gave a distinct French flavor to New Orleans's civic and institutional buildings." During the first half of the nineteenth century,

Fig. 7.12 Ursuline Convent, built in 1745 in New Orleans, Louisiana. A classic example of French colonial architecture, this building is the only remaining structure built during the time of French rule in New Orleans, thanks to several major fires that ravaged the French Quarter during the time that Spain ruled Louisiana

after the American annexation of Louisiana, French born architects continued to be employed to design the city's buildings and cemeteries, as well as prominent structures in cities upriver from New Orleans.

At the time of the first census undertaken after New Orleans's annexation by the United States, it was the nation's seventh largest city (or the fifth largest, if a couple of nearby urbanized places that were subsequently annexed are considered as part of Philadelphia), and by 1840 it had risen to third largest (Gibson 1998). Not only was it one of the nation's largest cities, but a significant share of its population spoke French, was foreign born (as New Orleans attracted large numbers of immigrants), and was non-white. Many of its non-white residents were free-blacks, with roots in the former French West Indian colony of Saint Domingue (See also Chap. 8), strengthening the city's French Creole character. While both black and white Creoles fled revolutionary Haiti, royalist refugees fleeing the French Revolution had also settled in Louisiana. Given their background, many residents of New Orleans viewed their community as being more sophisticated than those from which their American arrivals had come (Lewis 2003). Although New Orleans displays a street pattern influenced by the French long-lots and many buildings display French architectural detailings, only one unaltered colonial building, the Ursuline Convent built between the 1745 and 1750, remains from the first and longest period of French colonial administration.

Because of the city's rebuilding following several devastating fires, the ethnic landscape of New Orleans is far more complex, representing French, Spanish, and African influences. In explaining why "New Orleans is a foreign city", Robert Sauder notes,

> It was not absorbed into the American realm until nearly a century after its founding, and even today it exudes an ambiance that has no parallel elsewhere in America.... French, Spanish, and African influences blended into an indigenous Creole culture which today is manifest in the city's novel townscapes (Sauder 1996:7).

Although most of what we see today in New Orleans was built during Spanish rule rather than French administration (Fig. 7.13), many French Creole architectural traits are also visible in structures built long after the colonial period.

"Creole cityscapes" in New Orleans, as explained by Campanella (2006:221), "manifests itself in the hundreds of structures exhibiting Creole architectural styles, most of them built before the mid-nineteenth century." In parts of the city, historical detailings by Creole craftsmen, including "solid carpentry of a St. Roch frame house, the exquisite plasterwork of an uptown mansion, the intricate gingerbread on an Irish Channel shotgun house, [and] the iron-work of a French Quarter townhouse" (Campanella 2006:222), highlight the ethnic contributions, even though the structures were built well after "the glory days of Creole architecture."

Fig. 7.13 Manuel Lanzos House, built in New Orleans' French Quarter in 1788. While displaying many French colonial features, this French Creole cottage was built while New Orleans was administered by Spain. Note how the main level is raised off the ground and displays an extensive gallerie, with the space below enclosed by a wall

7.4.1 French Creole Architecture in New Orleans

It has been claimed that "Louisiana's Creole architecture is perhaps the most complex and elegant vernacular tradition surviving from the colonial period" (Edwards 1986:62). The oldest plantation mansions in rural southern Louisiana display the French Creole style, and even though many have disappeared, a number of prominent examples remain (Rehder 1999). Within New Orleans two distinct types of French Creole housing were constructed, the Creole cottage, the most common type of house in the French Quarter (Heard 1997), and the Creole townhouse. Indeed, one-third of the currently extant structures in the French Quarter illustrate the second-generation of French Creole architecture—that style which incorporates elements that date from the period of French colonial rule supplemented by architectural elements brought from the West Indies, Spain, Latin America, Canada, and Africa (Campanella 2008). As Campanella (2008:141) explains, "Dating mainly from the 1820s and 1830s, this tradition exhibits an indigenous New Orleans look and design that harks back to colonial antecedents, but with local modifications and variations." While many structures in the low-lying African-American portions of the city as well as numerous buildings in the swampy parishes south and east of New Orleans were destroyed by Hurricane Katrina in 2005, the historic French Quarter survived largely unscathed.

7.4.1.1 French Creole Cottages

French Creole cottages display a mixture of housing traits of Normandy, Quebec, and Haiti. Its origins were the Norman farmhouse with a galerie, a porch that runs the entire length of a wall, which was introduced to Haiti where two regional variations evolved. One version of the Creole cottage was based upon a two-room plan, while the other had three rooms. While the three-room plan was initially most popular in Louisiana, it was eventually supplanted by the four-room cottage, which was an expansion of the two-room cottage, creating a more boxy house or mansion than the elongated three-room cottage (Oszuscik 1992).

Although considerable variation in style exists among Creole cottages, they share many common features. These include a centrally located chimney, placed within inner walls rather than outside (Fig. 7.14). When additions were added, these outer rooms often lacked fireplaces. Likewise, Creole cottages typically had a galerie, especially in rural settings, that sometimes ran along all four sides of the dwelling. Steep roofs of over 45 degree slopes were common (Oszuscik 1992). Hip or pavilion roofs were commonplace, although gabled roofs are seen on Creole cottages, particularly within urban areas such as New Orleans. For well over a century builders of Creole cottages utilized the more complex interlocking Norman truss to support their roofs before adopting a simpler system that supported lighter weight roofs.

Regardless of house type, French Creole cottages and townhouses were built "flush with the sidewalk and typically

Fig. 7.14 House built in French creole style with front galerie in Plaquemine, Louisiana. Many houses in this style were constructed during the twentieth century

have plain facades, except for the delicate wrought-iron or cast-iron galleries which adorn the upper stories of the larger dwellings" (Sauder 1996:10). As Karen Kingsley explains, "Galleries on one or more sides of a building are one of the distinguishing features of Louisiana's architecture.... Galleries not only substituted for interior hallways by serving as a circulation path, but also provided outdoor living and sleeping space" (Kingsley 2003:21). Indeed, they were an integral part of the home, with detailings along their floors and on their ceilings similar to what one would expect within the interior of the house.

The floor and porch of Creole cottages were raised to avoid wetness. Walls were typically constructed upon a timber sills that either rested upon cypress blocks or brick (Edwards 1988). If the space below the superstructure was enclosed, it consisted of solid brick walls sixteen to twenty inches thick (Robinson 1983). The primary types of wall construction during the colonial period consisted of either *briquette entre poteaux* or *bousillage entre poteaux*, which involved filling the area between vertical posts with either brick or mud, respectively, and covering with clapboards or weatherboards (Edwards 1988).

Within the French Quarter of New Orleans many of the one and a half story Creole cottages are of half-timbered construction, covered with stucco and roofed with slate (Sauder 1996). Their window and door arrangements are also unusual, with "the street front of the Creole cottage [having] four shuttered openings, at least two of which are casement doors. The other two may also be casement doors or may be casement or double-hung windows.... The arrangement of openings is symmetrical—either door-window-window-door or window-door-door-window" (Heard 1997:28). At the rear of the cottage is a gallery or porch.

7.4.1.2 French Creole Townhouses

The French Quarter contains many two or three story Creole townhouses constructed around an interior courtyard. Although the two to three and a half story Creole Town House evolved from a Spanish Colonial house type, the style was modified by the numerous refugees from Haiti who arrived in the early nineteenth century (Heard 1997). The Creole townhouse, taller than the Spanish Colonial house, typically included a carriageway, a loggia, a courtyard, and one or more balconies.

The loggia is an open space within the townhouse adjacent to the courtyard. It frequently includes the primary stairwell within the structure, located just off the courtyard, with the arched opening to the stairwell acting to widen the open area. Thus, "Upper level loggias, glazed but quite open at their stairwells, provide one of the subtle 'in-between' degrees of enclosure which characterize French Quarter buildings" (Heard 1997:102). Typically, the loggia has only one of its four sides open to the outdoors.

Many, but not all, Creole townhouses include an entresol, which Heard (1997:40) defines as "a low-ceilinged storage or service floor placed above the commercial first level and connected with it." Both the first floor and the entresol are enclosed behind tall arched openings, with the doors or lower windows opening into the ground floor, while the "entresol's windows are the semi-circular tops of the arches" (Heard 1997:40). The residential floor or stories of the Creole townhouse are above the entresol. While found in Creole townhouses, it has also been argued that this style was actually introduced from Spanish Havana.

7.4.1.3 French Creole Plantation Mansions

The French Creole plantation mansion shares many characteristics with the French Creole cottage, such as having a central placement of chimneys, but it is considerably larger and more rectangular in shape, and much more likely to have a hip roof (Fig. 7.15). Like the Creole cottage, the front rooms of a French Creole mansion all opened onto a gallery, acting as an open-air hallway connecting anywhere from two to six rooms. All stairways within the Creole mansions were on the exterior, connecting with the galleries. Mansions are often 60–90 feet long by 35–50 feet wide, with the main floor raised a half to full story above the ground, with a brick wall enclosing the space at ground level. The walls of the main structure were either of all wood or a half-timbered construction called colombage. Those that had the French version of half-timbering had large cypress timbers, with either brick or bouisillage providing the nogging between the beams (Rehder 1999).

Examples of French Creole plantation mansions dating from the eighteenth and nineteenth centuries, including some built during the period of French colonial rule, can be found in the New Orleans metropolitan area as well as up the Mississippi River. The Pierre Menard house, built between 1818 and 1820 near Kaskaskia, Illinois, is another example of a French Creole mansion, with its wrap around galerie, poteaux-sur-sole walls, and double-hipped roof.

7.4.2 French Creole Architecture along Gulf Coast and Upriver

French Creole architecture is not just confined to New Orleans. Natchitoches, founded in 1714 and Louisiana's oldest permanent settlement, "became a center for Creole culture and architecture" (Kingsley 2003:305). Besides that city's commercial district with its cast-iron canopies and wrought-iron galleries and the older houses with their extensive galleries (porches running the length of the side, termed galerie in French), several nearby plantations also display French Creole architecture. Within the historic districts of Biloxi, Mississippi; Mobile, Alabama, and Pensacola, Florida (Fig. 7.16) one can also find

Fig. 7.15 Drestrehan Plantation House built between 1787 and 1790, facing the Mississippi River, in St. Charles Parish, Louisiana. The galleries that surround this house and the hipped roof, with a change in roof slope, can be clearly seen in this large plantation house, which resembles an oversized French creole cottage. Some of the detailings of this house were later modified to incorporate Greek Revival styles

Fig. 7.16 Charles Lavalle House, built around 1805 in Pensacola, Florida. This French Creole style cottage, located in the Seville Square Historic District, was erected while Pensacola was under Spanish administration

Fig. 7.17 The Jean Baptiste Valle House was built in 1792 in Ste. Genevieve, Missouri. It clearly displays front and side galleries, chimneys enclosed within the side walls, and a steep roof whose pitch changes towards the galleries. Its walls are of poteaux-sur-sole construction

examples of French Creole architecture, even though those within the last two cities were built after the French colonial period. There has been a recent revival of interest in French Creole architecture, particularly the Creole cottage, along the Gulf Coast (Oszuscik 1992).

7.4.2.1 French Creole Buildings in the Mississippi Valley

French Creole buildings dating from the eighteenth century are found in both Missouri and Illinois in the Mississippi River Valley. Walls of these cottages are of poteaux-en-terre, poteaux-sur-sole, or pièce-sur-pièce construction methods (Fig. 7.17, See also Fig. 7.8). The poteaux-en-terre walls consist of upright posts or planks set together, while vertical planks in the similar, but more common, poteaux-sur-sole walls rest on a wooden sill or stone foundation instead of directly on the ground. The pièce-sur-pièce construction method consisted of horizontal logs, with full dove-tail notching to fix the corners. Spanish moss was used in bousillage, the caulking of mud or clay and moss, to fill the cracks within the walls. The wall structure is considered the most unique characteristic of French houses (Johnson 1974).

Use of these three wall styles was not limited to the Creole cottage. The poteaux-en-terre, poteaux-sur-sole, and pièce-sur-pièce walls were an early construction style used throughout French America. Vertical log construction remained popular in the Mississippi Valley and the Great Lakes region much longer than in Canada. The Cahokia courthouse built around 1737 in Illinois is an example of such a building. Pièce-sur-pièce walls were constructed of hewn logs laid horizontally, and held in place by grooves cut into a vertical timber called a coulisse. The presence of the coulisse or vertical mortise posts, which often partly framed doors or windows, is a key diagnostic trait of early French Canadian structures (de Julio 1996). Pièce-sur-pièce buildings were the most common form in French Canada and were used in a majority of fur trading posts, including those of the Hudson Bay Company (Kniffen and Glassie 1966). Several surviving examples of French-Canadian pièce-sur-pièce houses remain in Wisconsin, both in the Green Bay and Prairie du Chien areas (Fig. 7.18). Because later clapboarding sometimes covered their original walls, the character of some of these buildings had been both hidden, yet better preserved.

7.4.3 French Cemeteries in Southern Louisiana

The cemeteries of New Orleans and southern Louisiana are noteworthy because of the extensive use of above ground tombs (Fig. 7.19). Although as Saunders (1996) indicates, this is partly an adaptation to the high water table in the low-lying area, it also reflects French and Spanish culture. Indeed Metairie Cemetery was patterned after a cemetery in Paris, and the above ground vaults in New Orleans are similar to those within the large historic cemetery in Port-au-Prince, Haiti. Not only were "the tightly packed straight avenues" of

Fig. 7.18 Fur trader's cabin built around 1800 in the French Canadian style, now preserved in Heritage Hill State Park in Green Bay, Wisconsin. The pièce-sur-pièce walls of hewn logs laid horizontally between vertical timbers, called a coulisse, are clearly visible

Fig. 7.19 Burials in many Louisiana cemeteries were in above ground vaults. These tombs with their French inscriptions are located in the cemetery adjacent to St. John the Baptist Catholic Church in Edgard, Louisiana

Paris's Père Lachaise Cemetery replicated in New Orleans's St. Louis Cemetery Number 2, but "[m]any of the most sought-after tomb designers in nineteenth-century New Orleans were French immigrants" (Kingsley 2003:77). Whether of whitewashed plaster covered brick or marble, the above ground tombs of New Orleans's older cemeteries display the city's ethnic heritage, both in their design and the use of the French language on the older tombs.

7.5 Cajun Landscapes and Architecture

Cajun farmsteads underwent considerable evolution in Louisiana, and their buildings differed substantially from French antecedents. Nevertheless, certain commonalities mark Cajun farmsteads, including the grouping of many small structures set far back from road. The farmhouse is typically closest to road (yet setback and it is usually the largest building, often larger than the barn). Farm buildings are often constructed of cypress, which is among those woods that are most rot resistant. All the buildings on the farmstead are arranged facing either the same direction as the house or they are perpendicular to it. Numerous fences, all parallel to the house, also mark the farmstead (Comeaux 1992). Cajun fences were traditionally cypress pieux or board fences. Although Cajun fences differ from traditional Acadian styles, one variety of gate is similar in both locations (Comeaux 1996).

7.5.1 Cajun House

The Cajun house is similar to a French Creole cottage (Fig. 7.20). R. Warren Robison (1983:157) argues that Louisiana's Acadian or Cajun cottage's "timber frame, except for the type of wood used, was almost the same as that used in faraway Acadia." In contrast, Malcolm Comeaux (1996) notes that the Cajun house was not an adaptation of Acadian styles, but modifications of a style imported from the French West Indies. The architectural historian Karen Kingsley likewise argues that the Cajuns "adopted the essentials of the Creole house…[with] a principal difference between a Creole and an Acadian house is that in the latter, the attic was used as a sleeping space and was accessed by a stair or ladder from the front gallery" (Kingsley 2003:24).

The Cajun house is of a heavy frame construction built on piers, elevating the structure two to three feet above the ground. It is typically two rooms wide and has two front doors with no central hall. Residents use galeries, or elongated front porches extending the length of the house, to connect the rooms. It has been claimed that it is the presence of the galerie that gives Louisiana country homes their character, with the galerie being used as "a weather protected, outdoor living room" (Rushton 1979:163), providing access to all the ground floor rooms. A steep exterior stairway extends from the front porch or galerie to a loft, or grenier, under the steep, gabled roof (Comeaux 1992).

Because both French Creole cottages and Cajun houses display conspicuous porches or galeries, the exterior staircase, or garconnier stairs to the grenier, sets apart the Cajun house. As Rushton explains, "The galerie staircase, so steep as to be almost a ladder, remains a distinctive Cajun feature throughout the nineteenth century, and one way of distinguishing the basic Cajun from the basic Creole house: the

Fig. 7.20 Cajun cottage built in 1867 in Grand Coteau, Louisiana. While both this house style and the Creole cottage both have the galerie, the Cajun cottage often displays an exterior staircase, or garconnier stairs, as is shown here

Creoles put their stairs out back" (Rushton 1979:163). Cajun houses are gradually becoming less common on the landscape. Although he notes that most Cajuns prefer to live in contemporary housing, Comeaux (1996:38) writes, "it is quite fashionable to build modern homes on concrete slabs to outwardly resemble 'the traditional Cajun house.'" An example of a modern Cajun house with garconnier stairs can be seen in the same community where the nineteenth century Cajun cottage shown in the figure was photographed.

7.5.2 Cajun Barns and Dance Halls

Cajun barns were actually derived from German traditions rather than the much larger Acadian structures that were more appropriate in a colder climate (Comeaux 1989). The characteristic Cajun barn (Fig. 7.21) measured about thirty feet square, with a recessed central corn crib entered from a gabled front end. Shed appendages, like lean-tos, were added onto the sides. Other farm outbuildings included chicken coops, a potato house, and an outdoor privy, among other structures (Comeaux 1992).

The Cajun communal dancing hall was the only other distinctive Cajun structure. It was basically a larger version of the Cajun house. Unfortunately, very few of these remain, yet several types of distinctive Cajun celebrations seasonally mark the Louisiana landscape.

7.5.3 Cajun Christmas Eve Bonfires and Mardi Gras Runs

During the Christmas season, constructions of elaborate stacks of lumber and sugar cane to be burned as Cajun Christmas Eve bonfires (Bragg 1995:8) line certain rivers. It has been a Christmas Eve tradition to light large bonfires atop river levees, with hundreds of bonfires being ablaze simultaneously along the Mississippi River (Fig. 7.22). Today twenty-foot high stacks of logs and sugar cane are prepared for the bonfires, while at one time stacks up to forty feet high were arranged.

The tradition of Christmas Even bonfires goes back over 200 years among the Cajuns. Rick Bragg (1995:8) describes this custom, which he notes "is one of the more unusual…of America's Christmas traditions." "Carefully constructed teepee shaped towers—built around a frame of willow and covered with cane reed" (Bragg 1995:8) are erected atop the Mississippi River levee and when set ablaze on Christmas Eve, provide a long line of glowing lights. While conical piles of lumber are most commonly built over a period of 3–4 weeks for the bonfires, a wide variety of other structures and shapes can also be found in southern Louisiana. Cajuns also celebrate the coming of Lent.

During the week or so before Fat Tuesday, Mardi Gras runs are held in about 18 communities in rural south-western Louisiana. Such runs attract anywhere from a dozen or so to hundreds of participants. Costumed and masked participants

Fig. 7.21 Cajun barn with central crib with flanking sheds, located in Grand Coteau, Louisiana

Fig. 7.22 Erection of wooden scaffolds atop levee for Christmas Eve bonfires in Lutcher, Louisiana

travel by horses or motorized wagons in a colorful and boisterous parade, stopping to visit residences and small communities. Participants are all male in some communities, in others both male and female runs take place, while others include children. Some of these Cajun Mardi Gras runs are comprised of white participants, while others are black. The scheduling of the runs likewise varies, with some occurring on the Saturday or Sunday preceding the start of Lent, while others occur on Fat Tuesday. The parade of costumed participants through the rural countryside is typically followed by a dance and a gumbo feast (Comeaux 2008).

7.6 French Canadian Barns

While we most often consider the Cajun and French Creole landscapes of Louisiana when the French ethnic landscape in the United States is reviewed, French Canadians also left their mark along the nation's northern border. While both the Upper Peninsula of Michigan and northern Minnesota have populations in which French Canadians are disproportionately represented—a legacy of both their role in lumbering and the fur trade, their greatest concentration is in New England and upstate New York. The St. John River, which forms part of the international boundary separating Maine and New Brunswick, divides the Madawaska region that was settled by Acadians who were first expelled from Acadia and subsequently relocated from the lower reaches of the St. John River in New Brunswick.

The Quebec long barn is primarily seen in Quebec (Fig. 7.23) and along the St. Lawrence, but a few examples have also been described in the United States. Typical features of the Quebec long barn include its elongated rectangular shape with a gabled roof. Its walls are of vertical board construction. The barn has multiple bays, facilitating a variety of functions, and a hayloft is located over the single story barn. Not only did this style influence the construction of a few barns constructed near the Canadian border in New England and far upstate New York, but Allen Noble (1984b) argues that it is related to Belgian barns of Door County, Wisconsin (see Fig. 10.14).

The Madawaska twin barn is closely related to the Quebec barn, but it also has English traits. This barn is found primarily in the St. John Valley of far northern Maine. Shaped like an "H", the Madawaska twin barn essentially consists of two long narrow barns, which are parallel and connected by a low passageway. This style is believed to be the result from influence by Quebec immigrants.

7.7 Ethnic Landscape Comparisons: Quebec and Louisiana

While both Quebec and Louisiana were part of the seventeenth and eighteenth century French colonial settlement of America, considerable ethnic, cultural, and political differences exist between French Quebec and Cajun Louisiana today. Their histories of settlement varied, as the two regions attracted settlers

Fig. 7.23 Quebec long barn northeast of St.-Jean-Port-Joli, Quebec. Note that this barn, sided in wooden shakes, is entered via a ramp bridge

from different parts of France and incorporated cultural traits from dissimilar peoples with whom they encountered. Indeed, one is tempted to ask: Why is Quebec threatening to leave Canada, yet the Cajun culture in Louisiana is assimilating into Anglo American culture? In speculating about the differences—which may be related to the fact that the Acadians and Quebecois differed culturally even in the eighteenth century, one can offer several other possible explanations.

In the centuries since their original French settlements, both Louisiana and Quebec have experienced immigration by other populations; however both the numbers and ethnic composition of these later arrivals differed greatly. Quebec saw the settlement of its Eastern Townships by both American Loyalists and many Scots, yet within the core of the French ethnic Quebec homeland, along the both the Gulf of St. Lawrence and the St. Lawrence River, focused upon Quebec City, the population remained overwhelmingly French. Many other Loyalists relocating from the United States and British immigrants went to Upper Canada, now called Ontario. In contrast, French settlement in Louisiana included not only displaced Acadians, but large numbers of French Creoles who fled revolutionary Haiti. Even during the days of French rule, Louisiana also received German immigrants and Canary Islanders. Unlike Quebec, Louisiana included a significant number of African slaves within its population. Louisiana was ruled for four decades as a Spanish colony after the French defeat in the French and Indian War. Following the American acquisition of Louisiana, the state was inundated by newcomers, particularly those of British heritage arriving from the Southern states. Later during the nineteenth century, New Orleans received large numbers of immigrants from Germany, Ireland, and Italy, while at the same time maintaining many trade and cultural links with Latin America. Thus, the French culture within Louisiana was more ethnically diverse than within Quebec.

Social scientists have long recognized that maintenance of language fluency is key to preservation of many aspects of ethnic culture. Louisiana differed significantly from Quebec with respect to the role of the French language. While the 1805 Louisiana constitution was written in French, in 1845 a new constitution authorized the use of French and English in public documents. However, the use of the French language in legal documents has been outlawed since the Civil War period. Public school instruction in Louisiana was in English only, with the speaking of French on school-grounds being banned in 1922. In contrast, parochial schools dominated within Quebec, and those schools played a major role in not only maintaining the Roman Catholic religious culture, but of preserving the French language. While the Cajuns and Creoles of Louisiana were predominately Roman Catholic, there most children attended public schools. Furthermore, while French priests kept the French culture and language alive within Quebec, just the opposite occurred within Louisiana. There, "to gain national acceptance, the Catholic church hierarchy in French Louisiana was determined to Americanize its members… and many priests as well as parishioners could not speak French" (Estaville 2001:95).

While there have been recent reversals in Louisiana, with a Cajun being elected governor and an effort at a French language revival beginning in 1968, schools had to import teachers from France to provide French instruction—hardly preserving the variety of unique French patois that had evolved in southern Louisiana. Around the same time Louisiana was encouraging the teaching of French as a second language, Quebec was enacting its language laws to prohibit the use of any languages other than French in outdoor advertisements and on road signs. Thus, in Quebec ties to both the French language and French culture remain connected and strong, while in Louisiana's they are weaker, being much more diluted by multiple other ethnicities. Indeed, David Kaplan (1994:595) explains, "Quebec's policy on language clearly demarcates a boundary that divides 'French' Quebec from 'English' Canada and that charges each territory with determining its social, economic, and cultural composition." Focusing inwardly upon the province of Quebec, the separate identity of the French Canadian, at least the Québécois identity, has grown within their ethnic homeland.

Today, Montreal is the second largest French speaking city in the world, and both it and Quebec City have numerous visible reminders of their French ethnicity. As Cole Harris eloquently explains, "Québec and Montréal were the most comprehensive transplantations from France in the New World…[and v]isitors likened them to French towns" (Harris 1990:68–69). These cities and others of Québec retain a strong French ethnic character, far stronger than anything in Louisiana. Nevertheless, as we have seen, many French ethnic reminders are clearly visible on the landscape of Louisiana today, as well as in small enclaves scattered along the Mississippi River Valley and the Great Lakes regions.

References

Allen, James Paul, and Eugene James Turner. 1988. *We the people: An Atlas of America's ethnic diversity*. New York: Macmillan Publishing Company.

Bragg, Rick. 1995. Cajun Christmas tradition refuses to die down. *New York Times*, December 24, 1995, p. 8.

———. 2001. Reported to be vanishing, Cajuns give a sharp 'Non'. *New York Times*, August 16, 2001.

Campanella, Richard. 2006. *Geographies of New Orleans: Urban fabrics before the storm*. Lafayette: Center for Louisiana Studies.

———. 2008. *Bienville's Delemma: A historical geography of New Orleans*. Lafayette: Center for Louisiana Studies, University of Louisiana at Lafayette.

Comeaux, Malcolm L. 1989. The Cajun barn. *The Geographical Review* 79 (1): 47–62.

———. 1992. Cajuns in Louisiana. In *To build in a New Land: Ethnic landscapes in North America*, ed. Allen G. Noble, 177–194. Baltimore: Johns Hopkins University Press.

———. 1996. The Acadians of Canada and the Cajuns of Louisiana: Cultural change over time and distance. In *Ethnic persistence and change in Europe and America: Traces in landscape and society*, Veröffentlichungen der Universität Innsbruck 213, ed. Klaus Frantz and Robert A. Sauder, 29–45. Innsbruck: University of Innsbruck.

———. 2008. The Cajun Rural Mardi Gras. (unpublished paper and handouts presented at Annual Meeting of Pioneer American Society and Eastern Historical Geography Association in Baton Rouge, October 17).

de Julio, Mary Antoine. 1996. The Vertefeuille House of Prairie du Chien: A Survivor from the Era of French Wisconsin. *Wisconsin Magazine of History* 80 (1): 26–56.

Dormon, James H. 1983. The Cajuns: Ethnogenesis and the shaping of group consciousness. In *The Cajuns: Essays on their history and culture*, ed. Glenn R. Conrad, 233–251. Lafayette: Center for Louisiana Studies, University of Southwestern Louisiana.

Edwards, Jay. 1986. French. In *America's architectural roots: Ethnic groups that built America*, ed. Dell Upton, 62–67. Washington: Preservation Press.

———. 1988. The Origin of the Louisiana Creole Cottage. In *French and Germans in the Mississippi Valley: Landscape and cultural traditions*, ed. Michael Roark, 9–60. Cape Girardeau: Center for Regional History and Cultural Heritage, Southeast Missouri State University.

Ekberg, Carl J. 1998. *French roots in the Illinois country: The Mississippi frontier in colonial times*. Urbana: University of Illinois Press.

Ens, Gerhard J. 1996. *Homeland to Hinterland: The changing worlds of the Red River Metis in the nineteenth century*. Toronto: University of Toronto Press.

Estaville, Lawrence E. Jr. 1993. The Louisiana-French homeland. *Journal of Cultural Geography* 13 (2): 31–45.

———. 2001. Nouvelle acadie: The Cajun homeland. In *Homelands: A geography of culture and place across America*, ed. Richard L. Nostrand and Lawrence E. Estaville, 83–100. Baltimore: Johns Hopkins University Press.

Faragher, John Mack. 2005. *A great and noble scheme: The tragic story of the expulsion of the French Acadians from their American homeland*. New York: W. W. Norton and Company.

Fischer, David Hackett. 2008. *Champlain's dream*. New York: Simon and Schuster.

Gibson, Campbell. 1998. *Population of the 100 Largest Cities and Other Urban Places in the United States: 1790 to 1990*. (Population Division Working Paper No. 27). Washington: U.S. Bureau of the Census. https://www.census.gov/population/www/documentation/twps0027/twps0027.html. Accessed 1 Dec 2016.

Griffiths, Naomi. 1973. *The Acadians: Creation of a people*. Toronto: McGraw-Hill Ryerson Limited.

Harris, Cole. 1990. French Landscapes in North America. In *The making of the American landscape*, ed. Michael P. Conzen, 63–79. Boston: Unwin Hyman.

Heard, Malcolm. 1997. *French Quarter Manual: An architectural guide to New Orleans' Vieux Carré*. New Orleans: Tulane School of Architecture.

Héroux, Andrée. 1992. The landscape as memory—The Laurentian Seigneuries. In *The heart of French Canada*, ed. Dean Louder, 85–110. New Brunswick: Rutgers University Press.

Johnson, C. 1974. Missouri-French houses: Some relict features of early settlement. *Pioneer America* 6 (2): 1–11.

Kaplan, David H. 1994. Two nations in search of a state: Canada's ambivalent spatial identities. *Annals of the Association of American Geographers* 84 (4): 585–606.

Kingsley, Karen. 2003. *Buildings of Louisiana*. New York: Oxford University Press.

Kniffen, Fred, and Henry Glassie. 1966. Building in wood in the Eastern United States. *The Geographical Review* 56 (1): 40–66.

Kornwolf, James D. 2002. *Architecture and town planning in Colonial North America*. Baltimore: The Johns Hopkins University Press.

Lewis, Peirce F. 2003. *New Orleans: The making of an urban landscape*. Second ed. Santa Fe: Center for American Places.

Meinig, Donald. 1986. *The shaping of America: A geographical perspective on 500 years of history. Volume 1, Atlantic America, 1492–1800*. New Haven: Yale University Press.

National Park Service. 2016. *Ste. Genevieve Final Special Resource Study and Environmental Assessment*. U.S. Department of Interior.

Noble, Allen G. 1984a. *Wood, brick, and stone: The North American settlement landscape, Volume 1: Houses*. Amherst: University of Massachusetts Press.

———. 1984b. *Wood, brick, and stone: The North American settlement landscape, Volume 2: Barns and farm structures*. Amherst: University of Massachusetts Press.

Noppen, Luc, and Lucie K. Morisset. 1999. The architecture of old Quebec, or the history of a palimpsest. *Material Culture Review/ Revue de la culture matérielle* 50 (Fall/Automne): 11–25.

Oszuscik, Philippe. 1992. French creoles on the Gulf coast. In *To build in a New Land: Ethnic landscapes in North America*, ed. Allen G. Noble, 136–156. Baltimore: Johns Hopkins University Press.

Price, Edward T. 1995. *Dividing the land: Early American beginnings of our private property system*, Geography Research Paper No. 238. Chicago: University of Chicago Press.

Radding, Lisa, and John Western. 2010. What's in a Name? Linguistics, geography, and toponyms. *The Geographical Review* 100 (3): 394–412.

Rehder, John B. 1999. *Delta sugar: Louisiana's vanishing plantation landscape*. Baltimore: Johns Hopkins University Press.

Roark, Michael. 1988. Imprint of the French in North America: Longlots in the Mid-Mississippi valley. In *French and Germans in the Mississippi valley: Landscape and cultural traditions*, ed. Michael Roark, 111–123. Cape Girardeau: Center for Regional History and Cultural Heritage, Southeast Missouri State University.

Robinson, R. Warren. 1983. Domestic architecture in Acadiana in the eighteenth and nineteenth centuries. In *The Cajuns: Essays on their history and culture*, ed. Glenn R. Conrad, 156–173. Lafayette: Center for Louisiana Studies, University of Southwestern Louisiana.

Rushton, William Faulkner. 1979. *The Cajuns: From Acadia to Louisiana*. New York: Noonday Press.

Sauder, Robert A. 1996. New Orleans' ethnic landscapes. In *Ethnic persistence and change in Europe and America: Traces in landscape and society*, Veröffentlichungen der Universität Innsbruck 213, ed. Klaus Frantz and Robert A. Sauder, 7–27. Innsbruck: University of Innsbruck.

Suchan, Trudy A., Marc J. Perry, James D. Fitzsimmons, Anika E. Juhn, Alexander M. Tait, and Cynthia A. Brewer. 2007. *Census Atlas of the Unites States*, Series CENSR-29. Washington: U.S. Census Bureau.

Trépanier, Cécyle. 1991. The Cajunization of French Louisiana: Forging a regional identity. *The Geographical Journal* 157 (2): 161–171.

Villeneuve, Paul. 1992. Québec—The old capital takes on new trappings. In *The heart of French Canada*, ed. Dean Louder, 111–135. New Brunswick: Rutgers University Press.

Wyckoff, William. 2014. Creating regional landscapes and identities. In *North American Odyssey: Historical geographies for the twenty-first century*, ed. Craig E. Colten and Geoffrey L. Buckley, 335–357. Lanham: Rowman and Littlefield.

8 African-American Landscapes

The African-American population has grown to represent slightly under one of every seven Americans, with the 2010 U.S. Census counting 42,020,743 individuals of Black racial ancestry, up from the 36,419,434 counted in 2000. Of all the ethnic groups that have contributed to the landscape and heritage of the United States, that of African-Americans is harder to define and for many Americans, it is underappreciated. Many ethnic populations sought a better life in the New World, whether deliberately fleeing persecution or economic deprivation in the Old World or perceiving greater opportunities in America. Yet, the ancestors of many of the nation's African-Americans arrived in chains. Unlike the thousands of white convicts and indentured servants that Britain shipped to America, those arriving aboard the slave ships could neither obtain their freedom, nor that for their descendents. Nevertheless, even as slaves African-Americans made numerous contributions in shaping the American landscape. Yet not all African-Americans in the antebellum South were slaves, and locally, such as in New Orleans, free blacks played a major role in the construction of the city. Later, the arrival of many thousands of black immigrants from the West Indies and others from Africa, starting in the later half of the twentieth century, has added to the diversity of the nation's African heritage. This diversity of backgrounds among those individuals of African ancestry have led some to question who should be considered as African-American—those who recently immigrated from one of the countries of Africa, but who might also be considered as Ethiopian-American or Nigerian-American or those whose ancestors were brought to America as slaves two or three centuries ago and whose African heritage has been trammeled by their enslavement and subsequent subordination by the dominant white culture (Berlin 2010).

8.1 African-American Population Definition

One question that has not only troubled the U.S. Census Bureau, but American society as a whole, is defining the African-American population. The question, "Who is a black?", has been answered differently not only over time, but by various segments of society. In the United States the definition of a person as being black or African-American historically ignored the various gradations between being all black and all white. Indeed, within certain Southern states until the 1980s, a person only needed one drop of black blood, or only some distant African-American ancestor, to be legally defined as being black (Davis 1991). Our concepts of races were shaped by the legacy of slavery and subsequent discrimination against those persons descended from slavery and are social constructions (Banton 2015). President Barack Obama is universally heralded as African-American, yet his lineage is half European. Indeed, some U.S. blacks have far more European genetic heritage than African, yet American society has considered them black if they visually look like they have African racial features, even though their skin may be relatively light, or if one of their parents—regardless of skin color—was considered black. The way race has been defined in the United States is considerably different from the recognition in many Caribbean and Latin American societies of numerous varieties of mulattos or colored persons, those individuals who are of mixed black and white ancestry. Furthermore, racism among Americans "is unique in its character" (Frazier 2016:25) and has been persistent, influencing politics, internal migration, and settlement patterns. The presence of racism constrains the processes of move-

J.A. Cross, *Ethnic Landscapes of America*, DOI 10.1007/978-3-319-54009-2_8

ment and assimilation that ethnic populations may otherwise experience.

Most African-Americans meet the definition of being racially mixed, as it is estimated that 75 to over 90% of African-Americans have some white blood (Davis 1991), even though fewer than one in ten persons reporting black or African-American ancestry on the 2010 Census indicated they had mixed black and white ancestry (Rastogi et al. 2011). Many African-Americans, possibly a quarter, also have American Indian blood. These mixtures are the result of miscegenation, or interracial intercourse, beginning when the first slave owners forced themselves upon their female slaves and continuing to intermarriage, which has only been legal in certain states since the last part of the twentieth century. Only 1.8 million persons reported on the 2000 Census as having both black and other racial ancestry—and 45% of these were black and white (McKinnon 2001), yet that number had grown to 3.1 million by 2010 (Rastogi et al. 2011), still far below actuality. Racism has long forced individuals of mixed racial ancestry to live their lives as African-Americans in segregated neighborhoods. It also negates the utility of such concepts as the melting pot and assimilation when considering African inclusion in American society (Berlin 2010).

While many individuals made considerable efforts to hide it, particularly during the period of strict segregation and racial discrimination which only began to break down following the passage of the 1964 Civil Rights Act, many American whites also have small amounts of black blood, with some estimates being that 20% of white Americans have some African racial heritage (Davis 1991). Regionally, the figure is far higher. Thus, the role of passing, in which light-skinned blacks moved to communities in which they were unknown and presented themselves as being white, contributed to the movement of African genetic traits into the white population. This can be seen in families where siblings were classified into different races, yet shared the same parents who were considered black. Thus, African cultural heritage may be more important than race or genetics in defining U.S. blacks.

8.1.1 Slavery and Spatial Distribution of African-Americans

The contemporary distribution of the African-American population (Fig. 8.1) reflects the heritage of slavery until the last third of the nineteenth century, the northward migration of blacks seeking industrial jobs during the twentieth century, and the legacy of racial discrimination and segregation that led to a racialized landscape in which blacks were concentrated into particular communities or neighborhoods. Blacks or African-Americans accounted for over half of the population in 106 counties nationwide in 2010, up from 96 in 2000, and all but one of these are a county that the U.S. Census Bureau considers to be located in the South, with the lone exception being St. Louis City, Missouri, not far from the South (McKinnon 2001; Rastogi et al. 2011). These counties are clustered along the Lower Mississippi Valley, particularly within the Delta Country of northwestern Mississippi, and extend across the Piedmont of Mississippi, Alabama, and Georgia into South Carolina, and then northwards into Tidewater Virginia and south-central Maryland. Five counties, three in Mississippi and two in Alabama, were over 80% black in 2010, as was one independent city in Virginia. Jefferson County, Mississippi, with 86.0% of its population of African-American ancestry, had the greatest proportion of its population that is black. The African-Americans across the South are the legacy of slavery in the plantation South and the large number of former black agricultural workers.

The largest numerical concentrations of African-Americans are outside the South. New York City had a black population of 2,228,145 in 2010, and Chicago had 913,009, down from 1,084,221 in 2000. These numbers represented 27.2% and 33.8% of their 2010 populations, respectively, smaller shares than in 2000. The cities with the third and fourth largest numbers of blacks, Philadelphia, Pennsylvania and Detroit, Michigan, had even greater dominance by blacks of their populations, 45.0% and 84.3%, respectively. These cities first began attracting large numbers of African-American workers to their factories during World War I when the stream of immigrants from Europe was cut off at about the same time that devastation from the boll weevil was destroying many of the Southern cotton fields where blacks had labored. This black migration north accelerated again during the Second World War, when job opportunities and less legislated racism lured another wave (Berlin 2010). The nation's capital also attracted large numbers of African-Americans, and the District of Columbia was 52.2% black in 2010, down from 61.3 in 2000. (It had reached 71.4% in 1970.) Several other southern cities also have large black populations, including New Orleans, Louisiana (61.2% in 2010, down from 67.9% in 2000, before the huge black exodus caused by Hurricane Katrina), Birmingham, Alabama (74.0%), Memphis, Tennessee (64.1%), Baltimore, Maryland (65.1%), and Atlanta, Georgia (55.2%).

8.1.2 History of Slavery

Slavery existed at times in all 13 British colonies as well as in French Louisiana and Spanish Florida, but it was outlawed the earliest in the North. Slavery began in 1619 in Virginia, when Africans first arrived as indentured servants and

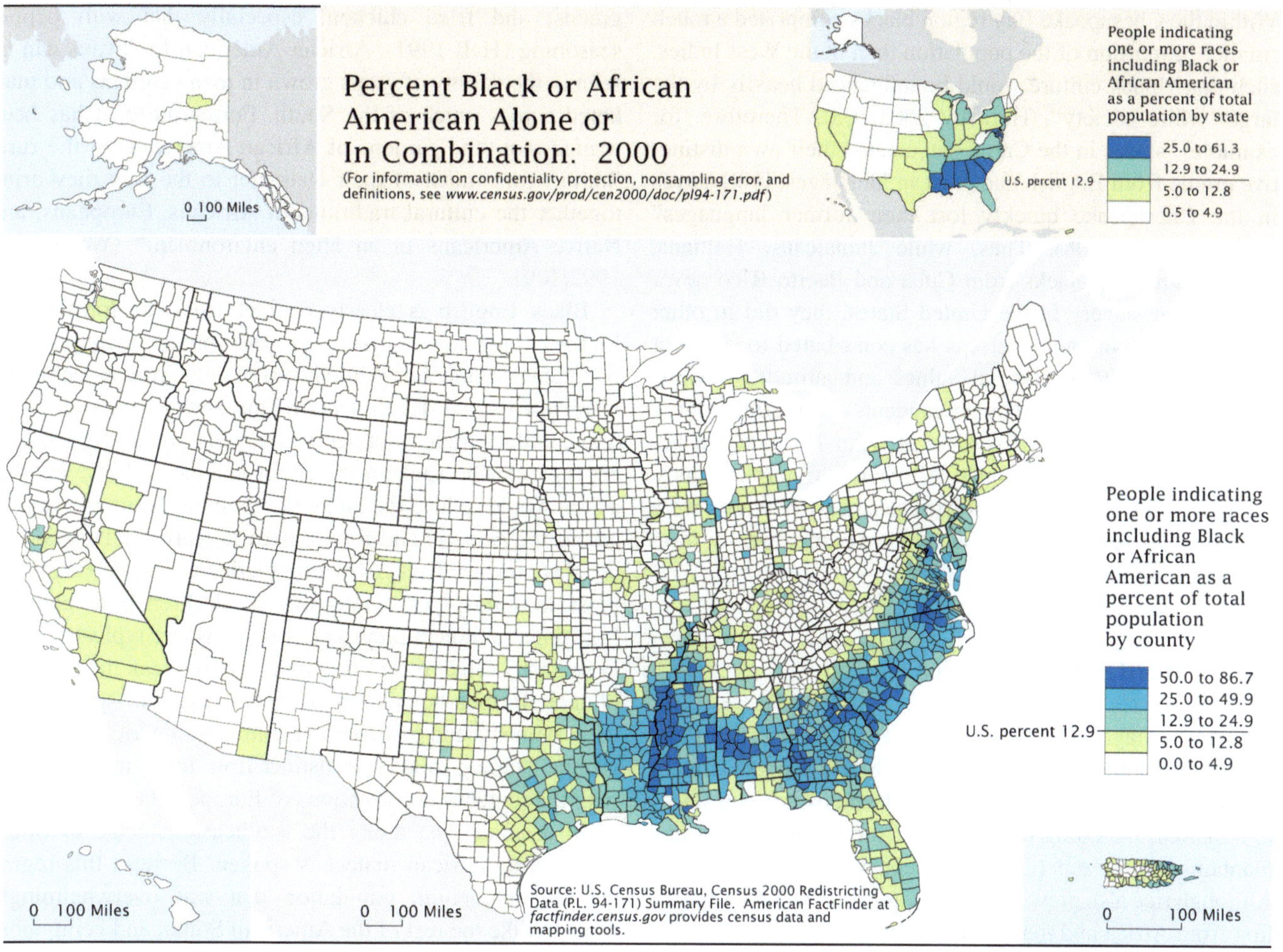

Fig. 8.1 Proportion of the population per county or parish that reported Black racial ancestry at time of the 2000 Census. Note that most of those counties with the largest share of their populations being Black are those areas of the South where slavery was widely practiced. Those scattered counties in the northern states with large percentages of their populations being Black attracted large numbers of industrial workers during the first half of the twentieth century (Reproduced from U.S. Census Bureau map in McKinnon (2001))

quickly lost their rights to freedom. Most slaves imported into the British colonies came from several coastal stretches between Senegal and the Bight of Biafra, plus Angola (Hornsby 2005), areas along the western and central coast of Africa. Yet, regionally in the South there were differences in the predominant African sources of their slaves (Berlin 2010). The importation of slaves was outlawed in 1819, yet the practice continued, supported by the breeding of slaves and smuggling. The system of slavery in the United States was only halted with the Emancipation Proclamation of 1863 and subsequent legislation at the conclusion of the Civil War. Of the millions of Africans shipped across the Atlantic Ocean as slaves, only about one in twenty wound up in the United States. Far more went to the West Indies or Brazil, where the life expectancy of a slave was 6 months. Even greater numbers of Africans were shipped as slaves to the Middle East (Sowell 1996). Unfortunately, slavery still exists in several parts of Africa.

Most whites in the South owned no slaves, and far fewer owned more than one or two. Only a small portion of the white population lived on plantations with large numbers of African slaves. Indeed, only one quarter of the white families in the South owned any slaves, and of those families who owned slaves, only 12% had 20 or more slaves (Aiken 1998). Yet because large plantations could own hundreds of slaves, many counties in the antebellum South had predominately black populations.

Not all African-Americans are descended from slaves, yet the vast majority are. There were even a few cases where African-Americans and Native Americans owned slaves. Most black immigrants from the West Indies have slave heritage from that region, an area where slaves and their descendants comprised a far greater proportion of the population than within the states of the U.S. South. Indeed, the landscape of slavery within the West Indies distinctly differed from that within North America because of the slave numbers.

Within the Chesapeake Bay region blacks comprised a much smaller proportion of the population than in the West Indies, such that "black culture would be influenced heavily by the larger white society" (Hornsby 2005:108). Therefore, for example, "slaves in the Caribbean created their own distinctive creole from English and African languages, [yet] slaves in the Chesapeake quickly lost their former languages" (Hornsby 2005:108). Thus, while Jamaicans, Haitians, Dominicans, and blacks from Cuba and Puerto Rico never experienced slavery in the United States, they did in other lands. This experience overseas has contributed to many of them having different social values and attitudes towards education than those held by descendents of American slaves (Sowell 1981). Some persons of African heritage are not descended from slaves, such as recent Somali refugees. Likewise, tens of thousands of Africans, many of whom were professionals or graduate students, have come directly from Africa during the past four decades.

8.2 Heritage from Africa

African heritage strongly influenced the foodways of the American South—not only the diet of blacks, but also that of whites. A variety of crops were introduced by African slaves to America, including okra, millets, sorghum, bananas and plantains, yams, red (Cayenne) pepper, peanuts, and rice. Although rice also grew in Asia, it was introduced to America first from Africa and many slaves were specifically obtained from rice growing areas of coastal West Africa for use in South Carolina and Georgia plantations (Hall 1991). Indeed, slaves were more knowledgeable about rice cultivation than their masters, and they worked in the malarial infested fields that the planters who chose to live in Charleston and Savannah avoided. While peanuts were consumed in a variety of forms, including peanut soup during antebellum days, later research by George Washington Carver, an African-American food scientist, revolutionized the uses for peanuts with his creation of peanut butter. Africans also influenced livestock raising. Slaves from Senegal and Gambia in West Africa brought knowledge about saddle horns, lassos, and open-range cattle herding and influenced the open-range cattle herding that later diffused north from New Spain into the Great Plains (Sluyter 2014).

The culinary skills of African-Americans, who not only fed their families but also cooked the meals for their owners during slavery and for white employers for a century and half thereafter, resulted in soul food. Indeed, African-inspired foods and dishes became an intimate part of Southern foodways and strongly influenced Southern white cooking. Certain foods clearly reflect African heritage, including gumbo, a soup or stew incorporating okra and chicken or seafood; stewed greens, particularly collard or mustard greens; and fried chicken, especially that with pepper seasoning (Hall 1991). African-American foodways can be seen in the variety of crops grown in many gardens and marketed across much of the South. Furthermore, it has been argued that "the gardens of African Americans in the rural South…owe much of their character to the way they bring together the cultural traditions of Africans, Europeans, and Native Americans in an alien environment" (Westmacott 1992:109).

Black English is clearly distinct from the form of the English language spoken by the predominately white American population. While much of their African languages were lost by the descendents of the slaves, some elements remain. Slave traders deliberately mixed Africans from different regions who spoke a variety of languages, believing that action would reduce the threat of rebellion during the voyage across the Atlantic and force slaves to learn the language of the slave owner. Thus, African-Americans learned English in most locations, yet a few still speak French in parts of Louisiana (although they, and the Cajuns, are typically fluent in English), while recent immigrants may speak the language of their former homelands. African-American speech contains elements and construction common to African language groups, helping to keep it distinct from those more familiar with the sentence constructions of European languages.

In a small area along the southeastern coast of South Carolina an African dialect is spoken. Because this region had an antebellum population that was overwhelmingly slave, unlike the rest of the American South, and being more akin to the West Indies, a creole language developed that is "an amalgam, drawing on vocabulary and grammar from several West African languages, pidgin used along the coast of West Africa and onboard English slaving ships, and Standard English" (Hornsby 2005:121). Gullah is the only remaining distinct language of the African-Americans, and it displays the influence of Bantu words in both its vocabulary and in African place names in South Carolina, Georgia, and Florida (Pollitzer 2005).

There has long been a lack of appreciation of the contribution of African-Americans in naming places. Lauret Savoy refutes the earlier work of George Stewart who wrote "the oppressed African left little mark upon the map," explaining that "as diaspora linguists like Annette Kashif have observed, languages and naming patterns from Africa crossed the Atlantic, too" (Savoy 2015:82–83). African inspired toponyms can be found in the tidewater region of the Southeast, and Savoy (2015:83) notes that there are names "with at least partial roots in Africa," which were long attributed to Native American sources, citing Suwannee as an example. Yet there are possible African sources for the word, and early residents two centuries ago along the river at Suwannee Old Town comprised a maroon community of Seminoles and fugitive slaves (Savoy 2015).

In contrast to Indian place names, "those of American blacks often conveyed a human situation, such as hardships in slavery, just as place-names in the Congo commemorate significant human experience, emotion, or action" (Pollitzer 2005:122–123). Even though enslaved, the "sea-island black was most fertile at naming everything, even the narrow salt-water drains interlacing the marshes.... Pindar Town preserved the Kongo word mpinda for peanut, but white men probably did the naming after the word had become current in local speech" (Pollitzer 2005:123). Ninety-three place names in South Carolina appear to have Bantu roots, as do 17 in Georgia and 24 in Florida (Pollitzer 2005). Other African inspired toponyms have been identified in Mississippi and North Carolina, and "hybrid words admixing tongues from two or more continents may be more common than once thought" (Savoy 2015:84).

8.2.1 African-American Contributions to the Antebellum South

African-Americans, both as slaves and as free persons, have made major contributions to the built environment and the patterns of settlement in the American South. "Indeed, the complexities and opportunities posed by their presence were major determinants of southern regional architecture" (Anthony 1976a:9). African-Americans not only built their own homes, dug the ditches that drained marshes that were then planted with rice or indigo, and cleared the fields that they cultivated, but they labored to construct many of the structures that were occupied by European-Americans.

Cultural baggage was brought to the South by both whites and blacks, and thus much of what is seen today displays elements of both. For example, blacks from Africa brought knowledge regarding "climatic adaption suitable for the rain-forested tropics, extended family compound arrangements, and ways of building in mud and thatch" (Anthony 1976a:9). Because slaves built both their own housing and dwellings for white plantation owners, their heritage is shown in both.

8.2.2 Plantations and Their Buildings

Plantation mansions and outbuildings were modeled after European architectural styles and plans, yet these structures were constructed by black masons and carpenters. While the floor plan and major construction design reflect European custom, African architectural elements were sometimes incorporated, particularly in decorative arts such as wrought iron grillwork or wood finishings. Within New Orleans the ironwork is best seen in the French Quarter and it is "attributed to black artisans and French and Spanish influences" (Campanella 2006:135).

Many outbuildings on plantations displayed African traits in their proportion, siting, and construction (Anthony 1976b). For example, within Williamsburg, Virginia "groups of outbuildings with their modest dimensions and pyramidal roofs create the visual effect of a piece of an African village with its multiplicity of dwelling units and granaries" (Anthony 1976b:11). Slaves and, in the case of Louisiana, free black contractors not only built, but they often designed outbuildings, blending European ideas about building materials with African concepts.

The skills of African-American slaves in constructing plantation houses, and how European and African architectural knowledge was amalgamated, can be seen within parts of the West African country of Liberia today. Established as a refuge for freed American slaves, some of these former slaves replicated the designs of Southern plantation houses in the structures they constructed in Africa. Some of these buildings remained visible on the landscape a century and half later (Belcher 1988).

8.2.2.1 Slave Communities

Because plantations had large numbers of slaves, they represented a sizable black community. The location of slave houses was nucleated, with rows of slave cabins, whose siting varied between large and small slaveholders (Fig. 8.2). Where arranged by the landlord, slave quarters were often in regular rows, while slave quarters were more irregularly spaced where plantation operators exerted less supervision over the slave villages. Yet, there was a "slave preference for a landscape marked by few overt boundaries and fixed sites, an environment open to and characterized by movement" (Vlach 1993:14).

Small slaveholders typically clustered slave homes near their owner's residence, to its side or rear (Aiken 1985). In contrast, large slaveholders arranged slave cabins in distinct patterns, often erecting "numerous cabins on 'streets' to form what appeared to be villages and small towns, [with the] most common form [being] a single street with houses on both sides" (Aiken 1985:386). These hamlets of slave quarters often accommodated 50–150 residents and were located near the overseer's house, enabling their location away from the mansion of the plantation owner. Typically, as a plantation increased in size, "when the population increased to more than 100, the slaves were divided between the old landholding and a newly created plantation" (Aiken 1985:386). For example, George Washington's Mt. Vernon plantation split its slaves into 4 units, with each group working in a separate part of his holdings.

Fig. 8.2 Ruins of slave houses constructed of tabby at Kingsley Plantation around the beginning of the 1800s. These slave quarters, built in a semi-circular pattern, are preserved in Timucuan Ecological and Historical National Preserve in Florida

8.2.2.2 Slave Housing

The type of slave housing built depended upon the plantation owner. Differences have been noted based upon the ethnic heritage of their owners, especially whether French or British (Fig. 8.3). Likewise, the financial investment of the owner was important, as sometimes little expenditure was made. Some slave housing strongly reflected native African housing styles, such as the use of wickerwork walls of woven branches plastered with mud (Fig. 8.4). Slave huts of this wattle and daub construction were commonly built in Georgia, South Carolina, and Louisiana. Furthermore, within South Carolina, the status of early African slaves "as a separate population was further emphasized by the fact that often their dwellings were actually African-style dwellings, small rectangular huts constructed with mud walls and thatched roofs" (Vlach 1993:155). Nevertheless, slave quarters, when considering the South as a whole, were far more commonly log cabins.

Typical British American slave housing consisted of single or double-pen (saddlebag) structure (See Chap. 9). The single-pen structure, consisting of one room typically 16 by 18 feet, was occupied by one family. Double-pen structures were occupied by two families. Most of the British colonial slave quarters were wood frame structures, with clapboards on a timber frame. Nevertheless many others were log cabins, with the space between the logs either chinked or unchinked. Although not as common, other slave quarters on the Atlantic seaboard were of brick, tabby (a crude poured concrete), or coquina (stone formed of compacted shells). It should be remembered that very little slave housing still exists in contrast to the many plantation mansions that have been preserved. Slave quarters were removed and many fell into ruin shortly after emancipation. Indeed, as John Vlach (1993:156) explains, "many of these structures were falling apart even while slaves were living in them." Thus, most of the slave housing was gone by the beginning of the twentieth century. As Dell Upton (1988:359) reminds us, "the surviving structures are misleading." Furthermore, only small elements of black settlement from this period remain visible on the landscape, such as slave burial grounds and sites of slave housing (Aiken 1985).

Within Antebellum South there was a huge social and economic chasm between plantation owners and poor white farmers. Thus, as John Vlach explains, "cabins of slaves and white yeoman farmers found first along the southeastern Atlantic coast and later in the frontier back country were so similar that it is often difficult to assign cultural origins" (Vlach 1986:43). A variety of other buildings designed to accommodate the work activities of the African-American slaves, or to enhance their welfare, were also constructed on many plantations. These include communal kitchens to prepare the meals of the slaves, smokehouses to preserve the pork that fed both the slaves and their masters, hospitals—which in many, but not all, cases were simply a cabin set aside to accommodate ill slaves and those giving birth, and even chapels for the religious instruction of the slaves (Vlach 1993).

Fig. 8.3 Brick slave quarters at Magnolia Plantation, located south of Natchitoches, Louisiana. These buildings, unusual in their permanency of construction, are preserved in part of Cane River Creole National Historical Park in northwestern Louisiana, which preserves an ethnic landscape shaped by the French, Spanish, and Africans

Fig. 8.4 Reconstruction of mud-chinked log slave quarters at Carter's Grove Plantation, near Williamsburg, Virginia

8.2.3 Differing Black and White Antebellum Landscapes

Blacks and whites had different perspectives on the landscape during the antebellum period. Furthermore, it has also been argued that many historic sites featuring preserved plantation buildings and antebellum mansions are often presented from the planter's perspective rather than from the slave's, or the African-American contributions are simply ignored (Dennis 2006; Hoelscher 2006). Even though both whites and black slaves lived in same area, they interacted differently and moved in dissimilar ways in that environment, and they viewed the landscape from different perspectives. Dell Upton (1988:364) notes that while "the planter's landscape offered the image of an orderly society that focused on himself and linked him to his peers, the slave's landscape took a different form… the slaves' responses to the planter's landscape." Utilizing Thomas Jefferson's Monticello as an example, he explains how the plantation owner's views of the landscape, "to see and not be seen," were the antithesis to the landscape envisioned by the slaves. Thus, Upton writes that the "slaves' landscape was described from the point of view of someone surrounded by other people's power, and its landmarks were plantation houses and fields differentiated by ownership" (Upton 1988:366). Another scholar similarly notes the existence of parallel landscapes in the antebellum South, so that "within the official, ordered landscapes established by planters, there was another system of definitions developed by slaves" (Vlach 1993:x).

While plantations were overwhelming black communities when considering their total populations, much of what remains today is the landscape ordered by their white owners. Yet, if one looks carefully, elements of the black oriented landscape can be identified on some plantations a century and half following emancipation. For example, while much of the African-American heritage on plantations in South Carolina's low-country has physically disappeared, with slave quarters and the structures in which they toiled gone, the canals dug across the former rice fields remain. While barely visible at best from the ground, they are conspicuous from the air. While Samuel Dennis (2006:91) is correct in lamenting that "low-county rice plantations will remain racial projects that materialize the ideology of white supremacy, through the discourse of planter paternalism, in the visible landscape," it must be remembered that without African-American slave labor there would have been no plantation, no mansion house, nor fields alternately watered and drained by those canals that do remain to this day.

8.2.4 African-American Slaves in the Antebellum Southern City

African-American populations were scattered throughout the cities of the South before the Civil War, yet their areas of movement were different from that of the white population. Blacks generally lived in several locations: black districts on out of way streets, which were especially common among the free blacks; shantytowns near a city's edge, which grew to become the largest of post-emancipation black settlements; and back-alley sites. As Kellogg (1977: 311) explains, "A prestigious avenue with many whites residing on it would be backed by a row of servants' quarters on a parallel alley." This interspersal of blacks in white residential areas facilitated the employment of black domestics who lived near the houses where they worked (Fig. 8.5).

Slave quarters for urban domestics were often in long narrow buildings behind their owner's main house, which was set on the street (Figs. 8.6 and 8.7). Thus, the narrow side of the black residence faced the street, such as can be seen today in Charleston and New Orleans, among other prominent southern cities. In addition, high masonry walls surrounded the lot, "and access to the street was through the main house or a side door" (Anthony 1976a:18). Furthermore, although some writers have noted servants' quarters facing alleys, they were often "omitted, because they might become gathering places for the slaves" (Anthony 1976a:18). Thus, alleys are only common in southern border cities, such as Richmond, Virginia, very unlike the prominence of alleys in Pennsylvania.

8.2.5 African-American Freemen in the Urban Landscape

While the vast majority of blacks in the Antebellum South toiled as slaves in the fields or in their masters' houses, there were scattered populations of free blacks, many of whom were of mixed racial ancestry. New Orleans, Louisiana, was particularly noteworthy for the size of its free black population and for their significant impact in shaping the urban landscape (Oszuscik 1992). The slave rebellion and revolution during the late 1790s in the French colony of St. Domingue, which became the independent country of Haiti in 1804, resulted in thousands of French whites and mixed race creoles fleeing the island. While many who fled also brought slaves with them, thousands of free blacks who also came settled primarily in Louisiana and the southern areas of Alabama and Mississippi. Thus, at the time of the 1810 Census, when New Orleans was 37% white and 63% black,

Fig. 8.5 Servants' quarters constructed behind Latimer House (erected in 1852) in Wilmington, North Carolina. Slaves were often housed near their workplace, with whites living in the opulent mansions and blacks living in structures placed to the rear of the properties

Fig. 8.6 Service building and courtyard wall along Dumaine Street in New Orleans, Louisiana. Slaves assigned to domestic duties were once housed in the narrow service building within the French Quarter of New Orleans, Louisiana

Fig. 8.7 Slave quarters in service building located behind French Creole cottage along Burgundy Street in New Orleans' French Quarter

43% of all blacks in New Orleans were free persons (Vlach 1976). These free blacks were very important in the building trades as carpenters and masons. Not only did the free blacks build and buy houses, but several of the prominent contractors employed by whites were free blacks.

Within New Orleans several historic neighborhoods have long been associated with African-American culture. For example, the suburb of Fauberg Tremé, located immediately inland from the French Quarter, has long been associated with its African-American residents. While established two centuries ago as a Creole enclave, for whites, mixed race free persons of color, and slaves, by the early twentieth century Tremé was almost entirely a black neighborhood, reflecting both its African and French Creole heritage. While that heritage is displayed in the architecture of the houses, churches, fraternal halls, and shops that were constructed, it was also shown by the music, food, parades, and street culture (Crutcher 2006). While urban renewal, freeway construction, and gentrification threatened the neighborhood, its greatest threat came with its inundation during Hurricane Katrina.

Among the prominent and distinctive decorative features not only in New Orleans's French Quarter, but also seen in Mobile, Alabama, are the elaborate wrought iron balconies (Fig. 8.8). This iron work was done largely by blacks, who dominated such work during the early 1800s. The decorative motifs displayed are very similar to wrought iron standards in the West African nation of Dahomey (Anthony 1976b:12).

8.2.6 The Shotgun House

The shotgun house is by far the most distinctive African-American house form. While the first shotgun houses in the United States were constructed at the beginning of the nineteenth century—over a half century before emancipation, thousands were constructed during the twentieth century, and houses influenced by this style continue to be built. An estimated 60,000 examples were built in greater New Orleans, not just in the French Quarter, but also throughout the metropolitan area (Fig. 8.9). Thousands of additional examples were constructed elsewhere in Louisiana and throughout much of the Mississippi Valley (Fig. 8.10). The shotgun house is one room wide, three or more rooms deep, with the doors all aligned, front to rear. The primary entrance to the single story shotgun house is at its gable end, which typically has a front porch with a sloping roof. While some viewers have commented that the house was named because its shape resembled a shotgun, or because if a shotgun was discharged through an open front door, its load would pass through each of the interior doors given their placement, the word is actually of African origin. Indeed, the West African word "to-gun" translates as "house" (Vlach 1986).

The style of the shotgun house was developed in and introduced from Haiti. There it reflected a blend of Yoruban (African) houses with Arawak structures as adapted by French buccaneers (Vlach 1976). Large numbers of shotgun houses remain in both Port-au-Prince and Gonaïves, Haiti.

Fig. 8.8 Intricate wrought iron grillwork displayed in many of the balconies of New Orleans's French Quarter is typically the handicraft of African-American blacksmiths, although its usage is most common within French and Spanish communities with ties to the Caribbean

Fig. 8.9 Shotgun house in New Orleans, Louisiana. Note that both the front door and front window are louvered and full length

Fig. 8.10 Shotgun houses in Louisville, Kentucky. Note that the width of the shotgun house approximates the length of the car

Indeed, in noting that "architectural links between Port-au-Prince and New Orleans cannot be denied," John Vlach (1976:60) explains that "the nonessential details that are associated with the shotgun in Haiti are also associated with the shotgun in Louisiana." Within New Orleans the oldest shotgun houses, near the northernmost corner of the French Quarter are found in an area once inhabited by mixed race Creole women from Haiti (Campanella 2006). In Louisiana, the shotgun house style continued evolving. A double front "door" style is common in both Haiti and New Orleans. The openings are full-length and shielded by louvered shutters, such that "the pattern of two tall, evenly spaced apertures cannot be missed" (Vlach 1976:57). The oldest shotgun houses had hip roofs, yet others are gabled. The front of the house is often embellished; and it may be covered by brick, whitewashed, or have wider clapboards than the sides.

Numerous variations of shotgun houses exist. The double shotgun is a duplex, side by side with a shared roof, whose peak demarked the separate quarters. The camelback shotgun has a second story over a rear room (Fig. 8.11), while the lateral wing shotgun has a side room by the rear room and a porch along front and side. Not only are there variations in this major subtype of the shotgun houses, including those without galleries to those with one, two, or three, but examples exist within Haiti, Louisiana, and elsewhere along the Gulf Coast (Oszuscik 1992). A few shotgun houses had a narrow side gallery providing access to rear rooms that were rented to lodgers.

The popularity of shotgun houses persisted long after the institution of slavery was abolished, and this style is not just associated with black residents. Shotgun houses became very popular in the Southern states. Until the mid twentieth century, it was the most common style of house built in Louisiana (Fig. 8.12). George Carney (1983:59) notes that in parts "of rural Louisiana, it is the only house type found…. Shotguns proliferated in river towns, including New Orleans, Vicksburg, Memphis, St. Louis and Louisville." Indeed, it has been estimated that there are 11,000 shotgun houses in Louisville, Kentucky alone. Additional shotgun houses are found in the Wabash Valley of southern Illinois and Indiana. The author has seen shotgun houses along the Ohio River in Ohio. In Oklahoma shotgun houses became common in oil boom towns, where prefabricated shotgun houses could also be moved by railroad flatcars. Shotgun houses were cheap, being easily and rapidly constructed. Simple in design, their construction required few tools, "usually only an axe and a few hammers," and the use of "[b]lueprints or drawings were unnecessary" (Carney 1983:62). Shotgun houses are widespread in certain neighborhoods along the Gulf of Mexico, ranging from Houston to Key West, as well as in the Mississippi Valley.

High houses are two story versions of shotgun houses (Fig. 8.13). They have a two-tiered front gallery with a gabled front, and were typically of wood frame construction.

Fig. 8.11 Double camelback shotgun house with a second floor room at its rear. This New Orleans, Louisiana duplex consists of two side by side residences, sharing a single roof, extending straight back from the front door. Note the arrangement of the windows with the louvered shutters opened

Fig. 8.12 Shotgun houses predominate within some African-American neighborhoods of Donaldsonville, Louisiana

Fig. 8.13 High houses in New Orleans, Louisiana

While they are much more common in Haiti than in the U.S., they can be found along the Gulf Coast from Florida to Texas. Architectural historians view the front porches, such as the gallery provides, as being of African heritage. As Vlach explains, "for almost 250 years the southern front porch has owed its existence mainly to the adaptive genius of local carpenters acting on African notions of good architectural form" (Vlach 1986:45).

8.3 Post-Emancipation African-American Landscapes

Following their emancipation from slavery, the African-American ethnic landscape began changing, rapidly in places. Upon emancipation many former African slaves made deliberate efforts to alter their landscape. The resulting "post-emancipation African-American community landscape was composed of the buildings that held importance to the freemen" (Wallace 2007:10). These buildings included the church and school, along with a scattering of houses. While no longer slaves, racial discrimination and segregation played a profound role in constraining how and where blacks altered the landscape. Landowners reorganized their former plantations, and a new system of sharecropping evolved. As sharecroppers, blacks farmed small plots of land and shared the crop with the landlord, rather than either earning wages or paying rent. They often built a new house on this farm plot, not wanting to live in their former slave quarters. In an 1877 account by a South Carolinian, it was explained that, "the negroes used to be kept…in cabins clustered together near the residence of their owner or overseer, [but] since their liberation they have shown quite a tendency to desert the relics of their former subjection" (quoted in Aiken 1985:388). Later, during the middle of the twentieth century, African-Americans likewise abandoned their sharecropper cabins.

Thus, the former nucleated slave communities disappeared. As Charles Aiken nicely explains,

> Seventy years after slaves have been freed, almost all traces of the nucleated settlement pattern of slavery had disappeared except in the Louisiana sugar plantation region. Although many of the exslaves created small hamlets at the back of the large estate-farms in the Kentucky Inner Bluegrass, throughout the cotton and tobacco plantation regions of the lowland South most rural blacks lived in isolated dwellings scattered across the landscape (Aiken 1985:387, Reprinted with permission of the American Geographical Society).

Yet, a half century following the Civil War, America's black population remained concentrated in the South. The 1910 Census found that 89% of nation's black population was in the South, and 79% of all Southern blacks lived in rural areas. Within many of these areas, African-Americans remain the majority population today.

Reminders on the landscape of the post-emancipation settlements include the abandoned general stores that once served sharecroppers, as most of the former sharecropper

Fig. 8.14 Greater New Hope Missionary Baptist Church located a mile from the hamlet of Bland in rural section of Alachua County, Florida. The church and its Sunday school building differ little from those of white congregations, but the church's name is key to the identity of its African-American congregation

shacks had disappeared even three decades ago. Thus, the most prominent reminders are those African-American churches (Aiken 1985). Rural Black churches were erected shortly following the Civil War, and were usually Baptist or Methodist (at that time African Methodist Episcopal, which is now merged with the United Methodist). The names of many of these churches spoke of freedom, such as Liberty, Greater Liberty, Fredonia, and New Caanan, among others (Fig. 8.14). Schools were often built associated with these rural black churches. Footpaths once connected the homes, church, and store, and these footpaths were more important than roads. These paths did not follow a grid, being more irregular, winding paths, than many roads. Today these paths have been converted to roads or have been abandoned, yet many rural black churches remain.

Sharecroppers' cabins have now largely disappeared from the Southern landscape, just as the number of black farmers has precipitously declined. A century ago most rural black families were tenant farmers, who worked lands for the landowner on a crop share basis. While some of the cabins were originally slave quarters, many were later replaced or covered with clapboards during the early 1900s. Sharecropper cabins were typically single-pen or saddlebags (see Chap. 9), and had a small porch along their front. Philippe Oszuscik (1992:176) notes that "most rural houses of freedmen, ex-slaves, and subsequent generations ...continued to be smaller, plainer, and to have fewer windows than the rural houses of Whites, and some later structures even had dirt floors (an African tradition)." Today, even the brick or stone chimneys of many abandoned sharecropper cabins have disappeared from the landscape.

8.4 African-American Landscape Evolution in the 20th Century

The African-American landscape underwent tremendous change during the twentieth century. At the beginning of the century, most African-Americans lived in the rural South, while by the end of the century several major cities in the North had become overwhelmingly black. Segregation was legally mandated throughout most of the nation as the century began, but with the Civil Rights Act of 1964, blacks who had sufficient financial resources could live within integrated neighborhoods by the end of the century. Nevertheless, major economic disparities among racial groups remained. During the twentieth century huge black ghettos developed in certain large metropolitan areas. While a great migration of six million blacks to the North marked much of the century (Berlin 2010), by the end of the century more African-Americans were returning to the South than leaving it (McHugh 1987). Change in the spatial distribution of African-Americans and how they shaped their landscape took place in both the rural South and in the nation's largest

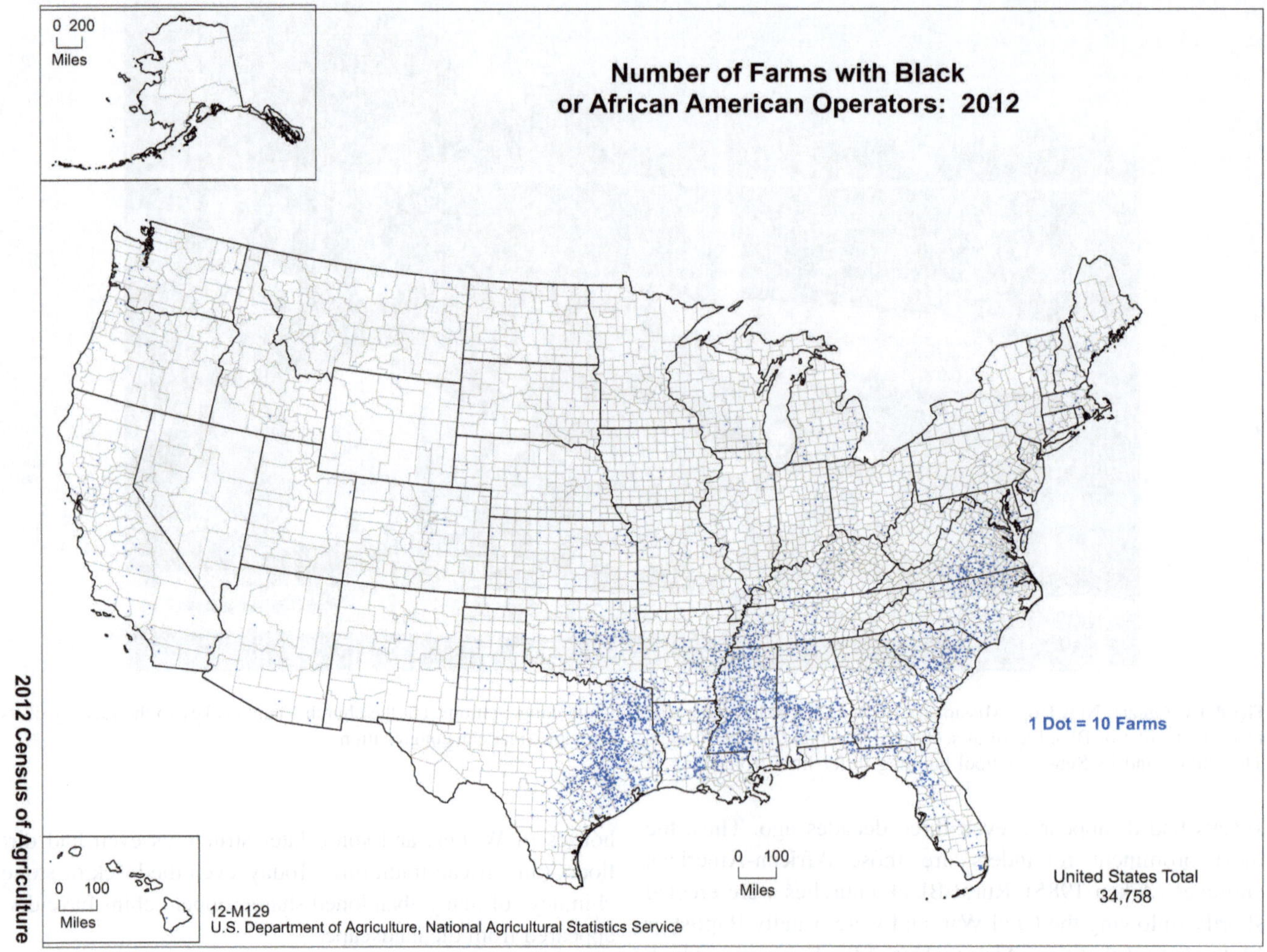

Fig. 8.15 African-American farmers are mostly found within the heart of the Old South, although a far smaller proportion of the region's farms are operated by Blacks today than a half century ago. Nevertheless, Blacks work on a far larger proportion of the region's farms than they either own or manage (Reproduced from the *2012 Agricultural Atlas* published as part of the 2012 Census of Agriculture NASS (2014b))

cities. Significant black populations remain in many rural districts, hamlets and cities of the South where slavery was practiced a century and half ago, yet other rural areas where many slaves were employed—such as parts of western Maryland (Wallace 2007)—had lost almost their entire Black population by the twenty-first century.

8.4.1 Rural African-Americans in the South

Few African-Americans live in rural areas outside of the South, yet as we saw earlier, they dominate the population throughout the Black Belt running from eastern Arkansas, across central Mississippi, Alabama, and Georgia, into the Carolinas. Blacks have abandoned farming in much of the South, with fewer than 3% of rural blacks living on farms. By the time of the 2012 Census of Agriculture only 34,758 African-American operated farms were enumerated nationally, yet only 7846 of these farmers produced more than $10,000 of crops or livestock annually, in contrast with 907,970 farms that were operated by whites. Yet of the nation's remaining African-American farmers, the vast majority are located in the South (Fig. 8.15). In Mississippi, the state which has the highest proportion of African-American farmers, blacks are outnumbered by over seven to one by white farmers. Only 1238 of that state's black farmers sold more than $10,000 in produce in 2012. In Alabama, African-Americans accounted for only 4.6% of all farmers selling more than $10,000 of crops annually. While Texas had the greatest number of black farmers (8551), their total sales were 0.5% of those of white farmers and also less than sales by both American Indians and Asians (NASS 2014a).

8.4.1.1 Rural African-American Communities

The loss of black farm operators does not mean that African-Americans have left the rural countryside. Many have found jobs as agricultural workers on white owned farms or work in nearby communities. Beginning around 1975 the rate of African-Americans returning to the South started exceeding those moving North, and this return migration has been both to the cities and rural countryside. Many African-Americans coming back to the rural South have chosen to locate near their former home place or near relatives (Cromartie and Stack 1989). There has also been movement from dispersed tenant farms to both cities and rural hamlets. Rural towns of fewer than 2500 residents and hamlets are both growing, resulting in both unplanned assemblages of dwellings and planned subdivisions.

While modern rural housing occupied by African-Americans is generally indistinguishable from dwellings lived in by other populations, many African-Americans occupy older structures that are essentially rural versions of shotgun houses or are patterned after single or double pen cabins. Although many rural dwellings are surrounded by a yard of cut grass, it has been observed that African-Americans are far more likely to have a yard swept clean of debris and grass than are their Caucasian neighbors. Sweeping the yard with a brush broom was historically not only common among many African-Americans—and it continues to be seen in areas of the rural South, but it perpetuates a style commonly seen in West African communities (Westmacott 1992).

Contemporary rural African-American hamlets consisting of unplanned assemblages of dwellings are typically associated with an extended family (Aiken 1985), and have been for several decades. The black owner of a farm or small tract of rural land allocates lots for the homes of various family members. Such homes often vary from well-kept modern houses to mixtures of various qualities. While these dwellings are usually located near the road, some of the houses may be behind that of another family member.

Planned rural subdivisions and hamlets, where the land is platted and sold with the idea of bringing profit to the developer, are also important. These housing developments have been created by both black and white developers with the African-American homeowner in mind. Some are sponsored by the Farmers Home Administration, especially since the passage of the 1960s civil rights laws. They are typically located in or at the margin of existing communities, yet a few are freestanding hamlets (Aiken 1985). These developments result in modern well-constructed houses in the countryside that are architecturally indistinguishable from new homes of nearby Whites. Nevertheless, much substandard African-American housing remains a half century after the passage of the civil rights acts.

8.4.1.2 Black Ghetto Towns

Within parts of the South, such as in the Delta Country of northwestern Mississippi, African-American expansion within the towns and small cities resulted in what Aiken (1990) described as the establishment of rural black ghetto towns. There are many non-metropolitan municipalities where the white population has decreased substantially and the black population has increased. For example, in Yazoo Delta County, Mississippi, since 1950 the proportion of communities that are over 50% and 75% black rapidly increased, with the white neighborhoods declining. Since the middle of the twentieth century movement of both blacks and whites "in the Yazoo Delta has resulted not only in concentration of blacks in particular municipalities but also in increased segregation among the region's municipalities" (Aiken 1998: 328–9). As the black population grew, white exodus occurred from transitional neighborhoods.

As the African-American population increased, it moved into both new neighborhoods of federally sponsored housing and into the margins of the white neighborhoods. Charles Aiken described the initial response of the residents of these white neighborhoods of predominately black rural communities who wished to maintain their white enclaves, where residential density was lower than within the black neighborhoods. Given that fewer whites desired to purchase residential properties than those which were on the market, houses might become vacant, "but they are not overtly 'for sale,' with signs in the yards and advertisements in newspapers" (Aiken (1990:236). Sometimes white residents would buy such vacant houses which were then "razed to protect the core from incursion by blacks" (Aiken (1990:236). The transitional area around this white core grew to have a mixed race composition, as "buying houses to prevent blacks from doing so eventually becomes an economic burden" (Aiken 1998: 333). At the time Aiken studied the Yazoo Delta municipalities in the 1980s and 1990s African-Americans comprised the majority of residents in most of them, yet sizable white neighborhoods remained. Overall, the African-American majority has grown since then, with few whites remaining. For example, Friars Point, Mississippi was 94.7% black in 2010, Shelby, Mississippi was 95.0% African-American, and Tchula, Mississippi was 97.6% black.

Within these ghetto towns the black residential areas have often been renovated with federal housing funds, new residential areas developed, and public housing built. While there have been improvements in the infrastructure, such as sewers, water, and roads paved, the public schools remain highly segregated, as the few white families remaining usually send their children to private academies. As these small cities become increasingly black and primarily inhabited by poor residents, their downtown retail districts decline, the number of stores decreases, and there is a "restructuring from one partly oriented to relatively affluent whites to one that almost exclusively serves low-income blacks" (Aiken 1998:334). For example, white customers have abandoned the historic downtown business district of Helena, Arkansas, and those businesses that remain open are clearly oriented to a poor African-American clientele, seen by the nature of wares and services offered and the advertising signs. Even more striking is the virtual abandonment of the historic downtown district of Cairo, Illinois. Not only are many blocks bereft of buildings, but during the summer of 2006 the author encountered the debris of one collapsed business building blocking part of the city's main street. Typically, these abandonments were not caused by African-Americans, but by white business owners who disinvested in their communities.

8.5 African-American Urban Landscapes in the South

African-Americans have long contributed to the growth of urban areas in the South, and with their decline in employment on farms, migration to cities in both the North and South ensued. Those blacks who moved to Southern cities continued to expand upon the residential segregation that had existed before emancipation. Newcomers settled into black neighborhoods, which had distinct boundaries with white neighborhoods, separated by such features as a railroad line, a river, or topographic features. Nevertheless, black and white districts were often in close proximity. David Lee clarifies, "As a black district was surrounded by white housing, a new black area, often with substandard housing, arose elsewhere" (Lee 1992:376). Thus, within many Southern cities blacks were found within a number of neighborhoods, separated by white neighborhoods. Yet, the process of white to black residential succession has also occurred in the urban South, starting several decades after it had begun in Northern cities (Tettey-Fio 2016).

It was within several Southern cities that a vibrant and increasingly politically active black middle class of business owners and professionals developed African-American oriented business districts and neighborhoods by the mid twentieth century (Fig. 8.16). Among the best know are Auburn Avenue in Atlanta, Georgia and Beale Street in Memphis, Tennessee. These concentrations of "black churches, the African American press, and black-owned barbershops and beauty parlors" (Inwood 2011:148) and numerous businesses provided what are termed black counterpublic spaces. As Inwood (2011:148) explains, "[t]he most vital aspect of these public spaces was the belief among African Americans

Fig. 8.16 Auburn Avenue, located just east of downtown Atlanta, Georgia, was the center of a vibrant African-American business district during much of the twentieth century before the passage of the Civil Rights Act. Big Bethel African Methodist Church (built 1891, rebuilt 1922) is one of three prominent African-American churches along the core of the Auburn Avenue. Churches have historically played very prominent roles within African-American communities

that they are exclusively in the presence of other black people," places where African-Americans could gather to discuss issues relevant to their community and plan actions that led to their obtaining civil rights. Thus, "African Americans have turned the landscapes created to segregate them into counterpublic spaces, places where black identities and cultural traditions flourished separate from and in opposition to the white-dominated public sphere" (Alderman and Modlin 2014:279). It can be argued that today's black majority colleges and universities, such as Morehouse College and Spelman College in Atlanta, also created, and in many aspects, perpetuate such spaces.

8.5.1 Atlanta's Auburn Avenue

Auburn Avenue during the first two-thirds of the twentieth century was not only the wealthiest African-American community in the nation, with "three of the most important African American-owned financial institutions in the U.S. South" (Inwood 2011:149), but also home of several elite social clubs and fraternal organizations (Fig. 8.17). It was also home of the Southern Christian Leadership Conference and several prominent churches, including Ebenezer Baptist Church, which Reverend Martin Luther King, Jr. served as co-pastor before his assassination in 1968. None of the prominent churches along Auburn Avenue, including also Wheat Street Baptist Church and Big Bethel African Methodist Episcopal Church, display architectural traits that conspicuously suggest any African ethnicity. Indeed Ebenezer Baptist Church displays a simplified Late Gothic Revival design (Fig. 8.18). It is described as having "the usual apparatus of a Baptist church—central lectern on a dais with seating for the choir and, at the very rear, a baptistry, with an illuminated cross above" (Williams 1997:123). These churches played a disproportionate role in shaping the community's cultural landscape, and their pastors took leadership roles in both the community and the civil rights movement. The original Ebenezer Baptist Church building is now part of the Martin Luther King, Jr. National Historic Site. Indeed, Auburn Avenue was situated "as a crossroads for African American culture, politics, and commerce during the time of segregation.... Auburn Avenue was a symbol that black people could be successful and govern themselves" (Inwood 2011:153).

Today African-American businesses in Atlanta are much less focused upon Auburn Avenue, yet one of the prominent black financial institutions occupies a new headquarters building just a block from its historic office. Businesses focused upon a black clientele remain, including several funeral homes and hairdressers, yet several buildings are boarded-up. Redevelopment to provide additional housing to enhance the African-American community has been undertaken (Inwood 2011). Yet with less discrimination and blacks

Fig. 8.17 Prince Hall Masonic Temple Building and former headquarter of the Southern Christian Leadership Conference, which led the effort opposing racial segregation. Professional and social institutions were prominent components of the Auburn Avenue African-American community, particularly before blacks gained admittance to those that served white professionals

Fig. 8.18 Ebenezer Baptist Church, constructed between 1914 and 1922 along Auburn Avenue in Atlanta, Georgia for an African-American congregation, was a center for the struggle for civil rights when Reverend Dr. Martin Luther King, Jr. served as its pastor. The visitors' center of the Martin Luther King, Jr. National Historic Site is located across the street from the church

assuming leadership roles in city government, African-American business activities have dispersed to other parts of Atlanta. Unlike so many other communities, the majority of land along Auburn Avenue remains in African-American ownership and it remains a strong symbol to African-American identity. As Inwood (2009:93) notes, "Although Auburn Avenue's economic prominence declined as African American capital moved to other parts of the city, its cultural significance increased as Martin Luther King's legacy received mainstream attention." Yet, Atlanta remains a segregated city, but the character of that racial division has changed over the past several decades.

Atlanta is widely viewed as one of the most progressive Southern cities, attracting many newcomers, including African-Americans relocating from Northern locations. Only the New York City metropolitan area has more blacks than the Atlanta metropolitan area, and the city is the nation's fourth largest black-majority city (Strait and Gong 2015). Whites have begun to repopulate the city and the suburbs, particularly the northern suburbs which had once been almost entirely white, are becoming increasingly more multiracial—with not only more African-Americans, but also Hispanics and Asians. But when one looks at its "neighborhood level residential space…Atlanta remains highly segregated at the neighborhood scale and in certain contexts is becoming even more residentially segregated with time" (Strait and Gong 2015:136).

8.6 Urban African-Americans in Northern Cities

Five million blacks moved North over a period of four decades beginning during World War I. This movement was particularly large between 1940 and 1960, when three million persons relocated (Berlin 2010). This migration was the result of the tremendous need for employees in Northern manufacturing industries, starting when European immigration was cut off during the World Wars. Furthermore, wages in the North were much higher than what blacks could earn in the South, even though they were lower than the average for whites. At the same time that jobs were available in the North, the collapse of cotton production in the South because of the boll weevil infestation acted as a push factor, as did the harshness of segregation and racial discrimination in the South.

West Indian blacks also began migrating to several northern cities beginning in the early 1900s. Thousands of blacks began immigrating to the United States, coming largely from Jamaica, Barbados, and Trinidad. While they came initially mostly from the British West Indies, more recently large numbers of blacks have arrived from Haiti and the Dominican Republic. At the same time many Puerto Ricans, some who were at least partly African in heritage, began arriving in the United States, but these individuals are not counted as immigrants, given American jurisdiction over Puerto Rico since

1898. In contrast to how others might view them, many of the Dominicans and Puerto Ricans "identify as 'White' or 'Other' [on the Census] to distinguish themselves from African American" (Boswell and Jones 2007:139). Because of historical differences in colonization that resulted in language differences, linguistic affiliations strongly influence the choice of neighborhoods among West Indian blacks, with blacks from the British Commonwealth countries favoring different neighborhoods than those from the Spanish speaking Caribbean. Both groups often settled in neighborhoods separate from those that attracted African-Americans from the South.

West Indian blacks largely settled in a few major cities, especially New York City. By the 1920s one-quarter of Harlem's population had Caribbean heritage. (See also Chap. 6.) The U.S. Census Bureau estimated that New York City was the home to 530,000 Dominicans in 2005, 7% of the city's total population. West Indian blacks also arrived in large numbers within Miami, where they now represent 48% of that city's black population (Johnson 2008).

West Indian blacks are culturally distinct from American blacks, which was recognized decades ago. West Indian blacks more likely to have urban rather than rural backgrounds. As the black sociologist Thomas Sowell reports, "West Indians were much more frugal, …entrepreneurial, … [whose] children…outperformed native black children in school" (Sowell 1981:219). He further noted that "West Indians in the United States had lower fertility rates and lower crime rates than either black or white Americans" (Sowell 1981:219). West Indian blacks are disproportionately represented among black professionals. Examples of prominent persons with black West Indian ancestry include Marcus Garvey, Shirley Chisholm, Malcolm X, Sidney Poitier, and General (and later Secretary of State) Colin Powell.

8.6.1 Black Ghetto Formation

African-American ghetto development differed from that experienced by European immigrants within America's large cities. As Ira Berlin explains, African-Americans moving into northern cities "confronted formal and informal prohibitions—restrictive covenants, zoning regulations, and so-called civic associations—employed by white residents to halt the 'invasion' of their homelands" (Berlin 2010:182). Thus, they became concentrated in ghettos that were predominately to overwhelmingly black. The urban geographer Richard Morrill, describing the difference between the experience of European immigrants and blacks, noted that many European immigrants lived in urban ghetto slums. But over time "as their incomes grew and their English improved they moved out into the American mainstream, making way for the next group" (Morrill 1965:340). At the time European immigrants lived in ghettos in the nineteenth and early twentieth centuries, few African-Americans lived in northern cities. As European immigrants or their children moved out of the ghettos early in the twentieth century, African-Americans often replaced them. Once established, the African-American ghetto has tended to persist. Indeed, segregation rates became so high as to result in hypersegregation, in which blacks comprised over 95% of the residents (Tettey-Fio 2016). Thus, while the ghetto represented a transient residence for many European ethnic groups, as Eugene Tettey-Fio (2006:72) explains, "for black migrants this same zone became a permanent ghetto, a racial geography produced by prejudice and discrimination."

The growth of the African-American ghetto was quick. For example, Chicago and Detroit had relatively few black residents before World War I, yet by 1950 they had 492,265 and 300,506 African-American residents, respectively. The Great Migration of African-Americans into northern cities continued through the 1950s and 1960s. Racism and discrimination limited those neighborhoods into which these non-white newcomers could reside. "While Jim Crow laws legally separated blacks and whites in the southern states, institutional mechanisms—restrictive deeds, redlining, loan denials and steering, became the tools of northern whites to restrict the living spaces of black Americans" (Tettey-Fio 2006:74), hypersegregated neighborhoods first developed in the North. Once Southern blacks began their rapid settlement of Northern cities, they quickly came to dominate broad parts of the city. Indeed, "The rapidity and magnitude of the Negro stream not only have increased the intensity and size of ghettos in the North but no doubt have also accelerated the white 'flight to the suburbs' and have strongly affected the economic, political, and social life of the central cities" (Morrill 1965:341). The African-American ghetto was almost always near the central business district, in a part of the city "often containing formerly elegant houses intermingled with commercial and light industrial uses" (Morrill 1965:342).

The placement of public housing with its large concentration of African-Americans only exacerbated the forces that promoted ghetto development. Since the 1930s the location of public housing has served as a "prime ghettomaker" as explained by Ford and Griffin.

> Unwilling to antagonize 'good neighborhoods' or political supporters by locating massive low-income housing projects open to blacks within their midst, planners crammed most such projects into existing ghettos, thus 'institutionalizing' them. At the same time, much of the community was often 'renewed' out of existence (Ford and Griffin 1979:144).

By the beginning of the twenty-first century many of the failed public housing complexes had been abandoned. Chicago demolished its notorious Cabrini-Green high-rise residential towers, favoring the dispersal of its former residents into low-rise structures within a variety of neighborhoods. Even with its acute housing shortages following Hurricane Katrina, New Orleans began razing some of its former public housing complexes. Yet it was not just the construction of public housing that disrupted minority neighborhoods. The location of many expressways in low income neighborhoods disrupted many ethnic minority communities.

8.6.2 African-American Ghetto Landscapes

The population of the African-American ghetto is almost exclusively black, with very few whites. The average income in these areas is less than within white areas, and the level of education of the population less than in white areas. The schools are often substandard, both with older and more crowded facilities and smaller proportions of teachers who are appropriately credentialed. Many public services are often neglected, including garbage pickup, street maintenance, and parks and playgrounds, as the community neither has the wealth nor political clout to receive the attention provided to wealthier, non-black, districts. Thus, within the "inner city ghettos, landscapes are characterized by visual blight" (Frazier and Anderson 2006). Both unemployment and crime are often much higher, and "home ownership is less and the houses are older, less valuable, more crowded, and more likely to be substandard" (Morrill 1965:342), and this remains true a half century after Morrill's observation. Furthermore, over half of the black households with children are "headed by women, many of whom are the poorest of the poor" (Walker and Wilson 2001:25). The lack of adult males within many black ghetto families is a significant factor contributing to both the poverty and crime that stigmatize many inner city ghettos, and the nation's high incarceration rate is a major contributor to the absence of adult males, particularly black males.

Houses in the black ghetto are often subdivided into smaller apartments, whose maintenance is often neglected by owners, who are frequently absentee. The architectural styles of the houses in the ghetto typically reflect the heritage or choices of their former occupants, as the African-American residents are occupying an already built environment. For example, New York City's Harlem was once an Italian neighborhood, while Chicago's vast south-side ghetto has swallowed numerous German, Czech, Dutch, Hungarian, Irish, Italian, Lithuanian, Polish, and Swedish neighborhoods (Cutler 1995). In Milwaukee, its north side German neighborhoods have become the heart of its huge African-American ghetto. While the houses, churches, and retail buildings constructed by these European immigrants largely remain, the presence of African-Americans within these former immigrant neighborhoods has transformed their appearances.

Alterations to the ghetto landscape include the creation of store-front churches, the opening of "used furniture and clothing stores, the independent grocery stores, the taverns, and … establishments found in low-income metropolitan areas dominated by blacks" (Aiken 1990:238). Because of the minimal offerings of fresh fruits and vegetables within many of the small groceries, ghettos are often considered "food deserts." Banks are often replaced by check cashing services, payday loan offices, and rent-to-own shops. At the same time many shops that once catered to upscale customers have been relegated to other uses and former ethnic associations and synagogues are now occupied by other groups.

While white flight from neighborhoods with growing numbers of black residents continues in many communities, the exodus of more prosperous blacks from blighted ghetto neighborhoods has also occurred. For example, within several suburbs of metropolitan Detroit, Michigan, "[w]hile there continues to be strong resistance to integrated neighborhoods—witness patterns of white flight to the suburbs in virtually every city in Michigan—blacks have acquired wealth and status to live outside the economically depressed areas" (Walker and Wilson 2001:24). The Detroit area is not alone. Prosperous middle-class black neighborhoods and suburbs are found within many metropolitan areas, including Atlanta, Chicago, Washington, and Los Angeles, among others.

8.6.3 Black Middle-Class Neighborhoods

Black middle-class neighborhood development and flight from the inner city ghetto remains different from that of white neighborhoods. Because both middle-class and working-class African-Americans have been moving from blighted inner cities, "middle-class status meant different things for blacks and whites; achievement of class status did not provide African-Americans the same freedom of residential choice" (Frazier and Anderson 2006:88). Thus, black neighborhoods, while often associated with the run-down, older inner city, also include newer black communities and subdivisions outside the city core. These can closely resemble white middle-class suburbs, as black middle-class and wealthier professionals have fled the crowded poorer central city ghettos for the same reasons that whites have departed. Unlike the inner city ghetto, in some of these neighborhoods African-Americans are the first to occupy their newly built houses, structures that are indistinguishable from new houses within white or integrated neighborhoods of residents having comparable incomes.

While arguments can be made that such middle class black neighborhoods should not be termed a ghetto, Larry Ford and Ernst Griffin view such developments in the San Diego, California metropolitan area as the "New Ghetto."

They explain that this type of "ghetto developed as largely middle-class, semi-suburban neighborhoods that are more akin to their white counterparts than to the core of the ghetto. Yet they remain black and largely isolated from the white community" (Ford and Griffin 1979:141). Similarly, black suburban development in the South since World War II has proceeded in a manner more alike that in California than in the North. In the South, to "meet the growing demand for black housing and to maintain the color line, southern white institutions politically accommodated blacks by providing new housing in acceptable but separate suburbs at the edge of existing cities" (Frazier and Anderson 2006:86). While predominately African-American, such neighborhoods are typically no more ethnically distinctive than most other modern American suburban developments.

8.7 Recent Immigration from Africa

During the past couple of decades direct immigration flows from Africa to the United States have grown. Between 1997 and 2006, a total of 145,000 Africans arrived in the United States as political refugees. Over 59,000 of these came from Somalia, while another 13,000 had fled from Ethiopia. Neither of these locations had figured prominently as source regions for African-Americans who were already in the United States, whether their roots were traced to slavery in North America or the West Indies. In addition, large numbers of other Africans have immigrated to the United States. In 2006 alone, the United States saw 112,108 immigrants from Africa (U.S. Dept. Homeland Sec. 2007). Among the three largest source regions of black African immigrants during 2006 were Ethiopia, Nigeria, and Somalia, and these African ethnics displayed major differences in their destinations within the United States. Thirty percent of the Somalis went to Minnesota and another 12% went to Ohio. Sixteen percent of the Nigerians went to Texas, while the others were widely scattered across the nation. Among the 16,157 Ethiopian immigrants, 26.8% went to the region around the District of Columbia, including Maryland and Virginia. Over 9% settled in Minnesota.

In 2014, the most recent year for which immigration statistics are available, the United States admitted 98,413 immigrants from Africa, less than 10% of global admissions. The greatest numbers of African immigrants came from Ethiopia, 12,300, and Nigeria, 12,828 (U.S. Dept. Homeland Sec. 2016). Refugees, of whom 17,501 were admitted from Africa in 2014, were less prominent among African immigrants than during the previous decade. In contrast to the number of blacks from Africa, the United States admitted far more persons of African heritage from the Western Hemisphere in 2014. For example, 44,577 immigrants arrived from the Dominican Republic, 19,026 came from Jamaica, and 15,274 entered from Haiti. Although 46,679 persons arrived from Cuba, a smaller proportion of these persons are likely to have African heritage than those immigrants from the other three countries, which are predominately black.

The recent arrival of African immigrants has begun to alter the American landscape. Some of these African ethnics have settled in areas which have little history of African-American settlement, such as the Somalis, Ethiopians, and Liberians within the Minneapolis-St. Paul, Minnesota metropolitan area. Others have focused upon regions of the nation which have long-established African-American communities, yet many of these recent African immigrants have ethnic heritages that are quite at variance from those existing black communities. Research within the Washington, DC metropolitan area indicates that African immigrants "are disproportionately more likely than other immigrants to settle in counties where the majority of the population is black" (Friedman et al. 2005:219). Thus, not only has the Washington metropolitan area attracted nearly 17% of recent African immigrants, but these "African-born blacks are not living in neighborhoods where they are highly concentrated among their coethnics. Nevertheless, they are living in neighborhoods that have a relatively larger share of blacks than of other immigrants" (Friedman et al. 2005:226). The arrival of the African immigrants can be seen by the establishment of neighborhood businesses, such as ethnic restaurants (Fig. 8.19) or shops and groceries displaying ethnic signage and colors, and churches or mosques reflecting their immigrant heritage.

The Ethiopian community within the Washington metropolitan area, the nation's largest, is too new to be marked with architecturally distinctive ethnic churches. While the majority of the Ethiopian immigrants are Christian, mostly from a Coptic (or Ethiopian Orthodox) tradition, the congregations are meeting in former commercial structures, only distinguished as churches by their signage (Chacko 2003a). With time, it is possible that these congregations may build structures more reminiscent of Ethiopia. It is in several neighborhoods of both the District of Columbia and Silver Spring, Maryland that the Ethiopian ethnicity of the immigrants is displayed in a variety of commercial establishments. Elizabeth Chacko (2003a:33) notes that "Ethiopian identity is clearly marked and discernible to the interested observer." She explains that many grocery stores, restaurants, and clothing stores display signs in Amharic. In addition she describes a "recurring image on storefronts and signs that has strong emotional and national appeal…, the Ethiopian flag in its entirety, or the flag's horizontal bands of green, yellow, and red" (Chacko 2003a:34). More ephemeral marks on the cultural landscape are the periodic use of neighborhood parks for the celebration of traditional Ethiopian holidays or religious festivals, which are linked to an area of Africa which had no connection to the trans-Atlantic slave trade.

Fig. 8.19 Ethiopian restaurant in Frederick, Maryland. The arrival of recent African immigrants has resulted in the establishment of ethnic restaurants in many communities, with the heritage displayed in the signage and window displays, and often in the selection of paint colors

Other African immigrants are arriving from parts of Africa that were intimately tied to the slave trade a couple of centuries ago, and these new arrivals have formed neighborhoods that show clear links to their natal states. For example, within New York City a Ghanaian community has developed in the Concourse area of the Bronx.

Large cities, as well as university cities, have often been the focus for blacks newly arriving from Africa. These immigrants have considerably broadened the diversity of America's black urban population, as within "America's major cities, 15% of the black American population is foreign born…a rich tapestry of brown-skinned people as culturally complex in their differences, backgrounds, and outlooks as those people lumped together under the all too convenient labels of 'Asian' or 'European'" (Johnson 2008:37). These individuals are far more closely linked to their ethnicity than the majority of whites and blacks in the United States, whose American roots may go back several centuries. In many respects, blacks and whites who have deep roots in the United States have more shared cultural values and experiences, albeit at times from dissimilar perspectives, than either group has with the culture of these newest African immigrants.

8.8 African Ethnic Landscapes of the Future

Undoubtedly, the African ethnic imprint upon the American landscape will continue to evolve in the future. While there is hope that Americans will become more appreciative of the tremendous contributions of African-Americans, both before and after emancipation, in shaping the cultural landscape of the American South, the African cultural imprint upon the nation is also becoming more regionally and ethnically diverse. Blacks whose ancestors left Africa several centuries ago and whose cultural heritage was formed by living many generations in a variety of West Indian countries have brought their mixture of African, Native American and European colonial traditions to several northern cities, some which previously had few black residents. Because of the legacy of slavery and centuries of acculturation through interaction with other cultures, the black American sociologist Thomas Sowell (1981:184) wrote "little or no African culture survives among American Negroes." Yet as we have seen in this chapter, while some of their landscape attributes were imposed upon them, African-Americans have contributed much to the American landscape, even though it is derived from only small fragments of their original culture. The Africa that they left is not the post-colonial Africa from which contemporary African immigrants have come, adding to the diversity of African ethnic elements we see today upon the landscape.

While some recent African immigrants are uncomfortable with the term "African-American," believing that their cultural heritage differs in many aspects from those blacks whose roots extend back two or more centuries in the United States and whose behavior has been shaped by a lengthy interaction with white European-American culture (Chacko 2003b), their presence contributes to the diversity of African cultures represented within the United States. Put simply, many recent African immigrants come from lands with very different heritages than those territories that contributed millions of slaves to the Western hemisphere. Thus, it is not surprising that many Ethiopian and Somali immigrants may think of themselves as Ethiopian-Americans or Somali-Americans, rather than African-Americans, just as German-Americans and Italian-Americans identify with their national origins more than their continental origins. Africa is far larger than Europe, and Americans are being increasingly reminded of its tremendous ethnic diversity as the variety of African heritages represented by America's black population is increasing.

References

Aiken, Charles S. 1985. New Settlement Pattern of Rural Blacks in the American South. *The Geographical Review* 75 (4): 383–404.

———. 1990. A New Type of Black Ghetto in the Plantation South. *Annals of the Association of American Geographers* 80 (2): 222–246.

———. 1998. *The Cotton Plantation South Since the Civil War*. Baltimore: Johns Hopkins University Press.

Alderman, Derek H., and E. Arnold Modlin Jr. 2014. The Historical Geography of Racialized Landscapes. In *North American Odyssey: Historical Geographies for the Twenty-first Century*, ed. Craig E. Colten and Geoffrey L. Buckley, 273–290. Lanham: Rowman and Littlefield.

Anthony, Carl. 1976a. The Big House and the Slave Quarters: Part I. Prelude to New World Architecture. *Landscape* 20 (3): 8–19.

———. 1976b. The Big House and the Slave Quarters: Part II. African Contributions to the New World. *Landscape* 21 (1): 9–15.

Banton, Michael. 2015. *What We Now Know about Race and Ethnicity*. New York: Berghahn.

Belcher, Max. 1988. *A Land and Life Remembered: Americo-Liberian Folk Architecture*. Athens: University of Georgia Press.

Berlin, Ira. 2010. *The Making of African America: The Four Great Migrations*. New York: Viking.

Boswell, Thomas D., and Terry-Ann Jones. 2007. Caribbean Hispanics: Puerto Ricans, Cubans, and Dominicans. In *Contemporary Ethnic Geographies in America*, ed. Ines M. Miyares and Christopher A. Airriess, 123–150. Lanham: Rowman and Littlefield.

Campanella, Richard. 2006. *Geographies of New Orleans: Urban Fabrics Before the Storm*. Lafayette: Center for Louisiana Studies.

Carney, George O. 1983. The Shotgun House in Oklahoma. *Journal of Cultural Geography* 4 (1): 57–71.

Chacko, Elizabeth. 2003a. Ethiopian Ethos and the Making of Ethnic Places in the Washington Metropolitan Area. *Journal of Cultural Geography* 20 (2): 21–42.

———. 2003b. Identity and Assimilation among Young Ethiopian Immigrants in Metropolitan Washington. *The Geographical Review* 93 (4): 491–506.

Cromartie, John, and Carol B. Stack. 1989. Reinterpretation of Black Return and Nonreturn Migration to the South. *The Geographical Review* 79 (3): 297–310.

Crutcher, Michael. 2006. Historical Geographies of Race in a New Orleans Afro-Creole Landscape. In *Landscape and Race in the United States*, ed. Richard H. Schein, 23–38. New York: Routledge.

Cutler, Irving. 1995. Ethnic Chicago. In *Geographical Excursions in the Chicago Region*, ed. Michael P. Conzen, 21–29. Washington: Association of American Geographers.

Davis, F. James. 1991. *Who Is Black? One Nation's Definition*. University Park: Pennsylvania University Press.

Dennis, Samuel F. Jr. 2006. Seeing Hampton Plantation: Race and Gender in a South Carolina Heritage Landscape. In *Landscape and Race in the United States*, ed. Richard H. Schein, 73–94. New York: Routledge.

Ford, Larry, and Ernst Griffin. 1979. Ghettoization of Paradise. *The Geographical Review* 69 (2): 140–158.

Frazier, John W. 2016. Racial Geography vs. Ethnic Geography: A Distinction. In *Race, Ethnicity, and Place in a Changing America*, ed. John W. Frazier, Eugene L. Tettey-Fio, and Norah F. Henry, 3rd ed., 25–45. Albany: State University of New York Press.

Frazier, John W., and Roger Anderson. 2006. People on the Move: African Americans Since the Great Migration. In *Race, Ethnicity, and Place in a Changing America*, ed. John W. Frazier and Eugene L. Tettey-Fio, 83–96. Binghamton: Global Academic Publishing.

Friedman, Samantha, Audrey Singer, Marie Price, and Ivan Cheung. 2005. Race, Immigrants, and Residence: A New Racial Geography of Washington, D.C. *The Geographical Review* 95 (2): 210–230.

Hall, Robert L. 1991. Savoring Africa in the New World. In *Seeds of Change: Five Hundred Years Since Columbus*, ed. Herman J. Viola and Carolyn Margolis, 158–169. Washington: Smithsonian Institution Press.

Hoelscher, Stephen. 2006. The White-pillared Past: Landscapes of Memory and Race in the American South. In *Landscape and Race in the United States*, ed. Richard H. Schein, 39–72. New York: Routledge.

Hornsby, Stephen J. 2005. *British Atlantic, American Frontier: Spaces of Power in Early Modern British America*. Hanover: University Press of New England.

Inwood, Joshua F.J. 2009. Contested Memory in the Birthplace of a King: A Case Study of Auburn Avenue and the Martin Luther King Jr. National Park. *Cultural Geographies* 16 (1): 87–109.

———. 2011. Constructing African American Urban Space in Atlanta, Georgia. *The Geographical Review* 101 (2): 147–163.

Johnson, Charles. 2008. The End of the Black American Narrative. *American Scholar* 77 (3): 32–42.

Kellogg, John. 1977. Negro Urban Clusters in the Postbellum South. *The Geographical Review* 67 (3): 310–321.

Lee, David. 1992. Black Districts in Southeastern Florida. *The Geographical Review* 82 (4): 375–387.

McKinnon, Jesse. 2001. *The Black Population: 2000*. Census 2000 Brief. Washington: U.S. Census Bureau (C2KBR/01-5).

McHugh, Kevin E. 1987. Black Migration Reversal in the United States. *The Geographical Review* 77 (3): 279–292.

Morrill, Richard L. 1965. The Negro Ghetto: Problems and Alternatives. *The Geographical Review* 55 (3): 339–361.

National Agricultural Statistics Service (NASS). 2014a. *2012 Census of Agriculture: United States Summary and State Data, Volume 1,*

Geographic Area Series, Part 51. Washington: U.S. Department of Agriculture. AC-12-A-51 http://agcensus.usda.gov/Publications/2012/Full_Report/Volume_1,_Chapter_1_US/usv1.pdf. Accessed 28 November 2016.

———. 2014b. Number of Farms with Black or African American Operators: 2012. In *2012 Census of Agriculture: Agricultural Atlas*. Washington: U.S. Department of Agriculture. https://www.agcensus.usda.gov/Publications/2012/Online_Resources/Ag_Atlas_Maps/Operators/Characteristics/12-M129-RGBDot1-largetext.pdf. Accessed 28 November 2016.

Oszuscik, Phillipe. 1992. African-American in the American South. In *To Build in a New Land: Ethnic Landscapes in North America*, ed. Allen G. Noble, 157–176. Baltimore: Johns Hopkins University Press.

Pollitzer, William S. 2005. *The Gullah: People and their African Heritage*. Athens: University of Georgia Press.

Rastogi, Sonya, Tallese D. Johnson, Elizabeth M. Hoeffel, and Malcolm P. Drewery, Jr. 2011. *The Black Population: 2010*. Census 2010 Brief. Washington: U.S. Census Bureau (C2010BR-06).

Savoy, Lauret Edith. 2015. *Trace: Memory, History, Race, and the American Landscape*. Berkeley: Counterpoint.

Sluyter, Andrew. 2014. African Arrivals and Transformations. In *North American Odyssey: Historical Geographies for the Twenty-first Century*, ed. Craig E. Colten and Geoffrey L. Buckley, 49–66. Lanham: Rowman and Littlefield.

Sowell, Thomas. 1981. *Ethnic America: A History*. New York: Basic Books.

———. 1996. *Migrations and Cultures: A World View*. New York: Basic Books.

Strait, John Byron, and Gang Gong. 2015. The Impact of Increased Diversity on the Residential Landscape of a Sunbelt Metropolis: Racial and Ethnic Segregation Across the Atlanta Metropolitan Region, 1990–2010. *Southeastern Geographer* 55 (2): 119–142.

Tettey-Fio, Eugene L. 2006. Black American Geographies: A Perspective. In *Race, Ethnicity, and Place in a Changing America*, ed. John W. Frazier and Eugene L. Tettey-Fio, 69–81. Binghamton: Global Academic Publishing.

———. 2016. Historical and Contemporary Black-American Geographies. In *Race, Ethnicity, and Place in a Changing America*, ed. John W. Frazier, Eugene L. Tettey-Fio, and Norah F. Henry, 3rd ed., 153–168. Albany: State University of New York Press.

United States Department of Homeland Security. 2007. *2006 Yearbook of Immigration Statistics*. Washington: U.S. Department of Homeland Security/Office of Immigration Statistics.

———. 2016. *2014 Yearbook of Immigration Statistics*. Washington: U.S. Department of Homeland Security/Office of Immigration Statistics.

Upton, Dell. 1988. White and Black Landscapes in Eighteenth Century Virginia. In *Material Life in America, 1600–1860*, ed. Robert B. St. George, 357–369. Boston: Northeastern University Press.

Vlach, John M. 1976. The Shotgun House: An African American Architectural Legacy. *Pioneer America* 8 (1): 47–56 and 8 (2): 57–70.

———. 1986. Afro-Americans. In *America's Architectural Roots: Ethnic Groups that Built America*, ed. Dell Upton, 42–47. Washington: Preservation Press.

———. 1993. *Back of the Big House: The Architecture of Plantation Slavery*. Chapel Hill: University of North Carolina Press.

Walker, Lewis, and Benjamin C. Wilson. 2001. African Americans in Michigan. In *African Americans in Michigan*, ed. Lewis Walker, Benjamin C. Wilson, and Linwood H. Cousins, 3–30. East Lansing: Michigan University Press.

Wallace, Edie. 2007. Reclaiming the Forgotten History and Cultural Landscapes of African-Americans in Rural Washington County, Maryland. *Material Culture* 39 (1): 9–32.

Westmacott, Richard. 1992. *African-American Gardens and Yards in the Rural South*. Knoxville: University of Tennessee Press.

Williams, Peter W. 1997. *Houses of God: Region, Religion, and Architecture in the United States*. Urbana: University of Illinois Press.

British Landscapes in America

9

British immigrants played a critical role in shaping the American landscape, with the rebellion of British colonial subjects leading to the establishment of the United States of America. While collectively those persons claiming ancestry from one or more of the ethnic groups inhabiting the British Isles rival those of German ancestry as being the largest ancestry group in the United States, British heritage belies considerable ethnic uniformity. Indeed, persons of British ancestry came from throughout the British Isles, representing a diversity of peoples.

9.1 Diversity of British Peoples

During the time period in which the greatest numbers of persons immigrated to the United States from the British Isles, that entire region was governed as one sovereign entity. Indeed, all of Great Britain has been administered as the United Kingdom since 1707, and Ireland was part of the United Kingdom from 1800 to 1921. Scotland, which was incorporated into the English monarchy in 1603, was formally annexed to form the United Kingdom in 1707 following a period of bloody resistance. Ireland, which the English crown began ruling in 1541, was formally annexed by Britain in 1800, but it left the United Kingdom in 1921 after an armed struggle for independence. Yet six counties in Northern Ireland resisted annexation to Ireland and remain part of the United Kingdom. Wales was conquered in 1284 and has been governed ever since as an English principality, yet it retains some of its original territorial integrity and Welsh is still spoken by a minority of its people. Likewise, Cornwall in far southwestern England was annexed by the English crown about the same time, yet the assimilation of its people resulted in the extinction of the Cornish language two centuries ago.

The British peoples are not as united as the United Kingdom's name would imply, and this was most recently demonstrated by the results of 2016's "Brexit" referendum to leave the European Union, with Scotland and Northern Ireland wanting to remain, while the English—particularly in the rural areas outside of London—seeking to leave. The British Isles had a long history of interaction among its Gaels, Picts, and English and received cultural and genetic transmission by various invaders, including the Romans, three Germanic groups—Angles, Saxons, and Jutes, and Norsemen or Danes (Miles 2006). French culture was introduced several times, not only by William the Conqueror, the Norman whose ascension to the English throne in 1066 introduced a Norman aristocracy to Britain, but centuries later by thousands of French Huguenots who fled persecution in France. Because each of these groups invaded or settled within only portions, or specific areas, of the British Isles, there are strong regional differences in their influence upon the local culture.

9.1.1 Cultural Differences Among the British

There are significant cultural differences among the population of the British Isles. For example, among the English language and prominence of the Church of England (or Anglican, Protestant Episcopal Church) are well established, even though strong regional accents remain and many English citizens no longer attend religious services. In contrast, while 99% of the Welsh speak the English language, Welsh, an ancient Celtic language, is still spoken by a fifth of the population. Furthermore, while the protestant church in Wales is prominent, many of the Welsh are Methodist rather than Anglican. Likewise, many of the Cornish are Methodist. The Scots (sometimes called the Scottish) incorporate a variety of Scots words into the English they speak, and their accent is distinct from the various English accents. Furthermore, Gaelic is still spoken by nearly 58,000 residents of the Scottish Highlands (Macfarlane 2016). Unlike the prominence of the Church of England in England, the established church in Scotland is the Church of Scotland, from which the Presbyterian churches in the United States have evolved. Yet many

J.A. Cross, *Ethnic Landscapes of America*, DOI 10.1007/978-3-319-54009-2_9

Highlanders remain Roman Catholic. To the north of Scotland lie the Outer Hebrides, where its language "is currently being lost" (Macfarlane 2016:23).

9.1.2 The Irish

In Ireland the majority of the population speaks English, by a greater proportion today than in the 1800s when much of the population fled the famine stricken land, yet Irish (Gaelic) is still spoken by small minority. While the population within the Republic of Ireland (Eire) is overwhelmingly Roman Catholic, that within Northern Ireland is split between Roman Catholicism and Presbyterianism. Within Northern Ireland a majority of the population are of Scots-Irish (or Ulster-Scots) ancestry. The Scots-Irish are persons of Scots heritage who were first forcibly settled into the northern part of Ireland by King James I beginning in 1609 in an effort to diminish the effectiveness of resistance to the English by both the Scots and the Irish. Additional Scots settlement in Ireland occurred throughout the seventeenth century. As T. M. Devine notes, the Scots-Irish are "an identifiable ethnic group, a hybrid people with their territorial roots in Ireland but also Scottish in terms of religious loyalties, culture, speech and intellectual heritage" (Devine 2003:141). The Scots-Irish emigrated in large numbers between 1717 and 1775, and their heritage is much more Scottish than Irish, but their ancestry was sometimes reported as Irish because they came to America from Ireland. Indeed, arrivals of Scots during the 15 years preceding the outbreak of the Revolutionary War greatly exceeded the numbers of arriving English settlers (Taylor 2016). Furthermore, Ulster-Scots "massively outnumbered direct migrants from Scotland not only because of their much higher and earlier rate of emigration but also because many Scots remained loyal to the crown in the American [Revolutionary] War and after 1783 left the USA for British North America [Canada]" (Devine 2003:154). Thus, from Ireland the United States received an infusion of both Irish and Scottish culture. Yet, not all immigrants from Ireland during the eighteenth century were Scots-Irish. Catholics, as well as Protestant Anglicans, arrived from both northern and southern Ireland. Yet, as Williams (2002:44) notes, a "very substantial number of Catholics" who came to America from Ireland "were predominately unchurched."

9.1.3 Populations of British Ancestry in the United States

At the time of the 2000 U.S. Census, 24,509,692 persons reported that they were of English ancestry (Fig. 9.1), a decline of over eight million from 1990. The 5-year American Community Survey estimates for 2000–2014 were 25,181,294. Because of differences in response to the ancestry question between the two censuses and the American Community Survey and the fact that those reporting simply "American" ancestry in 2000 increased by nearly as many as those reporting English ancestry fell, Table 9.1 reports the ancestry data from both censuses and the survey estimate.

The U.S. Census ancestry data indicates the first and second ancestry reported on the census returns. Because many individuals have multiple ancestries—and the ethnic mixing of the population increases each generation—many persons who are partly of English or Irish ancestry may not report it. Indeed, consider the quandary facing a person who is precisely one-quarter each of English, Irish, French, and German. What two ancestries would this person report on the census form? Of those surveyed for the 2000 Census, 58.0% reported a single ancestry, 22.0% reported two ancestries, while 19.9% simply ignored the question. In addition, the 5 year 2010–14 American Community Survey showed 22,365,250 persons indicating American ancestry, disproportionately overrepresented in the Upland South, a region whose settlement history is largely British. Thus, in all reality, the figures above underestimate the population that is of British ancestry (as likewise with other groups). Furthermore, there are also definitional problems in the Census, as persons of Scots-Irish ancestry are not consistent in how they report their ancestry, which is also variously reported as Irish, Scottish, or Scotch-Irish.

9.1.4 Diversity Among the English

The English themselves differed regionally with respect to religion, their attitudes towards education, and their economic class structure. David Hackett Fischer (1989) in his magisterial *Albion's Seed* distinguished four distinct British groups that immigrated to America, each coming primarily from a specific region of England. These included Puritans who came from East Anglia, that portion of eastern England north of the London area; a group he called the Cavaliers, who came from Southern England, west of the London area; Quakers who came from the North Midlands; and Backcountrymen who came from the Borderlands, that region including the England-Scotland border region, Scotland, and Northern Ireland. These groups differed significantly in a variety of social conditions: religious attitudes and fervor; level of literacy; social standing and wealth; social organization and class differences; and attitudes towards work. Furthermore, each group differed in the location of their settlements in North America and in their experiences after arrival.

These differences are eloquently described by Fischer:

> The character of Virginia's great migration…differed in almost every important way from the Puritan exodus to Massachusetts. From the start, immigrants to the Chesapeake colony were more highly stratified, more male-dominant, more rural, more agrarian, less highly skilled, and less literate (Fischer 1989:231).

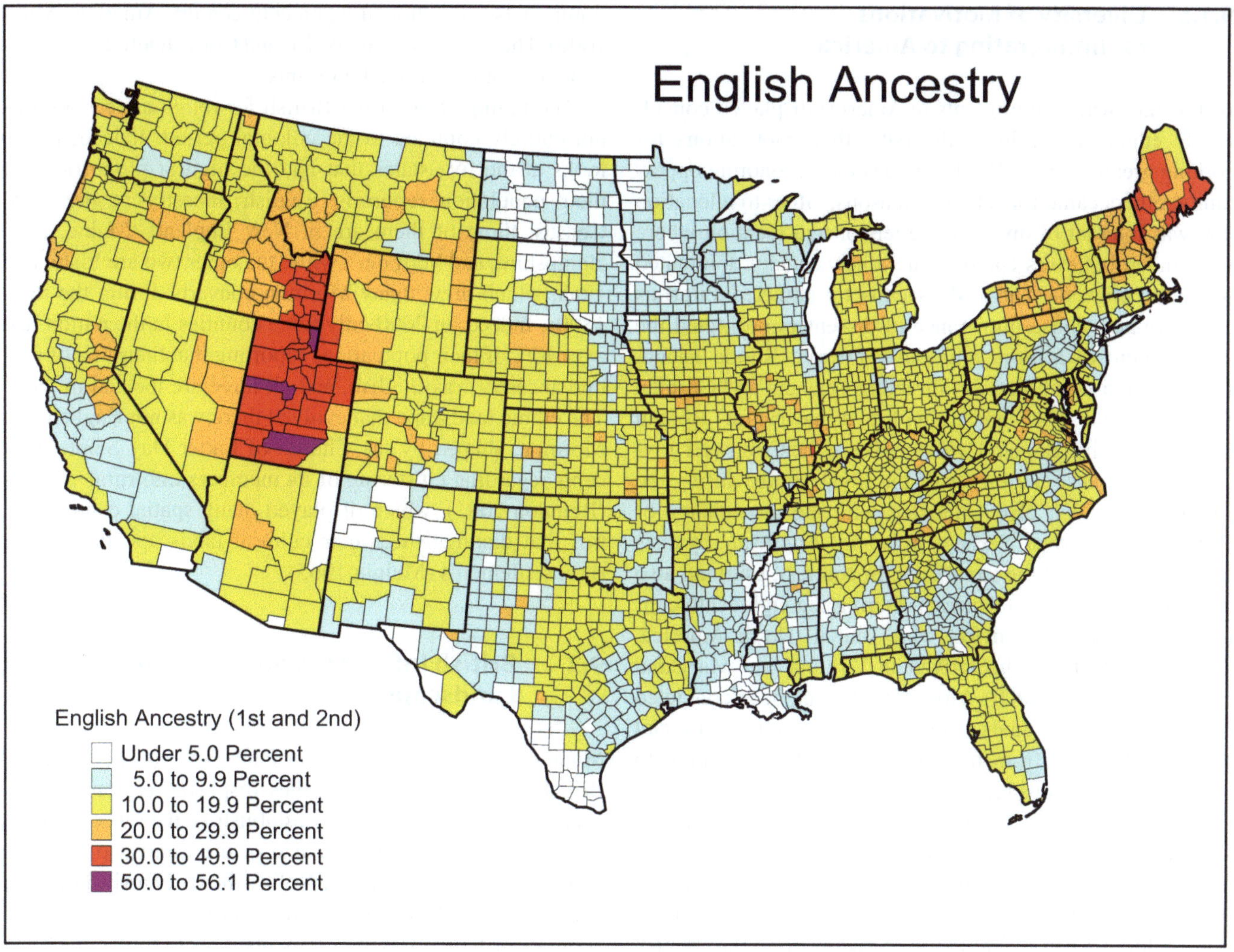

Fig. 9.1 Map showing the percentage of individuals who reported their ancestry on the 2000 Census indicating that English was their primary or secondary ancestry. English ethnics account for at least 10% of the population within over two-thirds of all U.S. counties, yet those counties with the largest percentages of their populations reporting English ancestry are in northern New England and Utah and adjacent portions of Idaho and Wyoming. The large proportions reporting English ancestry in northern New England are a legacy of English colonial settlement, while the large proportions in Utah reflect the fact that the Mormons attracted many English converts during the nineteenth century

Table 9.1 Population of British ethnics in the United States

Ethnic ancestry	1990 census	2000 census	2010–14 survey
British	1,119,154	1,085,718	1,261,179
English	32,651,788	24,509,692	25,181,294
Irish	38,735,539	30,524,799	33,918,058
Scotch-Irish	5,617,773	4,319,232	3,102,082
Scottish	5,393,581	4,890,581	5,401,397
Welsh	2,033,893	1,753,794	1,779,664
Celtic	29,652	65,638	48,939
Cornish	3991	Not reported	Not reported
Manx	6317	Not reported	Not reported
Northern Irish	4009	Not reported	Not reported

1990 and 2000 data from U.S. Censuses of those dates. 2010–2014 data from American Community Surveys conducted over 5 year period between 2010 and 2014. Data accessed utilizing American FactFinder on-line data acquisition tool operated by U.S. Census Bureau

Indeed, as David Miles (2006:327) explains, the settlers of Jamestown "transported their attitudes with them and half of them were 'gentlemen'." Cavaliers who settled the southern American colonies were far more likely to be followers of the Church of England and came from an area of England that not only had rebelled against Protestant innovations in the Church of England in 1549, but had also supported the Royalists during the English Civil War (Fischer 1989). Indeed, the English Civil War between 1640 and 1660 saw skirmishes fought on both sides of the Atlantic, and as Kevin Phillips (1999) describes in his *Cousins' Wars*, hundreds of men returned to England from Connecticut and Massachusetts to fight for the Parliamentary forces opposing the crown, while defeats of royalists supporting the Stuarts sent refugees to Virginia.

9.1.5 Diversity of Motivations for Immigrating to America

Just as the British immigrants to America displayed considerable ethnic and cultural diversity, their motivations for coming greatly varied. While some sought economic opportunity, others came for religious reasons, either to a location that was tolerant of divergent beliefs or to a location where likeminded believers could create a new society dominated by such believers. Thus, Massachusetts attracted Pilgrims and Puritans, whose intolerance of dissenters prompted the development of a more religiously tolerant Rhode Island. Pennsylvania was noted for attracting Quakers and German Pietists, while Maryland provided a haven for Roman Catholics fleeing England with the rise of the Cromwell government. Because the ouster of the royal family during the English Civil War also coincided with the ascendancy of Puritan control over Parliament, many Anglican royalists came to Virginia at the same time. In contrast, England also shipped its convicts to America. During the eighteenth century nearly 50,000 convicts were sent to North America, approximately half who arrived in Virginia and Maryland (Fischer and Kelly 2000). Most often having been convicted of theft, these individuals were typically drawn from a social class far different from that occupied by those coming to America for religious reasons.

Besides convicts, Britain sent thousands of indentured servants to its American colonies, particularly those south of New England. While some individuals had deliberately indentured themselves, many individuals—including numerous children—had been forcibly taken from the British and Irish streets and were "sold into servitude" (Jordan and Walsh 2008:16). During the period of indenture, which could be substantially lengthened if the servant tried to escape, the indentured servant was often treated similar to African slaves. As Jordan and Walsh (2008:111) explain, "huge numbers of white servants didn't live to see the day of freedom." Amongst the earliest indentured servants, the majority died before completion of their indenture, and those who lived "would emerge from bondage landless and poor" (Jordan and Walsh (2008:111). Because so many indentured servants were sold in the middle colonies, the presence of these British colonists accentuated the highly stratified social system that existed among the Cavaliers.

It is important to note that before the Revolutionary War immigrants were settling English colonies, and they viewed themselves as English even in North America (Taylor 2016). Strong ties with England were kept until the time of the Revolution, and even after the Revolutionary War the British continued to be a major trading partner with portions of the United States. Yet, while the English are sometimes viewed as dominating the immigrants coming to Britain's colonies, the Scots-Irish influx was strong, representing about 40% of all immigrants into colonial eighteenth century America (Miles 2006). They, as well as the Welsh and Cornish left distinct, but geographically restricted, imprints.

The immigration of the British for religious reasons was not entirely confined to the colonial period. Indeed, if one looks at the five states that had the highest proportions of their population reporting English ancestry in the 2000 Census, three of them are in New England (Maine, New Hampshire, and Vermont), while the other two are Idaho and Utah, which drew many Mormon converts during the mid-1800s. Indeed, in 2000 only three counties in the nation had over 40% of their populations reporting that their first or second ancestry was English, and all three are located in Utah. The ethnic landscape that evolved there was not English, but Mormon, shaped by the Church of Latter Day Saints' precepts regarding clustering of its members into rural villages (Bennion 2001), which displayed many spatial characteristics that distinguished their cultural landscape from other rural settlements (Meinig 1965).

9.2 British Influences on the American Landscape

Much of the generic American landscape can be traced to British origins. Even though they displayed regional differences, the English were early established, they were settling in English colonies, and the size of the English population simply overwhelmed others. Noting the early arrival of large numbers of English colonists in North America, the cultural geographer Peirce Lewis comments, "Inevitably, most early American folk houses were really English, or were quickly made to *look* English, just as our tastes in food, in clothes, and in landscape gardening were English" (Lewis 1975:3). He continues explaining that when "it came to the really important question of how a man's house should *look*, English standards prevailed from Labrador to the frontiers of Spanish Florida" (Lewis 1975:3). This remains true today. Green grass manicured lawns surrounding houses, which are located on unfenced lots, are a major contribution of the British to the American landscape.

9.2.1 English Influences Differ between Northern and Southern States

Strong regional differences developed between the northern and southern states regarding how the English colonists shaped their architecture and landscape. The South historically had stronger and more lasting ties with England than the rest of the country. These connections to England can be explained "perhaps because its economic links were traditionally so strongly forged to England, perhaps because it received so few immigrants and ideas from outside the British Isles, or perhaps, as in parts of Canada, merely to declare it independence from Yankee domination" (Lewis 1975:20,22). Indeed, before the American Civil War the output of southern plantations was frequently shipped directly to Great Britain for processing, and the South supplied the cotton necessary to operate Britain's textile mills. Likewise, many Southern planters sent their sons to be educated in Britain's colleges and universities.

The British cultural domination in the South was unique. New England, Pennsylvania, and the Mid-Atlantic colonies of the British had included large numbers of other ethnic groups from their earliest days. For example, the Dutch colony centered upon Manhattan once extended east into Connecticut and southwest into the Delaware River Valley, where a short-lived Swedish colonization effort was also made. Germans began immigrating into Pennsylvania during the later part of the seventeenth century, and by the time of the American Revolution large areas of Pennsylvania were German-speaking (Nolt 2002). Because of its religious tolerance, Rhode Island attracted a wide variety of colonists, including the first Jewish congregation in America. Later immigration into the former English colonies outside the South added large numbers from other ethnic groups, such as the Irish in the 1840s, French Canadians, and Italians beginning in the 1890s.

In contrast, the South was largely ignored by the great waves of immigrants from the 1840s through the time of the First World War. (It should be mentioned that while Germans who arrived during colonial times did influence the settlement of portions of North Carolina's Piedmont region and French Huguenots contributed to the settlement of the Charleston, South Carolina area, these were local exceptions to the dominant British cultural landscape that characterized the South.) Because of the institution of slavery and the resulting large African-American population, the South provided fewer employment opportunities for immigrants. It likewise had little land open to homesteading and few industrial jobs in its cities, which themselves were smaller and less numerous than those in the northern and Midwestern states. Thus, during the huge wave of immigration in the last half of the nineteenth century, most of the South had little to attract the multitudes of Germans, Eastern Europeans, and Italians who flooded into the United States. Even after slaves were emancipated, they and their descendants remained in the region, providing a cheap workforce with which few unskilled immigrants sought to compete. Furthermore, the Reconstruction Era South lacked the political institutions and the type of economic opportunities to attract many immigrants, and religious intolerance and bigotry discouraged settlement by Roman Catholics through the first half of the twentieth century.

Thus, British settlements in the American South differed from British settlements in both Pennsylvania and New England. We will begin by examining those British influences on the landscape in the American South first, simply because English settlement within that region began earlier than within New England. Although the first English settlement at Roanoke Island ended in failure, that at Jamestown, Virginia in 1607 led to the establishment by the mid seventeenth century of hundreds of plantations lining the shores of the estuaries flowing into Chesapeake Bay.

9.3 English Landscapes in the American South

The English who colonized the Southern colonies largely came from the south and west of England, an area not only noted for its strong support for the English crown and the traditional, liturgical Church of England, but a region of England that had practiced slavery on a large scale during the early Middle Ages (Fischer 1989). The English who settled in the American South came from the part of England where the manorial system was well established and "dominated by the great estates of the gentry, with their environing clusters of small houses inhabited by tenants and subtenants" (Fischer 1989:245). Those English who migrated came from both classes, the second or third sons among the gentry who had little prospect of inheriting their family estates because of the practice of primogeniture, and the masses of tenants, who lacked the wealth and education of the gentry (Fig. 9.2).

The respect that the Virginians had for the English crown was clearly demonstrated by the large number of places that were named after members of the royal family, not only at the time of the earliest settlements (such as Jamestown and Williamsburg), but even during the years leading up to the American Revolution. For example, in the middle of the eighteenth century many places in Virginia and North Carolina were named in honor of the House of Hanover from which the royal family came, naming counties or cities after the family name, after German localities associated with the House of Hanover, or after King George, his sons Prince Frederick and Prince William, or his wife Queen Charlotte. Thus, just a few examples include New Hanover, Lunenburg, Brunswick, and Charlotte (Phillips 1999:73). It is interesting to note that these names remain today, while those of several unpopular royalist governors and a Tory prime minister were removed. At the same time, names of Whig politicians in Britain, from which the Revolutionaries received considerable sympathy, were kept on the American landscape (Phillips 1999).

Royal names and titles were not the only source of English toponyms adopted in the South. Some counties were named after English counties, with the overall mix of names having "no parallel in England and reflect[ing] the mingling of settlers from different regional backgrounds in the Chesapeake"

Fig. 9.2 Carter's Grove Plantation house, built about 1750, east of Williamsburg, Virginia. Much of what is preserved from the colonial period was erected by the planter elite, who tried to replicate manorial houses erected in southern England. However, such structures (and this photograph only shows the right side of the mansion, with the left side being a mirror image of what is shown) are hardly representative of the type of quarters in which most white colonists in the South lived

(Hornsby 2005:99). In contrast, as we will see later, place names in parts of New England did correspond to how they were spatially arranged in England.

9.3.1 Landscape Legacy of Plantations

The Southern landscape shows influences of both the British elite and the African slaves who worked their fields and built their structures. While most whites in the colonial and antebellum South were not wealthy, the plantation system had a profound effect upon the landscape that evolved within the region. Plantation homes were conspicuous and ostentatious, yet such plantation mansions were homes to only a small percent of the white population. Because planters, particularly in Virginia and Maryland, typically lived on their plantations, unlike many West Indian sugar planters who remained in England and left plantation operations in the hands of overseers, the plantation landscape along the Chesapeake more closely replicated the manors of southern England. As Stephen Hornsby (2005:107) explains, "Unlike the great sugar planters, the tidewater gentry rarely, if ever, visited Britain and had to be content with recreating England in the colonies rather than enjoying England at first hand."

Plantations played a major role, socially, economically, and politically. Small farmers lacked the economic clout to directly market their cotton or tobacco overseas, so they often dealt with nearby plantations that simply added their product to that which was being exported. Likewise, because relatively few market towns developed where local farmers could obtain the skills of blacksmiths, coopers, or other artisans, many small farmers turned to nearby plantations, where slaves with the requisite skills could provide the needed services. Plantations became the symbol of the South. In his article, "Plantations created the South," Sam Hilliard explains that the plantation was simultaneously a symbol of "wealth and prestige," "agricultural production," and "a cruel and exploitative system… [that] played an important role in the development of southern agriculture, economy, and society" (Hilliard 1980:409). Furthermore, landscape scholars have readily identified visible differences in the landscape of plantations established by the English in Virginia, the Carolinas, and Georgia from those created by the French in Louisiana.

Because of the dominance of the plantations and the relative absence of villages or towns, "the neoclassical Anglican churches of Virginia and Maryland were sited not on village greens but rather at rural transportation nodes among widely scattered plantations" (Williams 1997: xiii). Even to this day, many rural areas of the South are dotted with many small churches (Fig. 9.3) along a highway away from any cluster of houses or other buildings.

Fig. 9.3 Wacahoota United Methodist Church in a rural area north of Williston, Florida. This church dates from 1852

Not only did the allocation of lands within the southern colonies not typically involve townships, but the granting of headrights to many early settlers and promoters who brought settlers resulted in a fairly irregular pattern of land subdivision. Unlike in New England, lands were far less likely to be surveyed before the arrival of settlers, who were often permitted to locate and survey their own parcels. While there are places were lands were surveyed into squares, rectangles, and other shapes, the pattern of individual choice, where lands were surveyed and marked by using the metes and bound system, could result in highly irregularly shaped parcels (Fig. 9.4). This was particularly true for those parcels that remained, fitting into those spaces not previously selected by those first staking their headrights (Price 1995).

9.3.2 South Country Cavalier Influences in Virginia

The Virginia landscape was intimately related to slavery, prompting the English to deliberately set themselves apart. Indeed, because the establishment of slavery in Virginia created distinct societies, the "European community saw itself as separate and superior to the slaves and indigenous peoples, and emphasized its Europeanness in defining itself" (Guelke 1992:143). At the same time, the existence of slavery meant that there were few jobs for European settlers in colonial Virginia, given that so many of the jobs were filled by slave labor. Thus, many of the "smaller planters and their children had difficulties making a living in the settled areas and gravitated to the frontier where it was easier to establish an independent livelihood" (Guelke 1992:143). Hence, the characteristics of the English landscape that evolved in the longer settled coastal communities of the South differed somewhat from the British landscape seen in the later settled interior settlements in the South.

9.3.3 English Influences on House Architecture

Housing types in British America varied by the origins of the British immigrants. In Virginia and Maryland building styles evolved from the vernacular architecture of southern England. As the architectural historian Cary Carson notes, "Wealthier planters found that an English west country plan divided a one-story house conveniently into two halves—a hall, inner room and attic chambers for a master's family and a separate work room or kitchen with quarters overhead for indentured servants and slaves" (Carson 1986:55,57).

It was not unusual to design plantation houses replicating the styles in England (Fig. 9.5). For example, "[s]ome Virginia planters either patterned their houses upon specific English country houses or availed themselves of architectural guidebooks published in London to ensure that their homes would conform to the latest British fashions" (Vlach 1993:4).

Fig. 9.4 Irregularly shaped fields in the Shenandoah Valley near Waynesboro, Virginia. Areas of early British settlement not only lacked the regularity of those lands that were later surveyed into townships and sections, but significantly differed from the patterns seen in areas of early French and Spanish settlement

Fig. 9.5 Wright's Chance Plantation house in Centreville, Maryland. This house, probably erected in the late seventeenth century, displays a rigid symmetry that was both typical of Georgian buildings in England and in the English colonies of North America

Fig. 9.6 Tidewater house located in East New Market, Maryland

The great houses were perfectly symmetrical, with a central passage running from front to back. Fischer notes that

> In most respects Virginia's plantation houses were exactly like middle-sized manors in south and west of England during the seventeenth and eighteenth centuries.... The great house was set far back from the road, with a cluster of small outbuildings for kitchens, stables, servants, and elaborate gardens in front. This description of manor houses in the south and west of England fits the great houses of Virginia exactly (Fischer 1989:267, Reprinted by permission of Oxford University Press, Inc.).

Yet differences also existed, prompted by climate and the local availability of building materials. Within tidewater Virginia the lack of local stone necessitated the far greater use of brick, yet many buildings were constructed of wood, which was far more plentiful in America than in England. Furthermore, while "the grand brick Georgian houses, and the manners and mores of the tidewater gentry were reminiscent of some aspects of the landscape and society of southern England,... far too much was different to consider the Chesapeake a replica society of England" (Hornsby 2005:109).

Within the coastal plain of Maryland and Virginia evolved a group of structures termed the Tidewater cottage or the Tidewater house, which had a wood frame construction covered with clapboards (Fig. 9.6). These Tidewater houses are viewed as having evolved from the English cottage, and are characterized as being "one-room deep; one story and often attic; steep roof; two rooms in length; exterior, end chimneys" (Kniffen 1965:565). Over time the style evolved both in size and height, with some being constructed atop a brick foundation, nearly a floor high beginning in the later 1700s.

9.3.4 Georgian Style Architecture

A variety of public buildings in Virginia are classic examples of Georgian style architecture (Fig. 9.7). This architecture "reflected Renaissance ideals made popular in England by architect Sir Christopher Wren" (Poppeliers et al. 1983:18). Not only was this architectural style utilized in rebuilding London following its Great Fire of 1666, but it was widely copied both elsewhere in England and in the colonies. The first building constructed in North America that was designed by Sir Christopher Wren is the Wren Building, built in 1695 on the campus of the College of William and Mary in Williamsburg. Also in Williamsburg, Virginia are the Governor's Palace and Capitol, similarly built in the Georgian style. This style remained popular into the Revolutionary period, with some Georgian style buildings being constructed into the early 1800s. While the work of English architects, as well as immigrant builders and craftsmen, can be seen in some Georgian buildings in the United States, far more commonplace was the use of "architectural building manuals known as pattern books" (McAlester and McAlester 1990:142) by American builders and carpenters.

Fig. 9.7 The Wren Building at the College of William and Mary in Williamsburg, Virginia. This structure, built in 1695, was designed by Sir Christopher Wren, and is a classic example of Georgian architecture, with its rigid symmetry

Hallmarks of Georgian design include a rigid symmetry, axial entrances, geometrical proportions, hipped roofs, and sash windows (Poppeliers et al. 1983). Thus, the window arrangements flanking a central doorway, marked by a classical pediment, were mirror images of each other (Fig. 9.8). By mid-Georgian period Palladian designs were often seen, with giant pilasters that marked building corners and a double or two-story portico (entry porch). Large Georgian style buildings were often constructed in five parts, with a large central area, flanked by low connecting structures leading to larger sections at the opposite ends of the buildings. Georgian Revival architecture appeared during the late nineteenth and early twentieth century, and that style is still utilized by some architects.

9.3.5 Virginia and the I-House

In both the tidewater and piedmont sections of Virginia, as well as in the valleys of the Appalachians, evolved the I-house (Fig. 9.9). It is two rooms wide, one room deep, and has two stories, with its front doorway centered on the long wall between the gable ends. While it may have been influenced by both British and German styles coming from Pennsylvania, it was, as Fischer and Kelly (2000:260) explain, "an American invention that spread through many parts of the United States." On the other hand, Kniffen (1965:555) notes "The 'I' house is one of several compounded from the old English unit consisting of one room and end chimney" and Lenney (2003:222–224) adds that the I-house type is "a loosely defined family of two-story, single-pile houses carried over from Britain." Widely favored by planters and farmers of more moderate wealth, the I-house design was carried west into Kentucky and Tennessee. From there it spread into the Midwestern states of Indiana, Illinois, and Iowa, from where it derived its name, the "I-house," because of its dominance during the later nineteenth century within those states whose names started with that letter. As John Rehder (2004:107) writes, "The I-house is North America's most ubiquitous folk house." Even to this day, the I-house remains one of the more commonly seen older house types in much of the rural Midwest, although they are far less common than they were a half century ago.

In the Upland South I-houses were often built of logs, but examples can be found built of almost any material—lumber, brick, or stone. Within both Tidewater Virginia and the Midwest most I-houses were of wood frame construction. The outside gable chimneys are said to show English traits. Indeed, while noting that considerable variation exists in the construction material and chimney location within I-houses, Kniffen (1965:555) describes those "qualities all 'I' houses unfailingly had in common: gables to the side, at least two rooms in length, one room deep, and two full stories in height." Over time, many received additions of porches or rearward extensions.

Fig. 9.8 Old Customs House in Yorktown, Colonial National Historical Park, Virginia. Note the position of the doors, both centrally located in their walls, flanked by windows displaying rigidly symmetrical fenestration

Fig. 9.9 I-house located north of Sharpsville, Maryland. This style of two story house, with a central front entry flanked by single a room at each side on the first floor, evolved in Virginia and spread throughout much of the Midwest, in particular within Indiana, Illinois, and Iowa

Fig. 9.10 St. Luke's Church in rural Benns Church, Virginia. This gothic building, similar in design to many parish churches in England, was constructed as an Anglican Church in 1632

9.3.6 Anglican Churches

Anglican churches, built in the colonial South, display a strong physical connection with those found in England (Fig. 9.10). Cary Carson explains, "The only specialized public buildings based on English or continental models in the seventeenth century were churches…. Each Protestant sect brought along a church plan and fittings suited to its own liturgy" (Carson 1986:57–58). In addition, Fischer notes that the "Anglican ideal was a small parish church, solidly constructed on a cruciform plan, with an altar and cross at the eastern end of the building and a pulpit tucked into a corner" (Fischer 1989:336). One such example is St. Luke's Church (1632), Isle of Wight County, Virginia, which displays gothic influence. Another can be seen in the ruins of the church built in 1639 at Jamestown. The Anglicans dominated the early religious landscape of colonial Virginia, and 37 Pre-Revolutionary Anglican Churches still remain in Virginia (Williams 1997).

The connection between church architecture in the South and in England has been noted by many scholars. Sir Christopher Wren, the English architect who also designed many church buildings in London following its Great Fire of 1666, sought to build "auditory churches." Thus, Anglican churches in Virginia were designed so that if "additional space was needed, wings were added in the form of a transept" (Williams 1997:104), creating a church in the shape of a cross. He cites Williamsburg's Bruton Parish Church as an example. The majority of the surviving Anglican churches in the South were constructed of brick, ornamented by the use of Flemish bond, in which the laying of the brick alternates in the placement of headers and stretchers.

While many Anglican churches lost members and declined following the American Revolution, when the church lost its official status and was shunned by many members, it made a comeback in the second decade of the nineteenth century and some buildings were restored. Furthermore, the High Church Ecclesiological movement of the mid nineteenth century influenced the construction of Episcopal and other churches. This movement, with Anglican roots, is most conspicuously displayed in the churches of the South (Williams 1997).

9.4 Puritans Create the New England Landscape

The Puritan migration from 1629 through 1640 brought approximately 21,000 persons to New England. Coming largely from East Anglia, itself an area that had witnessed an influx of Angles from western Denmark nearly a millennia earlier, which had significantly altered the culture of that area of east central England, the Puritans comprised the largest migration of immigrants until the arrival of Catholic Irish in the 1840s, fleeing the potato famine (Fischer 1989).

In addition, about 20,000 Puritans went to the West Indies, particularly to Barbados, and many of these individuals, or their descendents, later came to North America after the introduction of slaves to the West Indies reduced the economic prospects of smallholders.

Puritans often "emigrated as congregations seeking sites with ample room for expansion" (Meinig 1986:92). Thus, a village of families in East Anglia was transplanted across the Atlantic. The Puritans settled as nuclear families "more so than any major ethnic group in American history" (Fischer 1989:25), with their age-distribution fairly well balanced, including children through grandparents. Unlike the earliest settlers in Jamestown, who were all male, the Puritans had a more equal ratio between the sexes, approximately 1.5 males to each female. Furthermore, the majority of the Puritans were middle class, and two-thirds were literate. Coming from an area where household industry was commonplace, a far greater proportion of the Puritans were craftsmen or tradesmen than among the other British immigrant groups.

The Puritan migration overwhelmed the earlier arriving Pilgrims in overall numbers. Indeed, only several hundred persons lived at Plymouth, Massachusetts when the Puritan migration began, and thus the Puritans came to shape the landscape of Massachusetts and New England rather than the Pilgrims. Population growth in the centuries since the initial Puritan migration resulted in millions of descendents, estimated as "more than sixteen million by 1988" (Fischer 1989:17), which account from the large population of English ancestry in the northern third of the United States. Yankees of Puritan stock occupied much of New England, eastern New Jersey, and northern New York, and then spread west into northern Ohio, southern Michigan, northern Illinois, and southern Wisconsin.

The legacy of the English colonists can be seen through both the built environment and the place names that Yankee settlers took into the Midwest nearly two centuries after the initial Massachusetts colonization (Lenney 2003). Village greens, similar to those within New England villages, can also be found in some of the small communities established by New Englanders in southern Wisconsin. Likewise, styles of houses have been traced westward from New England across New York into the northern section of the Midwestern states (Kniffen 1965).

9.4.1 English Toponyms in New England

The Pilgrims and Puritans who settled Massachusetts and surrounding areas of New England drew on the names of many places in East Anglia and largely ignored Indian names when naming their towns, even though both Massachusetts and Connecticut are Algonquian. Relatively few towns were named for individuals before the Revolutionary War, as "the custom was un-English" (Lenney 2003:15). The seventeenth century settlers of Massachusetts often named towns after their previous residence or their birthplace (Lenney 2003). Not only did clusters of New England towns reflect the East Anglian, Midland, or West Country origins of the Puritans, but towns might be located in a county that had the same name as their namesake in England. Furthermore, towns in Massachusetts bearing English toponyms sometimes displayed the same spatial relationship as they did in England. For example, collocations "of two or more towns that bear not simply the names but also replicate the geographical proximity of their English originals give some grounds to think that the settlers hailed from that district" (Lenney 2003:37). Examples include Boston and Lynn, Worcester and Leomister, and Amesbury-Salisbury-Newberry-Andover, among others. While the Puritans also adopted a few Biblical names, use of names from the Puritan strongholds in England was more common.

The character of place naming in New England is explored in great detail by Lenney (2003). Besides copying names from their English homeland, New England place names were also bestowed by royal officials and even by royalty. After Captain John Smith had explored the coast of New England, before English settlement had actually begun, the future King Charles I was provided a map and permitted to choose the names, naming the Charles River for himself and capes after a couple of his relatives (Lenney 2003). By the late seventeenth and early eighteenth century, the titular seats of many English peers were utilized by the various royal governors to name newly established towns. While names of British prime ministers and royal governors were bestowed on some New England towns in the mid eighteenth century, some of those named after Tories were renamed during the American Revolution, while those of Whigs were typically retained.

9.4.2 Town Surveys and Land Division

The patterns of surveyed lands in New England are quite varied, ranging from long lots, rectangles and parallelograms to far more irregular shapes consistent with a metes and bounds cadastre. However, most land in New England was first placed into townships, which were then surveyed into parcels—both lots within a village (initially) and outlying farms—before they were sold or allocated to settlers (Price 1995). Because direct land grants to individuals were less common in New England than in the Southern colonies, town proprietaries who received the grant typically had the land surveyed into rectangular parcels that were then distributed to settlers. While not all parcels were equal in size, the regular lot lines were more easily located, deeds were less ambiguous, and litigation was less common (Lenney 2003). The parcels were measured using standard surveying tools, designed to measure area in acres and length in rods (16.5 feet). Thus many parcels, delineated by their long-abandoned stone fences and early road right-of-ways that can be seen today, reflect the English measures then in usage.

The arrangement of fields in New England was influenced by both the concepts of open-field farming, which "was a medieval holdover whose days were numbered" in England (Lenney 2003:105), and the "more modern one of enclosed fields from East Anglia" (Lenney 2003:104). Thus, while Plymouth, Cambridge, and Charlestown, among others, were founded as open-field villages, by the mid-seventeenth century farmers were enclosing their fields and consolidating their outlots. Among the later platted towns, such as those platted in New Hampshire in 1750, "such villages were by then largely a legal fiction meant to comply with royal charters" (Lenney 2003:119). Regardless, New England developed a strong village tradition.

Given the various patterns of land division among the towns and villages of New England, and it has been argued that these are related to both the source region of the English immigrants and the timing of settlement. Edward Price explains that

> much of the diversity among New England villages sprang from diversity among their ancestor villages in England. Recent studies have shown surprisingly close parallels between specific New England townships and the English villages or districts from which their settlers came. These parallels include sizes of landholdings, shapes and orderly arrangement of parcels, equality or inequality in land assignments, open-field or enclosed farming (Price 1995:33)

Yet even with similarity, fields and house lots were typically larger in New England than in England (Price 1995). Field patterns were far less irregular than in the American South.

Stone fences typically separated fields within New England, which once had over 150,000 miles of such walls (Lenney 2003). Today many of these stone fences remain, but they are often hidden in the forests that have reclaimed many of the fields and pastures that were abandoned beginning in the nineteenth century when cheaper crops from the Midwest began to be marketed within the urban centers of the region.

9.4.3 Town and Village Settlements in New England

Puritans established a landscape of villages in which farmers initially clustered, and they developed the concept of New England towns, which included that area within two or three miles of the centrally located village (Fig. 9.11). The village was the center of settlement within the town. As Allen Noble notes, "the meetinghouse or town hall and the adjacent tavern provided the focal point for the settlements" (Noble 1984a:17), from which the contemporary New England

Fig. 9.11 Church and post office facing village green of Angelica, western New York. The formation of villages with churches, shops, and houses focused upon a village green in Upstate New York provides clear evidence of the spread of ideas from Puritan New England towards the west. Similar examples, although less common, can be found as far west as southern Wisconsin

Fig. 9.12 Tombstone in Granary Burying Ground in Boston, Massachusetts. While the tombstones were rearranged into straight lines during the 1930s in an effort to bring more order to the old downtown cemeteries, the winged deathhead carvings on the tombstones still convey the attitudes that the eighteenth century Puritans from East Anglia had about the role of religion in their lives

village evolved. It has been suggested that towns were organized closely around a village for religious reasons. As John Stilgoe explains,

> The early Puritans had strong ideas on how the environment should be structured to safeguard society's moral fiber. Above all, they believed that members of the flock should stay close to the meetinghouse and each others for reasons of spiritual as well as physical safety (Stilgoe 1976:3).

The Massachusetts General Court went so far as to rule in 1635 that "'no dwelling house shall be built above half a mile from the meetinghouse in any new plantation'" (Stilgoe 1976:3). Yet as Lenney (2003:103–4) notes, the requirement that homes be so located "was intended in part as a measure of defense, in part to accord with the English village template, and in part as conducive to a certain type of ecclesiastical polity."

9.4.3.1 Meetinghouses to Churches

The meetinghouse was the center of community, and functioned as the site where both religious services and governmental meetings took place. Elaborate formal church buildings were generally absent from New England until the eighteenth century. The creation of the meetinghouse resulted from "the Puritan desire to house their Bible-oriented services in a place uncontaminated with the sacral associations of the term 'church'" (Williams 1997: xiii). Because the Puritans wanted a place of "worship that could not be regarded as itself a holy place," they created meetinghouses modeled after "late medieval guildhalls" (Williams 1997:5). Their church had no altar, but instead a centrally placed pulpit.

The graveyard was laid out before or by the meetinghouse, and it was typically not as orderly or as well maintained as Anglican graveyards. The presence of the cemetery served as a reminder about sin, and tombstone inscriptions contained warnings, such as death heads (Fig. 9.12). The character of the grave markers did not reflect East Anglian styles, for the simple reason that few gravestones were utilized in either location until the 1640s (Lenney 2003), yet the Puritan beliefs carried over from England influenced the types of markings.

While New England villages are often distinguished by their high steepled churches today, these mostly appeared after the nineteenth century began, as "the old plain-style meetinghouse was reoriented gable-end to the road and graced with a steeple" (Lenney 2003:247). While the Puritan Church remained the established church in Massachusetts until 1833, during the eighteenth century a variety of Anglican churches with towering steeples were built in urban centers, including Old North Church (officially named Christ Church) in Boston. Christ Church has a four-staged brick tower topped by a steeple. These Anglican churches display the neoclassicism espoused by Sir Christopher Wren and further developed by James Gibbs, whose "designs were printed in the 1720s and

Fig. 9.13 Old South Meeting house in downtown Boston, Massachusetts. While dwarfed by modern high rise buildings, this Congregational Church was built in 1729 as a Puritan Meeting House, and it is clearly marked by its 183 foot steeple

widely circulated in the colonies" (Williams 1997:9). Thus, similarly designed neoclassical churches were erected in England and its American colonies. While to the Puritans such steeples were originally viewed with skepticism, with time they and their Congregational successors embraced them (Fig. 9.13). As Stephen Hornsby notes,

> Congregational churches that began appearing in the New England landscape during the late eighteenth century were not mere replicas of their English Anglican counterparts… [as] the Congregational church had married English Georgian style and form with New England materials to create a distinctive New England vernacular building (Hornsby 2005:144-450).

It is these Congregational churches with their tall steeples that one associates with New England villages.

During the nineteenth century many New England villages developed their steeple rows that so distinguish their skylines today. Not only did churches have steeples, but schools had belfries and courthouses and town halls had bell towers or cupolas. As Lenney (2003:141–2) explains, with "the formal separation of church and state that occurred in Massachusetts in 1833 and earlier elsewhere: two buildings were required where one had stood before." At the same time, a schism among the former Puritans had some moving towards Unitarianism, while those who were more conservative became Congregationalists. Congregational churches "continued in the neoclassical Wren-Gibbs tradition, but differed in their degree of ornamentation" (Williams 1997:10). They typically had a large columnated porch or portico and a towering multi-staged steeple.

9.4.3.2 New England Settlement as Nucleated Village

Towns did not always develop highly compact settlements, as many of the farmers began building houses on their farms. Indeed, geographer Joseph Wood (1991:32) asserts that the "village tradition is invented. Puritan communities were commonly dispersed settlements. While constructing compact center villages in the nineteenth century, Romantic New Englanders invented a tradition of Puritan antecedence." He explains that the settlers of New England "organized and chartered towns as instruments of land division and ecclesiastical and political organization" (Wood 1991:34). Wood explains that

> Like towns, villages were instruments of settlement, and like town dwellers, villagers rarely formed compact settlements. Rather they laid out towns and villages as communities of freehold farmers on dispersed, single-family farmsteads, forming a corporation and an ecclesiastical parish--not dissimilar to what one could have found at the time in parts of England. … Towns and villages alike had a center, where were located a meetinghouse and occasionally a tavern. Community interaction was focused on the center, which was not itself a village in the contemporary sense of the term (Wood 1991:34–5, Reprinted by permission of the American Association of Geographers).

Yet by the 1800s the New England village took on its more contemporary definition, with greater compactness of its settlers. The New England village pattern still endures, "even though isolated individual farmsteads also have become a common feature" (Noble 1984a:17).

Fig. 9.14 Church and town hall facing village green in Bloomfield, Ohio. This area was originally part of Connecticut Western Reserve, lands in the Northwest Territory that were claimed by Connecticut until 1800 and which were largely settled by New Englanders. The subdivision of counties into rural townships is associated with settlement from New England. Town government did not evolve in areas settled by British ethnics from southern England, Scotland, or Ireland

Town government initially took place at the meeting house for both religious and secular issues. This provided a type of democratic involvement at a much more local level than the county level so common within the southern colonies. This concept came from East Anglia, where town and parish were interchangeable. Over time, the establishment of political towns and the holding of town meetings spread from Massachusetts into upstate New York, and then across the Upper Midwest into Michigan, Illinois, and Wisconsin (Fig. 9.14). Indeed, even as far away as southeastern Wisconsin the initial settlement was by New Englanders, giving the state its system of towns and a few early established settlements, such as Greenbush with its New England style church and tavern (preserved as Wade House State Historic Site) and Cookesville with its village green, even though subsequent settlement by thousands of Germans and northern Europeans reshaped the ethnic landscape of most of the state.

9.4.3.3 Common Lands and Village Commons

Village commons are frequently seen in New England, and some, such as the Boston Common, are well known (Fig. 9.15). Forty-two percent of the 306 towns in Massachusetts have commons today (Lenney 2003), although many have diminished in size in the centuries since their establishment. These commons, or greens as they are sometimes called, were derived from the tradition of holding lands in common ownership. As John Stilgoe explains, "If the group came from that part of England where common-field husbandry endured in practice or memory, and most groups did, they often laid out common fields around the nucleus of houselots" (Stilgoe 1982:47). Thus, because most of the English who settled in New England came from the lowlands of eastern England, that part of Britain best known for its village greens (Hoskins 1985), this landscape tradition was replicated in New England. The commons not only provided a safe central location to gather livestock together at night, but they also provided a field upon which the local militia could train.

The most widespread shape for a common is a triangular or wedge, typically placed between two roads converging within a village. Commons often cornered on a meeting-house, causing public buildings to be located in the middle of the street (Lenney 2003). Commons in western New England are more likely to be square, rectangular, or oval shaped, as is also true for remaining commons in the Midwest.

9.4.3.4 Urban Street Patterns

Urban street patterns in New England rarely followed a rigid grid (Fig. 9.16). As Peirce Lewis eloquently explains,

> Most New England cities grew in the old-fashioned European way, with main streets following old paths, and new streets and alleys added in haphazard bits and pieces as the need arose. The street plan of Boston is typical--a tangled skein of crooked streets that looks more European than American (Lewis 1990:98).

Fig. 9.15 Eastern side of Boston Common within downtown Boston, Massachusetts, with towering steeple of Park Street Church. Once a common area where animals were brought together and militia training occurred, today Boston Common functions as a downtown urban park. New England settlement, even outside of New England, is often marked by such green spaces

Fig. 9.16 The pattern of roads in older downtown Boston and Charleston, Massachusetts is far less orderly than the more grid-like arrangement within areas of the city settled within the past two centuries (Reproduced from U.S. Geological Survey 15 min topographic map of Boston, Massachusetts)

The pattern in Salem, Massachusetts is similar. Likewise, the pattern of rural roads remains more irregular than the grid that became so typical of the American Midwest.

9.4.4 Puritan Built Landscapes

The architecture in New England was dominated by the use of sawn lumber in timber-framed buildings. Wood-sheathing and clapboards were common siding materials, a continuation of folk custom from eastern England, in particular East Anglia, East Sussex, and Kent. In New England "house carpentry was much like that of eastern England in the many complex details of post and beam construction" (Fischer 1989:63). In areas of Britain outside of East Anglia wood was scarce, and settlers from those areas favored brick and stone, yet New Englanders continued their use of wood, even in areas favored with abundant stone or clay for bricks. Unlike other areas, log construction was very rare even among the earliest buildings in New England. As Fred Kniffen notes, New England "was essentially an area of frame buildings, log construction being regarded as a pioneer expedient hardly worthy of any considerable care" (Kniffen 1965:558). Several types of wood frame houses came to characterize New England settlements.

9.4.4.1 New England House Types

Many of the early houses in New England were simple one room or one bay structures, typically about 16 feet on each side, or with one length being slightly greater. It has been noted that the one room houses of the English are more likely to be square, while among the Scots-Irish and Germans, the bay was "usually at least 5 feet wider than it was deep" (Lenney 2003:186). With the size of the bay determined by the space between posts or crossbeams, houses could be expanded by adding additional bays. New England's Hall-and-Parlor house, with a central chimney located between two bays, was derived from a type "already well established in East Anglia" (Lenney 2003:202). While few such houses remain today, the style evolved into others that are still seen.

Salt-box houses are a simple rectangle of two stories, with a central chimney and a back room roofed by an extension of the roof line from the second floor. In essence, it was a two-story Hall-and Parlor house with a lean-to added to its rear, yet it was all constructed at the same time. With gables at each end of the house, the front eave was above the second floor, while in the rear the roof extended in a straight line from its ridge to back of house, often ending only five or six feet above the ground. Thus, the second floor provides limited headroom, with some of the limitations—at least towards the rear—of the Cape Cod house (Noble 1984a), which is discussed in the next paragraph. The first floor rear of the salt-box house often accommodated the kitchen and pantry. This style is viewed as clearly coming from East Anglia, where it was called an "outshut" (Lenney 2003).

The Cape Cod box (or Cape Cod house) is a structure of one and one-half stories, with a room or two or an attic tucked under the gabled roof (Fig. 9.17). Rectangular in shape, this style developed in eastern England in the late 1500s and early 1600s. This style was commonly seen throughout colonial New England, and it remains popular today "long after it was abandoned in England" (Fischer 1989:66). Indeed, within portions of Vermont over half of the houses were of the Cape Cod style during the early-mid nineteenth century (Lenney 2003), and the style reappeared throughout the northern section of the United States in the mid-twentieth century.

Garrison houses are another house type seen in New England (Fig. 9.18). These structures are of two to two and a half stories with their gables at the sides. The key diagnostic characteristic of the garrison house is their second-story overhang over the first floor wall, with the "jetty" projecting out one or two feet (Fischer 1989). The entrance door is positioned under this overhang on the eve side, or long side, of the house. Allen Noble (1984a:22) notes that the "second-story overhang is a relict feature stemming from house design in medieval Europe." Garrison houses are conspicuous today in older neighborhoods of East Anglia from which many Puritans emigrated, including the city of Norwich. The garrison house was a common house type, especially in the Connecticut Valley, and Paul Revere's house within Boston is also an example.

The New England Large House is typically wood framed and of two to two and one-half stories, and it is sometimes termed the central chimney colonial house (Fig. 9.19). This house was very symmetrical, with its chimney located in the exact middle of the house. Its window openings, all of the same size, are equally spaced. It was large, with ten rooms being typical, five bays on each floor, and had two stairways. Evolutionary variations of this house type have been traced by Kniffen (1965) across upstate New York into the Upper Midwest.

9.4.4.2 New England Barns

The New England English barn is also distinctive (Fig. 9.20). It is a small rectangular timber-framed barn, with two or three bays if one counts the center runways. The central threshing space was flanked by rooms or bays on both sides. The basic structure could be expanded into a larger barn by adding alternating bays and runways. As Noble explains, in keeping with "English tradition, New England barns are timber framed, held together by mortise-and-tenon joints" (Noble 1984b:17). They had a steep roof, often exceeding 45 degrees, and they were sided with vertical boards (Noble 1984b). As Yankee settlers moved into the Midwest, they built similar barns "with remarkably little change" (Kniffen 1965:558).

Fig. 9.17 Cape Cod style houses with clapboard siding, little roof overhang on their gabled ends, and one and a half stories remained popular in the twentieth century, yet the design originated four centuries earlier in eastern England

Fig. 9.18 Paul Revere House in Boston, Massachusetts. Built in 1680, it is not only the oldest surviving house within downtown Boston, but it is a classic example of a garrison house, with its second floor jutting outwards over part of the sidewalk

Fig. 9.19 New England large house near West Townsend, Massachusetts

Fig. 9.20 New England barn, with wind vane atop the square cupola ventilator on roof, near Ely, Vermont

Fig. 9.21 New England connected barn with stone house built in 1845 located near Plymouth Notch, Vermont

The New England connected barn was an early modification of the New England barn, in which the barn was connected to the farmhouse via a series of buildings (Fig. 9.21). Thus, between the barn and house one might find the tool shed, summer kitchen, wash house, stable, and an earlier house or addition. This style is found mostly in northern and eastern New England, basically north of the southern boundary of Massachusetts. In a mid-twentieth century survey, the connected, or connecting barn as it was sometimes named, comprised about one-fifth of the region's barns (Zelinsky 1957).

The gable-entry banked barn was built with its long axis parallel to the slope (Fig. 9.22). At one end the second story is entered, while on the opposite side its entrance leads into the basement. Allen Noble argues that the "gable-entry barn appears to be a derivative of the type that occurs through the rough country of the English Lake District" (Noble 1984b:39).

9.5 Quakers Shape the Pennsylvania Landscape

English settlement in the area between the Southern colonies (from Maryland south) and New England took place in a region of considerable ethnic diversity. New York had been established by the Dutch, who first relinquished their control of New Amsterdam to the English in 1664, and it had long attracted a wide variety of immigrants. While the establishment of Pennsylvania and West Jersey by the Quakers drew large numbers of this persecuted pacifist group from the North Midlands of England, where they were most numerous, it also drew Quakers from Wales and Ireland. Yet, it is even more difficult to trace English influence in this region along the Delaware River, because of the major influence of other ethnic groups upon the mixing of cultural values and traditions. Indeed, there were major Scandinavian (especially Swedish) settlements in Delaware Valley before the Quakers began arriving, and the Schuykill River that flows though Philadelphia was named by early Dutch settlers from New Amsterdam. Not all Quakers came from the British Isles, and Dutch and German Quakers came from the Rhine Valley, where William Penn had worked as a missionary (Fischer 1989). Other German Pietists were also attracted to Pennsylvania, including Mennonites, Moravians, and the Dunkards or Brethren. Out of this interesting mixture came the distinctive Pennsylvania Town, recognized for its cultural distinctiveness (Zelinsky 1977, Lewis 1990), resulting from a blend of cultural contributions from several ethnic groups, especially the Midlands English and German. New Jersey received Dutch settlers, Germans, Swedes, English Quakers, and even New Englanders. Indeed, Newark was established by Puritans from Connecticut. Thus, the resulting cultural diversity is displayed in New Jersey's "rich architectural legacy of patterned brick Quaker houses, bank-ended German houses, flared-eaved Dutch houses, and New England saltboxes" (Lenney 2003:183).

Fig. 9.22 Gable-entry barn near Ashby, Massachusetts

Yet, certain aspects of British culture did become imprinted upon the regional landscape of the Mid-Atlantic states. The Quakers, under the direction of William Penn, laid out the streets of Philadelphia in a grid pattern, unlike what was generally seen in Massachusetts or Virginia. Having an antipathy towards memorializing personal names, streets were numbered or named for objects. The regional architecture reflected a blending of both British and German styles, yet even that inspired by the English was distinct from that found within the British colonies both to the north and south. To this day one is far more likely to see buildings of brick masonry construction in the Mid-Atlantic Region than in either New England or the South.

9.5.1 Ethnic Architecture in the Delaware Valley

Many of the early houses of the Quakers in the Delaware Valley were built of fieldstone. David Hackett Fischer describes the links between the Old and New Worlds.

> These country houses of the Delaware Valley were similar in outward appearance to farm houses in the north of England. Methods of masonry in the Delaware resembled split-cobble and fieldstone farm buildings of the Lake counties, north Lancashire, west Yorkshire, east Cheshire and the Peak District of Derbyshire (Fischer 1989:476, Reprinted by permission of Oxford University Press, Inc.).

The Midlands region of England experienced a construction boom during the early seventeenth century, resulting in many stone and brick houses (Hoskins 1985). Thus, it should not be surprising that numerous brick houses were constructed in eastern Pennsylvania, given the origins of both its English and German settlers. The appearance of small pent roofs and door hoods above windows and doors, respectively, is also common on houses, meetinghouses, and barns in Pennsylvania and western New Jersey. This is similar to their usage in the north Midlands (Fischer 1989).

9.5.2 Quaker-Plan Houses

Architectural historians define a distinctive "simple cottage with three rooms on the first floor, a corner stair leading to a full second story, and a chimneystack with several fireplaces on one wall" (Fischer 1989:479) as the Quaker-plan house. Some scholars claim that both this house and a larger version with four rooms on both floors bear a strong resemblance with houses found in the north Midlands (Fischer 1989). Yet others, such as Noble (1984a:45) note that "the origins of this structure are diverse yet strongly related to the influence of Germans from the Rhine Valley." (See also Chap. 11.) Yet it also has characteristics of a two-story Georgian house in which the entrance door is off-center, thus being a "two-thirds Georgian" style house. Given the mixture of settlers

Fig. 9.23 Free Quaker Meeting House in Philadelphia, Pennsylvania. Constructed of brick in 1783, the plain simplicity of this house of worship is clear, and steeples and towers are lacking

within the greater Pennsylvania region, this house style probably displays elements from more than one area of Europe.

9.5.3 Quaker Meetinghouses

Quaker meetinghouses were, and remain, clearly distinguishable from the churches of other European Christians (Fig. 9.23). During the colonial period, "[m]eetinghouses in both England and the Delaware tended to be simple rectangular buildings, with massive stone walls and plain white shutters" (Fischer 1989:524), with double doors so that men and women had different entrances. Meetinghouses lacked steeples that were so common in Anglican churches and later in the Congregational churches, yet they typically had more windows and were better lit. The architecture of the Quaker meetinghouses also influenced the style of churches constructed by other Reformist congregations throughout the Delaware Valley. Indeed, the "exteriors of some Calvinist and Lutheran churches were sometimes indistinguishable from Quaker buildings" (Fischer 1989:526). Nevertheless, the interiors of the churches of the others did differ from that of the Quaker meetinghouses, having pulpits and either altars or communion tables. During the nineteenth century the Quaker meetinghouses often incorporated various Georgian decorative features, but they remained plainer than the churches of other groups.

9.6 Scotch-Irish Landscapes in the Appalachians and Ozarks

Settlement by Scots-Irish frontiersmen was directed towards the Upland South from both Philadelphia and Charleston, the ports in which they most frequently disembarked. Pennsylvania authorities encouraged the Scotch-Irish to settle away from Philadelphia where they could act as a buffer between the Quakers and Indians (Daniels 2002). The Scotch-Irish settlers then expanded into the Upland South following the Shenandoah Valley (Great Valley), from which they traveled west via the Saluda and Cumberland Gaps. While they were not the first to explore the Appalachians, they were intimately involved with the first effective settlement of the region (Fig. 9.24). As John Rehder explains,

> It was not until European settlers arrived to build blockhouses, clear land, and establish farmsteads and stations that the first effective settlement could be realized. That first permanent layer created by eighteenth-century Scotch-Irish, German, and English settlers in Appalachia established patterns that have endured through more than two centuries (Rehder 2004:73).

The interaction of the Scots-Irish with other Europeans, such as the Germans, resulted in the distinct landscape seen in the region today. Yet it were the Ulster Scots who "became typecast as the archetypical American pioneers" (Devine 2003:155).

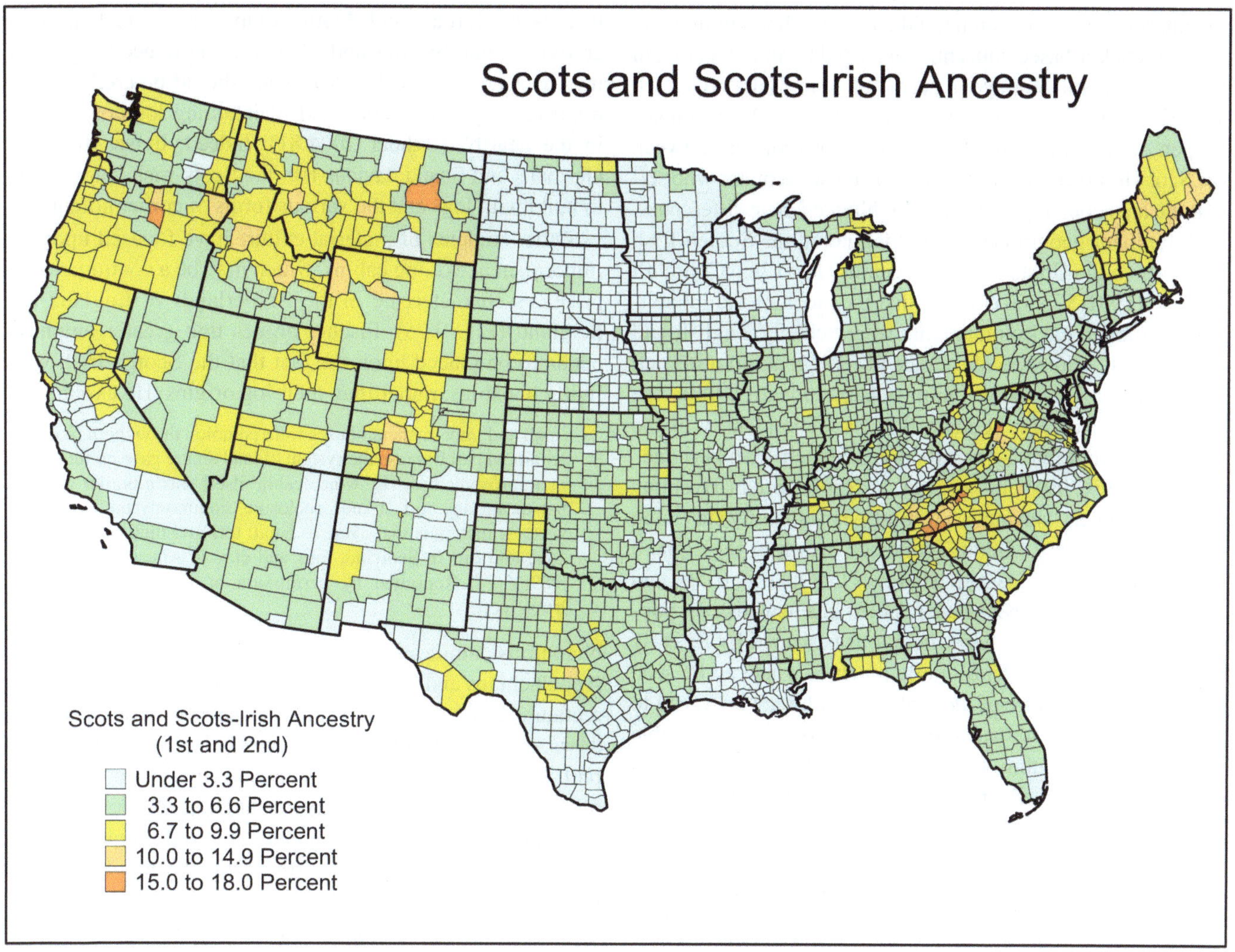

Fig. 9.24 Map showing the percentage of individuals who reported their ancestry on the 2000 Census indicating that Scots or Scots-Irish was their primary or secondary ancestry. Note the largest concentrations of ethnic Scots and Scots-Irish are within the central Appalachians and Upland South, the northern half of New England, and within many counties of the interior West

Unlike the settlements of Tidewater Virginia, Maryland, and the Southern coast of the Carolinas and Georgia, the Scots-Irish had little interaction with African slaves. Slavery was far rarer among the Scots-Irish, as they settled areas not appropriate for plantation agriculture. Over a century ago a well-know geographer, Ellen Churchill Semple, explained,

> If the mountains have kept out foreign elements, still more effectually have they excluded the negroes. This region is as free from them as northern Vermont. There is no place for the negro in the mountain economy, and never has been. In the days of slavery this fact had momentous results. The mountains did not offer conditions for plantation cultivation, the only system of agriculture in which slaves could be profitably employed. The absence of these conditions and of the capital wherewith to purchase negroes made the whole Appalachian region a non-slaveholding section. Hence when the rupture came between the North and South, this mountain region declared for the Union, and thus raised a barrier of disaffection through the center of the Southern States (Semple 1901:594).

Many areas of Appalachia and the Ozarks predominately settled by Scots-Irish have few African-Americans among their population today. Even some counties in northeastern Mississippi and northwestern Alabama lack significant black populations. This historical lack of slavery, lack of plantation development, and lack of settlement by immigrants from southern England, or later from elsewhere in continental Europe, have shaped a region whose cultural landscape stands in sharp contrast from that seen elsewhere in the South.

9.6.1 Settlement by Scots-Irish

Settlement by the Scots-Irish was typically spontaneous, not organized. Thus, an emphasis was placed on individual actions. As explained by Russel Gerlach, "Scotch-Irish settlement of the interior Ozarks was an individual effort, although in many cases it was kin-based with entire extended families locating in the same general area over a period of some years" (Gerlach 1992:291). The same generally holds for the Appalachians proper. As we will see in Chap. 11, the sharp contrast between the Scots-Irish settlers in Missouri and those Germans who settled the area contemporaneously highlighted their differences.

Prominent geographers have long noted the Scotch-Irish influence in the Upland South. A century ago Ellen Churchill Semple (1901:592, 601) noted that the stock of the inhabitants of the Kentucky mountains were "chiefly English and Scotch-Irish, with scarcely a trace of foreign admixture" and that their ballads show "unmistakable identity with the poem[s in]…Ancient Scottish Ballads." The identity of the Upland South remains "based upon a Scotch-Irish heritage" (Rehder 1992:99). In contrast, Terry Jordan-Bychkov (2003:9) argues, "any notion of European ethnicity surviving into the quaternary Middle Tennessee hearth of the Upland South is nonsense." Instead, he argues that intermarriage and acculturation with others have resulted in a new distinct population whose culture, nevertheless, was shaped by its early Scotch-Irish settlers. Thus, while much of what can be seen in the material culture of the Upland South may be traced to its Celtic roots, these structures have undergone considerable transformation within the Upland South. Henry Glassie similarly commented upon the mixing of heritages within the region:

> in the 18th century the Irish [Scotch-Irish], English, and Germans combined to build America. The Irish contribution is difficult to distinguish because it was integral to the formation of the American whole. The farm on the frontier in its solitary setting with its log house and scattering of outbuildings, whether erected by the Irish or not, meshed Irish, English, and German traditions into something new (Glassie 1986:75, Reprinted by permission of John Wiley and Son).

Thus, a variety of scholars have tried to explain the ethnic origins of log cabins, the vernacular housing of the region throughout the eighteenth and nineteenth centuries. Jordan-Bychkov (2003) notes that tens of thousands of these log cabins still remain, some of which remain occupied. Nevertheless, log cabins are far less conspicuous today than they were a half century ago.

9.6.2 Log Cabins in the Upland South

While log cabin styles became associated with the Scots-Irish, it has been argued that this construction method was learned from other ethnic groups living in southeastern Pennsylvania or the Delaware Valley. Indeed, neither the English nor Scots built log cabins in Great Britain. Claims have been made for both German and Swedish origins (via Delaware Valley settlements in the late 1630s) of log buildings. John Rehder argues that Terry Jordan and Matti Kaups are correct in their arguments that Swedes and Finns first introduced log buildings to America, but that "Germans should be credited with reintroducing, reinforcing, and maintaining log construction in the hearth" (Rehder 2004:81). Indeed, it was from the Germans and Scandinavians that the Scots-Irish learned how to build with logs and the various forms of corner notching.

Other students of material culture have focused upon the impermanence of log cabins, which may be a continuation of practices characteristic in the Borderlands of England and Scotland (Fig. 9.25). Fischer argues that greater emphasis needs to be placed upon the fact that the Scotch-Irish were living in cabins, rather than their log origins. He explains,

> Small and impermanent houses were common throughout North Britain, in part because the system of land tenure gave no motive for improvement… The word 'cabin' itself was a border noun that meant any sort of rude enclosure, commonly built of the cheapest materials that came to hand: turf and mud in Ireland, stone and dirt in Scotland, logs and clay in America (Fischer 1989:656-7, Reprinted by permission of Oxford University Press, Inc.).

Numerous similarities in the design of the cabin in the Appalachians and in northern Britain are itemized, including rectangular single room dwellings, earthen floors, few windows, and both front and rear doors (Fischer 1989). Cabins built by the English were typically square, while those built by Scots-Irish were typically rectangular. As John Rehder (2004:99) notes, "The rectangular cabin represented a direct descendent of dwelling dimensions from western Britain and northern Ireland."

Several types of log cabins are found throughout greater Appalachia. The single-pen cabin consisted of one room, which averaged 20 by 17 feet, with a fireplace located at one gable end and doors in both the front and rear. This one room cabin was sometimes called the "English-plan" cabin, and in many areas of the Upland South it was the most common style (Jordan-Bychkov 2003). The double-pen cabin consisted of two rooms, and three varieties of the two room cabin are found. The basic double-pen house, called the Cumberland House, had two front doors, with a fireplace at each end of the house. The saddlebag house had a central chimney between the two pens. Noting that its "origins are almost certainly British," Terry Jordan-Bychkov (2003: 34) describes its introduction into the Upland South via Tidewater Virginia and the South Carolina-Georgia Low Country. From there the dwelling style was carried inland and noticed by "southwestward-migrating Midlanders, who quickly adopted it and rendered the plan in notched-log construction" (Jordan-Bychkov 2003:34). The dog-trot house had an open air passageway between the two pens, with the rooms having their fireplaces located along their outer wall opposite their doorways that opened onto the central walkway (Fig. 9.26). While Fischer (1989) notes that the breezeway arrangement

Fig. 9.25 Single-pen log cabin along Blue Ridge Parkway in North Carolina. Jess Shields cabin was constructed about 1815 and it was occupied until the middle of the twentieth century

Fig. 9.26 Log dog-trot cabin located in Land Between the Lakes National Recreation Area in Tennessee. An open air passageway runs from the front porch between the two rooms, each which are entered from the passageway and have a fireplace along their outer wall. Note the dove-tail notching on the corners

of the dog-trot had long been used in Britain, Jordan-Bychov (2003) argues for a Finnish origin, brought to New Sweden, and then spread down the Great Valley by settlers from Pennsylvania. Because dog-trot houses sometimes resulted when a "English-plan" cabin was enlarged, and because of the predominance of the Scots-Irish among the settlers in the Upland South, their role in adopting the dog-trot house which eventually spread from the Upland South across the Piedmont and into the northern section of Florida needs to be recognized.

9.6.3 Farm Buildings and Their Arrangement in the Upland South

The arrangement of outbuildings in the farmsteads of the Upland South has been attributed to the British Borderlands region. John Rehder notes that the "dispersed settlement patterns and the sprawling arrangement of outbuildings on individual farms find identity with Northern Ireland and Scotland" (Rehder 1992:118). While certain types of farm buildings became associated with Appalachian areas settled by the Scots-Irish, their vernacular barn and crib-barns and snake fences are not directly related to the Old World, but are associated with development by the Scotch-Irish in the U.S.

In some respects the transverse crib barn (Fig. 9.27) so common in the Upland South reminds us of the dogtrot cabin, given that in the barn a runway separates cribs or stalls on either side, their doors all facing onto this interior runway. In contrast with the dogtrot house, the gables face to the front and rear within the transverse crib barn, with the central passage runway situated under the roof ridge. While most of these barns are wood frame structures, some the cribs are of notched log construction, permitting dual passageways to bisect the barn both under the gables and from eave to eave. Sheds could be added to the sides of the barn, providing additional space as the barns were laterally expanded. In this respect, as Jordan-Bychkov (2003: 53) notes, the ethnicity of their builders can be seen.

> Lateral rather than vertical expansion typified the houses and barns of the Celtic peoples of the British Isles, including the Scotch-Irish. Germans in Europe did precisely the opposite, enlarging upward (Jordan-Bychkov 2003: 53).

Because of the dominance of the transverse-crib barn, Jordan-Bychkov (2003) views it as "diagnostic" of the Upland South, defining the region.

Because extensive free range grazing of cattle and hogs typified the nineteenth century farms of the Upland South, as had been the case in Celtic Britain, fences were often utilized to keep livestock out of fields rather than to keep them penned in. As Jordan-Bychkov (2003:44) explains, "when the farmer abandoned a field, the 'stake and rider' rail fence, which had no posts, was typically disassembled and taken to the new clearing." There were several types of wooden split rail folk fences utilized in Appalachia besides the stake and rider, including the snake, worm, or zig-zag fence and the

Fig. 9.27 Transverse crib barn southwest of Bland, Virginia. Such barns are common in the Upland South where the populations were dominated by persons of Scots-Irish ancestry

post and rail fence (Rehder 1992). Although the actual design of the fences cannot be fully attributed to the Scots-Irish ancestry of their builders, their use and the continuance of open-range livestock grazing in parts of the South into the last half of the twentieth century clearly demonstrates a continuance of the population's Celtic heritage.

9.6.4 Scotch-Irish Arrangement of Space

The arrangement of the farmsteads and settlements of the Scots-Irish and their descendents was distinctive. Henry Glassie argues that the "The Irish dimension in the architecture of the American frontier is not a matter of a few discrete forms but of an approach to space generated from social values" (Glassie 1986:76). He notes the influence of enclosure that reshaped the British landscape creating "dispersed, independent and profitable farmsteads." Yet, he explains that "the Irish had always lived on separate farms" and "the independent family farm, so fundamental to American social history, can be read as a direct translation to the New World of an elder Irish tradition" (Glassie 1986:76,79). These ideas eventually became incorporated in the law governing the distribution of homestead lands in the American Midwest and Great Plains, requiring the homesteader to both live on his farm and to plant a portion of it into crops to gain the free land.

While it is in the rural landscape that most ethnic influences of Scots-Irish settlers can be seen, "Even some towns in the Appalachians bore the mark of Ulster urban tradition in the elongated market-place known as 'the Diamond' which is found to this day in several places in Northern Ireland" (Devine 2003:158).

9.6.4.1 Land Tenure and Land Surveys

Lands within those areas of the South that were settled early by the Scots-Irish were surveyed using the metes and bounds system. Following a system used elsewhere in the South, lands were not surveyed in advance of settlement, but individuals marked off their holdings by the use of landmarks, such as prominent trees or rocks, roads, and streams, and directions and distances from those landmarks. The result was not only a patchwork of highly irregular shaped lands, but also considerable litigation, given the impermanence of some of the landmarks and frequent contradictory boundaries.

Among the Scots-Irish land tenure was even more confused, given their propensity to squat upon lands without the benefit of a formal title. Russel Gerlach (1992) notes that the "Scotch-Irish were prone to occupy land without the formality of ownership, a tradition with strong Ulster antecedents. Squatting, as the practice was known, was widespread throughout the Scotch-Irish settlements of the interior Ozarks" (Gerlach 1992:292), and it persisted into the twentieth century.

The lack of order in their farmsteads extended to their promiscuous dumping of trash and broken-down equipment in their landscape. Throughout the Upland South, extending south into regions of northern Florida where the Scots-Irish and their descendents settled, trash is frequently dumped alongside country roads and into ditches and streambeds. McWhiney (1988:233) comments, "In Ireland and Scotland just as in the South, …numerous instances of one or more inoperable automobiles or rusting farm machines parked in a householder's front yard" can be seen.

9.6.5 Scotch-Irish Place Names

Regions of the United States that were first settled by the Scots-Irish can be clearly distinguished by their toponyms. The Scots-Irish rarely used names referring to royalty, unlike their common use by the Virginia Cavaliers. Place names were often taken from locations in North Britain. For example, Cumberland is a northwestern English border county, and it was used to name the Cumberland Mountains, the Cumberland Plateau, the Cumberland River, and the famous Cumberland Gap, as well as to name communities and counties in most states of the Appalachian region. In addition the Scots-Irish frequently used family and clan names for naming settlements. Indeed, as McWhiney (1988:4) explains, "All but 6 of the 112 Kentucky counties named for persons have Celtic names." Family and clan names were rarely used in Puritan New England or in Anglican Virginia.

9.7 Welsh Ethnic Landscapes in America

Nearly two million persons reported Welsh ancestry on the 2000 U.S. Census, yet most members of this ethnic group are widely dispersed across the nation. While the Welsh contributed to the early Quaker settlement in Pennsylvania, within only three states, Pennsylvania, Idaho, and Utah, did as many as 1.5% of the population report Welsh as their first or second ancestry in 2000. The 5-year American Community Survey estimates for 2010–2014 were lower, with only one of the three states showing that large of a proportion. In Utah 2.0% of the population indicated Welsh ancestry. The presence of the Welsh in Utah and Idaho, as with the greater than average proportion who are English, is associated with the large Mormon population within the region, given that the Mormons attracted many converts from those two ethnic groups in the 1800s. Indeed, 20.8% of Idaho's Oneida County was of Welsh ancestry in 2000, the largest proportion in the nation. Yet, the ethnicity of the Welsh in this area is overshadowed by the dominance of the Mormon Church on the landscape, forming the Mormon Culture Region that Donald Meinig (1965) has eloquently documented.

9.7.1 Distribution of Population of Welsh Ancestry

There are three counties in northeastern Pennsylvania where Welsh ancestry is reported by over 5% of the population, and it exceeds 2.5% in an additional four counties of Pennsylvania, plus three in upstate New York and two in Ohio. The Welsh settlements in these areas, plus southern Wisconsin, were initially primarily of farmers while later Welsh settlement was related to mining, quarrying, and tinplating, particularly near Scranton and Wilkes-Barre, Pennsylvania (Allen and Turner 1988).

Initially the Welsh settlements were distinctive, standing "out as people who spoke a foreign language very different from English and who tended to cluster together in clannish rural and industrial settlements" (Knowles 1995:247). However, subsequent arrivals of other Americans or new immigrants from elsewhere diluted Welsh influence in many locations. Some of the Welsh settlements were specifically placed on less desirable lands, simply because there they faced less competition from other groups in acquiring lands. As Knowles explains,

> Sustained access to affordable land was essential process because, unlike the Germans and Scandinavian immigrants who quickly established communities of several thousand residents in the rural Midwest, the smaller numbers of Welsh typically needed ten to twenty years to gather even one thousand settlers in a single community. The Welsh tended to settle in hilly areas not so much because they felt an affinity for rugged landscapes but because American and wealthier European immigrants bought up the best land on the agricultural frontier. The leftover islands of unclaimed or lightly developed inferior ground posed challenges that were familiar to Welsh hill farmers who were glad to be able to buy land adjoining friends' and relatives' farms" (Knowles 1995, "Immigrant Trajectories through the Rural-Industrial Transition in Wales and the United States, *Annals of the Association of American Geographers*, 247, Reprinted by permission of American Association of Geographers).

The Welsh, as a group, established rural enclaves where they maintained close-knit communities, maintaining "their religion, their national traditions, and their language" (Davies 1982:3).

9.7.2 Religious Buildings in the Welsh Ethnic Landscape

Religious buildings are the most distinctive feature of the Welsh ethnic landscape in America (Fig. 9.28). Many of the Welsh were Calvinistic Methodists, a large number who affiliated with the Presbyterians in 1919, while others joined the United Church of Christ. In Wisconsin there was a separate Welsh Presbyterian presbytery until the 1950s. Churches were often the first buildings constructed in new Welsh

Fig. 9.28 Peniel Community Church and ty capel near Peniel, Ohio. The Welsh church on the left has two front doors, common among Welsh ethnics, and the ty capel is located to the right

settlements, sometimes built even before houses. As Michael Struble and Hubert Wilhelm (1992) explain, the most visible features of the Welsh ethnic landscape in their settlement in southern Ohio are religious buildings, their small chapels and the ty capel. The ty capel, or house chapel, was a small building near a rural Welsh church. It provided a place to serve dinner to their minister following services and a meeting place to conduct church business. While ty capels are only known in southern Ohio, small distinctive wooden Welsh churches, often with two doors in their front gable end, can be found in the rural ethnic islands settled by the Welsh (Struble and Wilhelm 1992). Typically, Welsh congregations, such as those in Wisconsin, were small, with fewer than 75 members (Davies 1982). Within these small chapels, Gymanfas, great meetings or preaching festivals with associated hymn singing in the Welsh language, are held. Tombstones in Welsh cemeteries are frequently inscribed in Cymraeg, also perpetuating the memory of the Welsh language. In contrast with their churches and ty capels, Welsh ethnic farm buildings are rare. Most Welsh farms in Ohio and Wisconsin lack ethnically distinct buildings.

9.8 Cornish Ethnic Landscapes in America

The Cornish, a Celtic population that occupy Cornwall—the southwesternmost county of England, are a much smaller ethnic group than the Welsh in the United States. Indeed, on the 1990 Census only 3991 persons declared a Cornish ancestry, and one-quarter of this group were in Wisconsin. The 2000 Census did not report statistics for this group. A far larger number of Cornish had immigrated to United States, but most list English as their ancestry, as indeed Cornwall has been governed as part of England for well over a half millennia. Indeed, the Cornish diaspora, not only to the United States, but also to South Africa, Australia, and parts of Latin America has been well documented. It was fueled both by poverty at home (the Cornish suffered from the potato famine at the same times as the Irish and the development of overseas mines made many mines in Cornwall uneconomical) and by opportunities for Cornish miners overseas (Payton 2005).

9.8.1 Cornish Ethnic Population Distribution and Mining

The skills that the Cornish learned working in Cornwall's many tin mines were highly sought in America's mining districts of the nineteenth century, and the Cornish filled a disproportionate share of the skilled and managerial positions. The presence of Cornish miners was particularly significant in developing the lead mines of southwestern Wisconsin and the copper mines of the Keweenaw Peninsula of Michigan's Upper Peninsula. From these locations Cornish miners participated in the California Gold Rush and the establishment of the iron ore mines in Minnesota's Vermillion Range (Rowe 2004).

The largest remaining ethnic island of the Cornish in the nation is found in the vicinity of Mineral Point, Wisconsin. Between 1835 and 1850 an estimated 8500 Cornish miners came to Iowa, Lafayette, and Grant Counties, Wisconsin (Holmes 1944). While contemporary figures are unavailable, it was estimated in 1940 that 20,000 persons of Cornish descent lived in Wisconsin, centered in the southwestern region of the state where the Cornish had immigrated to work in lead mines. Thus, the Cornish were among the earliest European ethnics in Wisconsin.

9.8.2 Cornish Ethnic Buildings

The most conspicuous aspect of the Cornish ethnic landscape in southwestern Wisconsin is displayed by the stone buildings they constructed in Mineral Point, Dodgeville, Linden, and on numerous farms within the area (Fig. 9.29). A Cornish historian, Philip Payton, writes, "In Mineral Point, Cornish architectural influence is reflected to this day in the solid limestone cottages in Shakerag Street" (Payton 2005:137). Cornish buildings were typically of stone masonry construction, given that many of the immigrants had been stone masons and had immigrated to work in underground mines. In Mineral Point and Dodgeville can be seen Cornish miner's cottages. These are one-story stone buildings, whose chimneys are placed at their gable ends. In a few homes an additional room was added by using double-gabled design, with side-by-side gables, a style similar to what one sees in many of the older churches of Cornwall today. A noted architectural historian describes these houses the Cornish miners built around 1830.

> The fronts are cut limestone blocks laid with very thin joints and the backs are random ashlar. A typical one-story miner's cottage is the Ingraham house which was similarly faced with dressed stone on the front and ashlar on the sides and back (Perrin 1981:114).

In addition, the native stone walls of the Cornish "houses were two feet thick, the windows deeply sunken" (Holmes 1944:12). While the use of limestone in Wisconsin rather than the granite that was common in Cornwall resulted in some differences in external appearances of the structures, the masonry was similar. Not all buildings are of a single floor, and the heritage of the Cornish builders is now preserved at Pendarvis House, which abuts a two-storey masonry house, both of which comprise a state historic site.

The Cornish were overwhelmingly Protestant, as displayed by their old masonry churches. Indeed, Trinity Episcopal Church in Mineral Point, which they built 1839 between 1845, is the oldest church building in continuous use in Wisconsin.

Fig. 9.29 Pendarvis House in Mineral Point, Wisconsin. The Cornish miners who settled Mineral Point in the early nineteenth century were highly skilled at constructing stone masonry buildings, including these two houses, in which carefully cut blocks are closely fit

While a greater proportion of the Cornish are Methodist, some are Primitive Methodist, with very plain structures.

Although many commercial buildings in Mineral Point display Cornish heritage through their stone masonry construction, the Cornish also show their ethnicity by the presence of specialized food shops and restaurants advertising Cornish saffron buns and pasties, which contain vegetables and meat in a triangular pie crust, having first been developed in Cornwall to serve as a meal for miners working underground.

The legacy of the Cornish, as well as several other ethnic groups in the copper mining of Michigan's Keweenaw Peninsula, is also showcased in Keweenaw National Historical Park (Liesch 2011). There, the Cornish were the first emigrant group who developed the mines and built many of the mine buildings, although relatively few of their descendants live in the area today. The Cornish also left a lasting legacy in Grass Valley and New Almaden, California, where many Cornish miners came to seek gold and eventually settled into farming (Payton 2005).

9.9 British: American Interactions

Given its history as a former British colony, it is not surprising that a strong British influence can be seen upon much of the American landscape. Yet that influence itself is quite varied, not only reflecting the regional differences of the various British groups that settled in America, but the historical interactions of those settlers with both other ethnic groups and environmental conditions that they encountered in America. Furthermore, over time Americans influenced what was British, as many American ideas were conveyed back across the Atlantic to the British Isles. In its over two and a quarter centuries as a nation—and indeed for over a century before independence—American conceptualizations of the land and how it should be shaped were simultaneously diverging from British notions and striving to emulate the latest architectural styles coming from the United Kingdom.

References

Allen, James P., and Eugene Turner. 1988. *We The People: An Atlas of America's Ethnic Diversity*. New York: MacMillan Publishing.

Bennion, Lowell C. Ben. 2001. Mormondom's Deseret Homeland. In *Homelands: A Geography of Culture and Place across America*, ed. Richard L. Nostrand and Lawrence E. Estaville, 184–209. Baltimore: Johns Hopkins University Press.

Carson, Cary. 1986. English. In *America's Architectural Roots: Ethnic Groups that Built America*, ed. Dell Upton, 55–61. Washington: Preservation Press.

Daniels, Roger. 2002. *Coming to America: A History of Immigration and Ethnicity in American Life*. 2nd ed. New York: Harper Perennial.

Davies, Phillips G. 1982. *The Welsh in Wisconsin*. Madison: The State Historical Society of Wisconsin.

Devine, T.M. 2003. *Scotland's Empire and the Shaping of the Americas 1600–1815*. Washington: Smithsonian Books.

Fischer, David Hackett. 1989. *Albion's Seed: Four British Folkways in America*. New York: Oxford University Press.

Fischer, David Hackett, and James C. Kelly. 2000. *Bound Away: Virginia and the Westward Movement*. Charlottesville: University of Virginia Press.

Gerlach, Russel L. 1992. A Contrast of Old World Ideology: Germans and Scotch-Irish in the Ozarks. In *Ideology and Landscape in Historical Perspective*, ed. Alan R.H. Baker and Gideon Biger, 289–302. Cambridge: Cambridge University Press.

Glassie, Henry. 1986. Irish. In *America's Architectural Roots: Ethnic Groups that Built America*, ed. Dell Upton, 75–79. Washington: Preservation Press.

Guelke, Leonard. 1992. Ideology and Landscape of Settler Colonialism in Virginia and Dutch South Africa: A Comparative Analysis. In *Ideology and Landscape in Historical Perspective*, ed. Alan R.H. Baker and Gideon Biger, 289–302. Cambridge: Cambridge University Press.

Hilliard, Sam B. 1980. Plantations created the South. *Geographical Magazine* 52: 409–416.

Holmes, Fred L. 1944. *Old World Wisconsin: Around Europe in the Badger State*. Minocqua: North Word Press, Inc. (reprinted 1990).

Hornsby, Stephen J. 2005. *British Atlantic, American Frontier: Spaces of Power in Early Modern British America*. Hanover (NH): University Press of New England.

Hoskins, W.G. 1985. *The Making of the English Landscape*. London: Penguin Books.

Jordan, Don, and Michael Walsh. 2008. *White Cargo: The Forgotten History of Britain's White Slaves in America*. Edinburgh: Mainstream Publishing.

Jordan-Bychkov, Terry G. 2003. *The Upland South: The Making of an American Folk Region and Landscape*. Santa Fe: Center for American Places.

Kniffen, Fred. 1965. Folk Housing: Key to Diffusion. *Annals of the Association of American Geographers* 55 (4): 549–577.

Knowles, Anne Kelly. 1995. Immigrant Trajectories through the Rural-Industrial Transition in Wales and the United States. *Annals of the Association of American Geographers* 85 (2): 246–266.

Lenney, Christopher J. 2003. *Sightseeking: Clues to the Landscape History of New England*. Hanover (NH): University Press of New England.

Lewis, Peirce F. 1975. Common Houses, Cultural Spore. *Landscape* 19 (2): 1–22.

——. 1990. The Northeast and the Making of American Geographical Habits. In *The Making of the American Landscape*, ed. Michael P. Conzen, 80–103. Boston: Unwin Hyman.

Liesch, Matthew. 2011. Partnership and Photographs: Community Conceptions of Keweenaw National Historical Park. *The Geographical Review* 101 (4): 497–517.

Macfarlane, Robert. 2016. *Landmarks*. London: Penguin Random House UK.

McAlester, Virginia, and Lee McAlester. 1990. *A Field Guide to American Houses*. New York: Alfred A. Knopf.

McWhiney, Grady. 1988. *Cracker Culture: Celtic Ways in the Old South*. Tuscaloosa: University of Alabama Press.

Meinig, Donald W. 1965. The Mormon Culture Region: Strategies and Patterns in the Geography of the American West. *Annals of the Association of American Geographers* 55 (2): 191–220.

——. 1986. *The Shaping of America: A Geographical Perspective on 500 Years of History. Volume 1: Atlantic America, 1492–1800*. New Haven: Yale University Press.

Miles, David. 2006. *The Tribes of Britain*. London: Phoenix.

Noble, Allen G. 1984a. *Wood, Brick, and Stone: The North American Settlement Landscape. Volume 1: Houses*. Amherst: University of Massachusetts Press.

——. 1984b. *Wood, Brick, and Stone: The North American Settlement Landscape. Volume 2: Barns and Farm Structures*. Amherst: University of Massachusetts Press.

Nolt, Steven M. 2002. *Foreigners in Their Own Land: Pennsylvania Germans in the Early Republic*. University Park (PA): Pennsylvania State University Press.

Payton, Philip. 2005. *The Cornish Overseas: A History of Cornwall's 'Great Emigration*. Fowey: Cornwall Editions.

Perrin, Richard W.E. 1981. *Historic Wisconsin Buildings: A Survey in Pioneer Architecture 1835–1870*. Milwaukee: Milwaukee Public Museum.

Phillips, Kevin. 1999. *The Cousins' Wars: Religion, Politics, and the Triumph of Anglo-America*. New York: Basic Books.

Poppeliers, John C., S. Allen Chambers Jr., and Nancy B. Schwartz. 1983. *What Style Is It? A Guide to American Architecture*. Washington: Preservation Press.

Price, Edward T. 1995. *Dividing the Land: Early American Beginnings of Our Private Property System*. Geography Research Paper No. 238. Chicago: University of Chicago Press.

Rehder, John B. 1992. The Scotch-Irish and English in Appalachia. In *To Build in a New Land: Ethnic Landscapes in North America*, ed. Allen G. Noble, 95–118. Baltimore: Johns Hopkins University Press.

——. 2004. *Appalachian Folkways*. Baltimore: Johns Hopkins University Press.

Rowe, John. 2004. *The Hard-Rock Men: Cornish Immigrants and the North American Mining Frontier*. St. Austel: Cornish Hillside Publications.

Semple, Ellen Churchill. 1901. The Anglo-Saxons of the Kentucky Mountains: A Study in Anthropogeography. *Geographical Journal* 17 (6): 588–623.

Stilgoe, John R. 1976. The Puritan Townscape: Ideal and Reality. *Landscape* 20 (3): 3–7.

——. 1982. *Common Landscape of America, 1580 to 1845*. New Haven: Yale University Press.

Struble, Michael T., and Hubert G.H. Wilhelm. 1992. The Welsh in Ohio. In *To Build in a New Land: Ethnic Landscapes in North America*, ed. Allen G. Noble, 79–92. Baltimore: Johns Hopkins University Press.

Taylor, Alan. 2016. *American Revolutions: A Continental History, 1750–1805*. New York: W. W. Norton & Company.

Vlach, John M. 1993. *Back of the Big House: The Architecture of Plantation Slavery*. Chapel Hill: University of North Carolina Press.

Williams, John Alexander. 2002. *Appalachia: A History*. Chapel Hill: University of North Carolina Press.

Williams, Peter W. 1997. *Houses of God: Region, Religion, and Architecture in the United States*. Urbana: University of Illinois Press.

Wood, Joseph S. 1991. 'Build, Therefore, Your Own World': The New England Village as Settlement Ideal. *Annals of the Association of American Geographers* 81 (1): 32–50.

Zelinsky, Wilbur. 1957. The New England Connecting Barn. *The Geographical Review* 48 (3): 540–553.

——. 1977. The Pennsylvania Town: An Overdue Geographical Account. *The Geographical Review* 67 (2): 127–147.

Dutch, Belgian, and Luxemburger Landscapes: Ethnicity from the Low Countries

10

At the time the European scramble for colonies in North America began in earnest in the early seventeenth century, The Netherlands had recently gained its independence. Following the split between the seven predominantly Protestant northern provinces, considered the United Provinces that included the province of Holland, and the Spanish Netherlands, the Catholic provinces to the south were ruled by Spain until 1713. While these southern provinces were reunited with the northern Dutch provinces between 1815 and 1831, after being ruled successively by Austria and France, they became the independent country of Belgium in 1831. The Flemish spoken by residents of western Belgian is a dialect of Dutch, and Dutch is spoken throughout The Netherlands, although Frisian is the native language of the north central coastal province of Friesland. Given the intertwining histories of Belgium and The Netherlands, which together with Luxembourg comprise a region near the mouth of the Rhine River called the "Low Countries," it is only appropriate that the discussion of the ethnic imprint of Dutch, Belgian, and Luxemburger immigrants be considered in the same chapter. Initially, it was the Dutch from the United Provinces, which comprises today's The Netherlands, who were the first group from the Low Countries to make their mark on the American landscape.

10.1 Dutch Ethnic Landscapes in America

10.1.1 17th Century Dutch Settlement in New York

Americans often overlook the crucial role that the Dutch played in their nation's early colonial history. They frequently fail to acknowledge the lasting impact of Dutch attitudes about tolerance and free trade (Shorto 2004). Not only was New York City established as Nieuw Amsterdam by the Dutch West India Company in 1624, but Dutch territory extended from Connecticut through Delaware, separating the northern and southern English colonies. Unlike other colonies that were established by European monarchies, Holland was not only a republic, but its colony was a mercantile venture.

Centered upon the Hudson River, the Dutch settlements traded with Native Americans in the interior and with the West Indies. Nieuw Amsterdam, founded on the southern tip of Manhattan Island, maintained such superlative trading links with Europe that much of the trade of Virginia went through its harbor. Because of its mercantile orientation and the greater Dutch toleration of religious diversity, the colony came to include a highly diverse and dynamic population, and its residents achieved certain rights before they were received by those living in the English colonies of North America. Even following its take-over by the English, the Dutch residents remained, Dutch immigration continued, and the residents of the region were guaranteed political rights not found in the other English colonies (Shorto 2004).

The Dutch ethnic imprint on Manhattan Island and the Hudson River Valley can be seen in the many Dutch names in New York City as well as in buildings constructed in the Dutch colonial style upriver in the Hudson Valley. While fires, including those set during the siege of New York City during the Revolutionary War, had removed the last buildings dating from the period of Dutch Nieuw Amsterdam well over two centuries ago, downtown New York City still has its Canal Street, along the route of a long-filled canal dug by the Dutch, and Wall Street marks the location of the former defensive wall at the Dutch settlement's northern limit. Brooklyn, Harlem, The Bronx, and Staten Island all bear Dutch names, as do many communities up the Hudson Valley, such as Amsterdam, Rensselaer, Rotterdam, and Watervliet. In Staten Island's Richmondtown Restoration, Voorlezer's House, constructed around 1696 by a Dutch settler, remains a reminder of the architectural heritage brought by New York's first colonizers (Kornwolf 2002).

While examples of Dutch colonial style architecture and toponyms remain, Russell Shorto argues that the most important Dutch colonial contribution was the setting of "Manhattan on course as a place of openness and free

J.A. Cross, *Ethnic Landscapes of America*, DOI 10.1007/978-3-319-54009-2_10

trade…, something utterly alien to New England and Virginia" (Shorto 2004:310). Thus, it has been claimed that New Amsterdam was the "birthplace of the American melting pot" (Shorto 2004:312), given the tolerance of the Dutch colony and later the city for individuals practicing many religions and speaking numerous languages.

The Dutch settlements in the Hudson Valley brought only a few thousand colonists to America. Within none of the counties of New York, with the exception of Wayne County along Lake Ontario, did as many as a tenth of the population claim Dutch ancestry on the 2000 Census. Nevertheless, for nearly two centuries after the Dutch arrived their imprint upon the landscape of the Hudson Valley was very conspicuous, and many of the oldest surviving structures within the region today display Dutch architectural characteristics, even if they were built well after English colonial rule had been established. As James Kornwolf (2002:417) explains, "Dutch families living in northern Bergen and Passaic Counties [of New Jersey] and in adjoining Orange County in New York continued to build in traditional Dutch ways into the nineteenth Century."

10.1.2 Dutch Cultural Landscapes in the Hudson Valley

Few, if any, buildings remain from the period of Dutch colonial rule, yet a variety of buildings, including urban and rural houses, churches, and barns that were built during the eighteenth and nineteenth centuries incorporating many of the Dutch architectural elements remain. Dell Upton (1986:48) notes that the "most durable contribution to New World Dutch architecture was a framing system distinguished by parallel H-shaped bents or transverse frames." Utilizing massive wooden beams that were a foot or more thick and attached to upright posts several feet below their connection with the roof rafters, these formed the characteristic H-bent. Because this framing system differed from that used by other colonists, the Dutch-style buildings were distinctive. Furthermore, this framing system was used for both wood sheathed and masonry buildings.

Within the Hudson Valley of New York, along with portions of its Mohawk and Schoharie River tributaries, and in the Raritan Valley of New Jersey, distinctive Dutch barns can be seen. A key difference between the Dutch barn and the English barn is that the wagon doors of the Dutch barn are under the gable and that at least one is a Dutch door, which has "upper and lower portions [that] swing independently of each other" (Noble 1984b:20). Typically, these Dutch barns are square or are slightly rectangular with the gable ends being longer than the sides, with the central aisle being wider than the side ones. In his pioneering study of the New World Dutch barns, John Fitchen notes that "most of these Dutch barns are recognizable from a distance by their proportions generally, and particularly by the fact that their *width almost always exceed their length,* and their *height to the ridge is always more than twice the height of their side walls*" (Fitchen 1968:21–22). While there were other gable-fronted barns in the Mid-Atlantic states, the Dutch barns were distinctive because they had "a distinctive framing method that makes use of heavy H-bents with through-tenoned beams that divide the interior into a number of equal-sized, structurally defined bays" (Lanier and Herman 1997:189). The exterior walls of the Dutch barns are usually covered with horizontal clapboards, unlike most other contemporary barns that had vertical sheathing. While the Dutch barn had relatively small eaves, "unlike the English barn, the Dutch barn often has a small pent roof over the wagon doors" (Noble 1984b:21).

During the first several decades of the 1800s side-wall wagon entry barns, which are considered as either Dutch-

Fig. 10.1 Crailo House in Rensselaer, New York. This house was, probably constructed in the early eighteenth century and later remodeled, although one of the stones in its basement is inscribed with the date of 1642. Now preserved to illustrate Dutch colonial culture in Crailo House State Historic Site, this house was built on what was once the Patroonship of Rensselaerswyck

Anglo barns or Anglicized Dutch barns, appeared. While both varieties have the classic H-bents, the Anglicized barns were built with their entrance at the gable end, but were remodeled to have their entrance along the side or eve-wall, while the Dutch-Anglo barns were erected with the side entrance. Examples of both varieties have been identified in New Jersey and New York (Huber 2014).

As Fitchen (1968) noted 50 years ago, the number of surviving Dutch barns in New York and New Jersey had fallen considerably over the previous generation. Many of the surviving structures had been considerably modified to accommodate new uses or changing technologies. Nevertheless, several hundred Dutch style barns are estimated to have survived into the twenty-first century, according to the Dutch Barn Preservation Society's website.

Because many Dutch colonists and their descendents were urban oriented, several examples of Dutch architecture are found within cities, such as the eighteenth century Crailo House in Rensselaer (Fig. 10.1), which is preserved as a New York state historic site. Several other Dutch colonial houses remain within the greater Albany area (McAlester and McAlester 1990), while others are found in the area along Delaware Bay that also saw Dutch colonization in the seventeenth century (Fig. 10.2). Their houses, which typically had steep roofs, included the H-frames that distinguished their barns. Use of this framing system resulted in the upstairs being a half story. Dutch houses often had jambless fireplaces, in essence a hood leading to the chimney, which negated the need for an exterior brick chimney. Because their roofs typically lacked overhangs, a brick parapet was often constructed atop the gable wall which rose above the roof line, as shown in Fig. 10.1. Many Hudson Valley Dutch houses built after the middle of the eighteenth century had gambrel roofs, in which the slope of the roof differs from near the eaves to the roof's centerline. Typically, the roof's "shorter upper slope lies at an angle of thirty-five degrees and the longer lower slope inclines almost seventy-five degrees from the horizontal" (Noble 1984a:35), thus providing far more usable living space on the house's upper floor.

The Dutch also left a lasting legacy upon the religious landscape, including the survival of several colonial period Dutch Reformed churches within the Hudson Valley (Williams 1997). Largely of stone masonry construction, "most of the handful of surviving eighteenth-century Dutch churches in New York feature large square towers, either set into the body of the church itself or abutting it" (Williams 1997:61). While many of the current church buildings were constructed long after the church congregations were founded (Fig. 10.3), the continued presence of the Reformed Churches in the area contributes to the region's lasting Dutch ethnic legacy.

Fig. 10.2 Dutch house constructed in the late 1600s in New Castle, Delaware. This building, facing the village green surveyed by the Dutch in 1651, and remnants of a nearby dyke constructed by Dutch settlers in the mid-seventeenth century are reminders that Dutch colonial settlement not only occurred in New York, but also along Delaware Bay

Fig. 10.3 Flatbush Reformed Church, Saugerties, New York. This church, dating to 1809, is one of six Reformed Churches in the village of Saugerties

10.1.3 Dutch Cultural Landscapes in Michigan, Wisconsin, and Iowa

The 2000 Census reported that 1.6% of the nation's population claimed Dutch ancestry, and those areas with the largest proportions of their populations having Dutch ancestry were the destinations of immigrants from The Netherlands during the nineteenth century. Sioux County, Iowa had a greater proportion of its population claiming Dutch ancestry than any other county in the nation, 57.7% (Fig. 10.4). Besides this county in far northwestern Iowa, five counties nationally had populations with over 25% Dutch ancestry in 2000, including Lyon County, Iowa (37.2%), located adjacent to Sioux County; Douglas County, South Dakota and Ottawa County, Michigan, both with 35.1% Dutch ancestry; Marion County, Iowa, 25.9%; and Pipestone County, Minnesota, 25.1% Dutch. The American Community Survey 5-year estimates for 2010–2014 showed little change.

During the mid-nineteenth century large scale immigration from the Low Countries began, centered upon three Midwestern states. The religious affiliation of Dutch immigrants influenced both the pattern and character of their settlements. In Wisconsin, Dutch Roman Catholics were far more likely to settle in or migrate to cities that offered industrial jobs than were Dutch Protestants. The rural countryside where Dutch Protestants settled maintained its distinct ethnic identity far longer than the urban areas settled by Dutch Catholics, who were far more likely to be assimilated into the greater Roman Catholic community, particularly when nationality-based parishes were discouraged (Schreuder 1997).

The staunch Calvinism of many of the Dutch remains visible in the landscape, not surprising given that the Dutch immigrants of the mid-nineteenth century were disproportionately drawn from the Seceders, who rejected efforts to modernize the established Dutch church. Thus, tulips, decorative wooden shoes, and windmills mark a portion of southwestern Michigan as having many residents of Dutch heritage, as do the many Dutch toponyms, such as Bentheim, Borculo, Drenthe, Graafschaf, Holland, Noordeloos, Zeeland, and Zutphen, that are scattered across the landscape (Harmsel 2002). It is the absence of other activities that also highlight their ethnic religious heritage. As Larry Ten Harmsel explains,

> In these towns, where on Sunday nothing is bought or sold and the sound of lawnmowers is nowhere to be heard, where taverns and movie theatres have never stood, life strolls along to the peaceful rhythms of the seasons and the church bells (Harmsel 2002:28).

Similar imprints of Calvinism have been noticed near the Dutch settlement of Prinsburg in Minnesota's Kandiyohi County.

> People are expected to keep homes in good repair and yards well-groomed… There are no beer parlors, no liquor stores, no dance halls and no billiard parlors (quoted in deGryse 1981:197).

Because of the early split between the Reformed Church of America and the Christian Reformed Church, even small Dutch ethnic communities sometimes have two churches, rather than just one. Likewise, Christian schools are favored by members of the Christian Reformed Church, while members of the Reformed Church of America rely upon public schools (Harmsel 2002). Although many of the churches in these ethnic Dutch communities are viewed by as being less ethnically distinctive, on its buildings (Fig. 10.5) the Reformed Church's Hope College in Holland, Michigan "features the stepped gables and polychrome brickwork characteristic of the mother country's building style" (Williams 1997:168).

The architectural characteristics of the barns in southwestern Michigan also distinguish the Dutch ethnic region, just as the absence of drinking establishments. Allen Noble (1984b:137) notes that a conspicuous marker of "the Dutch areas is the occurrence of large, tripartite, transverse gable barns—the Michigan Dutch barn." Alike the barns in the

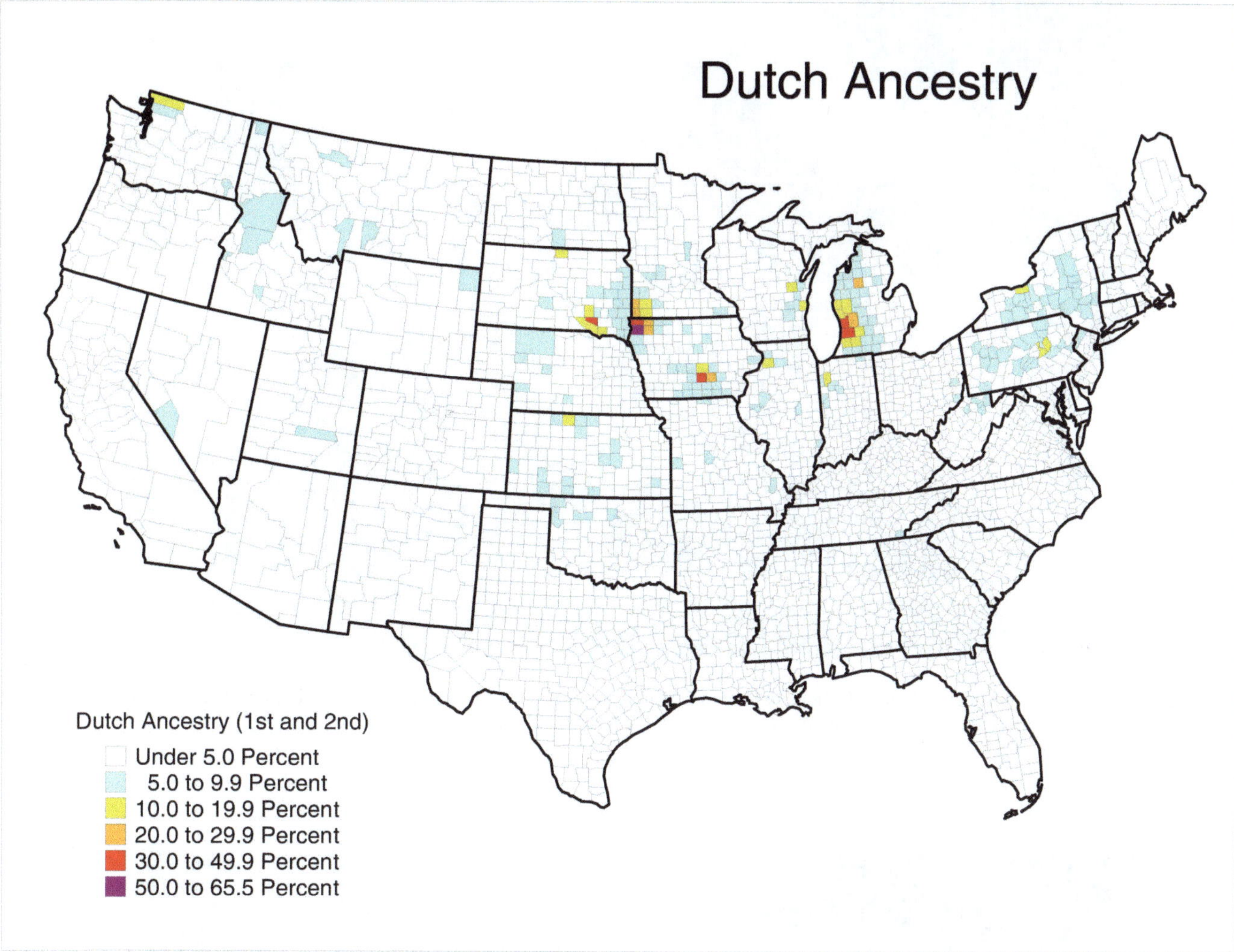

Fig. 10.4 Map showing the percentage of individuals who reported their ancestry on the 2000 Census indicating that Dutch was their primary or secondary ancestry. Within most U.S. counties Dutch ethnics account for under 5% of the population, and even in those counties of the Hudson River valley which were part of New Netherlands Dutch ethnics account for under 10% of the population. The largest islands of Dutch ancestry are found in Michigan and Iowa, with smaller groupings in Pennsylvania and Wisconsin

much older Dutch settled region of New York, these Michigan Dutch barns are wider, with a less steep roof, yet the barn doors remain under the gable, which may rise to a height that is several times that of the side walls. The walls are vertically sheathed, more in common with barns constructed by many other Americans.

The houses built by ethnic Dutch settlers within southwestern Michigan and northeast Wisconsin, while often rather indistinguishable from those built by other Americans, were most distinctive when built of brick (Fig. 10.6). What sets apart the Dutch brick buildings from those of other ethnic groups is their use of patterned brick work, in which most "bricks are a deep red and the decorative details are provided by employing a few bricks of a contrasting tan or brownish-yellow color" (Noble 1984b:139). Designs are typically geometric, under the eaves and around the window frames. While older houses displaying the Dutch decorative brick detailing (Fig. 10.7) can be found in the communities and rural region near Holland, Michigan and in Wisconsin's Lower Fox River Valley, modern houses lack such ethnic identification.

In south central Iowa Dutch settlement centered around Pella, while later immigrants settled in northwestern Iowa. Within these communities the Dutch have not only maintained some of their visible material culture, but they have also enhanced their identity by deliberately incorporating Dutch styles, motifs, and symbols in new construction. Thus, not only are Dutch style windmills seen in Holland, Michigan and in both Pella and Orange City, Iowa, but they are also seen in Dutch-settled Cedar Grove and Little Chute, Wisconsin (Fig. 10.8). In Pella a new canal was recently dug downtown, landscaped with Dutch flowers, and lined with the Molengracht buildings, which display characteristic Dutch architecture and stair-stepped brick gables, housing

Fig. 10.5 Voorhees Hall on campus of Hope College in Holland, Michigan. This building, displaying intricate diagonal designs in its brickwork and stepped gables, was erected in 1907 on a college associated with the Reformed Church of America

Fig. 10.6 Dekker Huis in Zeeland, Michigan. This brick house and adjacent store building were erected in 1876, and they display Dutch brickwork and the stepped gable. The building now houses the local historical society museum

Fig. 10.7 Brick farmhouse near hamlet of Drenthe, southeast of Zeeland, Michigan. Note the characteristic Dutch decorative detailing in the brickwork under the gable and eaves and above the windows

Fig. 10.8 Dutch style windmills have been erected in several communities to showcase their Dutch ethnicity. This Dutch windmill was constructed in The Netherlands, disassembled, and re-erected in Little Chute, Wisconsin. Note the stepped gables on the adjacent visitors' center

local businesses, restaurants, and a hotel. Such ethnically oriented construction has been used within many communities to enhance tourism.

10.2 Belgian Ethnic Landscapes in American Midwest

The Belgian ethnic population in the United States is relatively small. The 2000 Census showed that 360,642 persons reported Belgian ancestry, just 0.1% of the population, almost exactly what the American Community Survey estimated for 2014. Not only are Belgian-Americans few in number, but this group speaks two different languages. Those who are Walloons speak French, while the Flemings speak Flemish, a dialect of Dutch.

Numerically few, Belgian-Americans are highly concentrated into just a few areas of the Upper Midwest, where they have created a distinct ethnic landscape (Fig. 10.9). Indeed, nearly a third of the nation's Belgian-Americans are found in Wisconsin and Michigan. Those counties with the largest proportions of their population reporting Belgian ancestry on the 2000 Census included Kewaunee County, Wisconsin (20.6%), Door County, Wisconsin (9.8%), Lyon County, Minnesota (9.4%), and Brown County, Wisconsin (8.6%). Within six towns of Wisconsin's adjacent Door and Kewaunee Counties ethnic Belgians accounted for over 40% of the population in 2000, with the Town of Red River having the greatest proportion, 55.8%. As James Allen and Eugene Turner (1988:59) note, "the rural area between Sturgeon Bay and Green Bay became almost exclusively Belgian and the largest Walloon Settlement in the country." The percentages of the population who were estimated by the 5-year American Community Survey for 2010–2014 to be of Belgian ancestry were actually higher in Wisconsin's Door and Kewaunee Counties (12.7% and 22.3%, respectively) than reported in the 2000 Census.

Belgian-Americans are predominately Roman Catholic. It was the presence of a French-speaking priest in a small French Canadian settlement southeast of the Bay of Green Bay that prompted Walloons to settle the region east of Green Bay between 1853 and 1857 (Laatsch and Calkins 1992). Similarly, a Catholic bishop promoted settlement in Minnesota initially for the Irish, but those Belgians who arrived proved more successful. The religious devotion of many Belgian-Americans has contributed to their distinctive ethnic landscape.

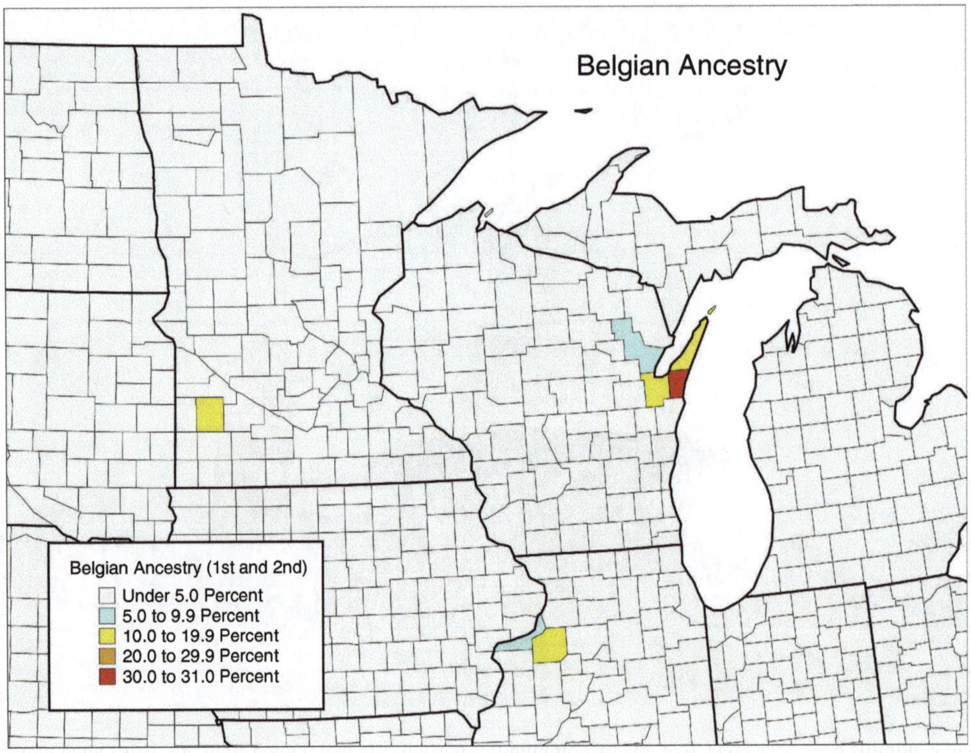

Fig. 10.9 Percentage of individuals who reported their ancestry on the 2000 Census noting Belgian ancestry in 2000. The vast majority of Americans of Belgian ancestry are located in the Midwest, and even within that region their ethnicity is most prevalent in the southern part of Wisconsin's Door Peninsula

10.2.1 Rural Belgian Settlement Landscapes

Small clusters of houses surround brick Roman Catholic churches in several ethnic Belgian hamlets in Wisconsin's southern Door Peninsula, including Namur, Rosiere, and Thiry Daems. Not unlike the ethnic German villages in Wisconsin's "Holy Land" east of Lake Winnebago, several farmsteads of the adjacent farms crowd the hamlet edges, bringing farmhouses and barns to be located near these churches. The single-towered brick St. Francis Catholic Church just north of Brussels, Wisconsin is even more impressive, and is adjoined by a cemetery and ornate grotto, with two dairy farmsteads and fields across the highway (Fig. 10.10). Yet, the most distinctive religious structures that set apart the Belgian-American ethnic landscape are small roadside chapels.

10.2.1.1 Roadside Chapels

Roadside chapels or Catholic votive chapels are constructed near the road at the edge of farmyards (Fig. 10.11). These small structures, typically seven by nine feet, were originally built to revere specific saints or to display gratitude. Their appearance has been described by Charles Calkins and William Laatsch.

> Most commonly, the chapels were of frame construction, sided with either clapboards or composition shingles.... Inside walls and ceilings were finished with lath and plaster or wallboard, usually painted pale blue, green or yellow (Calkins and Laatsch 1986:102).

Being built without windows, the roadside chapels could easily be mistaken for an outhouse, except for their placement and the display of a wooden or metal cross placed above their door or attached to the roof peak. Sometimes a religious inscription in French is placed above the door. Inside, a small altar opposite the door is the focus of the structure, and adorning its walls are "religious pictures and various birth, baptismal, first communion, marriage and death certificates" (Calkins and Laatsch 1986:102).

Two dozen roadside chapels remained in Wisconsin's southern Door Peninsula as recently as the mid-1980s, although at that time only ten were noted as being used on a regular basis (Laatsch and Calkins 1986). Yet even then their importance to the community was recognized, and they are often maintained by the "family whose ancestors built the structure" (Laatsch and Calkins 1986:119). While the roadside chapels are open to all who care to use them, being viewed as community structures more than the private part of farms, William Tishler and Erik Brynildson (1986:81) wrote that they "now are disappearing, no known replacements under construction." While it is true that several have been lost over the past decade, including one that was located on the right of way of a recently constructed divided highway, some remain visible, particularly along the side roads, such as east of Dykesville. Several have been relocated to the edges of churchyards where they can be more easily maintained, such as at Namur where a state historical marker was erected to commemorate the Belgian settlement; and this author has noted the appearance of a new roadside chapel by a farmstead near Brussels.

Fig. 10.10 St. Francis Catholic Church and cemetery with grotto located just north of Brussels, Wisconsin

Fig. 10.11 Belgian ethnic roadside chapel located southeast of Dyckesville, Wisconsin

10.2.1.2 Distinctive Rural Housescapes

Many other distinctive elements of rural housescapes mark the ethnic Belgian settlement. As Laatsch and Calkins (1986:117) note, "The most visible examples of the imprint were found in the forms of folk architecture…[which] survive as relics and continue to serve as a cultural index for understanding the original settlements." These Belgian ethnic structures include red brick houses, brick masonry over logs, bake ovens, and long three-bay barns.

Over 200 red brick houses, of modest size and unpretentious, are found throughout the Belgian ethnic region (Fig. 10.12). They "contrast with the frame and clapboard structures that characterize most of northeastern Wisconsin" (Calkins and Laatsch 1986:100). The brick houses of the Belgians are often nearly square and are two stories or one and one half-story, with two or three windows on each side. These Belgian brick buildings have a strong European precedent, and "are a prevalent characteristic of the Belgian-American rural Wisconsin landscape" (Tishler and Brynildson 1986:41). Some of the brick houses are actually massive log buildings that have been covered with a brick veneer, a house type "rarely found in other parts of the United States" (Tishler and Brynildson 1986:73). Besides the brick masonry houses, stone masonry houses are also relatively common in Wisconsin's Belgian ethnic region.

Many of the Belgian ethnic houses, be they constructed of brick or stone, have gabled roofs aligned with a door. A circular or semi-circular window is often centered under the roof peak of the gable facing the road. It has been described as being a "mason's trademark" (Calkins and Laatsch 1986:100). This round or semicircular attic window is described as "the real diagnostic trait of these houses" (Hart (2008:88).

A bake oven attached to the summer kitchen (Fig. 10.13) was at one time found on nearly every farm within the Belgian ethnic area, although by the mid-1980s only 15 remained (Tishler and Brynildson 1986). The distinctive outdoor oven was attached to the gable end of the summer kitchen.

> The oven was constructed on a platform of local dolomite, which stood four feet high and extended beyond the gable end of the summer kitchen. The oval, domed baking chamber was constructed of bricks made from local clays and rested on the platform. To protect the oven platform from weather damage, a frame gabled roof was built over the entire structure (Calkins and Laatsch 1986:100, Reprinted with permission of John Wiley and Son).

The bake oven's function was superseded by wood-fueled cook stoves and later by gas or electric ovens. Thus, the "remaining ovens are cultural relics, used only during religious holidays" (Calkins and Laatsch 1986:100). While many of the surviving bake-ovens are partly hidden in the backsides of farmsteads, a very conspicuous example, marked by a historic plaque on its wall, is located close to the main highway between Belgium and Namur at a farmstead that also contains a classic example of the three bay gabled roof barn.

Fig. 10.12 Belgian ethnic brick farmhouse located southwest of Brussels, Wisconsin. Note the characteristic circular window below the front gable

Fig. 10.13 Characteristic bake oven attached to gable end of summer kitchen near Belgian ethnic farmhouse north of Thiry Daems (Kewaunee County), Wisconsin

Fig. 10.14 Belgian ethnic barn near Dyckesville, Wisconsin. Many of the barns were built of logs in the region, more of an adaption to an abundant local resource than a replication of cultural tradition

The Belgian dairy barn is distinctive (Fig. 10.14). While it may have frame, log, or masonry walls, it most often has three bays with a gabled roof. Long, low, log hay barns are also common, with the log-work being unchinked, allowing free passage of air. These elongated Belgian ethnic barns are "roughly two or three times as wide as they are deep" (Noble 1984b:141). One of the most distinctive characteristics of some of the Belgian barns is a white star painted on the gable end of the barn just below the roof (Laatsch and Calkins 1992). While the star is possibly a local builder's trademark, it is distinctive and generally absent from the barns of other ethnic populations in the northeastern Wisconsin.

10.2.2 Belgians in Midwestern Cities

Belgian immigrants were also drawn to several Midwestern cities, even though their rural heritage is far better preserved and studied. In particular, many Belgian immigrants settled in Detroit, Michigan; LaPorte, Michigan City, and Mishawaka, Indiana; and Moline, Illinois.

The urban ethnic landscape of the Belgians in Mishawaka has received geographic study (Hume 2003). The St. Bavo Flemish Catholic Church, named for a seventh century Belgian saint, became the focus of Mishawaka, Indiana's Belgian population following its dedication in 1905. Ethnic Belgian retail establishments and homes were concentrated in the blocks near the church. Today, the twin-towered brick church remains the most prominent feature in the Belgian landscape of Mishawaka, with only a few businesses bearing Belgian names or advertising Belgian pastries remaining (Hume 2003). Yet today Belgians are a minority within Mishawaka, accounting for only 2.6% of its population, according to the American Community Survey for 2010–2014, fewer numerically and proportionally than Belgians in Green Bay, Wisconsin, where they comprise 6.6% of its urban population.

10.3 Luxemburger Ethnic Landscapes in the Midwest

Within the heart of the Belgian ethnic region of Wisconsin lies the Town of Luxemburg, which surrounds the village of Luxemburg. Notwithstanding its name, the village and town of Luxemburg's population were 0.1% and 0.0% of Luxemburger ancestry, respectively, while over 40% were Belgian in 2000. Likewise, the Town of Belgium in Ozaukee County, north of Milwaukee, had only 0.5% of its population reporting Belgian ancestry, but 19.9% were of Luxemburger ancestry. While such statistics clearly demonstrate the limitations of using toponyms to locate an ethnic population, many persons of Luxemburger heritage who immigrated to America came from the Belgium province of Luxembourg (deGryse 1981). The number of persons reporting

Luxemburger ancestry in the United States is quite small. Only 45,139 persons claimed Luxemburger ancestry in 2000, with four-tenths of these persons residing in Iowa, Minnesota, and Wisconsin. Even fewer numbers were estimated to have Luxemburger ancestry on the 5-year American Community Survey for 2010–2014. Nevertheless, Luxemburgers have left a distinct mark upon the cultural landscape of several small communities, including St. Donatus, Iowa; Rollingstone, Minnesota; and Dacada, Wisconsin.

In these settlements the Roman Catholic church with its tall central tower over the entryway is prominent, located in the center of the hamlet of Dacada and on a hillside overlooking the village of St. Donatus. It has been observed that "Rollingstone's Holy Trinity Church [is] reminiscent of the stone edifices in the Grand Duchy" (deGryse 1981:201). St. Donatus, with 34 buildings listed on the National Register of Historic Places, "including the church, rectory, outdoor way of the cross, pieta chapel, stonehouse bakery, smoke house, and restored limestone residences" (Bunkers 1999:15), is a particularly well-preserved Luxemburger village. It touts this heritage by highway signage denoting "Historic Luxembourg Village" at the village's edge (Fig. 10.15). Hewn stone masonry construction, often covered with stucco plastering, is common in both houses and barns there and in Dacada. While some of the farm buildings are now in ruins, their massive stone walls remain. The older one and one-half or two story houses are rectangular shaped with their long side facing the road, walls are plastered stone masonry, and their jerkin-style roofs have a truncated gable or half-hipped appearance.

There are examples of houses with this jerkin-style roof in St. Donatus, Dacada, and near Rollingstone (Fig. 10.16). With the wall on the gable end rising half to two-thirds of the distance from the side walls to the roof peak, where it met the hipped roof, even the roofless ruins of the stone walls of long-abandoned houses demonstrate their Luxemburger heritage. These houses typically lacked eves, having little or no overhang (Institut Grand-Ducal: n.d.).

Some of these houses are good examples of the Quereinhaus style of Luxembourg architecture. These structures originally were constructed as house-barns, with the barn or stable at one end (Fig. 10.17). Nevertheless, in contrast with the Old World villages, many other Luxemburgers in the American Midwest built barns separated from their houses. Stone walls, "so finely joined they required no mortar" (deGryse 1981:204) both fenced fields and stabilized steep slopes near Rollingstone.

The preservation of traditional Luxemburger buildings in several of their ethnic communities in the Upper Midwest has received support from the Grand Duchy of Luxembourg, which has sent money and artisans to restore stone houses in Elba, Minnesota; St. Donatus, Iowa; and near Fredonia, Wisconsin. Similarly, the Grand Duchy funded the preservation of a stone barn with arched entryways south of Belgium, Wisconsin and the historic lighthouse in Port Washington, Wisconsin.

Fig. 10.15 Sign proclaiming the Luxembourger ethnicity of St. Donatus, Iowa. Note the two stone masonry buildings with their jerkin-style roofs

Fig. 10.16 House with jerkin-style roof with St. Nicholas Catholic Church in distance in Dacada, Wisconsin. While several examples of Luxembourger farm buildings are located near this hamlet astride the Ozaukee and Sheboygan County line, a far larger population of Luxemburger ancestry lives in nearby Port Washington

Fig. 10.17 Back side of Gehlen barn within Luxembourger village of St. Donatus, Iowa. This limestone barn was built in 1839 and originally housed both livestock and people, with the narrow slots providing ventilation on this side and windows being located on the opposite side

10.4 Ethnic Persistence and the Future

The Dutch and their neighbors, the Belgians and Luxemburgers, have established small ethnic settlements that are best seen in the rural countryside of the Midwest. While far less numerous than their German, Slavic, or Scandinavian neighbors, their distinctive architecture and religious devotion ensures the persistence of these reminders of their ethnicity. In particular, the continued financial support of the Luxembourg government guarantees the preservation of key structures that not only chronicle an important part of the Grand Duchy's history but the ethnic landscape that these immigrants created in America.

References

Allen, James P., and Eugene Turner. 1988. *We The People: An Atlas of America's Ethnic Diversity*. New York: Macmillan.

Bunkers, Suzanne L. 1999. Reflections of Luxembourg in the Rural Midwestern United States. *Bulletin Linguistique et Ethnologique de l'Institut Grand-Ducal. Fascicule* 29: 12–31.

Calkins, Charles F., and William G. Laatsch. 1986. Belgians. In *America's Architectural Roots: Ethnic Groups That Built America*, ed. Dell Upton, 100–105. Washington: Preservation Press.

deGryse, Louis M. 1981. The Low Countries: Belgians, Netherlanders, and Luxembourgers. In *They Chose Minnesota: A Survey of the State's Ethnic Groups*, ed. June Drenning Holmquist, 185–210. St. Paul: Minnesota Historical Society Press.

Fitchen, John. 1968. *The New World Dutch Barn*. Syracuse: Syracuse University Press.

Harmsel, Larry Ten. 2002. *Dutch in Michigan*. East Lansing: Michigan State University Press.

Hart, John Fraser. 2008. *My Kind of County: Door County, Wisconsin*. Chicago: Center for American Places at Columbia College Chicago.

Huber, Greg. 2014. Side-Wall Wagon-Entry Barns in Monmouth County, New Jersey: Their Genesis and Types. *Dutch Barn Preservation Society Newsletter* 27 (1–2): 1–16.

Hume, Susan E. 2003. Belgian Settlement and Society in the Indiana Rust Belt. *The Geographical Review* 93 (1): 30–50.

Institut Grand-Ducal. n. d. Luxembourg Traditional Architecture in the United States. http://www.institutgrandducal.lu/Joomla/index.php?option=com_content&task=view&id=129&Itemid=240. Accessed 29 Nov 2016.

Kornwolf, James D. 2002. *Architecture and Town Planning in Colonial North America*. Baltimore: The Johns Hopkins University Press.

Laatsch, William G., and Charles F. Calkins. 1986. The Belgian Roadside Chapels of Wisconsin's Door Peninsula. *Journal of Cultural Geography* 7 (1): 117–128.

———. 1992. Belgians in Wisconsin. In To Build in a New Land: Ethnic Landscapes in North America, ed. Allen G. Noble, 195–210. Baltimore: Johns Hopkins University Press.

Lanier, Gabrielle M., and Bernard L. Herman. 1997. *Everyday Architecture of the Mid-Atlantic: Looking at Buildings and Landscapes*. Baltimore: Johns Hopkins University Press.

McAlester, Virginia, and Lee McAlester. 1990. *A Field Guide to American Houses*. New York: Alfred A. Knopf.

Noble, Allen G. 1984a. *Wood, Brick, and Stone: The North American Settlement Landscape. Volume 1: Houses*. Amherst: University of Massachusetts Press.

———. 1984b. *Wood, Brick, and Stone: The North American Settlement Landscape. Volume 2: Barns and Farm Structures*. Amherst: University of Massachusetts Press.

Schreuder, Yda. 1997. Americans by Choice and Circumstance: Dutch Protestant and Dutch Catholic Immigrants in Wisconsin, 1850–1905. In *Wisconsin Land and Life*, ed. Robert C. Ostergren and Thomas R. Vale, 320–330. Madison: University of Wisconsin Press.

Shorto, Russell. 2004. *The Island at the Center of the World: The Epic Story of Dutch Manhattan and the Forgotten Colony that Shaped America*. New York: Doubleday.

Tishler, William H., and Erik Brynildson. 1986. *The Architecture and Landscape Characteristics of Rural Belgian Settlement in Northeastern Wisconsin*. Madison.

Upton, Dell. 1986. Dutch. In *America's Architectural Roots: Ethnic Groups That Built America*, ed. Dell Upton, 48–53. Washington: Preservation Press.

Williams, Peter W. 1997. *Houses of God: Region, Religion, and Architecture in the United States*. Urbana: University of Illinois Press.

11 German-American Landscapes

Germans are the nation's largest or second largest non-Hispanic ethnic group, depending upon how the numbers are totaled. If all of the ethnic groups from the British Isles, including persons reporting British, English, Irish, Scotch-Irish, Scottish, and Welsh ancestry, are considered together, they outnumber the Germans, whether one looks at recent American Community Survey estimates or ancestry data from the 2000 Census. However, with 42,885,162 Americans reporting their first or second ancestry as being German when queried by the 2000 Census, German-Americans greatly outnumbered the 24,515,138 individuals reporting primary or secondary English ancestry. Five-year American Community Survey estimates for 2010–2014 showed 48,874,293 Germans, in contrast with 33,918,058 Irish, 25,181,294 English, and 17,330,315 Italians, the three largest groups of other Europeans. Since the earliest days of the republic, Germans have figured prominently. Indeed, at the time of the 1790 U.S. Census, Germans comprised 9% of the nation's population, the most numerous group other than those of English heritage.

11.1 History of German Immigration to America

Germans began migrating to America long before there was either a united Germany or a United States. During the late 1600s and the 1700s southeastern Pennsylvania was a center of German immigration. Not only had William Penn actively recruited German Quakers to join his colony, but many other German Pietists were attracted by the colony's pacifist settlers. Thus, Pennsylvania received thousands of German Reformed and Lutheran settlers, as well as Amish, Brethren (Dunkards), and Mennonites, giving it an overwhelmingly Protestant German population. Indeed, the first Bible published in the United States was printed in German. At the time of the American Revolution many communities in Pennsylvania were predominately German speaking and remained so well into the next century. With a population that was 38% German at the time of the 1790 Census, Pennsylvania was noted for its German heritage. Corrupting the spelling of the Deutsch language or because they came from the northern Rhine Valley near the Dutch border, these Germans became known as the Pennsylvania Dutch, with 255,807 reporting this ancestry in 2000. The American Community Survey between 2010 and 2014 estimated the Pennsylvania German population numbered 307,785.

11.1.1 Pennsylvania German Culture

The Pennsylvania Germans were the largest non-British population within England's 13 American colonies and remained so within the United States until the early 1840s. Furthermore, as the historian Steven Nolt explains, "As the largest group of white, non-British-stock people in the new nation, Pennsylvania Germans pioneered the process of ethnicization-as-Americanization" (Nolt 2002:5). Thus, the identity of the Pennsylvania Germans was evolving simultaneously with the creation of a new independent American identity. Nolt states that among the Pennsylvania German "the negotiation of ethnic and national identity took place in the context of a fragile young republic whose cultural majority was itself struggling to define" (Nolt 2002:5). The Pennsylvania Germans developed a culture that was distinctly different from that of German immigrants who began arriving in the mid nineteenth century, an identity that they viewed as American rather than German, with religion being a key to their identity. Hence, "encounters with mainstream American political and religious cultures did more than foster a Pennsylvania German sense of separation and sectarianism. They actually provided resources for building and maintaining identity in terms suited to the American scene" (Nolt 2002:7).

Because few German immigrants came to Pennsylvania in the nineteenth century and as a result of its particular German Protestant background, Pennsylvania German culture remains distinct from that of other German settlements. Indeed, even in the 1800s the arrival of newer German immigrants "prompted the 'Pennsylvania

J.A. Cross, *Ethnic Landscapes of America*, DOI 10.1007/978-3-319-54009-2_11

Germans' toward a greater self-consciousness of their own group identity" (McMurry and Van Dolsen 2011:3). At the same time, newly arrived Germans noted how the Pennsylvania Germans had become Americanized and sought "new and isolated" locations farther west for new settlements where they "could safeguard their German culture and create a more favorable environment for future generations" (Schroeder 1988:157).

Other Germans arrived during colonial days, particularly in New York and North Carolina. During the Revolutionary War, thousands of German mercenaries fought for the British. Most came from the German state of Hesse-Cassell, and following the war nearly a quarter of these Hessian soldiers, a total of 3200, remained in the United States, while others returned later bringing their families (Fischer 2004). The total number of German mercenaries who became Americans has been estimated between 5000 and 12,000 (Sowell 1981).

11.1.2 19th Century German Immigration

The period from the 1850s through the 1880s was the time of the greatest German immigration to the United States, although sizeable immigration actually began a couple of decades earlier. The Germans arrived in three distinct waves. From 1845 to 1855 939,000 Germans came, with 97% of these coming from southwestern Germany (then largely the German states of Baden, Bavaria, Hesse, Nassau, the Palatinate, Pfalz, and Württemberg). Following the conclusion of the American Civil War, from 1865 to 1873 another 1,066,000 immigrants arrived, most of whom were from northwest Germany (primarily Hanover, Oldenburg, Ost Friesland, Schleswig-Holstein, and Westphalia). The final wave came between 1880 and 1893 bringing 1,849,000 immigrants, with the largest numbers coming from northeastern Germany, especially Prussia, but also Mecklenburg, Pomerania, and Upper Silesia (Conzen 1980; Sickles-Taves and Allsopp 2005; Bawden 2006). Given the spatial distribution of religious denominations among the German states, large numbers of Catholics were represented by those Germans from southern Germany, in contrast to more Protestants from other German lands. Thus, "unlike other immigrant groups, German-speakers did not comprise a single homogeneous religious group" (Max Kade Institute 2005:19).

Many German immigrants came as farmers and settled throughout the Midwest and into the Great Plains. Wisconsin was particularly attractive and the state employed German immigration commissioners and published publicity material in German to attract immigrants. A lasting legacy of the success of this German immigration is that Wisconsin, North Dakota, and South Dakota had more than 40% of their residents reporting German ancestry on the 2000 Census (Fig. 11.1). Nearby Minnesota, Iowa, and Nebraska all reported populations whose German ancestry exceeded 35%. In four North Dakota counties Germans comprised over two-thirds of the population in 2000. They exceeded 60% of the population in four additional North Dakota counties, three each in South Dakota and Nebraska, two in Minnesota, and one each in Iowa, Wisconsin, and Ohio.

German colonization efforts also took place in Texas and Missouri. German settlement was widely encouraged by the government of the Republic of Texas, and many Germans continued immigrating following Texas' statehood in 1845. The Hill Country near Austin and north of San Antonio retains a strong German character. Three counties in that area boasted populations in 2000 where over one-third claimed German ancestry. Within Missouri, not only did many Germans move to St. Louis, but many settled within the Lower Missouri River Valley. Although St. Louis's African-American population now outnumbers its German-Americans, persons of German ancestry still comprise over half of the population in one county along the Missouri River west of St. Louis and in one county of Illinois just downriver from St. Louis.

German settlements were often planned colonization efforts, both by domestic and overseas groups. A variety of factors prompted the German immigration, including overpopulation at home, and a lack of land, especially in the Palatinate where farmlands were divided among all heirs, resulting in farms that were too small to be economically viable. In addition, the elimination of village common lands hurt farmers, and unemployment of artisans was widespread with the rise of the Industrial Revolution's factory system and British exports (Sowell 1981). While most German immigrants were lower middle-class farmers, the arrival of the "Free-thinkers" in the late 1840s brought many professionals. Political conflict and related religious intolerance also prompted immigration, with the situation varying among the various German states and principalities. Germany itself was not unified until 1871, after much of the mass migration of Germans to America had occurred.

Germans were considered the last of the "Old Immigrants" (Ward 1971). They shared some characteristics with the British. English is a Germanic language as is German, and initially the German immigrants were largely Protestant as were most of the British. Germans (and Scandinavians) had close ties with Lutheran, Reformed, or Evangelical churches, while Catholics comprised only about one-third of their number. Germans were viewed as assimilating rapidly with other Americans. Hubert Wilhelm explains, "The German immigrant readily adopted these American forms, thereby contributing to the creation of a midwestern American cultural hearth" (Wilhelm 1992:66). Nevertheless, there were efforts to create staunchly German settlements, such as in the vicinity of Hermann, Missouri, where Germans from Pennsylvania established a planned settlement.

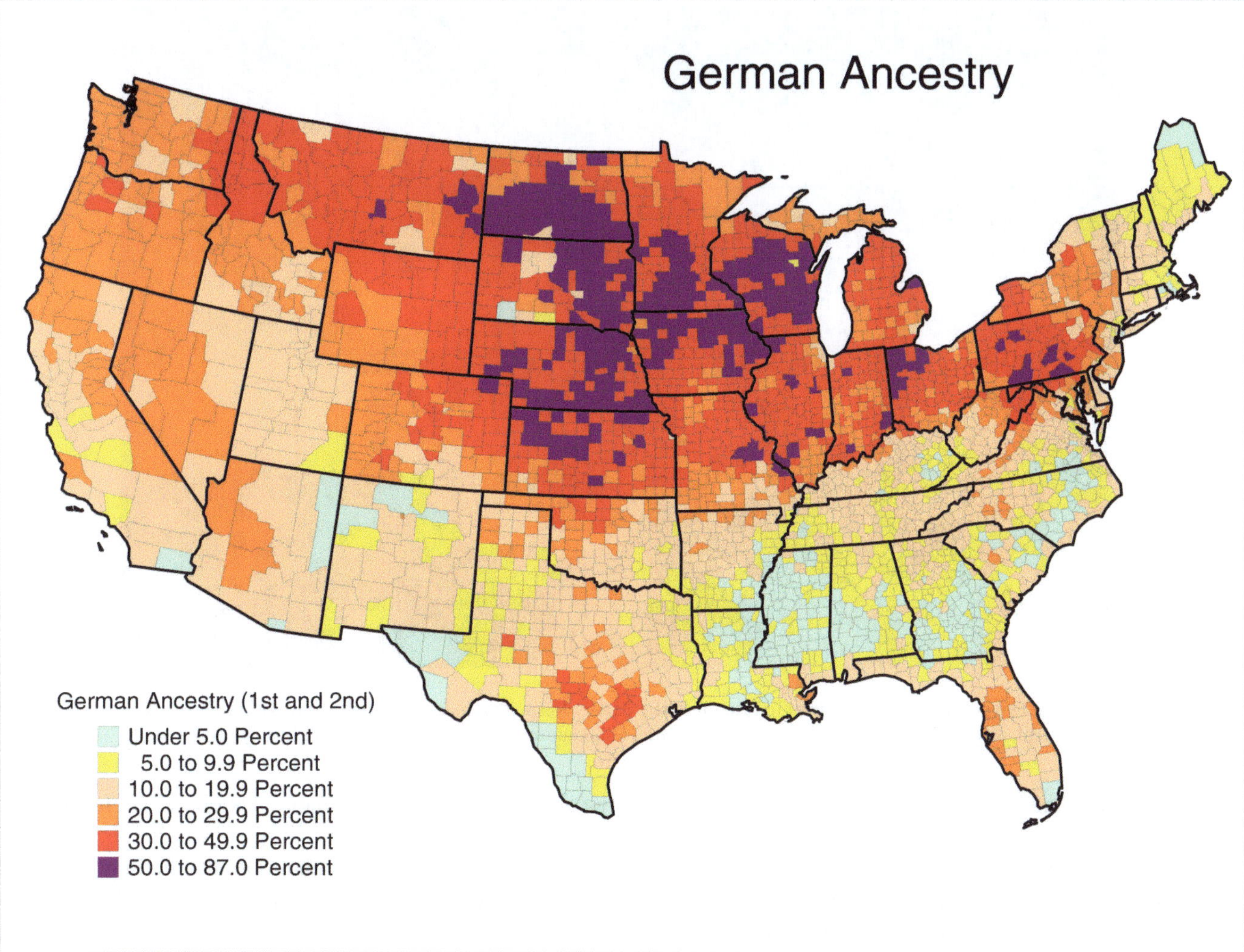

Fig. 11.1 Map showing the percentage of individuals who reported their ancestry on the 2000 Census indicating that German was their primary or secondary ancestry. Note the broad swath of the northern half of the nation, including Pennsylvania, the Midwest, the northern Great Plains, and the northern Rocky Mountain states in which the at least 30% of most counties reported German ancestry. Furthermore, many counties are over half German. A smaller concentration of German-Americans can be seen in central Texas

11.1.3 German-American Assimilation in the 20th Century

Both World Wars greatly accelerated the rate of assimilation by German-Americans. During World War I German-Americans were widely mistrusted, and in the case of the large ethnic community around St. Cloud, Minnesota, national guardsmen patrolled the community during the duration of America's involvement. Discrimination and harassment of Germans was widespread. Teaching the German language was dropped in many schools, and the use of German in commercial signs, on facades of churches, and in everyday conversation nearly ceased, and the majority of German-language newspapers stopped publication (Kirschbaum 2015). The German identity of many items was deemphasized, and toponyms changed. For example, while Berlin, Wisconsin kept its name (unlike Berlin, Ontario, which became Kitchener; Berlin, Michigan, which became Marne; and New Berlin, Ohio that became North Canton), its pronunciation was changed. Potsdam, Kansas was renamed Pershing, Germania, Iowa became Lakota, and East Germantown, Indiana was also renamed Pershing (Fuchs 2015a). In Milwaukee, where its German heritage was so strong that it was referred to as the "German Athens" and "the most German city in America" (Max Kade Institute 2005:25), there was removal of German architecture and art. The still-standing Germania Building in Milwaukee saw the removal of its Germania statue, and its name was chiseled off the stone façade of the building, and there were demands that the towers atop the building, referred to as 'Kaiser's helmuts', be removed (Hoelscher et al. 1997). The popularity of copying the contemporary works of German architects in designing houses, retail establishments, breweries, and churches collapsed with the war, and preservationists today lament the

Fig. 11.2 German style maypole erected along street in front of Berlin Family Restaurant in Berlin, Wisconsin. Within Europe maypoles are most commonly found in Germanic countries, such as Germany and Austria, although some are also found in Scandinavian countries

loss of so many buildings simply because they are eclectic (City of Milwaukee 1994). So effective was the drive to de-emphasize German heritage that in rural Missouri "even descendents of the immigrants developed a reticence about their cultural heritage and tended to discount the importance of their ancestors' contributions" (van Ravenswaay 1977:viii).

11.1.4 German-American Heritage in the 21st Century

In recent years a new appreciation has been shown for German heritage. In Wisconsin "annual festivals celebrate, and some would say market, German heritage" (Bawden 2006:79). For example, in Milwaukee the Grand Avenue Mall includes a Speisegarten and the newly constructed high-rise building on the site of the demolished Pabst Building "explicitly mimics the hallmark motifs of this unique architectural heritage" (Hoelscher et al. 1997:399). In New Ulm, Minnesota, the 102-foot monument to Hermann, the German warrior who defeated the Roman Empire's attempt to annex German lands, was dedicated in 1897 by "German immigrants as a vehicle of fostering and expressing unity, ethnic pride, and cultural ties to their homeland" (Fuchs 2015b:16). It has been renovated, and the city "refurbished its business street to reflect its German origins" (Conzen 2003:78). Thus, "community leaders and local entrepreneurs have actively developed the 'German Town' into a regional 'place brand'… and invite visitors to 'Discover Germany in Minnesota' at local festivals (e.g. the annual Oktoberfest), restaurants, shops, museums, and historic sites" (Fuchs 2015b:20). Other communities also showcase their German heritage to attract tourists, including Gutenberg, Iowa; Hermann, Missouri; and Fredericksburg, Texas, among many others. Berlin, Wisconsin conspicuously displays a German-style maypole along its main business street (Fig. 11.2). Wineries have reopened in areas where German immigrants had established extensive vineyards in the nineteenth century, and those along the Missouri River near Hermann showcase their German ethnicity (Fig. 11.3). Additional communities have affected images of German ethnicity, such as Leavenworth, Washington, which was made-over to imitate a Bavarian village (Frenkel and Walton 2000).

11.2 Diversity of German Population

German ancestry includes a variety of peoples, including Prussians, Bavarians, Palatinates (from the Rhineland along the Rhine River), and Bohemians (from the border areas with the Czech Republic), among many other groups. As we saw in our earlier discussion of the three waves of German immigration during the last half of the nineteenth century, each successive wave brought Germans from different German lands. Even a region such as the Rhineland was highly diverse, as the area was divided among over 50 different principalities in the late eighteenth century. Thus, "the Rhineland has never been a homogenous area politically, historically, socially, or geo-

Fig. 11.3 Vineyard and nineteenth century brick winery building at Stone Hill Winery in Hermann, Missouri. Before Prohibition this was the nation's second largest winery, and it and many others nearby wineries along the Missouri River were established by German immigrants

graphically" (Langer 2006:59). Not only did various German groups migrate to America at different times, but many American regions that received large German ethnic populations drew disproportionately from certain specific German states or principalities. For example, northwest Germans from Hannover, Oldenburg and Brunswick formed a disproportionate share of the German population of St. Charles and Warren counties of Missouri (Kamphoefner 1987). Within specific townships, concentrations from specific German regions are even stronger. Within eastern Wisconsin separate areas settled by Baden Germans, Hanover Germans, Hessian Germans, Hünsruck Germans, Lipper Germans, Schleswig-Holstein Germans, Pomeranian Germans, and Saxony Germans have been mapped by the developers of Wisconsin's Ethnic Settlement Trail, along with those settled by German Russians and undifferentiated mixtures of Germans.

Germany only united in 1871, yet many German-speaking populations lived outside of its boundaries. Thus, many ethnic German immigrants came from areas of Europe that were not part of Germany either then or now, including Switzerland, Austria, plus parts of present-day France, Poland, the Czech Republic, and Russia. Significant numbers of Germans lived in scattered communities in central and eastern Europe that were administered by the Austro-Hungarian Empire before the First World War, while others lived along the Baltic Sea, east and north of present Germany. German ethnics also include German-Russians (persons of German heritage who settled along Russia's Volga River—also called Volga Germans), Black Sea Germans, and German Jews (Sowell 1996). Indeed, the abundance of German surnames in central North Dakota is related to German-Russian immigrants (Fuchs 2015a). Germans from certain other areas of Europe were associated with various pietistic groups, such as the Moravians and Westphalians.

Religious identity was more important than regional ethnicity for many Germans. Indeed, among most immigrant "Germans religion was at least as important an aspect of ethnic identity as language, and immigrants tried to live near other Germans of the same religious denomination" (Allen and Turner 1988:51). Thus, many German-American settlements, particularly in rural areas, can be differentiated based upon their religious orientation. Indeed, denominational differences are not only much older than the presently understood German identity, but they are conspicuously displayed through religious architecture.

Each of these groups of Germans had many distinct cultural traits. As William Tishler (1986a:142) explains, given their diverse regional and religious heritage "German settlers brought diverse building skills and traditions… [and] German-American vernacular architecture varied extensively." Different German groups developed separate areas as settlers from specific German states, and only later did these groups establish their German-American identity. As Richard Zeitlin (1977:14) clarifies, acculturation of German immigrants from diverse areas of central Europe resulted in "the creation of a 'German' identification" which "developed only after they had reached America." Thus, German-American identity and culture are not homogeneous. Because of their different origins and resulting chain migration, the German populations in many cities and areas of the United States represent descendents of differing German groups. For example, although many Germans settled rural areas as farmers, others went to cities, the selection of which was strongly related to their origins. Thus, "Each of the East Coast cities was dominated by a different group: Baltimore by Hessians, Philadelphia by Württenbergers, Boston by Badenese, and New York by Bavarians" (Kamphoefner 1987:81).

Because Germans themselves had many diverse traditions, coming from a wide area of central Europe over a period of three centuries, regional differences can be seen in the German-American ethnic landscape. Distinctions are partly dependent upon the time of immigration. Germans immigrating to America during the colonial period were largely coming from the Rhine River Valley, often termed the Palatinate, extending south into Switzerland, and were largely pacifists fleeing religious persecution. While these earliest German immigrants largely settled in Pennsylvania, later arriving German immigrants settled elsewhere. Yet at later times immigrants came from different areas of Germany, regions that had not previously contributed to the trans-Atlantic migration. Germans also had a diverse religious background, ranging from Lutherans, German Calvinists, and Roman Catholics to Mennonites, the Amish, and other conservative groups to the liberal Free-Thinkers, many who were non-believers. Because the material culture of Germany evolved over time, immigrants brought knowledge of what was then current or popular at home.

German ethnic landscapes in the United States developed both in locations where immigrants arrived directly from Europe and in places where secondary migrations of German-Americans took place. Geographers have traced distinct movements of German cultural traits from Pennsylvania west across southern Ohio and into Indiana as well as a large German immigration directly into Ohio and the rest of the Midwest. Geographers have also shown a movement of German-Americans, descendents of those who initially settled southeastern Pennsylvania, who moved west and then southwest down the Shenandoah Valley into western Virginia and North Carolina (Zelinsky 1977). On the other hand, German immigration into Texas was more direct, although many Germans briefly lived in port cities such as New Orleans or Galveston before heading into interior Texas.

11.3 German Heritage Influences American Culture

The fact that Germans have largely assimilated into the dominant American culture and that the conspicuous outward display of German heritage ceased with the two world wars does not negate their profound influence upon contemporary American culture. In his *Ethnic America*, the sociologist Thomas Sowell (1981:43) summarized their diverse contributions, noting that "American industry, education, military defense, eating and recreational patterns all reflect the contributions and influence of German Americans." Germans contributed many words, concepts, and traditions to the generic American culture that developed. Even the Christmas tree, which so many today associate with the American celebration of the holiday, was introduced by German immigrants (Sowell 1981). Marching bands, organized orchestras, and gymnasium sports all were shaped by German-Americans. The organization of American universities is largely based upon German models, particularly the degree arrangements and system of faculty ranks. German immigrants later led the American race into space.

Beer and beer gardens have long been associated with Germans, even though there are now far fewer beer gardens than there were before World War I and Prohibition, yet there has been a resurgence in number of brewpubs. The best beers are often associated with Germans (Fig. 11.4), and the centers of brewing were long those cities with large German populations, such as Milwaukee and St. Louis. Neighborhood taverns catering to the entire family, serving both food and drink, are commonplace today in communities with large German-American populations. The tavern had a major role in the community and social life of German-Americans. As Richard Zeitlin (1977:24) explains, "While not uniquely German, it was the way Germans used the beer hall that made this informal institution function as an integral part of community affairs." It was a German meeting place, rather than a place for a quick drink. Even today, many of the communities with the greatest number of taverns per capita are located in states with large populations of German ancestry.

Sausages are just one of many foods associated with various German populations, and there are regional favorites, such as bratwurst among Wisconsin Germans, mettwurst among Cincinnati Germans, and knackwurst among others. Celebrating with both their beer and sausages, Octoberfest celebrations are held annually within many cities throughout the American Midwest where so many German-Americans continue to reside. While not all Octoberfests today are as German heritage centered, being more a promotion of a local culturally diverse identity, the Max Kade Institute (2005:28) notes "the more specifically German-themed Octoberfests nationwide reflect an American phenomenon that is striking to European Germans, namely the predominance of symbols specific to traditional Bavaria, which strive, misleadingly, to evoke a single 'German' culture."

11.3.1 German-American Ties to the Land

Germans developed stronger ties to their rural farmlands than did many other ethnic groups or Americans of Yankee heritage. As William Tishler (1986a:142) notes, "With a strong love for the soil, they [Germans] regarded their new land with a deep sense of permanency. Stability and longev-

Fig. 11.4 Former brewery of the Dubuque Brewery and Malting Company in Dubuque, Iowa. This Romanesque structure, in which 200 employees once manufactured beer following the merger of four local breweries in 1892, housed one of the Midwest's more prominent breweries before Prohibition. Germans played a major role in establishing this and many other large breweries throughout the Midwest

ity became hallmarks of their built environment and land ethic." Historian Richard Zeitlin (1977:24) likewise described the German settlers "feeling of proprietorship towards their land." Germans are characterized as looking upon their "land as the foundation, the basis for an estate that could be passed on to succeeding generations, a viewpoint that assumed locational stability" (Zeitlin 1977:25). The prominent cultural geographer, Carl Sauer, noted nearly a century ago that "stability remains the most distinguishing characteristic of the German stock. Where Germans have located in most cases they have remained" (Sauer 1920:173).

German-American inheritance practices have led to stability and persistence of land ownership that can readily be seen today (Fig. 11.5). Research findings of Sonya Salamon, a rural sociologist working largely in Illinois, demonstrate that "German farmers are obsessed with land; obtaining, keeping, or transferring it is the stuff of community discourse" (Salamon 1992:100). She finds that German ethnic farmers are fixated upon owning and working their lands, and while profits are necessary for them to remain in farming, "profits are not the end desired but a means to maintain continuity of the family name in conjunction with a particular farm within a German community" (Salamon 1992:101). Not only have German-Americans been focused upon maintaining their farmlands, but they were also more likely "to invest in fine barns rather than grand houses" (Conzen 2003:28), an action which enhanced the likelihood that their farm would be profitable. As Sauer explained, based upon his studies in Missouri,

> Because of this stability of ownership their improvements surpass in durability, if not in elegance, those of any other group of farmers. They build by preference of stone or brick. Where good stone is available, as at Westphalia and Hermann, one sees not only stone houses, but stone barns, sheds, and fences (Sauer 1920:173–174).

Much of what Sauer noted early in the twentieth century is still visible. Because of their numbers and success, German-Americans dominate the contemporary Midwestern agricultural landscape.

German-American farmers, in comparison with Yankees (Americans of British heritage) own a greater proportion of their farm land, are less likely to rent land, have farms which are smaller but more similar in size to those of their neighbors, operated farms whose size is limited by family capabilities, and are less likely to have heavy debt (Salamon 1992:93). Work done in Wisconsin indicates that dairy farmers of German ancestry are more likely to obtain their farmland from parents or relatives, used fewer technologies oriented towards maximizing production, but "were significantly more inclined to use input-cost minimizing approaches such as management-intensive rotational grazing" (Cross et al. 2000:480), all actions that facilitate greater success for entering farmers.

German-Americans account for a disproportionate share of Midwestern rural farm population, a greater share than they had a century ago (Conzen 2003). Overall 59.1% of Midwestern farmers are German. In the early 1990s it was highest in Wisconsin where 65.6% were of German descent,

Fig. 11.5 Classic example of a German bank barn south of Glen Rock, Pennsylvania. Clearly distinguished by its forebay projecting over the entrance to its lower level, this large German bank barn typified the fact that many German ethnics invested more in constructing their barns than their houses

followed by Ohio (64.7%), Nebraska (64.0%), Iowa (61.4%), Minnesota (59.0%), and Illinois (57.9%), according to Salamon (1992:34). Furthermore, Salamon (1992) reported that rural settlements of German-Americans (especially if they were of the same religious denomination) were more likely to survive as economically and socially vibrant communities than those of mixed ethnicity and religion, such as those villages and hamlets characterized by Yankees. Thus, as Kathleen Conzen summarized, "descendents of German immigrants… have not only preserved distinctive rural communities and cultures; they have also retained their commitment to a rural way of life to a greater extent than almost any other ancestry group" (K. Conzen 1990: 5). It is not surprising that many of the best examples of German ethnicity on the land will be found on the farms and in the villages of rural American.

11.4 Types of German Ethnic Landscapes in Rural America

Ethnic Germans brought many notions regarding how to construct houses, barns, and other buildings with them. Germans were far more likely to utilize brick or stone masonry construction than were many other European immigrant groups, such as the English, the Scots-Irish, or the various Scandinavian peoples, and they also introduced fachwerk construction, utilizing half-timbering techniques. Germans were also highly skilled in the construction of log buildings, which were often among the first structures built in their communities.

11.4.1 Rural German-American Landscapes

The rural landscape and farms of German-Americans have long been distinctive. Even in the 1700s German-American farmers in Pennsylvania were distinguishable from other Americans. An historian notes, "Eighteenth- and nineteenth-century travelers often marveled at the cathedral-like bank barns, skillfully tended fields, carefully managed woodlots, and orderly farmscapes that they encountered in German-settled parts of the Philadelphia backcountry (Lanier 2011:11). Over two centuries later, Russel Gerlach, in studying the German ethnic landscape in the Ozarks and comparing it with that of their Scotch-Irish neighbors, noted the greater permanence of the German-American farm buildings. Put simply, he noticed that "[t]he majority of structures in German areas were constructed in the nineteenth century while very little remains of the nineteenth century architectural heritage of the Scotch Irish" (Gerlach 1992:298). Not only had they survived, but the German structures were larger. Indeed, throughout much of the nation Germans were responsible for erecting the most impressive farm buildings.

Fig. 11.6 St. Lawrence Catholic Church towers over hamlet of St. Lawrence, Wisconsin

11.4.1.1 German Churches and Cemeteries

Many rural areas settled primarily by German-Americans are clearly marked by their distinctive churches, however their character is strongly influenced by denomination. While Anabaptist groups erected relatively plain structures—if any at all, lacking a tower or steeple, German Lutherans and Catholics constructed towering edifices. Referring to the area north of Cincinnati, Ohio, Peter Williams, an expert on religious architecture, describes the numerous red brick churches that German Catholics built on a "grand scale" during the nineteenth century. He explains, "Most of their churches were adaptations…of styles familiar in the old country: the Gothic, the Renaissance, and the *Rundbogenstil*, the German variety of Romanesque revival" (Williams 1997:163). The situation in Ohio, where towering steeples—sometimes rising over a hundred feet—marked the location of various rural village and hamlet, was also replicated in east central Wisconsin. The area east and south of Lake Winnebago became known as "Wisconsin's Holy Land" because of the proliferation of Roman Catholic churches scattered amongst its German-American hamlets, many which were named after their parish churches (Schlemper 2004). Here also, the brick or stone masonry churches (Fig. 11.6), with either a high central tower or twin-towered façade, were typically the community's largest building, often located atop a low hill giving them even greater visual presence.

Protestant churches are conspicuous in those areas where German settlers were members of Lutheran or German Reformed churches (many of which are now associated with the United Church of Christ). While the exterior architecture of many Lutheran Churches is not that different from German Catholic churches, both often displaying conspicuous gothic steeples, their interiors were quite different. In German settlements "many Catholic churches had elaborate Gothic altars and vivid stations of the cross; Reformed churches, echoing their Calvinist heritage, would have simpler, pulpit-centered interiors, though often featuring dark wooden Gothic-style furnishings" (Williams 1997:166). Churches in the German settlements of the Texas Hill country, "regardless of denomination, differ in important architectural aspects from those of the neighboring Anglo-Americans" (Jordan 1980:119). The Germans distinguished their churches by constructing them of stone or fachwerk in a Gothic style (Fig. 11.7). As Jordan explains, this style includes

> stepped towers, jagged steeples and pointed windows, and the elaborate use of stained glass. One church, the original St. Mary's Catholic at Fredericksburg, has the rare Gothic feature known as a helmet tower (Helmturm), a convex-sided crown atop the vestibule tower (Jordan 1980:119) .

Within the Texas Hill country, German church styles in both rural and urban settings are similar.

While a few early Roman Catholic churches in the Midwest had their names and cornerstones inscribed in German, such practice was strongly discouraged by many of the Irish-American bishops. On the other hand, inscriptions in German (Fig. 11.8), giving the name of the "Kirche" high on the front façade or above the door, were very common-

Fig. 11.7 Towering Gothic stone towers of adjacent German churches in Fredericksburg, Texas. The tower on the left is of the new St. Mary's Catholic Church, built in 1901, while the tower on the right is of the old St. Mary's Catholic Church, built in 1863. The older church displays a helmturm, or Gothic helmet tower, while the newer church features a far more common type of Gothic tower

place on Lutheran and various German Reformed churches constructed before World War I and are a strong reminder of the ethnicity of their builders.

Cemeteries providing the final resting places for German ethnics, and these sites, particularly those established before World War I, also display a distinct ethnic landscape. Even within the panhandle of north Texas, not an area known for a dominant German ethnicity, German cemeteries can be distinguished from those of other groups. For example, Roper and colleagues (2006) note that the cemetery boundaries are delineated by fences or lines of trees, that the older tombstones were inscribed in German, and that iron crosses marked some of the graves (Fig. 11.9). Similarities have been noted by this author in German ethnic cemeteries in the Texas Hill County, within the Missouri Valley, and in Wisconsin. While almost all of the headstone inscriptions were carved in German Gothic lettering before the First World War, afterwards a

Fig. 11.8 German name carved in stone on tower above entrance of Deutsch Evangelische Lutheraner Zions-Kirche in rural Longtown, Perry County, Missouri. This church, built in 1912, is now called Zion Lutheran Church

smaller proportion were so marked, such that by the middle of the twentieth century virtually all German ethnic gravestones were carved in English. In some German cemeteries, such as one adjacent to a rural German Lutheran Church in east central Wisconsin, delicate carvings of a heavenly city adorned many of the granite tombstones from the early twentieth century, with the buildings displaying a clear German medieval character (Fig. 11.10). In life and in death, the Germans displayed their ethnic character on the landscape.

11.4.2 Rural German-American Villages

Most German immigrants dispersed on rural homesteads like other immigrant groups or Yankees, given the requirements of the Homestead Act mandating such siting of the farmers' homes in exchange for free land. Nevertheless, several utopian communities were established in more traditional European village style. Examples include New Braunfels and Fredericksburg, Texas; Hermann, Missouri; and the Amana Colonies in Iowa. Begun in 1855, the Amana "settlements reflected the layout of medieval European agrarian villages,… [featuring] unadorned communal residences and artisan shops around a centrally located church and were dominated by barns and other agricultural buildings" (Tishler 1986a:146). While many aspects of the communal living of the Amana residents are now history, the Amana settlements today attract many visitors to their shops and restaurants in century-old buildings. The houses and barns remain in the villages (Fig. 11.11), which are surrounded by fields and pastures, contrasting sharply with the dispersed farmsteads of those other farmers surrounding the Amana Colonies (M. Conzen 1990). In contrast, while Fredericksburg attracts many tourists today, its early German settlers quickly built

Fig. 11.9 Metal grave markers in cemetery enclosed by iron fence in New Braunfels, Texas. Note that the two older inscriptions are in German, while the more recent one on the right is in English

Fig. 11.10 German ethnicity displayed in language and decorations on tombstones in Zion Evangelical Lutheran Cemetery northeast of Zittau, Wisconsin. Use of the German language in tombstone inscriptions continued at this cemetery through the period between the two world wars

homes on their farm plots, only returning on weekends to their "Sunday houses" that were constructed on their town lots (Jordan 1966; Leiding 1992).

It was not just utopian German immigrant groups who sometimes were able to at least partially replicate their Old World village structure, even with dispersed farmsteads. From the Interstate Highway between Milwaukee and the Lake Winnebago area another example is readily visible (Fig. 11.12). Sitting atop a hill is the small community of St. Anthony with its stone masonry church with its high tower, surrounded with a cluster of farmhouses—some of German style log construction—and a variety of German bank barns.

The immigration of Rhenish Prussian Catholics to the area immediately east of Wisconsin's Lake Winnebago

Fig. 11.11 Barns were erected at the edge of the cluster of housing within the communal villages of the Amana Colony. These barns remain in use in East Amana, Iowa

Fig. 11.12 Farm buildings and houses cluster near the now closed St. Anthony Catholic Church in the German ethnic hamlet of St. Anthony, Wisconsin. Note the log building (one of several in the community) to the left and the large German bank barn with its forebay to the right

beginning in the late 1830s resulted in a cluster of rural villages focused upon the parish churches and parochial schools (Schlemper 2004). This cluster and another in Wisconsin's Dane County have been described as "adaptations of the centralized village pattern—the "*dorf* culture" found in heavily Catholic Bavaria" (Zeitlin 2000:39). Arriving from small villages in the Eifel region of Rhenish Prussia, Catholic churches and schools were the first public buildings constructed in these villages of east central Wisconsin, being erected before the first stores and taverns. With the establishment of a nearby Capuchin monastery at Mt. Calvary, not only were the local parishes supported, but it became a focus of religious pilgrimages (Schlemper 2004). The resulting settlement, known for over a century as Wisconsin's "Holyland," is "[a]mong the most distinctive of these large clustered German settlements in Wisconsin" (Schlemper 2006:189). The monastery and parish churches dominate the rural landscape today, one and a half centuries after their establishment. Likewise, around Roxbury and Martinsville in northwestern Dane County, German Catholics created hamlets around a central church, with shops, taverns and farm buildings clustered nearby (Zeitlin 2000).

Within the area of the Missouri Ozarks settled by Germans are a number of elongated "string-shaped village(s), or *Strassendorfer*, [that were] a form a settlement that the Germans were acquainted with" (Gerlach 1976:67). These villages originally consisted of a single street of houses flanking a church, with a long narrow strip of land extending away from the dwellings. While growth might result in a parallel street, most was focused down the street, creating long narrow communities. For example, Altenburg, Missouri is described as being "two miles in length and two lots in width" (Gerlach 1976:67). In some of the string-shaped villages the homes are placed close to the road, without much of a front yard, as in the case of Westphalia, Missouri.

Volga Germans, such as those who settled villages in Ellis County, Kansas, also attempted to establish nucleated farm villages, which initially were centers of communal farms. Although that form of land tenure is long gone, the limestone farm houses and towering rural churches, the center of the nucleated villages, remain (Kloberdanz 1980). While other examples can also be cited, the vast majority of German farmers lived on dispersed farmsteads.

11.4.3 Distinctive German-American House Architecture

Germans who settled in Pennsylvania were widely noted for their substantial stone masonry construction even in the eighteenth century. As McMurry (2011:37) explains, "Scholars have identified several characteristic plan types that were strongly associated with German-speaking residents in colonial Pennsylvania." The ernhaus is among the earliest German house types in America, with examples such as the stone masonry Hans Herr house remaining in southeastern Pennsylvania (Fig. 11.13). The ernhaus, which has also been called a Continental German house, a Flürkuchenhaus, or a stoveroom type house, is characterized by having its entry door opening directly into the kitchen, or küche (McMurry 2011). The fireplace and chimney were central to the house, with an entertaining room on the opposite side of chimney from the kitchen.

11.4.3.1 German-American House Characteristics

Many architects argue that it is often difficult to recognize German houses.

> From New York to North Carolina, 18th-century German houses still exist that are recognizable by their parts rather than by a completely familiar form. Most significant, the kitchen sometimes lost its position as the principal entry space and was relegated to the cellar. In other cases, the kitchen remains on the first floor but the overall plan is barely distinguishable from an English or Dutch two-room house (Chappell 1986:69, Reprinted with permission of John Wiley and Son).

Although the floor plan may not necessarily be distinct, the method of construction and the choice of building materials frequently distinguish German ethnic communities. In areas of rural Pennsylvania "the rural landscape is still dominated by these buildings" (McMurry 2011:56), even though they were constructed well over a century ago.

While German influences upon vernacular architecture were quite varied, and indeed the German immigrant communities underwent assimilation while in America, there are some distinctive elements. Lauren Sickels-Taves and Philip Allsopp (2005:87) identify three key distinguishing features of German ethnic houses: "the overall configuration is rectilinear; the placement of the main entrance door is on the long side; [and] the main door leads directly into the kitchen or Küche." Germans not only brought distinctive ideas about how to build their houses, but also their barns, and even occasionally, housebarns.

11.4.3.2 German-American Stone Buildings

Germans who located in other regions, even though immigrating to America a century or more later, constructed a wide variety of buildings of stone. Within the Texas Hill Country, such as at Fredericksburg, stone was very frequently utilized for construction, being used for houses, barns, churches, commercial buildings, and fences (Fig. 11.14). Within the Missouri River German settlements stone was widely used to construct barns and houses, particularly outside the larger communities, while substantial stone churches were constructed. Within Wisconsin many examples of well-built stone masonry barns, farmhouses, and churches remain in the counties immediately north of Milwaukee.

Fig. 11.13 Hans Herr House, constructed 1719, near Willow Street, Pennsylvania. This well-preserved stone house, now roofed with slate but once roofed with thatch, shows the classic German propensity to build of masonry, to place the entrance doorway along the long side, and to have a centrally located hearth

Fig. 11.14 Stone masonry barn near Anhalt, Conal County, Texas. German ethnics were far more likely to construct their barns and farmhouses with stone masonry than Texans of British or American ancestry

These structures were typically built by stonemasons who had learned their craft in Germany, and while the Germans were not unique in building with stone, the frequency with which they utilized stone and certain architectural detailings produced a distinctly German landscape. For example, during the mid-nineteenth century the Germans in Missouri frequently constructed brick arches above windows and doors in structures otherwise built of stone and installed a Hausspruch, a carved stone giving the date of construction and the name or initials of the homeowner, above the doorway or in a gable of the house (van Ravenswaay 1977). While Germans in the Texas Hill Country often followed southern styles, they were far more likely to erect their structures of stone rather than wood. Thus, their "small, sturdy stone houses…displayed construction methods brought from Germany, including half-timbering (Fachwerk), casement windows, exterior plastering and whitewashing, and thatched roofs" (Jordan 1966:166). While the thatched roofs have long been replaced with sheet metal or shingles, many of these century and a half old buildings remain, even though German ethnics largely ceased constructing stone houses in the 1890s. In places such as Castroville, Fredericksburg, and New Braunfels, Texas, along with the many rural farmsteads established by German immigrants, they remain a conspicuous part of the ethnic landscape.

> Because of solid construction, many of the original stone and Fachwerk houses are still standing and in use, and their unique blend of native architecture and European construction methods will continue to be prominent in the cultural landscape of the German Hill Country for years to come (Jordan 1966:167).

What Terry Jordan wrote a half century ago remains true today.

11.4.3.3 German-American Log Houses

Germans often built their first houses of logs. Log buildings were particularly common among immigrants from western and southern Germany, especially among those who settled in the Upper Great Lakes region. American styles strongly influenced the types of log cabins that Germans built in the Missouri Valley, particularly with respect to fireplace location and the pitch of the roof. Nevertheless, the log houses of Missouri Germans provided a higher and more spacious attic by extending the log walls well above the ceiling of the first floor. Furthermore, the "Germans also brought the ceiling joists through the walls of their houses" (van Ravenswaay 1977:115) and arranged their rafters differently. In addition, throughout the Missouri Valley, "even in the single-room log houses, the stairway to the attic is boxed in, a remnant of late medieval form" (van Ravenswaay (1977:115). Many German log cabins were subsequently replaced by more substantial "stone, half-timbered, or brick houses similar to their homes in Europe" (van Ravenswaay (1977:143).

11.4.3.4 Fachwerk in German Ethnic Buildings

Fachwerk, a German half-timbered style of construction that is sometimes termed deutscher Verband in Wisconsin, was utilized in many early buildings constructed in German-American settlements (Fig. 11.15). Fachwerk has a wood timber framing, with brick being used as a filler or nogging between the timbers. A variant style of half-timbering, called Dreckhaeuser, used mud covered wooden slats placed between timbering, and examples are found in eastern Ohio. While mud or other nogging materials were sometimes used, brick was especially prominent after the 1860s. Most of these walls were covered with boards or plaster for weather-proofing; thus one often cannot readily see evidence of fachwerk buildings until the structures have fallen into disrepair. Fachwerk was particularly common among immigrants from northern and eastern Germany, and surviving examples are found in Pennsylvania, western Ohio where over 100 remain (Fig. 11.16), central North Carolina, the Texas Hill Country, the Missouri Valley in eastern Missouri, and in particular east central and southeastern Wisconsin where 300 can be found, including houses, barns, taverns, and churches. This cluster of German half-timber buildings in southeastern Wisconsin is "the largest known concentration in the United States" (Tishler 1986b:278).

Fig. 11.15 Saxonia House in rural farming hamlet of Fillmore, Wisconsin. The fachwerk construction of this building, built in 1855, can be seen by the timberwork bulging outward in the plastered side walls. Plastering such walls, while largely obscuring the timbers, greatly enhances the longevity of the wooden support beams

Fig. 11.16 Building of fachwerk construction in Zoar, Ohio, showing use of brickwork between the timbers framing the structure. This structure was built in 1825 by the utopian communal Society of Separatists of Zoar

Houses of fachwerk construction in Wisconsin have distinctive floor plans. These dwellings "have remarkably uniform room arrangements within a simple, rectangular plan...[with] front and rear entries at or near the center of the axial walls which opened into a central hallway" (Tishler 1986b:280).

11.4.3.5 German-American Brick Masonry Construction

Brick masonry construction is far more common for houses and churches in many German-American communities (Fig. 11.17). Both the manufacture of brick and their use in construction were familiar to most German immigrants, and brick was used frequently for larger buildings, such as churches, mills, breweries, and gymnastic halls. Thus, the German settled area of west central Ohio was marked by its "great number of tall, brick church spires" (Wilhelm 1992:69). Likewise, within all but one of the German settlements in Missouri the most common building material was brick (van Ravenswaay 1977).

The "German House" in Ohio was typically constructed of brick and was often called a "chapel style house." As Hubert Wilhelm explains, this was because local surveyors

> felt that these small houses had been influenced by the slightly earlier construction of brick churches. ... The form of the house, however, is the same—story-and-a-half with steeply pitched roof, gable-end chimneys, and entrance on the long side. On the first floor there are always four rooms, including the kitchen, placed toward the right rear. The upstairs is divided into two rooms (Wilhelm 1992:69, Reprinted with permission of Johns Hopkins University Press).

German ethnicity was also displayed in the manner that brick was laid. A common bond among Missouri Germans was to use five to seven rows of stretchers, brick with its long side exposed, followed by a row or layer of headers, in which the short edge is exposed (van Ravenswaay 1977). Brick was used in the building cornices. By the late 1800s "the segmental arch became rounder, and so commonly was its sprightly form used that it is one of the signatures of German construction in Missouri, as elsewhere" (van Ravenswaay 1977:231).

11.4.3.6 Later German-American Dwellings

Later German-American builders integrated many other cultural influences into their houses. Yet, as Kathleen Conzen (2003:57) explains, even though "Germans might adopt the temple-and-wing shape of the Midwestern farmhouse, but inside they retained the traditional continental *küche-stube-kammer* (kitchen, heated living space, bedchamber) arrangement, with interior chimneys and the kitchen dominant." Yet Germans were often noted for constructing houses similar to their neighbors, that had "cottage floorplans [that] varied very little and were easily built" (Meyer 1988:197). Many of the cottages they built in southern Illinois "portray the German interest in efficient and economical living space" and "German building traditions from the old country were reinforced and modified upon contact with French creole cottages" (Meyer 1988:197 and 199). Thus, the housing of Germans, particularly later arrivals or second or third-generation houses are not necessarily good indicators of their ethnicity, as German ethnics adopted elements from other groups, including Americanized Georgian style house plans. However, their barns are far better in displaying German ethnicity.

Fig. 11.17 Brick masonry construction dominates buildings within the German ethnic community of Washington, Missouri, located to the east of Hermann. The large building is the Louis Wehrman House, built before 1860 in the German Classic-Revival style

11.4.4 Midwestern Barns Display German Influence

Germans had a tradition of three field agriculture, a mixed farming operation whereby several crops were grown in rotation, along with the grazing of animals (and the use of their manure). This type of agriculture required barns to shelter animals, necessitating "a sizeable barn with sufficient space to store hay and sheaves of grain awaiting threshing, and to accommodate a threshing floor and animal stalls" (Wilhelm 1995:66–7). Thus, many observers have noted that barns on farms of German ethnics are large, far larger than the typical farmhouse. Indeed, unlike most others (excluding the French), German barns were designed to provide animals shelter besides a place to store crops and thresh grain. Because an investment in their barn furthered the entire profitability of the farm, German-Americans often spent more resources building their barns than their houses, exactly the opposite of the behavior of many Anglo-American farmers.

11.4.4.1 Heritage from Germany and Switzerland

In Germany and Switzerland farm animals were traditionally sheltered on the lower floor of houses, and many structures were built several centuries ago that combined the features of a house and barn. While such German housebarns with animals and people in the same structure, under the same roof, and separated by an interior wall are commonly seen in Germany and Austria, they are quite rare in United States (Fig. 11.18). Yet, a few German immigrants built them and surviving examples can be found in Wisconsin, Missouri, and the Dakotas. Unlike the spread of German inspired barn styles from the early colonized Pennsylvania, these housebarns were erected by individuals who immigrated from Germany before the American Civil War. Several remaining examples of these early housebarns, including examples in Missouri and eastern Wisconsin, are of fachwerk construction. Nevertheless, by far the vast majority of German-American farmsteads had a wide variety of distinct, separate farm buildings, yet sometimes arranged contiguously around a courtyard.

11.4.4.2 German Bank Barns

German bank barns are very commonly seen throughout the Middle Atlantic Region of Pennsylvania and New Jersey extending through the Midwestern states into the northern Great Plains. The style was derived from barn types introduced into southeastern Pennsylvania in the late 1600s and reinforced by ideas brought by later German immigrants. Although German bank barns in the United States visually look like barns from southern Germany and Switzerland, Allen Noble (1984:22) explains that "considerable debate exists, however, over whether the barn designs utilized in the New World were in fact evolved in this country, or whether

Fig. 11.18 Housebarn in German-Russian Mennonite community of Sommerfeld, Manitoba, just across the border from North Dakota. A few housebarns still remain in Ohio and Wisconsin

they are transplants from Europe." He reasons that it is most likely that "German, Swiss, and Austrian immigrants introduced several barn types, probably all related to one another and demonstrating an evolution from simple to complex" (Noble 1984:22). Research by Robert Ensminger (1992), both in Pennsylvania and Europe, strongly supports a Swiss origin of these barns, given similarities in design to early barns in Pennsylvania and sixteenth century barns in the Prätigau region of Switzerland that are still standing. Furthermore, "the early use of the term *Sweitzer* for this barn strongly suggests a Swiss origin" (Ensminger 1992:15). After reviewing the scholarship describing the ethnic links to Pennsylvania barns, McMurry and Ritchie Garrison (2011):107) concluded, "Research to date suggests that certain land tenure patterns, cultural foodways, decorative motifs, and labor organization occurred primarily among Pennsylvania Germans, and had specific architectural manifestations." Large bank barns are far more common among the Pennsylvania Germans.

The German bank barn is often located on sloping ground, with its basement area used to house animals (Fig. 11.19). The upper level of the barn is used for storage, such as of feed, hay, and farm implements, and it is entered on the side opposite the basement opening. If the barn is constructed on level ground, an earthen ramp is built leading to the second level. A key diagnostic feature is that the upper floors of the barn overhang the entrance to the basement area. This forebay "projects over the front stable wall for a distance that varies from four feet to even twenty or more feet in some barns" (Ensminger 1992:1). The German bank barn has a gabled roof, and its main entrances are usually on the longer sides, as is its typical forebay.

11.4.4.3 German Bank Barn Varieties

Over a dozen types and subtypes of the German bank barn have been identified. The Sweitzer Barns (also Schweitzer or Swisser) are the most common variety of German bank barns, and indeed one of most common of all barn types (Fig. 11.20). It was typically constructed from 1730 until about 1850 and was the prototype of the standard Pennsylvania barn, built from 1790 until around 1890 (Ensminger 1992). The Sweitzer barn can be viewed as a two and one half story structure, with the lowest level typically constructed of stone masonry, while the upper portion is often of a wood frame construction (Noble 1984). The lower level of the Sweitzer barn, which is often partly excavated, provides space to stable the farm animals. The barn's main upper floor, which is reached by an inclined earthen ramp—assuming the barn was not erected on sloping land, typically contains either three bays or two bays separated by a central drive space. The central section of the barn's upper level was originally designed as a place to thresh grain. It had doors over the forebay that could be thrown open to facilitate a

Fig. 11.19 German bank barn in village of Railroad, Pennsylvania. Animals are confined to the lower level, hay and threshing occurred on the upper level

Fig. 11.20 Sweitzer style German bank barn north of Mount Aetna, Maryland

draft, to assist in the removal of straw from the grain. The Sweitzer barn is found from Pennsylvania into Ohio and Ontario and south into the Shenandoah Valley of Virginia, and is "almost always an indicator of original German settlement" (Noble and Cleek 1995:90). The derivative standard Pennsylvania barn was built not only in the areas where Sweitzer barns are located, but also in northern Indiana and Illinois, southern Wisconsin, and across the Mississippi River into Iowa and Nebraska, with even a few examples constructed in Oregon and Washington.

The forebay, the major distinguishing characteristic of these barns, is usually supported by cantilevered beams, and posts or columns are rare. The barn may be constructed of logs, timber frame, brick, or stone. Many of those of logs or timber frame are set on a foundation of stone walls that enclose the entire lower level. Closed-forebays were con-

Fig. 11.21 Variety of German bank barn with projecting foundation walls, located in hamlet of St. Anthony, Wisconsin

structed in some varieties of the Sweitzer and Pennsylvania barns, in which the lower-level wall extended underneath the end post at each end of the forebay (Fig. 11.21). The forebay itself is always of timber framing, with plank siding.

The positioning of the roof ridge line is the primary difference between Sweitzer and standard Pennsylvania barns. Within the Sweitzer barn the ridge is placed farther back from the forebay, resulting in an asymmetrical gable in which the eave of the forebay is lower than the eave on the barn's bank side. However, the ridge line is centrally placed over the lower level foundation walls of the Sweitzer barn. With the standard Pennsylvania barn, as Ensminger (1992:67) explains, "the roof ridge is centered over the whole barn structure…[producing] a *symmetrical* gable end for the upper, framed part of the barn and a forebay front wall height equal to that of the rear bank wall."

Another variety of the Pennsylvania barn, the extended barn, was built from 1790 until about 1920, and these had the longest projecting forebays. Unlike the other varieties of Pennsylvania barns, the forebay beam did not extend into the center or rear of the barn, but was instead supported by posts. Furthermore, the slope of its shed roof was less than that of the main barn. While it has been proposed that Wisconsin porch barns are another variant of German bank barn that should be called a Pomeranian barn as it was thought they were brought to the Marathon County area by Pommeranian immigrants from northern Germany (Bastian 1975), Ensminger (1992) argues that this is erroneous, as the style of the upcountry posted-forebay Pennsylvania Barn traveled across the country. These timber frame barns are longer and lower than Sweitzer barns, often with several pairs of threshing doors.

While the Pennsylvania barn derived from Swiss-German bank barns constitutes the most common variety of barn presently seen throughout much of the United States, barns built by German immigrants and their descendents varied regionally. For example, within east central Missouri numerous bank barns of two levels with a hayloft were constructed, yet the forebays so characteristic elsewhere are missing. In addition, some of the two level barns had their entryway in their gable ends, rather than the more common side entry. As with other German barns, these were frequently of stone masonry construction for at their lower level, while the upper level might be either stone masonry or wood. The stone masonry barns had numerous slits in their walls to provide ventilation (van Ravenswaay 1977). Yet not all the German built barns were of these designs. In the Texas Hill Country their barns, many which were constructed of stone, are smaller. Indeed, in Texas Germans patterned their farmsteads after those of Southern yeomen (Jordan 1966), as the area's climate did not necessitate the type of structure typical in their homeland. Yet even in Texas German barns are distinctive because of their stone construction.

While various German bank barns are fairy widely distributed, the Grundscheier, or ground barn, is found primarily in southeastern Pennsylvania and adjacent north central Maryland. Relatively rare today and often of stone construction, these German barns are unbanked and lack the type of forebay associated with the various bank barns (Noble and Cleek 1995). Older versions may have narrow slits in the stone walls that provide ventilation.

11.4.4.4 Decorations on German Ethnic Barns

Decorations were often placed on German ethnic barns. Sometimes white painted decorations around barn doors resemble stylistic arches and columns. Owl holes, small decorative cut-outs under gable of barns in the form of stars, diamonds, half-moons, and hearts, are most common in areas of German-Americans. For example, it has been noted that in Berks County, Pennsylvania, "many Pennsylvania German barns displayed these large discs painted with star patterns" (McMurry and Garrison 2011:108). In many of the German brick barns of Missouri decorative openings in the brickwork under the end gables provide ventilation (van Ravenswaay 1977). Because German-Americans are such a dominant force among American's family farmers, barns displaying German architectural heritage are so prevalent, and barns remain standing for a century or more following their construction, much of America's agricultural heartland displays a strong German ethnic imprint.

In some areas of German settlement, such as in parts of central Wisconsin, the separate buildings that comprise a farmstead are clustered around a central courtyard. This mitteldeutsches Gehöft provides shelter from the wind to animals that are enclosed by the barn and other farm buildings which abut at their corners. In some examples, the farmhouse is placed sufficiently close to the courtyard to form its fourth side.

11.5 Urban German-American Landscapes

German immigrants influenced the development of many American cities, yet it is sometimes difficult to separate German contributions from contemporary styles. Some of the German-American urban landscape did not differ much from prevailing American patterns. For example, Hubert Wilhelm (1992:77) observed that "Germans built the kind of structures popular in America at the time and…'there was no articulate expression of a German architecture.'" Nevertheless, many cities did develop distinctive German districts. For example Cincinnati, Ohio attracted thousands of German immigrants to its "Over the Rhine" district in the mid nineteenth century. While that area today is largely an African-American ghetto, many of its buildings still display a German architectural imprint, and Germans still account for nearly one-fifth of the city's population. Likewise, Germans left a strong visual imprint on St. Louis, Missouri, yet there they now comprise only about a sixth of the city's population. In both cities, German-Americans account for a far greater proportion of the suburban population.

11.5.1 Milwaukee as America's Most German-American City

One of the best large cities in which to see a remnant of a strong German imprint on the landscape is Milwaukee, Wisconsin, which has been described as "America's Most German-American city" (Gurda 1996). While German-Americans in 2000 accounted for only 19.9% of the city's population and 28% of Milwaukee County's population, given the large number of both Polish-Americans and African-American who resided there, during the late nineteenth century the city was termed the "German Athens," literally a "city transplanted from the Rhine" (Bawden 1996). In 1890 over 90% of Milwaukee's population were either foreign born or the children of immigrants, and over two-thirds of these first or second generation residents were German ethnics. This dominance by a German born population translated into the creation of a city that was remarkably German in appearance, so much so that by the end of the nineteenth century Milwaukee was viewed as the most foreign of America's cities, with its numerous breweries, taverns, "elaborate gothic church spires," and German-speaking population (Hoelscher et al. 1997). Indeed, the number of Germans was so dominant that Tim Bawden notes, "perhaps only in Milwaukee, did the Germans attain a level of local hegemony strong enough that they could create a civic, or public, landscape so overtly based on German, not American, architecture" (Bawden 1996:3). During the quarter century leading up to the First World War, Milwaukee showcased its German heritage through its houses, churches, business buildings, and other structures that were constructed. At the time "being German was something to be proud of, and Milwaukee's by then well-established and prosperous German-Americans had the resources to celebrate their ethnicity" (City of Milwaukee 1994). Germans constructed "a large number of residential, religious, commercial/industrial, and public buildings reflecting design elements of native German architecture that gave the city a distinctive Old World character" (U.S. National Park Service 2005:18).

11.5.1.1 German-American Houses

The prosperous families of Milwaukee began copying the plans of German houses when building their homes beginning in the early 1890s (Fig. 11.22). While the interior plan resembled American Victorian structures, the exterior displayed

> motifs borrowed from the architecture of the sixteenth and seventeenth century German renaissance. The distinguishing features were shaped gables, bellcast helmet-domed towers and oriels, deep arcaded porches, and timbered gables. Abundant bold sculptural detailing, intricate woodcarving, and massive ironwork enlivened walls laid-up in the preferred brown or earthtone colored brick, trimmed with terra cotta or stone (City of Milwaukee 1994).

Fig. 11.22 Henry Harnischfeger House built in 1905 in Milwaukee, Wisconsin. This house, constructed in the German Renaissance Revival style, includes statues built into the columns supporting its second story arcade on the right

While many of these houses were German Renaissance Revival or German Baroque Revival mansions, such as the Pabst Mansion, others just displayed considerable German detailing. Even on far less pretentious houses, the front porch and massive front gables, either shingled like those in Eastern Europe or covered with half-timbering were commonplace reminders of German ethnicity. Today many surviving German inspired houses can be seen in neighborhoods near Lake Michigan on the north side of Milwaukee (Fig. 11.23).

11.5.1.2 German-American Churches

German ethnicity is also seen in many of the older churches in Milwaukee. Beginning in the 1870s Milwaukee architects borrowed the plans of churches in Germany that were copied onto the Milwaukee landscape. Many of these churches had a symmetrical, "soaring central spire with low flankers in the front that almost completely conceal the nave of the church" (City of Milwaukee 1994:3), yet some German churches were twin towered. German architectural embellishments on Milwaukee churches include intricate brick detailing and repetitive gabled forms on the façade (City of Milwaukee 1994). As the city architectural survey notes, the interior vaulted ceilings, carved woodwork, and church furnishings also showcased German culture. Milwaukee's Ethnic Architecture Inventory lists eight surviving classic German churches built in the years up to 1901.

11.5.1.3 German-American Commercial Buildings

Commercial and industrial buildings also showcase the ethnicity of their builders, such as in the Pabst and Schlitz breweries (Fig. 11.24). These huge cream-colored brick structures "took on a Germanic character by employing crenulated parapets, shaped and stepped gables, deep corbel table friezes, pinnacles, massive arched portals and Gothic arched windows trimmed with label moulds" (City of Milwaukee 1994:14). Pabst Brewery owned many taverns, both in Milwaukee and elsewhere, which displayed a castellated style, reminiscent of the brewery. Indeed, German brewers "established vast outdoor beer gardens and erected hundreds of distinctive taverns and beer halls" (Bawden 2006:84). The Germania Building, complete with its exact replica of the Germania statue overlooking the junction of the Rhine and Moselle Rivers at Rudesheim, Germany, was built to house the nation's largest circulation German-language newspaper and is marked with copper domes topped with spiked finials at each corner of the building. It is just one of many remaining commercial buildings highlighting the Milwaukee's German heritage.

11.5.1.4 German-American Public Buildings

Milwaukee's public buildings remind the visitors of the ethnicity of the city's residents of a century ago (Fig. 11.25). The German Baroque Revival style Pabst Theatre has a

Fig. 11.23 Frederick Weinhagen House built in 1901 in Milwaukee, Wisconsin. This brick house was built in the German Renaissance Revival style in a neighborhood of many other houses displaying German architectural features

Fig. 11.24 Stepped decorative gable of the former Schlitz Brewery in Milwaukee, Wisconsin

"German convex mansard roof and walls overlaid with elegant shallow arcading enriched with curvaceous foliated decorations" (City of Milwaukee 1994:16). Milwaukee's City Hall (Fig. 11.26), built in 1895 and at the time of its construction the third tallest building in the nation, was clearly designed to highlight the city's German heritage. In its National Historic Landmark nomination, mention is made that it was "designed in a style commonly called German Renaissance Revival, its curved gable dormers and terra cotta ornamentation are reminiscent of the Northern Renaissance Revival architecture of Germany" (U.S. National Park Service 2005:4). Milwaukee's City Hall was "patterned

Fig. 11.25 German architectural embellishments displayed in the Pabst Theatre (*left*) and the Milwaukee City Hall (*right*). These buildings in Milwaukee, Wisconsin display German Baroque and German Renaissance styles, respectively

Fig. 11.26 City Hall, built in 1895 in Milwaukee, Wisconsin. This building displays both German Renaissance architecture and design elements copied from sixteenth and seventeenth century German rathauses

after features from actual sixteenth- and seventeenth-century German *Rathäuser*, or city halls, that [Milwaukee architect Henry] Koch had seen in Germany, his City Hall casts a distinctively Old World look on Milwaukee's downtown streetscape" (Hoelscher et al. 1997:391). In particular, similarities between the Hamburg, Germany Rathaus and Milwaukee's City Hall have been described, including their similar towers, "tripled arched-window arrangement, Flemish gables silhouetted against a steep roof, and finials or flagpoles" (U.S. National Park Service 2005:14). Furthermore, "many sculptural details, such as cherubs and the heads of foxes and wolves, were incorporated to indulge the German taste for richly carved ornament" (Hoelscher et al. 1997:391). The older portion of Milwaukee's Federal Building, with its numerous towers and gables, likewise evokes the Old World character of the city, as does the city's towering gothic-styled stone masonry water tower.

11.6 The German-Speaking Old Order Amish

While the majority of the German ethnics in the United States have become highly integrated into American society, reshaping the dominant American culture with their strongest ethnic imprint to be found in buildings constructed a century or more ago, a small minority of German ethnics has not only persisted in maintaining their separation from society, but are growing in numbers and spatial significance (Lamme 1993). Several groups of ultra-conservative Swiss-German Anabaptists have created distinct landscapes as they have separated their communities from mainstream American society (Kraybill et al. 2013). These groups include the Old Order Amish, the Old Order Mennonites, the German Brethren, and the Hutterites. While the Hutterites have established communal settlements in the western Great Plains, and in particular, within Canada's Prairie Provinces (see Chap. 18), the rapidly growing numbers of the Old Order Amish and Old Order Mennonite are expanding those areas of the United States where one sees their plain communities. While many new Amish settlements represent an infill within their traditional east central homeland, "distant settlement attempts are thriving more than ever" (Anderson and Kenda 2015:504), and the Amish are increasingly attracted to rural areas that have experienced declining populations.

The Old Order Amish share many religious beliefs with the Mennonites, from which Jakob Amman and his followers broke over the issue of shunning (excommunication) during the 1690s. While the Old Order Mennonites have created a landscape that visually is not too different from that of the Old Order Amish, the Amish are much more conservative than the mainstream Mennonites. The Amish were prompted to immigrate to America because of religious persecution at home. The first wave of Amish immigrants came between 1717 and 1750 from the Palatinate of Germany and from Switzerland and settled in Pennsylvania. The second wave came chiefly from the Alsace and Lorraine between 1817 and 1861, and established settlements primarily in Ohio, Illinois, Iowa, and southern Ontario (Fig. 11.27).

The Amish believe that they must live separate from unbelievers. As the well-known sociologist John Hostetler (1987:283) explained, "Living in a redemptive community, separated from the world, is essential to salvation." The Amish have a strong desire to live apart from others, and separate themselves from the world of the English, as they call all others, by their distinct dress, beards among their

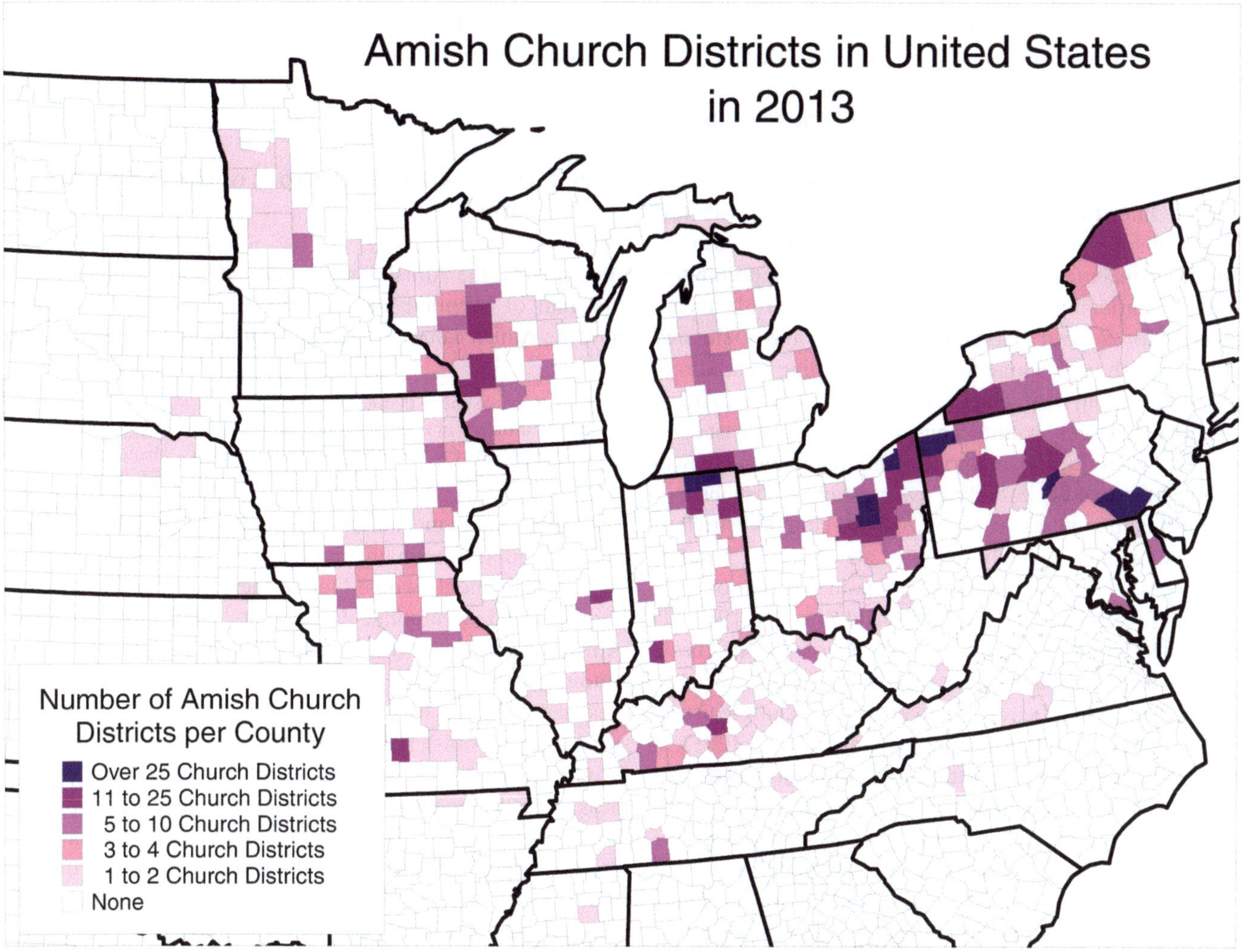

Fig. 11.27 Amish settlements have rapidly expanded over the past several decades to extend from Pennsylvania through the Midwestern states, with small communities now located from Maine to Montana, Wyoming, and Colorado. Amish settlements are also found in Ontario, New Brunswick, and Prince Edward Island. The impact upon the landscape is most conspicuous in the larger, and longer established, settlements with multiple church districts. An Amish church district or congregation typically includes 25–35 households

married men, speaking in German, traveling by horse-drawn buggy, and avoiding many varieties of modern technology. The Amish have traditionally been tied to employment in farming. As Hostetler (1987:285) explains, "The charter of Amish society requires that members make their living from farming; if not from farming, then from occupations of a rural or semi-rural character." Even though many Amish are no longer farmers today, particularly in their largest settlements, they strive to maintain a rural lifestyle, and many have turned to sawmill work, woodworking, and furniture or woodcraft making—all within a rural setting.

11.6.1 Population and Distribution of the Old Order Amish

The population of the Amish, who continue to speak the Pennsylvania German (or Pennsylvania Dutch) dialect within their communities, has increased significantly, given the large size of their families and their high fertility rate. With a doubling in their numbers every two decades, the Amish have experienced many long-distance moves from their original settlement regions as they seek new lands to farm. Within the area of their traditional settlement, particularly in Lancaster County, Pennsylvania and near Holmes County, Ohio, their population is such that there are few lands available for children who want to continue in farming. Thus, the Amish face a quandary—look to job opportunities outside of agriculture or move to areas where farmlands are available for sale at reasonable prices. While Ohio and Indiana Amish have a long history of spreading in all directions, establishing new settlements, the Pennsylvania Amish had remained largely within Pennsylvania until the past three decades, when high land prices and population growth prompted them to begin settling as far away as Wisconsin.

Amish communities are organized into church districts, a congregation that typically includes about 25–35 families who gather every other week at the home of one of their members for Sunday services. As Donald Kraybill (1989:11) notes, "The congregation, or church district, is the basic organizational unit above the family in Amish society." Because Amish families are relatively large, a church district often includes 150–200 persons. As Kraybill et al. (2013:5) explain, "Church districts that share a common history in a given geographic area constitute a *settlement*." By 2013 there were 2012 church districts nationally, scattered among 472 communities. The largest Amish settlement, with 238 church districts, was centered upon Holmes County, Ohio, overlapping into several adjacent counties. The second largest and oldest, having been established about 1760, covered portions of Lancaster and Chester Counties, Pennsylvania, with 187 church districts. The third largest Amish community was in Elkhart and LaGrange Counties, Indiana, with 153 church districts (Donnermeyer and Luthy 2013). In contrast to these huge settlements, most settlements in the states receiving the greatest number of Amish newcomers are considerably smaller. Thus, in Wisconsin, which has the fourth largest number of Amish settlements, the two largest settlements each have 13 church districts.

The total Amish population within the United States has more than doubled in the past quarter century, bringing their total to over 250,000 (Donnermeyer et al. 2013). Thus, there has been a tremendous change in the spatial distribution of Amish settlements since the 1970s. In 1974 49% of all Amish settlements and 75% of Amish church districts were located in Indiana, Ohio and Pennsylvania. By 2003 only 32% of Amish settlements were in those three states (Luthy 2003). New York and Wisconsin now have the third and fourth largest number of Amish settlements, even though nearly two-thirds of the total Amish population remains in Indiana, Ohio and Pennsylvania (Donnermeyer and Anderson 2015). Other states with many new Amish settlements include Michigan, Missouri, and Kentucky. Amish settlements now extend from Florida to Maine and as far west as Montana and Idaho.

11.6.2 Distinctive Landscapes of the Old Order Amish

As a society, the Amish became more distinctive during the first half of twentieth century. William Crowley describes how the Amish came to stand out from their neighbors.

> Their clothing, general plainness, and religious beliefs had previously made them distinctive. Now their farming techniques and their means of transportation—the horse and buggy—began …to represent pockets of retention of methods abandoned by more progressive agriculturalists (Crowley 1978:257–8).

The Amish avoid mechanized equipment, especially tractors and automobiles, to maintain their lifestyle. Use of a tractor to plow their fields would "tempt their members to expand acreages and accumulate large debts while driving other Amish off the land" (Hostetler 1987:279), while use of an automobile would take members away from their communities and enhance their opportunities to interact with the worldly, non-Amish population.

Old Order Amish landscapes are characterized by plainness and a non-mechanized lifestyle. Farming makes use of horse-drawn plows, harrows, and harvesters, and horse-drawn buggies, usually painted black, are used for travel. Plow horses with their huge hooves are seen around all Amish farms, and within nearby communities one sees facilities for horses, while the droppings of the horses mark the highways. Plainness, rather than an ostentatious German ethnic architecture, marks the landscape of Amish communities (Fig. 11.28). Thus, there is a lack of electrification, no power

Fig. 11.28 Large farmhouses on Amish farmstead south of Groffdale, Lancaster County, Pennsylvania. The Lancaster County Amish settlement, established in 1760 and now the nation's second largest Amish settlement, is dispersed over the rural countryside. Given their large family sizes and that multi-generations live together, Amish farmhouses are often large, and a "grandfather house" is often present. Note the large barn and the absence of electrical power lines and television antennae

lines running to Amish houses, no electric lights in their barnyards, and no television antennae atop their houses. Windmills are still frequently used to pump water. Because telephones are banned from houses, telephone booths may be placed at the end of farm lane and in their shops.

Amish buildings are not necessarily distinctive other than being plain, frequently painted white, and lacking electrification. Because placing lightning rods would question their faith, these are missing from their farmsteads, even though many of their English neighbors have them. Amish expansion in many areas has resulted in their occupation of buildings built by others, prompting the observation that "Amish settlements take on the character of the area where they are founded" (Lamme and McDonald 1984:90). Within Wisconsin, new Amish residents occupy farmsteads built by German, as well as Norwegian and Welsh immigrants (Cross 2016).

A disproportionate number of Amish farmers operate small dairy herds. Cross has noted,

> The landscape of the Amish dairy farms, particularly of those Grade B operations using cans, is conspicuously different from that of their "English" neighbors, even when the Amish farms had been previously managed by an "English" farmer. The Amish dairy farm is typically much smaller, both in acreage and number of cows. Cows are more likely to be seen grazing in a pasture, unlike so many other dairy operations where the cows are now fed within the barn year-round. Amish cornfields following harvest are sometimes marked by rows of shocked corn (Cross 2004:93–94).

To facilitate the receipt of their milk cans, which most milk processors will not accept, small cheese factories have been established by the Amish near their settlements. Those who utilize milk cans have changed little in their avoidance of technology, while the Amish who use bulk tanks have become more accepting of certain technologies, and as a result, have larger herds (Cross 2014).

Other visual clues indicate the presence of Amish populations. "Within many, but not all, Amish settlements farms commonly display signs including the words "No Sunday Sales" that advertise their vegetables, baked goods, quilts, or wood products. Signs warning of "Horse Drawn Vehicles" are increasingly found along rural highways" (Cross 2004:95). Buggies, always pulled by one horse, and painted black or occasionally gray, travel between Amish farmhouses as well as to nearby market towns (Fig. 11.29). What one does not see are Amish church buildings, but instead on Sunday one sees up to two dozen buggies parked at a farmstead, where the faithful are worshiping in either the barn or the farmhouse.

Amish farming communities have higher population densities than other rural areas, with a family size well above the national norm and a lack of mechanization limiting farm size, resulting in a greater number of farms per square mile.

Fig. 11.29 Amish buggies parked near Amish bakery at farmstead southwest of Dalton, Wisconsin. Most buggies are black, although certain church districts favor gray buggies or open black buggies. The placement of the orange reflective warnings on the backs of the buggies has been vigorously opposed by some Amish communities as infringing on their religious rights to maintain plainness

Thus, one sees "more life in the Amish country—more natural life, more animal life, more human life" (Hostetler 1987:279). As their population increases, estimated to exceed 308,000 in the United States and Canada in 2016 (Young Center 2016), Amish landscapes are increasingly appearing in locales where other ethnics—be they German Lutherans, German Catholics, or other European groups—previously shaped the ethnic landscape.

11.7 Other Germanic Landscapes

Not only did Germans immigrate to the United States from a wide variety of different German states and principalities, but immigrants from several parts of central Europe brought a blended heritage, strongly German, yet also steeped in a separate national heritage. Both the Austrians and Swiss are good examples, for while both nations have a long history of a separate national identity, German culture and language have long dominated. While we often identify the Amish as a distinct German group because of their language, their earliest origins were Swiss, although they lived for several centuries in the German Palatinate before migrating to America.

11.7.1 Swiss Ethnic Landscapes

Other immigrants from Switzerland came directly to the United States, bringing a heritage that was partly German, but the Swissness of their heritage is often displayed. One of the best examples is within south central Wisconsin where the community of New Glarus touts its Swiss heritage to bring tourists to attend its William Tell festival, to shop in stores selling Swiss style chocolates, cheeses, and sausages as well as cowbells and other curios, and to patronize a variety of restaurants (Fig. 11.30) featuring Swiss style fare (Hoelscher 1998). Within the region centered upon Green County, Wisconsin are a variety of cheese factories producing Swiss cheese, as well as the only manufacturer of Limburger cheese in the nation, and dairy farm silos can be spotted that display the Swiss flag. The same region is the home to several Reformed or United Church of Christ congregations serving the Swiss ethnic population. With 23.3% of its population reporting Swiss ancestry at the time of the 2000 Census, Green County led the nation in the proportion of its population claiming Swiss ancestry. The county with the second highest proportion of its population having Swiss heritage was Adams County, Indiana, which was 11.5% Swiss.

Fig. 11.30 New Glarus Hotel located in the heart of New Glarus, Wisconsin. Erected in 1853, the restaurant within this building serves Swiss style foods including cheese fondue, kalberwurst and other sausages, spatzli or Swiss noodles, and roesti within a building painted with Swiss inspired decorations and displaying Swiss style flowering window boxes

11.8 Précis

Germans, alike their British cousins who also speak a Germanic language, brought a large variety of architectural and material cultural elements to the American landscape. Not only are these seen by immigrants who came directly from the numerous German states and principalities, but also by German ethnics who arrived from various other lands of central and eastern Europe. Because of their vast numbers, and because they settled across so much of the country, German ethnic influences have shaped much of what is now considered the generic American landscape.

References

Allen, James P., and Eugene J. Turner. 1988. *We the People: An Atlas of America's Ethnic Diversity*. New York: Macmillan.

Anderson, Cory, and Loren Kenda. 2015. What Kinds of Places Attract and Sustain Amish Populations? *Rural Sociology* 80 (4): 483–511.

Bastian, Robert W. 1975. Southeastern Pennsylvania and Central Wisconsin Barns: Examples of Independent Parallel Development? *Professional Geographer*. 27 (2): 200–204.

Bawden, Timothy. 1996. Milwaukee's German Renaissance Landscape, 1890–1918. Paper presented at annual meeting of Wisconsin Geographical Society, Oshkosh, Wisconsin, Oct 12.

———. 2006. A Geographical Perspective on Nineteenth-Century German Immigration to Wisconsin. In *Wisconsin German Land and Life*, ed. Heike Bungert, Cora Lee Kluge, and Robert C. Ostergren, 79–91. Madison: Max Kade Institute for German-American Studies, University of Wisconsin.

Chappell, Edward A. 1986. Germans and Swiss. In *America's Architectural Roots: Ethnic Groups That Built America*, ed. Dell Upton, 68–73. Washington: Preservation Press.

City of Milwaukee. 1994. *Milwaukee Historic Ethnic Architecture Resources Study, Final Report*, 1994. Milwaukee: Department of City Development. August.

Conzen, Kathleen Neils. 1980. Germans. In *Harvard Encyclopedia of American Ethnic Groups*, ed. Stephan Thernstrom, 405–425. Cambridge: Harvard University Press.

———. 1990. *Making their Own America: Assimilation Theory and the German Peasant Pioneer. German Historical Institute Annual Lecture Series No. 3*. New York: Berg Publishers.

———. 2003. *Germans in Minnesota*. St. Paul: Minnesota Historical Society Press.

Conzen, Michael P. 1990. Ethnicity on the Land. In *The Making of the American Landscape*, ed. Michael P. Conzen, 221–248. Boston: Unwin Hyman.

Cross, John A. 2004. Expansion of Amish Dairy Farming in Wisconsin. *Journal of Cultural Geography* 21 (2): 77–101.

———. 2014. Continuity and Change: Amish Dairy Farming in Wisconsin over the Past Decade. *The Geographical Review*. 104 (1): 52–70.

———. 2016. Dairying Landscapes of the Amish in Wisconsin. *Material Culture*. 48 (2): 16–31.

Cross, John A., Douglas Jackson-Smith, and Bradford Barham. 2000. Ethnicity and Farm Entry Behavior. *Rural Sociology* 65 (3): 461–483.

Crowley, William K. 1978. Old Order Amish Settlement: Diffusion and Growth. *Annals of the Association of American Geographers* 68 (2): 249–264.

Donnermeyer, Joseph, and Cory Anderson. 2015. A Mid-Decade Update on Amish Settlement Growth. *Journal of Amish and Plain Anabaptist Studies.* 3 (2): 222–235.

Donnermeyer, Joseph, Cory Anderson, and Elizabeth C. Cooksey. 2013. The Amish Population: County Estimates and Settlement Patterns. *Journal of Amish and Plain Anabaptist Studies.* 1 (1): 72–109.

Donnermeyer, Joseph, and David Luthy. 2013. Amish Settlements across America: 2013. *Journal of Amish and Plain Anabaptist Studies.* 1 (2): 107–129.

Ensminger, Robert F. 1992. *The Pennsylvania Barn: Its Origin, Evolution, and Distribution in North America.* Baltimore: Johns Hopkins University Press.

Fischer, David Hackett. 2004. *Washington's Crossing.* New York: Oxford University Press.

Frenkel, Stephen, and Judy Walton. 2000. Bavarian Leavenworth and the Symbolic Economy of a Theme Town. *The Geographical Review.* 90 (4): 559–584.

Fuchs, Stephan. 2015a. An Integrated Approach to Germanic Place Names in the American Midwest. *The Professional Geographer.* 67 (3): 330–341.

———. 2015b. History and Heritage of Two Midwestern Towns: A Toponymic-material Approach. *Journal of Historical Geography* 48 (1): 11–25.

Gerlach, Russel L. 1976. *Immigrants in the Ozarks: A Study in Ethnic Geography.* Columbia: University of Missouri Press.

———. 1992. A Contrast of Old World Ideology: Germans and Scotch-Irish in the Ozarks. In *Ideology and Landscape in Historical Perspective*, ed. Alan R.H. Baker and Gideon Biger, 289–302. Cambridge: Cambridge University Press.

Gurda, John. 1996. America's Most German-American City: Milwaukee Looks Back on Its Legacy as the Deutsch-Athen. *German Life.*: 42–45.

Hoelscher, Steven D. 1998. *Heritage on Stage: The Invention of Ethnic Place in America's Little Switzerland.* Madison: University of Wisconsin Press.

Hoelscher, Steven, Jeffrey Zimmerman, and Timothy Bawden. 1997. Milwaukee's German Renaissance Twice-Told: Inventing and Recycling Landscape in America's German Athens. In *Wisconsin Land and Life*, ed. Robert C. Ostergren and Thomas R. Vale, 376–409. Madison: University of Wisconsin Press.

Hostetler, John A. 1987. A New Look at the Old Order. *Rural Sociologist.* 7 (4): 278–292.

Jordan, Terry G. 1966. *German Seed in Texas Soil: Immigrant Farmers in Nineteenth-Century Texas.* Austin: University of Texas Press.

———. 1980. A Religious Geography of the Hill Country Germans of Texas. In *Ethnicity on the Great Plains*, ed. Frederick C. Luebke, 109–128. Lincoln: University of Nebraska Press.

Kamphoefner, Walter D. 1987. *The Westfalians: From Germany to Missouri.* Princeton: Princeton University Press.

Kirschbaum, Erik. 2015. *Burning Beethoven: The Eradication of German Culture in the United States during World War I.* New York: Berlinica Publishing.

Kloberdanz, Timothy J. 1980. Plainsmen of Three Continents: Volga German Adaptation to Steppe, Prairie, and Pampa. In *Ethnicity on the Great Plains*, ed. Frederick C. Luebke, 54–72. Lincoln: University of Nebraska Press.

Kraybill, Donald B. 1989. *The Riddle of Amish Culture.* Baltimore: Johns Hopkins University Press.

Kraybill, Donald B., Karen M. Johnson-Weiner, and Steven M. Nolt. 2013. *The Amish.* Baltimore: Johns Hopkins University Press.

Lamme, Ary J. III. 1993. The 'North County' Amish Homeland. *Journal of Cultural Geography* 13 (2): 107–118.

Lamme, Ary J. III, and Douglas B. McDonald. 1984. Recent Amish Settlement in the North Country. *Material Culture.* 16 (2): 77–91.

Langer, Ute. 2006. Using Archival Resources in Germany for Research Focused on Emigration from the Rhineland in the Nineteenth Century. In *Wisconsin German Land and Life*, ed. Heike Bungert, Cora Lee Kluge, and Robert C. Ostergren, 57–76. Madison: Max Kade Institute for German-American Studies, University of Wisconsin.

Lanier, Gabrielle. 2011. Landscapes. In *Architecture and Landscape of the Pennsylvania Germans, 1720–1920*, ed. Sally McMurry and Nancy Van Dolsen, 10–31. Philadelphia: University of Pennsylvania Press.

Leiding, Gerlinde. 1992. Germans in Texas. In *To Build in a New Land: Ethnic Landscapes in North America*, ed. Allen G. Noble, 362–378. Baltimore: Johns Hopkins University Press.

Luthy, David. 2003. Amish Settlements Across America: 2003. *Family Life.* 36 (October): 17–23.

Max Kade Institute. 2005. *How German Is American?* Madison: Max Kade Institute.

McMurry, Sally. 2011. Rural Houses. In *Architecture and Landscape of the Pennsylvania Germans, 1720–1920*, ed. Sally McMurry and Nancy Van Dolsen, 30–65. Philadelphia: University of Pennsylvania Press.

McMurry, Sally, and J. Ritchie Garrison. 2011. Barns and Agricultural Outbuildings. In *Architecture and Landscape of the Pennsylvania Germans, 1720–1920*, ed. Sally McMurry and Nancy Van Dolsen, 94–124. Philadelphia: University of Pennsylvania Press.

McMurry, Sally, and Nancy Van Dolsen. 2011. Introduction: Architecture and Landscape of the Pennsylvania Germans, 1720–1920. In *Architecture and Landscape of the Pennsylvania Germans, 1720–1920*, ed. Sally McMurry and Nancy Van Dolsen, 1–9. Philadelphia: University of Pennsylvania Press.

Meyer, Douglas K. 1988. German Cottage Structure Types in Southwestern Illinois. In *French and Germans in the Mississippi Valley: Landscape and Cultural Traditions*, ed. Michael Roark, 191–208. Cape Girardeau: Center for Regional History and Cultural Heritage, Southeast Missouri State University.

Noble, Allen G. 1984. *Wood, Brick, and Stone: The North American Settlement Landscape, Volume 2: Barns and Farm Structures.* Amherst: University of Massachusetts Press.

Noble, Allen G., and Richard K. Cleek. 1995. *The Old Barn Book: A Field Guide to North American Barns and Other Farm Structures.* New Brunswick: Rutgers University Press.

Nolt, Steven M. 2002. *Foreigners in Their Own Land: Pennsylvania Germans in the Early Republic.* University Park: Pennsylvania State University Press.

Roper, Stephanie Abbot, Scott C. Roper, and Cindy Ferraresi. 2006. Cemeteries of the Texas Panhandle. *PAST: Pioneer America Society Transactions.* 29: 24–41.

Salamon, Sonya. 1992. *Prairie Patrimony: Family, Farming, and Community in the Midwest.* Chapel Hill: University of North Carolina Press.

Sauer, Carl O. 1920. *The Geography of the Ozark Highland of Missouri.* Chicago: University of Chicago Press. Reprinted 1968 by Greenwood Press, New York.

Schlemper, M. Beth. 2004. The Regional Construction of Identity and Scale in Wisconsin's Holyland. *Journal of Cultural Geography* 22 (1): 51–81.

———. 2006. The Borders of the Holyland of East-Central Wisconsin. In *Wisconsin German Land and Life*, ed. Heike Bungert, Cora Lee Kluge, and Robert C. Ostergren, 189–205. Madison: Max Kade Institute for German-American Studies, University of Wisconsin.

Schroeder, Adolf. 1988. German Folklore and Traditional Practices in the Mississippi Valley. In *French and Germans in the Mississippi Valley: Landscape and Cultural Traditions*, ed. Michael Roark, 155–177. Cape Girardeau: Center for Regional History and Cultural Heritage, Southeast Missouri State University.

Sickles-Taves, Lauren B., and Philip D. Allsopp. 2005. Making a Mark in America: The Architectural Ingenuity of Germanic Settlers. *Material Culture.* 37 (1): 85–106.

Sowell, Thomas. 1981. *Ethnic America: A History*. New York: Basic Books.

———. 1996. *Migrations and Cultures: A World View*. New York: Basic Books.

Tishler, William H. 1986a. Midwestern Germans. In *America's Architectural Roots: Ethnic Groups That Built America*, ed. Dell Upton, 142–147. Washington: Preservation Press.

———. 1986b. Fachwerk Construction in the German Settlements of Wisconsin. *Winterthur Portfolio.* 21 (4): 275–292.

United States National Park Service. 2005. National Historic Landmark Nomination: Milwaukee City Hall. NPS Form 10–900. https://www.nps.gov/nhl/find/statelists/wi/Milwaukee.pdf. Accessed 3 Dec 2016.

Van Ravenswaay, Charles. 1977. *Arts and Architecture of German Settlements in Missouri: A Survey of a Vanishing Culture*. Columbia: University of Missouri Press.

Ward, David. 1971. *Cities and Immigrants: A Geography of Change in Nineteenth Century America*. New York: Oxford University Press.

Wilhelm, Hubert G.H. 1992. Germans in Ohio. In *To Build in a New Land: Ethnic Landscapes in North America*, ed. Allen G. Noble, 60–78. Baltimore: Johns Hopkins University Press.

———. 1995. Midwestern Barns and their Germanic Connections. In *Barns of the Midwest*, ed. Allen G. Noble and Hubert G.H. Wilhelm, 62–79. Athens: Ohio University Press.

Williams, Peter W. 1997. *Houses of God: Region, Religion, and Architecture in the United States*. Urbana: University of Illinois Press.

Young Center for Anabaptist and Pietist Studies. 2016. Amish Population Change 1992–2016. Elizabethtown College. https://groups.etown.edu/amishstudies/files/2016/06/Population_change_1992-2016.pdf. Accessed 10 Sep 2016.

Zeitlin, Richard H. 1977. *Germans in Wisconsin*. Madison: State Historical Society of Wisconsin.

———. 2000. *Germans in Wisconsin*. 2nd ed. Madison: State Historical Society of Wisconsin.

Zelinsky, Wilbur. 1977. The Pennsylvania Town: An Overdue Geographical Account. *The Geographical Review.* 67 (2): 127–147.

12 Scandinavian Landscapes in America

Scandinavians came from five countries surrounding the Baltic Sea or facing the North Atlantic and largely settled within the northern portion of the United States, where they often found a climate and extensive coniferous forests similar to that of their Old World homeland. Scandinavians, regardless of their national origins, are "closely related in religious heritage, language and other aspects of culture" (Allen and Turner 1988:69). While the Vikings from Iceland via Greenland were the first known Europeans to attempt a settlement in North America, establishing a short-lived outpost along the shore of the northern peninsula of Newfoundland over a thousand years ago, their only lasting legacy upon the landscape is the barely discernable earthen ruin of their village site at L'Anse aux Meadows, only rediscovered by archaeologists in 1961. Between 1638 and 1656, both the Finns and Swedes were involved in efforts to establish the colony of New Sweden along the Delaware River, but it was integrated into the surrounding British colonies. Because of the presence of these early Swedish and Finnish settlers, some cultural geographers have argued that the extensive use of log structures by those who settled Pennsylvania, the Appalachians, and those areas farther west are a legacy of these Finns (Jordan and Kaups 1989). Several Swedish churches remain in Philadelphia and Delaware to mark the landscape of these early Scandinavians.

The greatest numbers of Scandinavian immigrants arrived in the United States during the second half of the nineteenth century. Of the Scandinavians, Norwegians and Swedes arrived in the largest numbers, and these ancestries were claimed by 4,477,725 and 3,998,310 Americans, respectively, on the 2000 Census. Danish ancestry was reported by 1,430,897, Finnish ancestry was indicated by 623,573, while Icelandic ancestry was cited by 42,716. The 5-year American Community Survey for 2010–2014 showed similar numbers, estimating the presence of 4,451,906 persons of Norwegian ancestry, 3,984,136 Swedish ethnics, 1,323,182 Americans of Danish heritage, 643,821 persons of Finnish ancestry, and 49,022 Icelanders.

12.1 Icelandic Ethnic Landscapes

While relatively few Icelanders came to America, they represented a large proportion of their island's total population. Although Icelanders typically settled in areas that attracted Norwegians, they comprised an important part of the population only in several small areas, such as Washington Island, at the tip of Wisconsin's Door Peninsula, and the vicinity of Walhalla in Pembina County, North Dakota. Indeed, 8.9% of Pembina County's population listed Icelandic ancestry on the 2000 Census, a higher figure than in any other county in the nation. Near Walhalla is Icelandic State Park that preserves an Icelandic homestead. While the population of nearby Thingvalla Township—note the Icelandic name—in Pembina County is small, 45.6% of its residents were of Icelandic ancestry in 2000. The Icelandic-American numbers were greater in the Town of Washington, Door County, Wisconsin, but there they accounted for only 9.3% of the town's population. Most of the Icelandic ethnics from Washington Island have reportedly relocated in Racine, Wisconsin (Hart 2008).

12.2 Norwegian Ethnic Landscapes

Migration from Norway to American began in earnest in 1825, although more came from the 1840s to the outbreak of the Civil War, by which time 70,000 Norwegians had arrived (Fapso 2001). Wisconsin was initially one of their primary destinations. Far more Norwegians immigrated following the Civil War, as Norway was second only to Ireland in the percentage of its citizens emigrating to America. Norwegian immigrants typically settled in small communities and agricultural settlements. Indeed, "Norwegians were statistically the most rural of all major nineteenth-century immigrant groups" (Lovoll 2006:xiii).

J.A. Cross, *Ethnic Landscapes of America*, DOI 10.1007/978-3-319-54009-2_12

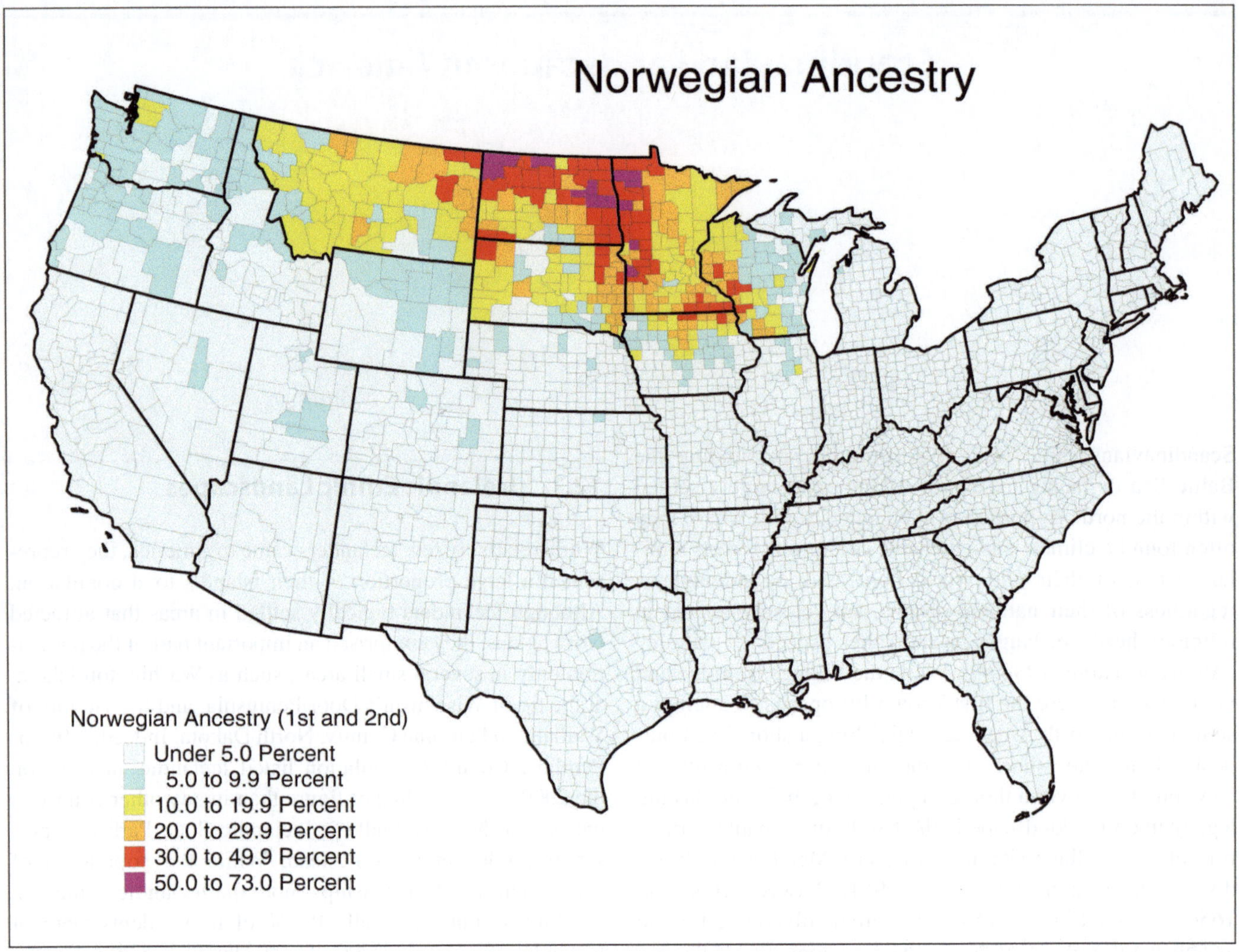

Fig. 12.1 Map showing the percentage of individuals who reported their ancestry on the 2000 Census who reported that Norwegian was their primary or secondary ancestry. Note the broad swath of counties running from western Wisconsin, across Minnesota and the Dakotas into Montana in which 20% to over half of the population reported Norwegian ancestry

12.2.1 Norwegian Ethnic Population Distribution

The largest concentrations of Norwegian-Americans are found in the Upper Midwest and the northern Great Plains (Fig. 12.1). Numerically, the state with the largest number of persons reporting Norwegian ancestry in 2000 was Minnesota with 850,742 Norwegian-Americans, who comprised 17.3% of its population. Wisconsin had the second largest number, 454,831, and persons of Norwegian ancestry accounted for 8.5% of its population. North Dakota, with 193,158 Norwegian-Americans had the greatest proportion of its population with that ancestry, 30.1%. While California, with 436,128, had the third largest number of persons of Norwegian ancestry, given that state's huge population, Norwegian-Americans only comprised 1.3% of its residents. Similarly, Washington had 367,508 Norwegian-Americans, comprising 6.2% of its citizens.

Persons of Norwegian ancestry accounted for over half the population within eight counties nationally, six in North Dakota and two in Minnesota. Those two counties with the greatest proportions of their population reporting Norwegian ancestry are Divide County, North Dakota with 64.7% and Steele County, North Dakota with 62.0%. With a large margin of error, the 5-year American Community Survey for 2010–2014 estimated that individuals of Norwegian ancestry approximated half of the population within these two counties. Nationally, the 2000 Census shows that persons of Norwegian ancestry comprised over a third of the total population within 49 counties nationally, 23 of which were in North Dakota, 18 in Minnesota, three in Iowa, two in both Montana and Wisconsin, and one in South Dakota. Given their prominence in this extensive area, Norwegians have had the opportunity to imprint their ethnicity upon the cultural landscape of much of the upper Midwest and adjacent northern plains. Although more limited in numbers,

Norwegian immigrants also left their imprint on Bosque County, Texas.

Norwegian ethnicity is displayed by numerous Norwegian toponyms, both for places and features, and by the character and arrangement of their many farm buildings. For example, within southwest Wisconsin's Coulee region, "Norwegian place names abound.... Towns have names such as Bergen and Christiana, and physical features have such names as Rundahl Coulee, Norwegian Hollow, Gronvold Ridge, Lindvig Ridge, and Rongstad Ridge" (Raitz and Mather 1971). Within these rural Norwegian ethnic settlements, we will first consider the built environment.

12.2.2 Norwegian Ethnic House and Farm Landscapes

Within early Norwegian immigrant communities in the Midwest, log construction was typical, although their construction methods often differed from that in Norway. For example, in Norway logs typically had a long groove and were double-cut notched for a tight fit, while in the American Midwest it was much more common to see dovetail corner notches with mortar placed between the logs (Henning 1986). Wooden pins placed through vertical holes in the logs protected buildings from warping. While sod roofs were used at first, these were replaced first by wood shingles, and then later by more standard American roofing materials. Nevertheless, the Norwegian log house preserved at Wisconsin's Little Norway outdoor museum retains its sod roof.

Norwegian immigrant houses typically display one of three main forms: a single room one and half story house; the two or three room Akershusisk house; and the one and half to two story house. The most typical Norwegian-style house was the two to three room Akershusisk house (Henning 1986). This house distinctly shows its connection to a medieval Norwegian dwelling, with the house's main room, termed a stue, being of log construction, while a smaller room of wood frame construction was appended to its gable end. Above the stue was a second floor built of logs. At the gable end of the house was a storeroom (kleve) of wood frame construction, which had a separate entryway or forstue (Henning 1986).

The distinguishing characteristics of the Akershusisk house are described by William Tishler.

> These rectangular houses are easily distinguished by a slightly off-center chimney constructed adjacent to the interior partition wall. They are symmetrical in outward appearance—the main entrance is usually centrally located in the axial wall, flanked by one of two windows, with the same window arrangement frequently repeated on the opposite side of the house (Tishler 1992:235, Reprinted by permission of Johns Hopkins University Press).

Additional rooms were added to the house, forming a T or L shaped floor plan in the United States, differing from the linear additions that were most common in Norway.

Some of the one and a half to two story houses of Norwegian immigrants are characterized by an enclosed side stairway or sval that provided access to a sleeping loft (bod) or a second floor (Fig. 12.2). While these are rare, examples remain in Wisconsin. The loft cottage with its characteristic overhanging front upper gable providing a loft sleeping space, is viewed as representing "Americanized simplifications of medieval loft cottages of Norway" (Tishler 1992:240).

One of the largest concentrations of one-room log houses built by Norwegian immigrants is within Wisconsin's Coon Valley, southeast of La Crosse, where Tishler (1992) noted that 54 still remained in the early 1990s. While describing their log construction, he also noted how the design deviated from that found within the Old Country.

> All were built with oak logs joined with full dovetail corner notches and were chinked with wood strips, clay, or a combination of both. With one exception, all have horizontal logs laid up to the eaves with board-and-batten or clapboard siding running to the ridge on the gable ends. Only one, a very early dwelling, exhibits the traditional Scandinavian full log gable end construction (Tishler 1992:234, Reprinted by permission of Johns Hopkins University Press).

The propensity of Norwegian-Americans to rapidly incorporate many new elements into the construction of their homes resulted in many of their homes differing, at least in their outward appearances, from those built in Norway (Bakken 1994). Norwegian architectural historian Reidar Bakken (1994), in describing examples from Wisconsin's Coon Valley, notes that the floor plans of many Norwegian-American houses more closely follow traditional ethnic patterns than what would be expected by looking at the buildings' exterior, which appear American.

12.2.2.1 Farmstead Arrangements

Many Norwegian immigrants preferred locations on wooded hills near a stream, typically avoiding prairies where possible, even though many Norwegians settled in the Dakotas. The coulees of southwestern Wisconsin and southeastern Minnesota (Fig. 12.3) provided an ideal location for the Norwegian immigrant farmsteads, and this area to this day retains many reminders of the farmsteads nestled in the narrow valleys.

The farm buildings in Norwegian ethnic farmsteads in the Upper Midwest were usually scattered in either a courtyard arrangement of buildings or in a linear or double linear arrangement. Both of these farmstead arrangements display Old World antecedents (Henning 1986). Within the typical "paired linear plan, the farm lane often divided the house, summer kitchen, well and shops on the up-slope side of the land and the barn and other livestock buildings on the down-slope side"

Fig. 12.2 Norwegian immigrant house preserved at Old World Wisconsin outdoor museum near Eagle, Wisconsin. Note the presence of a sval to the right, providing access to the sleeping loft, in this log house displaying dovetail notching

Fig. 12.3 Norwegian ethnic farmstead located along slope at edge of the valley in the Root River Valley near Peterson, Minnesota. Norwegian ethnics included many buildings in their farmsteads, with this example still including, between the farmhouse to the left and the tobacco barn to the far right, three small outbuildings and a barn, largely obscured by the conifers

(Henning 1986:152). Examples of this linear arrangement of farm buildings can be seen in both southeastern Minnesota and southwestern Wisconsin.

The variation in Norwegian-American farm layouts has been attributed to local differences related to the region of Norway from which the immigrants came. In the Coon Valley area of southwestern Wisconsin, some of the surviving farms show "indications of a double farmyard" (Bakken 1994:89) and many of the immigrants came from Gudbrandsdalen, where that style was still prevalent at the time of emigration. In these farmyards "a granary [and/or a horse barn, now often gone] situated in the middle of the farmyard separates the courtyard around the house from the barnyard and machine shed" (Bakken 1994:84). In contrast, within Bosque County, Texas, where Norwegian immigrants came from the province of Hedmark, "the placement of the stone fences, the house, and several of its nearest outbuildings…demarcate a rectangular yard, similar in form to the…traditional Norwegian *tun*" (Breisch 1994:100).

Norwegian immigrant farmsteads initially included a large number of buildings, with up to 16 separate buildings being noted on some of their farmsteads in Wisconsin (Fapso 2001). These structures could include the house, summer kitchen, springhouse, corncrib, granary, chicken coop, stable, barn, and various storage buildings. These buildings included both those typical of the Old World and those common among other American farmers. Those structures transplanted from Norway include the stabbur, designed for the storage of specific household supplies, and the eldhus or fire house, used to "bake flat bread, wash clothes, and heat water for butchering" (Fapso 2001:33). Because Norwegian-Americans rapidly adopted Yankee building styles and construction modes, it is not unusual to find a mixture of both ethnic Norwegian and more generic American architecture on the same farmstead.

In contrast to what was reported in Wisconsin and Minnesota, the Norwegian immigrant farmsteads in Bosque County, Texas "tended to be much less complex than those which the immigrants had left behind" (Breisch 1994:99). Norwegian ethnic farms in Texas are usually "composed of little more than the main house and a small group of log or stone outbuildings…[including] bank barns, vegetable cellars, corn cribs, springhouses, detached, or 'summer' kitchens, and storage or work sheds" (Breisch 1994:99). Thus, while fewer than in the Old Country, as well as fewer than in Wisconsin, the Norwegian ethnic farms in Texas still frequently displayed more structures than those of other groups.

The barns found on many Norwegian ethnic farms are basically a Norwegian version of the bank-barn that we previously described in Chap. 11 discussing German ethnic landscapes. The lower portion of the barn was typically built of stone or logs, or a combination of both. The upper portion of the barn was of post and beam construction, with a ramp and bridge providing access to the hay storage area. The entrance ramp was often covered by either a shed or gable roof (Henning 1986).

Norwegian-Americans distinguished themselves in Wisconsin, and in the past to a lesser extent in Minnesota, by their involvement in tobacco production. While scholars have argued for both a cultural (Raitz and Mather 1971) and economic (Ibarra and Strickon 1989) explanation of the connection, over two-thirds of the tobacco grown in Wisconsin came from farmers of Norwegian heritage. Although there are fewer Norwegian-American farmers in this region today and Amish farmers have acquired hundreds of farms in the region, it is the Norwegian settled area that continues to dominate Wisconsin's tobacco production, although the acreage is down, just as it has nationally. What tobacco farming remains is largely around the Norwegian settlement of Stoughton in Dane County.

12.2.3 Norwegian Heritage on Display

Norwegian names remain throughout the Driftless Area of Wisconsin and Minnesota, as do the Lutheran churches (Fig. 12.4), Norwegian fraternal organizations in the villages (Fig. 12.5), and Norwegian traits on some of the farm buildings. Within settlements, such as Westby, Norwegians display their heritage, just as they did nearly a half century ago when Karl Raitz and Cotton Mather described its ethnicity.

> Westby, a town of about 1,500 persons, was the focus of early Norwegian settlement in western Wisconsin...replete with Norwegian cultural symbolism…. The traveler entering Westby is greeted by a large, colorful, highway sign which welcomes him in the Norwegian language. Merchants on the main street proclaim their Norwegian heritage or their empathy with Norwegians by employing Norwegian motifs in the decoration of the building facades, by flying Norwegian flags, or by using Norwegian store names (Raitz and Mather 1971:689, Reprinted with permission of American Association of Geographers).

Around the time Raitz and Mather described the Westby area, linguists from Norway visited the area to rediscover its "pure" form of Norwegian, a version that had not been altered by a century of change in Norway. Today, Westby is 50% larger and among a number of communities that rely upon their Norwegian heritage to promote tourism. It has erected a stabbur with an overhanging second floor to serve as a tourist office (Fig. 12.6). The Norskedalen Nature and Heritage Center in nearby Coon Valley, Wisconsin showcases their heritage, preserving buildings of several Norwegian ethnic farmsteads. Stoughton, Wisconsin displays the Norwegian flag along its Main Street (Fig. 12.7), where a Norwegian Heritage Center has opened near its Sons of Norway Lodge and one sees Norwegian decorative

Fig. 12.4 Springdale Lutheran Church east of Mount Horab, Wisconsin. The steeple of this church was made in Norway and shipped to Wisconsin. While few architectural features distinguish Norwegian ethnic churches from those of other Lutherans, the Norwegian Lutheran synod in the U.S. was controlled at one time by the state Lutheran church in Norway

Fig. 12.5 Sons of Norway Lodge in Lanesboro, Minnesota. Ethnic based fraternal organizations help members of many ethnic groups both maintain and showcase their heritage

Fig. 12.6 Stabbur in Wesby, Wisconsin. This Norwegian styled building was erected on the farm of a Norwegian-American following a visit to Norway. It was purchased by the local Chamber of Commerce and moved to Wesby to enhance its promotion of local tourism

Fig. 12.7 Norwegian flags line Main Street in Stoughton, Wisconsin, where one can also see Norwegian ethnic decorations displayed on some of the buildings

designs on some of its buildings. Across the Mississippi River in Minnesota one can spot former tobacco barns, Norwegian fraternal halls, and Norwegian Lutheran Churches. Spring Grove, Minnesota, promotes itself as America's Norwegian community.

Although towering stave-wood churches mark the Medieval religious landscape of Norway, this style has not been utilized within the Norwegian ethnic churches in America, largely because it was long outdated even in Norway when the immigrants arrived. Nevertheless, small replica stave-wood structures can be seen as ethnic monuments in some Norwegian-American communities, ranging from Little Norway in Wisconsin to Rapid City, South Dakota.

12.3 Swedish Ethnic Landscapes

Nearly 1.25 million Swedes immigrated to America, the third highest rate of emigration of any European country (Palmqvist 1983). Nevertheless, Swedish-Americans dominate the ancestry in relatively little of the nation (Fig. 12.8). In only six counties did individuals of Swedish ancestry account for over 20% of the population in 2000. Kittson County, Minnesota had the largest proportion of its population reporting Swedish ancestry, 33.6%, while Isanti County, Minnesota, with 25.2%, had the second highest proportion. Swedish ancestry exceeded 20% within two additional counties in Minnesota, plus two in Nebraska. In a broad swath of northern Minnesota, plus scattered counties in

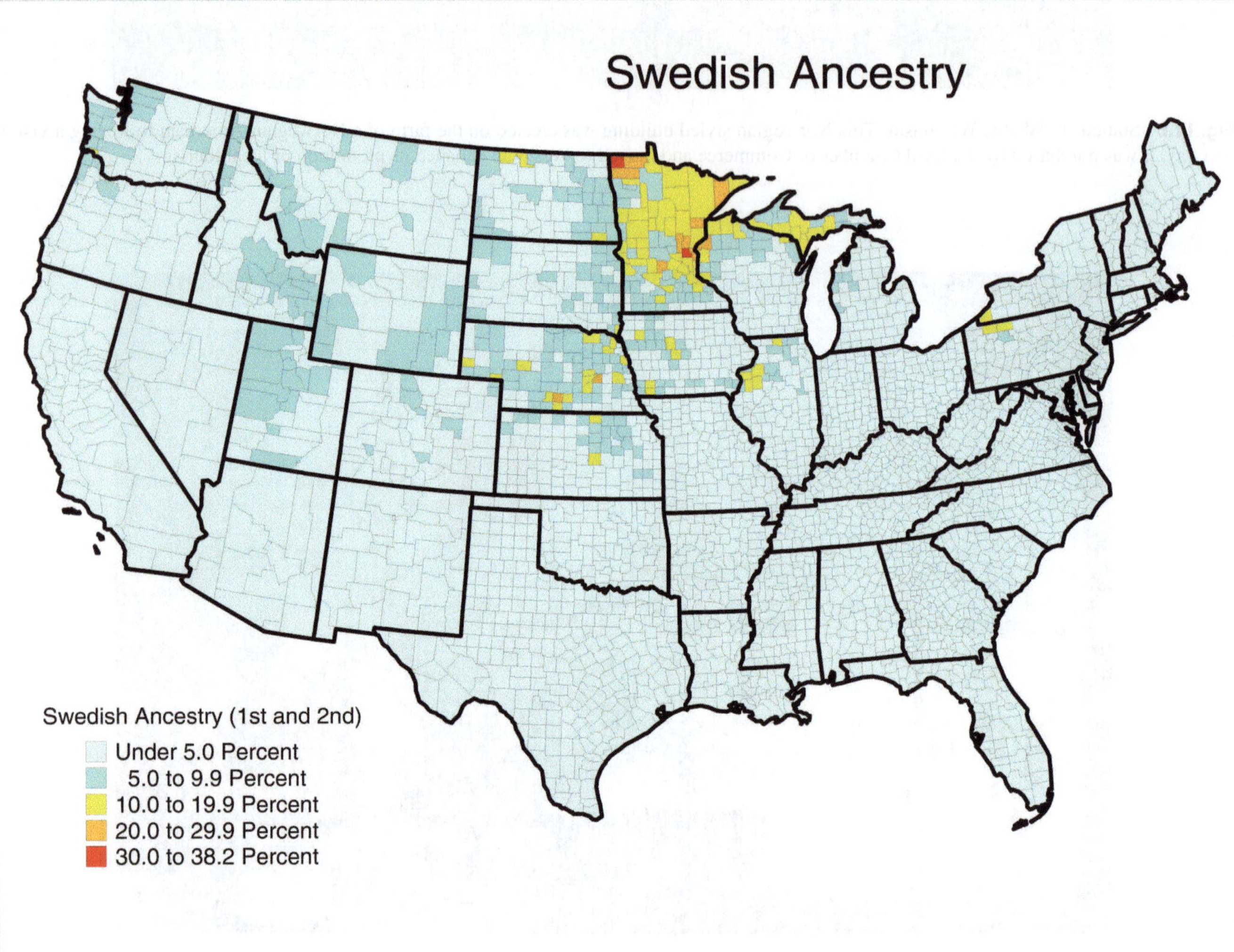

Fig. 12.8 Map showing the percentage of individuals who reported their ancestry on the 2000 Census who reported that Swedish was their primary or secondary ancestry. Note the clusters of counties near Lake Superior, along with a scattering of counties in the Great Plains between Nebraska and the Dakotas, in which over a tenth of the population reported Swedish ancestry. Swedes comprise over 5% of the population in many of the counties of the Pacific Northwest

northern Wisconsin, Michigan's Upper Peninsula, Illinois and the Great Plains Swedish-Americans account for over a tenth of the population. The 5-year American Community Survey for 2010–2014 indicates that Isanti County and Kittson County's Swedish population had fallen to 30.3% and 19.4%, respectively.

The Swedes left only a minimal ethnic imprint upon the landscape. After reviewing their impact, Frederick Hale (2002:40) notes that "those from Sweden might well be termed the transparent Scandinavians of Wisconsin" as they "quickly became nearly indistinguishable from its Yankee stock." He explains that they made little "determined effort to preserve their distinctly Swedish heritage; indeed, their geographical distribution rendered such a step nearly impossible" (Hale 2002:40).

While the Swedish familiarity with constructing houses and barns with logs or wood timber in the Old World was brought to America, and it has been argued that the Swedish stuga av celltyp and portlider evolved into the American saddlebag and dog-trot houses, respectively (Palmqvist 1986), cultural and historical geographers have also attributed those two styles to other immigrants. Nevertheless, the fact that many Swedish immigrants to New Sweden who located in the Lower Delaware Valley in the mid-seventeenth century, and in the Upper Great Lakes region two centuries later, built with logs using the distinct Swedish corner timbering is not disputed. Such construction with logs is not surprising, given that "log contruction was the most widespread building technique in Sweden at the time when Swedes arrived in Minnesota" (Palmqvist 1987:19).

Buildings constructed by Swedish immigrants in Minnesota retained certain elements of Old World architectural style, but were neither direct copies nor always easily distinguished from those built by other Americans in the areas they settled. Lena Anderson Palmqvist (1983:46) explains, "When comparing the rural architecture in Minnesota with contemporary Swedish buildings one is more apt to find similarities in the building technique, the type of house plan, the style of the building rather than direct copies of Scandinavian houses." She describes how some of the Swedish influences can be seen in some of the older houses in the Chisago County, Minnesota area. There it is noted that Swedish settler houses "preserved the characteristic lines, the long narrow building with a steep roof and a very narrow roof projection" (Palmqvist 1983:46), as well as the use of half-timbered walls, even though they were filled with cement, unlike the straw, hay, or brick commonly used in Sweden. Furthermore, Swedish-American vernacular buildings in Minnesota sometimes display vertical siding on timber framing, whereas most other groups more favored horizontal siding (Palmqvist 1983).

The identification of Swedish ethnic housing presents challenges, as Lena Palmqvist explains.

> 'Typical' Swedish houses are difficult to discern among old Midwestern houses. The traditional floor plan was still used, and wood was the predominant building material, but the lumber was sawn, often precut. The floor plan, the location of the door and symmetrical placement of windows on the facade have direct parallels in Swedish vernacular architecture (Palmqvist 1986:155, Reprinted with permission of John Wiley and Son).

Nevertheless, she argues that "the continuous changes in the rural architecture with more or less radical modifications of the original buildings give almost no intact example of Swedish pioneer dwellings" (Palmqvist 1983:51). While Swedish-Americans favored houses with symmetrical facades and parallel windows, which were typical of Swedish homes, such characteristics were also favored by other Americans, and are seen in the I-houses so typical across much of the American Midwest. Nevertheless, Swedes often marked their ethnicity (Fig. 12.9) by displaying the dala horse, often painted bright red, by their front door or hanging from the ceiling of their front porch.

Examples of more-refined Swedish architectural styles can be seen in some of the houses remaining at Bishop Hill State Park in Illinois, which preserves buildings that were erected at a Swedish religious colony in the 1840s, patterned after Swedish mansions. The few remaining examples of its early construction followed Swedish styles of wood framed buildings with wood cladding, with "long eaves and soffit returns with smaller top-story windows placed between them, very similar to those…of central-Swedish farmhouses contemporary to them" (White 2012:57). Yet even there, while the colony's initial buildings display Swedish architectural connections, within a decade construction had ceased to display such ethnicity (White 2012).

Barns in Swedish-American communities, alike those in other Scandinavian settlements, display greater connections with the Old World than do houses. Indeed, the Germanic influence that resulted in a variety of two-story bank barns in America can also be seen in both Swedish- and Norwegian-American communities. Within Swedish ethnic settlements around Chisago County, Minnesota can be seen what have been described as "elongated half-log barns" (Noble and Wilhelm 1995:11). In actuality, these Swedish ethnic barns resemble two structures built side by side with a small doorway connecting them through a shared wall. The differences in their construction are conspicuous, "with the cowbarn in stone- or log-construction and the haybarn built of heavy studs and generally with vertical siding" (Palmqvist 1983:51). Not only do these styles have "direct parallels in contemporary Swedish barns" (Palmqvist 1983:51), but Swedish ethnic barns "show stronger cultural influences than the dwellings, many of which have been rebuilt or replaced" (Palmqvist 1987:24).

As with many groups, Swedish immigrants frequently built churches patterned after those at home. For example, the "Chisago Lake Lutheran Church has been compared to

Fig. 12.9 Wood-framed house in Stanton, Iowa. Although architecturally very similar to wood-framed houses constructed by Americans of many ethnic backgrounds, the presence of the dala horse on the front porch distinguishes many of the houses occupied by Swedish-Americans within the community

the church building in Madesjö parish in Småland, Sweden" (Palmqvist 1983:86), given that the parishioners in Chisago Lake, who immigrated from the same parish in Småland, built a church inspired by their former place of worship. The Swedish-American Lutheran Churches were often wood frame neo-gothic structures, with a square wooden tower containing the entrance door at the gable end topped with a steeple. Others, such as the Sabylund Lutheran Church located northeast of Stockholm, Wisconsin (Fig. 12.10) and the historic Vasa Lutheran Church in Minnesota are of brick construction (Fig. 12.11), yet sporting conspicuous neo-gothic towers and steeples.

12.4 Finnish Ethnic Landscapes

Nationally, 623,573 Americans reported to the 2000 Census as being of Finnish ancestry, a decrease from the 658,870 persons who so reported in 1990 (Fig. 12.12). Nearly two-fifths of the total are found in the three states that adjoin Lake Superior. Michigan had the largest number, 101,351, concentrated in its Upper Peninsula, Minnesota had 99,388, while Wisconsin reported 36,047. Finns immigrated as miners and lumbermen, arriving in the 1890s and early 1900s, settling the Upper Great Lakes area where the physical landscape of rocky soils, lakes, and coniferous forest was similar to that in Finland. As Arnold Alanen (1995:122) explains, "Once the Finns arrived in the Lake Superior region…they also were culturally 'preadapted' to cope with the environment because of their previous experiences in a marginal agricultural region of Europe." Washington State had 40,290 Finnish-Americans in 2000, similarly reflecting the attraction of Finns to work in that state's lumbering industry around the beginning of the twentieth century.

12.4.1 Distribution of Finnish Ethnic Population

Although far less numerous than any Scandinavian group with the exception of the Icelanders, those Finns who settled near Lake Superior were that area's dominate ethnic group and have left a distinct ethnic imprint on the landscape. More persons reported Finnish ancestry within six counties of Michigan's Upper Peninsula than any other ancestry. While nationally Finns comprised only 0.2% of the population in 2000, within Houghton County, Keweenaw County, and Ontonagon County they were 34.1%, 37.9%, and 31.9%, respectively. While 5-year American Community Survey estimates for 2010–2014 were slightly lower, the proportion of the population reporting Finnish ancestry still

Fig. 12.10 Sabylund Lutheran Church located northeast of Stockholm, Wisconsin. Like those among the Norwegians and Danes, Lutheran churches mark the ethnic landscapes settled by Swedes

exceeded 30% within two of these three counties. Finns were over 20% of the population in three additional counties in 2000. Numerous toponyms throughout the region reflect this Finnish ancestry. Finnish-Americans today account for 2–5% of the population within several counties near the mouth of the Columbia River and near Butte, Montana where they were far more prominent 80 years ago. There they left their ethnic imprint on the cultural landscape of several Pacific Northwest rural communities that is similar to what is more widely seen in the Upper Great Lakes region, although far fewer locations bear Finnish names (Kilpinen 1994).

While Finns dominated the residential neighborhoods near several of the Lake Superior region's copper and iron ore mines, with streets bearing Finnish names, many of their communities were planned company towns. Finns were also sharply divided politically between conservatives, who sought to maintain their ethnic identity and language, and socialists or Red Finns, who not only discouraged settlement of the countryside, but sent thousands of their number to the Soviet Union in the early 1930s (Kaups 1995). The distinctive character of these conservative rural Finnish-Americans was first reported by Eugene Van Cleef in 1917, who viewed "the log constructed saunas and houses, and meadow hay barns [as serving] as diagnostics of the nationality of the builders" (Kaups 1995:6). While there are a few Finnish socialist workers halls remaining, it is the rural ethnic landscape of the Finns that continues attracting the greatest attention of geographers.

Fig. 12.11 Vasa Evangelical Lutheran Church, erected 1869 in hamlet of Vasa, Goodhue County, Minnesota. This church serves what is considered "the most intact, unchanged of the original Swedish colonies in Minnesota and remains one of the most representative," according to its nomination for the National Register of Historic Places

12.4.2 Rural Finnish-American Landscapes

Finnish-American farmsteads traditionally consisted of a large number of buildings, somewhat similar to what was found among Norwegians. The outbuildings were generally arranged in "an L-shaped or nearly square pattern," of which there were many variations (Kaups 1986:127). Their buildings were small, partly a result of poverty. Log construction was very common, especially at first, yet these buildings were constructed more recently than many of those within the Norwegian ethnic communities because of the later arrival of Finnish immigrants.

12.4.2.1 Finnish Log Buildings

Log construction was the characteristic Finnish style for all farm buildings, yet the association between Finns and log construction may have also been promoted to enhance the image of Finnish-Americans (Kaups 1995). As William Tishler (1984:17) observes, "Until quite recently, entire farmsteads of mostly log structures, including the characteristic sauna, existed in Bayfield County's Oulu Township and a few areas in Douglas and Iron counties." Finns showed a preference for using logs of conifers (Knipping 2008).

The form of log construction used by ethnic Finns was different from that of other ethnic groups. The logs were hewn flat to have an outward appearance of being wood planks. The top of the log was convex, while the bottom was concave, permitting them to lock together tightly when stacked. A cut, or chamfering, was made along the bottom edges of the stacked logs to form a V-shaped notch, marking the connection of the logs. Finnish immigrants used full dovetail corner notching like other Scandinavians (Noble 1993).

Finnish-American farm houses ranged from small log structures to Three-Room-Baltic-Houses, such as the one preserved at Old World Wisconsin (Fig. 12.13), with bedrooms flanking a central hearth. Chimneys usually had a central location, and roofs were usually gabled and covered with wood shingles (Kaups 1986). Such houses were typically entered through an enclosed porch or stoop. Having a stoop sheltering the door is an adaptation against severe winters, yet it was more commonly present on Finnish-Americans homes than the houses of other nearby ethnic groups.

Houses of Finnish ethnics were more likely to be unpainted than those of their neighbors, yet they often had "flower-filled window boxes which offset drabness" (Mather and Kaups 1963:501). In addition, "the presence of ladders along the sides and on the roofs of buildings" (Mather and Kaups 1963:500) often distinguished them from the houses of their neighbors.

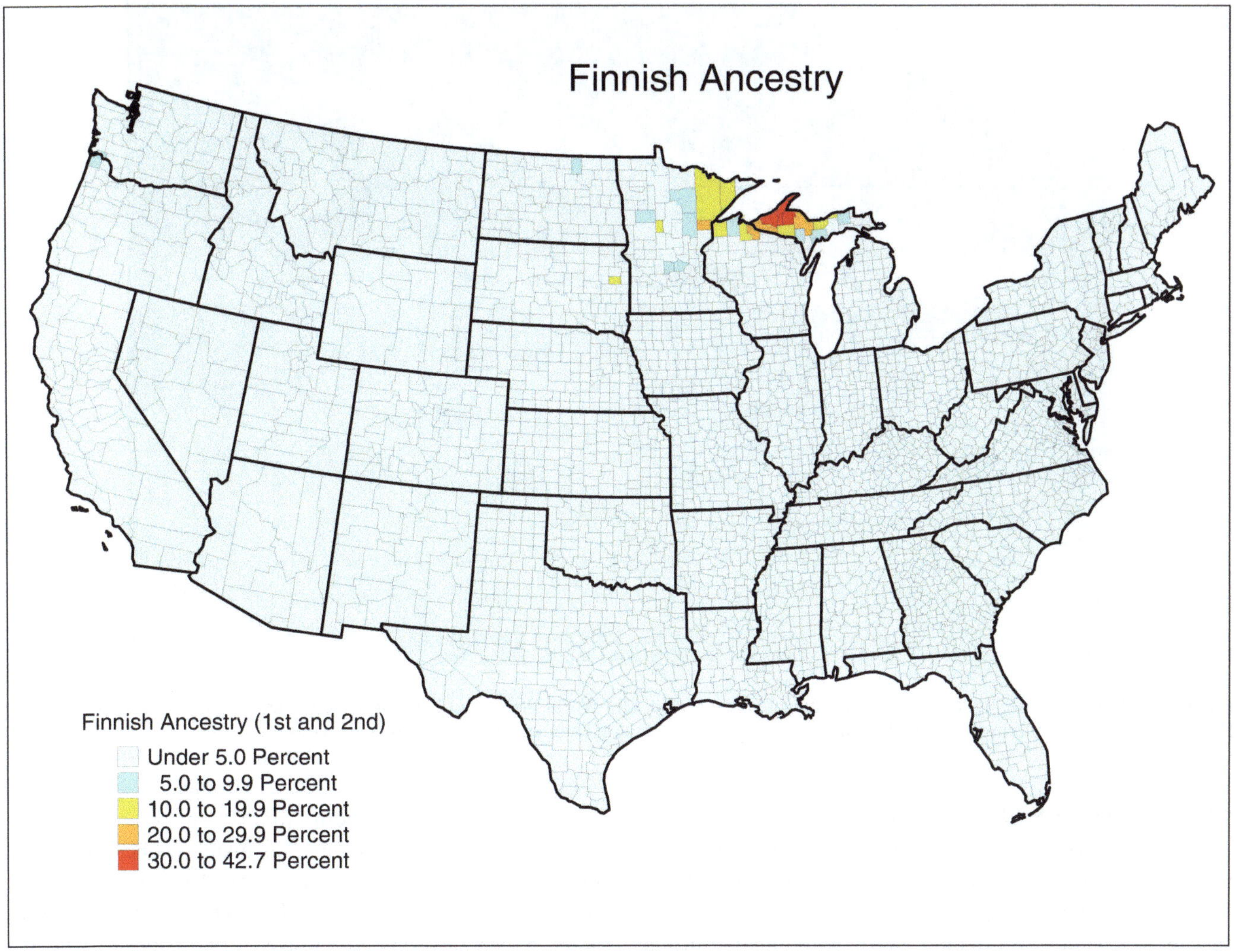

Fig. 12.12 Map showing the percentage of individuals who reported their ancestry on the 2000 Census who reported that Finnish was their primary or secondary ancestry. Note the swath of counties near Lake Superior, running from Michigan's Upper Peninsula to Minnesota's Arrowhead Country, in which over a tenth of the population reported Finnish ancestry

12.4.2.2 Finnish Saunas

The sauna has long been a key to the identification of Finnish farmsteads. Five decades ago Cotton Mather and Matti Kaups (1963:494) explained, "the sauna, or Finnish bath, is an example of a cultural index, and when one views a sauna in the field…there is indeed a high mathematical probability that one is in a Finnish cultural complex." The sauna was the most persistent of Finnish farm structures, and Mather and Kaups (1963) noted a study in Michigan's Upper Peninsula that found that 88% of Finnish rural homes had saunas. In her South Dakota study, Linnea Sando (2014:3) more recently explained, "The sauna became the most distinctive and longest lasting structures that Finns contributed to the Hamlin County landscape." Even though over time most Americans of Finnish heritage have replaced their more traditional houses with those virtually indistinguishable from those of other Americans, many saunas remain. Indeed, Finnish ethnic populations "continue to build and use saunas, incorporating them into their daily lives" (Sando 2014:3).

The Finnish sauna was typically a small one-room building of about 120 square feet centered around a hearth. Log construction was common in Michigan, Wisconsin, and Minnesota, and the sauna typically has a low ceiling, with entry through a small doorway. Sometimes the sauna has two rooms, both with a window, with the second room being used as a dressing room (Kaups 1992). In the more open prairies, saunas were built using similar floor plans, but of "the least expensive wood available…[and] stud frame construction of the saunas was much more common in South Dakota" (Sando 2014:9). While many saunas remain as separate detached buildings near the house, Sando (2014) reports that the construction of saunas attached to the house is becoming more common.

Fig. 12.13 Ketola House, built in 1894, relocated and preserved at Old World Wisconsin outdoor museum near Eagle, Wisconsin. Note the characteristic stoop sheltering the front door, ladder on the roof, and central chimney in this Finnish ethnic house

12.4.2.3 Finnish Ethnic Barns

There are a variety of Finnish ethnic barns, typically constructed for specific functions. Several small hay barns (latos), of rounded logs that are separated rather than tightly spaced to enhance ventilation, were traditionally located both at the farmstead and scattered in the various hayfields (Noble 1993). These meadow hay barns, often measuring 16 by 19 feet to 25 by 32 feet, were viewed as one "of the significant components of the cultural landscapes that the Finns molded" (Kaups 1992:262) in the northern Great Lakes region. Within parts of the Upper Midwest and the Pacific Northwest, such as in Long Valley, Idaho, Finns built hewn log cattle barns, with haylofts (Kilpinen 1994). Yet regional differences exist among the Finnish settlements. As Jon Kilpinen (1994:30) notes, "the slanted-wall hay barns common in western Finland and the upper Midwest" are absent from the Pacific Northwest. The connecting barn built in several stages is common in the Great Lakes region.

12.5 Danish Ethnic Landscapes

Danes did not cluster together to the degree that other Scandinavian peoples did, and they failed to leave a strong visible ethnic imprint on the landscape. For example, while the 2000 Census found that individuals of Norwegian ancestry accounted for over 60% of the population in eight counties within the northern Great Plains, there was only one county in the nation where as many as one-fifth of the population was of Danish ancestry. That was Audubon County, Iowa, where individuals of Danish ancestry comprised 32.9% of the population. The two counties with the next largest proportion of Danish-Americans (16.5% each) were Sevier and Sanpete Counties, Utah, testament to the participation by early Danish converts in the Mormon trek to the beehive state. In his discussion of housing built by Danish immigrants in Iowa and Minnesota, Signe Betsinger (1992:225) notes that "Danish influences quickly disappeared on the exterior of early houses," with ethnicity more visible in the interior, by the type of decoration of walls and the style of furniture.

Danish immigrants were far less numerous than those coming from Norway and Sweden. Thomas Carter (1986:118) explains that "while Scandinavian ethnic building traditions have figured prominently in the recent vernacular architecture studies, the Danish contribution has gone unnoticed." Yet a few examples may be found.

While the Danes initially favored long rectangular houses, symmetrically arranged with a central dormer perpendicular to the axial gable, Danish-Americans readily incorporated the styles of their American neighbors (Betsinger 1992). Building styles were changing in Denmark at the time that many Danes came to America, and while some of the older styles were brought, "more often it was the newer bourgeois

Fig. 12.14 Al Johnson's Swedish Restaurant in Sister Bay, Wisconsin. This structure blends the architectural traditions of both Swedish and Norwegian ethnics, and together with the goats grazing contentedly on the roof, uses its ethnicity to market itself

architecture that was brought to America" (Carter 1986:120). The result was that a common type of house in Danish-American communities was a scaled-down version of the Danish herrgard or country house that in America became the three-room pair house, where a large central room was "flanked by a pair of smaller rooms to either side" (Carter 1986:121).

Examples of large long-gabled masonry barns, "strikingly similar" to those built in the last half of the nineteenth century in Denmark, have been identified in South Dakota (Carter 1986). Likewise, Danish heritage influenced the architecture of a few churches in the Great Plains. Citing the example of St. Peder's Lutheran Church in Nysted, Nebraska, Carter (1986:122) explains that it exhibits "the square entrance tower and stepped gables of the Dutch style prominent in fifteenth and sixteenth century Danish church architecture."

12.6 Précis

Scandinavians left their mark upon portions of the American landscape, particularly within the Upper Great Lakes region and in the Pacific Northwest. While the similarities in their housing with popular American styles and in their farm buildings with numerous German bank barns have reduced the distinctiveness of much of their built environment, elements of their material culture are conspicuous reminders of their ethnic presence. For example, saunas can be found within the Finnish-American rural countryside of northern Michigan and Wisconsin, and stabbur buildings in Norwegian communities are used to promote their ethnicity. The Swedish restaurant in Sister Bay (Fig. 12.14), Wisconsin advertises its ethnic distinctiveness not only by its ornate log structure, but by the goats that graze its sod roof. Yet the same building is also indicative of the cultural interplay of the various Scandinavian ethnicities, inasmuch as it was built of logs obtained from Norway by Norwegian craftsmen, yet it employs Swedish waitresses (Hart 2008).

References

Alanen, Arnold R. 1995. Back to the Land: Immigrants and Image-Makers in the Lake Superior Region. In *Landscape in America*, ed. George R. Thompson, 111–140. Austin: University of Texas Press.

Allen, James P., and Eugene Turner. 1988. *We The People: An Atlas of America's Ethnic Diversity*. New York: Macmillan Publishing Company.

Bakken, Reidar. 1994. Acculturation in Buildings and Farmsteads in Coon Valley, Wisconsin, from 1850 to 1930. In *Material Culture and People's Art Among the Norwegians in America*, ed. Marion John Nelson, 73–91. Northfield: The Norwegian-American Historical Association.

Betsinger, Signe T. Nielsen. 1992. Danes in Iowa and Minnesota. In *To Build in a New Land: Ethnic Landscapes in North America*, ed. Allen G. Noble, 211–225. Baltimore: Johns Hopkins University Press.

Breisch, Kenneth A. 1994. Good Building Stone and a 'Lay of the Land' that Makes a 'Hominess': Norwegian-American Settlement Patterns and Architecture in Bosque County, Texas. In *Material Culture and People's Art Among the Norwegians in America*, ed. Marion John Nelson, 92–117. Northfield (MN): The Norwegian-American Historical Association.

Carter, Thomas. 1986. Danes. In *America's Architectural Roots: Ethnic Groups that Built America*, ed. Dell Upton, 118–123. Washington: Preservation Press.

Fapso, Richard J. 2001. *Norwegians in Wisconsin*. Madison: State Historical Society of Wisconsin.

Hale, Frederick. 2002. *Swedes in Wisconsin*. 2nd ed. Madison: Wisconsin Historical Society Press.

Hart, John Fraser. 2008. *My Kind of County: Door County, Wisconsin*. Chicago: Center for American Places at Columbia College Chicago.

Henning, Darrell D. 1986. Norwegians. In *America's Architectural Roots: Ethnic Groups that Built America*, ed. Dell Upton, 148–153. Washington: Preservation Press.

Ibarra, Robert A., and Arnold Strickon. 1989. The Norwegian-American Dairy-Tobacco Strategy in Southwestern Wisconsin. *Norwegian American Studies*. 32: 3–30.

Jordan, Terry G., and Matti Kaups. 1989. *The American Backwoods Frontier: An Ethnic and Ecological Interpretation*. Baltimore: Johns Hopkins University Press.

Kaups, Matti. 1986. Finns. In *America's Architectural Roots: Ethnic Groups that Built America*, ed. Dell Upton, 136–141. Washington: Preservation Press.

———. 1992. Finns in the Lake Superior Region. In *To Build in a New Land: Ethnic Landscapes in North America*, ed. Allen G. Noble, 242–267. Baltimore: Johns Hopkins University Press.

———. 1995. Cultural Landscape: Log Structures as Symbols of Ethnic Identity. *Material Culture* 27 (2): 1–19.

Kilpinen, Jon T. 1994. Finnish Cultural Landscapes in the Pacific Northwest. *Pacific Northwest Quarterly* 86 (1): 25–34.

Knipping, Mark. 2008. *Finns in Wisconsin*. Madison: Wisconsin Historical Society Press.

Lovoll, Odd S. 2006. *Norwegians on the Prairie: Ethnicity and the Development of the Country Town*. St. Paul: Minnesota Historical Society Press.

Mather, Cotton, and Matti Kaups. 1963. The Finnish Sauna: A Cultural Index to Settlement. *Annals of the Association of American Geographers* 53 (4): 494–504.

Noble, Allen G. 1993. The Finnish Landscape of the Upper Great Lakes. *The East Lakes Geographer* 18: 14–24.

Noble, Allen G., and Hubert G.H. Wilhelm. 1995. The Farm Barns of the American Midwest. In *Barns of the Midwest*, ed. Allen G. Noble and Hubert G.H. Wilhelm, 1–23. Athens: Ohio University Press.

Palmqvist, Lena Anderson. 1983. *Building Traditions among Swedish Settlers in Rural Minnesota*. Uddevalla: The Nordiska Museet – The Emigrant Institute.

———. 1986. Swedes. In *America's Architectural Roots: Ethnic Groups that Built America*, ed. Dell Upton, 154–159. Washington: Preservation Press.

———. 1987. Buildings on Swedish-American Farms in Two Minnesota Communities. *Swedish-American Historical Quarterly* 38 (1): 13–28.

Raitz, Karl B., and Cotton Mather. 1971. Norwegians and Tobacco in Western Wisconsin. *Annals of the Association of American Geographers* 61 (4): 684–696.

Sando, Linnea C. 2014. The Enduring Finnish Sauna in Hamlin County, South Dakota. *Material Culture* 46 (2): 1–20.

Tishler, William H. 1984. Built from Tradition: Wisconsin's Rural Ethnic Folk Architecture. *Wisconsin Academy Review* 30: 14–18.

———. 1992. Norwegians in Wisconsin. In *To Build in a New Land: Ethnic Landscapes in North America*, ed. Allen G. Noble, 226–241. Baltimore: Johns Hopkins University Press.

White, Janet. 2012. Swedish to American: Built Form at the Bishop Hill Colony. *Material Culture* 44 (1): 50–68.

13 Slavic Landscapes in America

The largest numbers of Slavic peoples who came to the United States immigrated from the northern Slavic nations, in particular those lands that are now parts of the Czech Republic, Poland, and Russia. When queried about their ancestry for the 2000 Census, 1,703,930 persons reported being of Czech or Czechoslovakian ancestry, 8,977,444 cited Polish ancestry, and 2,652,214 claimed Russian ancestry. Slovakian ancestry was cited by 797,764 Americans. The American Community Survey for 2014 estimated that 1,419,630 Americans had Czech ancestry, 9,249,392 had Polish ancestry, 2,762,830 were of Russian ancestry, and 706,662 had Slovakian ancestry. The increases in numbers of Poles and Russians were partly driven by immigration, as 203,470 Russians and 144,043 Poles gained permanent residency status within the United States between 2000 and 2013 (U.S. Dept. of Homeland Security 2014).

13.1 Slavs from Central European Cultural Transition Zone

Culturally, the nations east of Germany display a transition between Western Europe and Eastern Europe. Germans had a strong influence upon the culture of both Poland and the Czech Republic. Indeed, before World War II Germany included the lands around the present Polish city of Gdansk (formerly Danzig), and before World War I German Prussia covered a broad swath of territory between the present German border and Gdansk. Germany annexed the Sudentenland of Czechoslovakia, long settled by their ethnic brethren, in 1938 in an action that British Prime Minister Nevil Chamberlain mistakenly believed would appease Germany's quest for territory. Given the long history of interaction between Germans and both Czechs and Poles, it is not surprising that many Czechs and Poles who immigrated to the United States not only spoke German, but settled near German immigrants with whom they could communicate.

The Eastern European lands settled by the Slavs represent a transition zone in both language and religion. While all of the major languages in the area are considered Slavic, those in the west, including Czech and Polish, are written using the Roman alphabet, while those in the east, including Russian and Ukrainian, utilize the Cyrillic alphabet. Within Poland and the Czech Republic, the majority of the population is Roman Catholic, although the Czech Republic also includes some Calvinist Protestants. Russia, on the other hand, is predominately Russian Orthodox. This religious difference is displayed in the style of their crosses and the architecture of their churches, particularly the utilization of domes on towers, with even Catholic churches in Poland displaying some of the eastern influences. In addition, before World War II millions of Jews lived in Poland and far western Russia.

Slavic persons, as a group, tended to migrate to the United States later than Western Europeans. Nevertheless, there were exceptions, such as several prominent Poles who came to fight in the American Revolution, including Count Casimir Pulaski and Thaddius Kocziewski. Poles from German controlled areas and Czechs typically were the first Slavic peoples to immigrate in large numbers to the United States, with many Poles arriving by the middle of the nineteenth century and Czechs arriving after the American Civil War (Allen and Turner 1988). In contrast, Poles from the Russian controlled areas of their homeland along with Russians and Ukrainians came in the largest numbers after 1890. Not only did time of arrival influence where the Slavic populations settled, but it also shaped the type of ethnic material culture that was displayed in their communities. Our discussion will start with immigrants from the westernmost Slavic nations and move east, roughly approximating their arrival in America.

13.2 Czech Ethnic Landscapes in America

Czechs are natives of Bohemia and Moravia, areas that were ruled as part of the Austro-Hungarian Empire until 1918. Czechs, particularly those from the region of Bohemia, were

J.A. Cross, *Ethnic Landscapes of America*, DOI 10.1007/978-3-319-54009-2_13

nearly evenly divided between those who migrated into rural settlements and those who moved to urban centers in the United States, particularly Chicago, Cleveland, and New York (Allen and Turner 1988). Many of the Czechs who migrated to the United States during the 1850s were professionals or craftsmen, while the preponderance of those arriving a couple decades later were farmers.

Wisconsin was the earliest focus of Czech immigration, and before the Civil War a Czech population had settled in Milwaukee. Not only did Wisconsin actively solicit immigrants, but it had "a reputation for political and religious freedom" (Allen and Turner 1988:84). Farm families settled in Racine, Manitowoc, and Kewaunee Counties, and later in Price and Langlade Counties. Wisconsin also received a secondary migration from Chicago. Within the Great Plains, eastern Nebraska south of the Platte River similarly received many Czechs who had previously settled in Illinois, Wisconsin, and Iowa. In contrast, the area of Texas south of a line extending from Dallas to San Antonio was the destination of many Czechs immigrating directly from Europe.

13.2.1 Czech Ethnic Population Distribution

Today, those rural areas that attracted Czech immigrants during the nineteenth century continue to have the largest proportions of their populations reporting Czech ancestry (Fig. 13.1). Butler County, Nebraska, with 32.6% of its population claiming Czech ancestry and an additional 3.5% reporting Czechoslovakian in the 2000 census, had the highest proportional Czech ethnicity. It is surrounded by three of the nation's other six counties where Czechs accounted for at least one-fifth of the population in 2000, including Saline County, Nebraska with 23.2%, Colfax County,

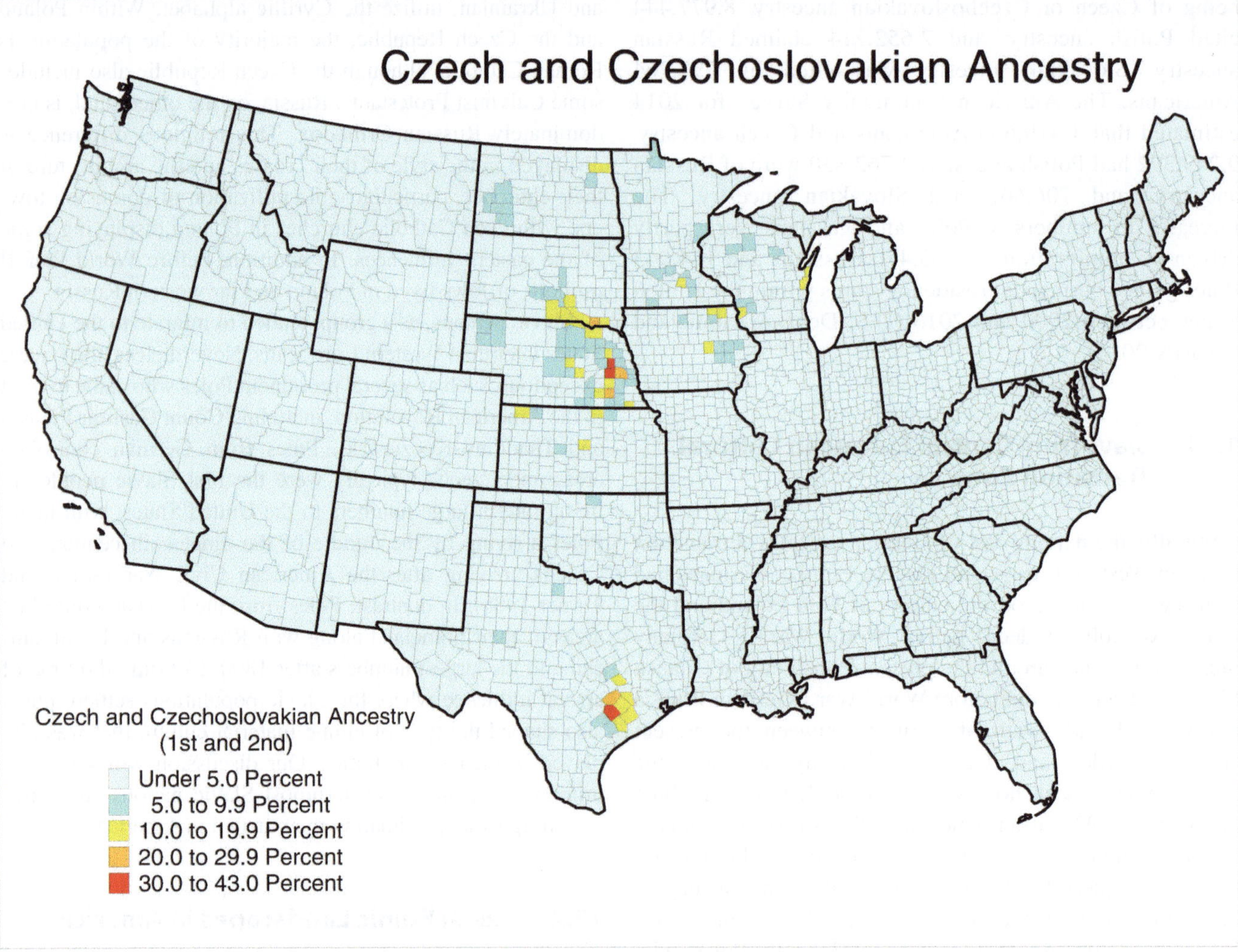

Fig. 13.1 Map showing the percentage of individuals who reported their ancestry on the 2000 Census indicating that Czech or Czechoslovak was their primary or secondary ancestry. Note the clusters of counties in southern Texas, Nebraska, and several Midwestern states in which over a tenth of the population reported Czech ancestry

Nebraska with 26.3%, and Saunders County, Nebraska with 20.9% Czech or Czechoslovakian ancestry. Bon Homme County, South Dakota was 21.1%. In southern Texas Lavaca County was 27.4% Czech or Czechoslovak while adjacent Fayette County, Texas was 20.0%. Wisconsin's Kewaunee County, which is also noteworthy for its large Belgian population, was 15.7% Czech. American Community Survey 5-year estimates for 2010–2014 had fallen in all of these counties except Lavaca County, Texas, where those of Czech and Czechoslovakian ancestry had risen to 30.7% of the population. Butler County, Nebraska remained the county with the highest Czech and Czechoslovakian proportion, 33.0% of its population.

While larger numbers of ethnic Czechs (including Czechoslovakians) reside in large urban counties like Cook County, Illinois and Cuyahoga County, Ohio (which had 67,690 and 29,814, respectively in 2000), they represented only 1.2% and 2.2% of the population, respectively. The 2010–2014 American Community Survey estimates for these counties, comprising the centers of the Chicago and Cleveland metropolitan areas, showed drops in both the total numbers and proportion of their populations reporting Czech or Czechoslovakian ancestry, with only Cuyahoga County exceeding 1 percent (1.7%).

13.2.2 Czech Religious Landscapes

While distinctive ethnic churches are a conspicuous feature of the cultural landscape for many European immigrant groups, this was often less so among Czechs. Many Czechs altered or abandoned their religious practices on immigration, and many Czechs were known as freethinkers, having no beliefs. As James Allen and Eugene Turner note, "The move to America led many [Czechs] to abandon Catholicism and join Protestant denominations, and many urban families gave up any religious practice. Czech communities here were often bitterly divided between Freethinkers and Catholics" (Allen and Turner 1988:84). Although the Czech Republic is chiefly Roman Catholic, this was largely imposed by conquest. There were Protestant activities there in the early 1400s, and Calvinist churches are still found in the Czech Republic. Churches in Bohemia typically display either Eastern or Western European architecture. Even where the Czechs have remained devoutly Roman Catholic in the United States, their church buildings may not necessarily be distinctly Czech. For example, within Kewaunee County, Wisconsin, the tall brick masonry Roman Catholic churches with their towering steeples were often built by German masons (Rucker 1996). Yet, it has also been observed that Czechs often emphasized their national heritage in their church buildings, and "their edifices contain liturgical icons using Czech script and are often named to commemorate great national folk heroes of their pre-Austrian past" (Rau 1992:287).

A distinctive religious landscape feature within the Czech community of Kewaunee County, Wisconsin is the presence of Czech roadside religious shrines (Fig. 13.2). While not as large or numerous as those seen in Central Wisconsin's Polish region, three to four foot high shrines and religious markers remain along the highway near Pilsen and Stangelville, Wisconsin. Typically these shrines were "topped by a crucifix and included a scripture verse in Czech" (Rucker 1996:45). A local historian explains that "many parishes had shrines at nearby intersections, and such markers were points of veneration during rogation days" (Rucker 1996:45). Among the extant roadside shrines, examples can be found carved of stone and crafted from medal. Similarly, the Czechs, along with some early German settlers, sometimes utilize iron cross grave markers within their cemeteries (Rau 1992).

As with many ethnic groups, Czechs typically inscribed tombstones in their cemeteries using their native language and utilizing familiar emblems or decorations. The gateway into the cemeteries were conspicuously labeled in Czech (Fig. 13.3). Based upon her study of several cemeteries around Praha, Texas, Eva Eckert (2002:201) wrote, "The stones provide a measure of continuity in the community, and also show through the content and culture of messages how the immigrants of the second and third generations changed their identity." Initially, the inscriptions were all in Czech, yet English words appeared mixed with Czech between the two world wars, and the layout of the words and dates came to parallel a more standard American layout. Shifts in grammar of the Czech inscriptions were noticeable, until finally, the "Czech used was formulaic and revealed nothing about the actual capacity to speak the language" (Eckert 2002:203). In many of the rural communities of the Plains such tombstones are the most conspicuous indicator of the Czech ethnicity of the settlers, given that most of the farms have consolidated and the descendents of the settlers have moved to distant cities.

Czechs built fraternal lodge-halls or sokols within many of their communities. Within South Dakota they displayed masonry construction and were labeled with Czech writing. While architecturally most of these structures are virtually indistinguishable from other plain American brick or frame buildings, "the use of the lunette-shaped front parapet" (Rau 1992:304) is a key to their ethnic identity. Examples of wood-framed sokols can also be found in Wisconsin (Fig. 13.4).

In some Czech communities the ethnicity is more shown by the signage on the buildings and the types of ethnic foodstuffs that are advertised, rather than any unusual

Fig. 13.2 Czech Catholic shrine at intersection of rural road one mile north of the Catholic church in Stangelville, Wisconsin. Note that the rural farmhouse is located close to the roads, which is more common among Czech ethnics than many other groups. Subsequent to the taking of this photograph, the highway was widened, resulting in the removal of the house and the relocation of the shrine by about twenty feet, where it is now flanked by a historical sign explaining the role the shrine played in the church's rogation ceremonies

Fig. 13.3 Entrance to Czech cemetery in Prague, Nebraska with the Czech words that translate as National Cemetery. Almost all of its pre-World War II tombstones are inscribed in Czech

Fig. 13.4 Zapadni Ceska-Bratrska Jednota or Western Czech Fraternal Association Hall north of Cadott, Wisconsin. This hall or sokol was constructed in 1907 with pressed metal siding over a wood frame

Fig. 13.5 Restaurant in Prague, Nebraska, with its name in Czech extolling its kolaches and displaying the Czech flag by the American flag

architecture. For example, the local restaurant (Fig. 13.5) in Prague, Nebraska showcases the community's ethnic heritage by its name highlighting the availability of kolaches, a type of yeast pastry containing a variety of semi-sweet fruit fillings, as well as the display of the Czech flag. Similar reliance upon the kolache to showcase a community's Czechness can be found within Hillsboro, Texas, among other communities.

13.2.3 Czech Farmscapes

Both in their settlements in the Midwest and in the Great Plains, Czech immigrants built houses that were distinctive, displaying elements of their Old World culture. Because they found different environmental conditions and building materials in Wisconsin compared with Nebraska and South Dakota, they often built their houses of logs or wood in the former, whereas stone masonry structures were more common in the latter. Nevertheless their buildings are still identifiable as being Czech (Rau 1992).

In his observations of the Czech settlement in southeastern South Dakota, John Rau (1992) describes the dominance of masonry construction among the Czechs, who utilized the local chalkrock that was very easily cut. This stone was used in most of their buildings, even those which were largely wooden. In addition, "with few exceptions, the Czech covered the exterior facades of stone walls with a thick, smooth veneer of plaster or mud" (Rau 1992:290). Such plaster covered walls are frequently seen in the Czech Republic today.

Log buildings were more common in areas of forests, yet the log walls were usually covered with clapboards. Similarly, planking covered the logs in the interior of the houses. Besides South Dakota, numerous examples of Czech built vernacular log structures remain in Kewaunee and Manitowoc counties, Wisconsin. There they have been described as representing "a remarkable continuation of the centuries-old Bohemian skill in fabricating sturdy log structures, [which] appears to be unique in America" (Tishler 1984:16). Although four types of log houses have been identified in these counties, in many cases clapboard siding now covers the walls. Nevertheless, their Czech identity is displayed by "traditional characteristics in their log construction details, widespread use of roof overhangs and forebays, gable exterior refinements, and close proximity to the adjacent road" (Tishler 1984:16). Furthermore, just as among their neighboring Belgian settlers, Czechs in Wisconsin sometimes clad their log houses with brick (Rucker 1995).

13.2.3.1 Traditional Czech House Plan

The houses built by Czechs, whether of logs or stone, had several distinctive arrangements, as explained by David Murphy.

> Only a few of the numerous Czech house types were used in America. All had a single story, a narrow front gable wall and varying depth. Most popular were single-room cabins, two-room cottages and triple-pen houses, the smaller usually designed to be expanded into one of the larger types. The largest traditional house to be built in any number was a tripartite, central-passage house with spatial similarities to the northern European parstuga (Murphy 1986:114, Reprinted with permission of John Wiley and Son).

Czech folk houses in America were typically long and narrow or were constructed in an L-shape. The three-bay house, with two side rooms flanking a central hall (termed a sin), was common. The central hall often served as a kitchen, with the chimney located in the inner partition wall between two rooms and the stairway leading up from the central hall. If the structure was built in an L-shape, the stairway and chimney were still located in the central room that connected the two rooms at right angles to each other (Fig. 13.6).

Fig. 13.6 Abandoned Czech ethnic farmhouse in Kewaunee County, Wisconsin. Located near the road, the upper floor of the L-shaped house is a typical half story, and under the front gable, partly obscured by the tree, is a decorative wooden fan-shaped ornamentation

The upper floor was usually a low half story, only having its ceiling at a normal height under the center of the gable.

Farmhouses built in Czech settlements were often built close to roads, sometimes only a few feet from the edge of the right-of-way. The proximity of these rural houses to the highway was sometimes like what one would expect within the center of a village or small city. In contrast, farmhouses of most other ethnic groups are set much farther back from the roadway. Many Czech farmsteads also had "little houses" or second houses for the use by grandparents.

13.2.3.2 Czech Barns and Outbuildings

The barns and various farm outbuildings, such as summer kitchens and granary or machine sheds, of the ethnic Czech settlements differed from those of their neighbors. For example, in eastern Wisconsin it was noted, that the large barns on Czech farmsteads "are remarkably similar in size and shape, and incorporate low foundations, massive logs, two cribs separated by a central threshing floor, and identical upper-level structural systems" (Tishler 1984:16). These barns are viewed as being derived from Old World types (Murphy 1986).

Ethnic Czech barns in Wisconsin, typically of log or timber construction, are often set at right angles to each other (Fig. 13.7). In other cases the barns were built in an L-shape. Silos could be attached, or built in the form of an L. One barn in Kewaunee County has a square stone masonry silo along one side, forming a short L. While rare, the square silo is disproportionately more common in the area (Rucker 1995). Czech immigrants also built long linear barns, resembling the design of housebarns. At the beginning of the twenty-first century, one such housebarn was still standing in Manitowoc County, Wisconsin.

In South Dakota Czechs were know for their masonry barns, of rubble or hewn chalkrock, with the masonry walls along the long sides extending to the eaves. Entry was through the gable end. These barns were viewed as containing a mixture of traditional and American designs (Rau 1992).

13.3 Slovakian Ethnic Landscapes in America

Two and a half times as many persons claim Czech ancestry than those claiming Slovakian ethnicity in the United States. Over 11 times more Americans report Polish ancestry. Thus compared to these other Slavic peoples, little study has been made of the Slovakian ethnic landscape in the United States.

The greatest numbers of Slovakian ethnics are found in a swath running from the Mid-Atlantic states to Illinois, with the greatest numbers in Pennsylvania and Ohio. The American Community Survey for 2014 estimated that 204,923 individuals of Slovakian ancestry lived in Pennsylvania, 29.0% of the nation's total.

Fig. 13.7 Czech ethnic barns were often arranged in an L-shape, such as on this farmstead south of Kellnersville, Wisconsin

13.4 Polish Ethnic Landscapes in America

Polish ethnicity can be claimed by many persons whose ancestors came from lands outside the present boundaries of Poland. Indeed, when many of the ethnic Poles came to America, there was no independent Poland, but instead an area that was split between German, Russian, and Austrian-Hungarian rule. While an independent Poland was reestablished following World War I, its borders were farther east than the current borders. Since World War II Poland includes many lands that had previously been German, with its border with the Soviet Union being shifted approximately 140 miles to the west. Thus, many ethnic Poles came from areas that are now part of Russia. Ethnic Poles who immigrated to America during the nineteenth century and before the First World War came from a variety of German states, the Russian Empire, and the Austro-Hungarian Empire. Immigration again rose during the 1980s, when many Poles fled Communist-ruled Poland before the fall of the Iron Curtain, and it is continuing.

13.4.1 Polish-American Population Distribution

Polish ancestry was reported by 3.2% of the United States population at the time of the 2000 census, a total of 8,977,444 persons, yet nearly 400,000 fewer than reported a decade earlier. The American Community Survey for 2014 estimated that the population of Polish ethnicity had risen to 9,249,392.

The largest numbers of Polish-Americans in 2000 were found in several Northern cities (Fig. 13.8). New York City, with 213,447 persons of Polish ancestry, had the most, yet there they represented only 2.7% of the population. In contrast, within several industrial cities along the Great Lakes Polish-Americans were not only numerous, but they comprised a larger share of the population and distinctive Polish ethnic neighborhoods can be found. For example, Chicago counted 210,421 persons of Polish ancestry, comprising 7.3% of its population. Cook County,

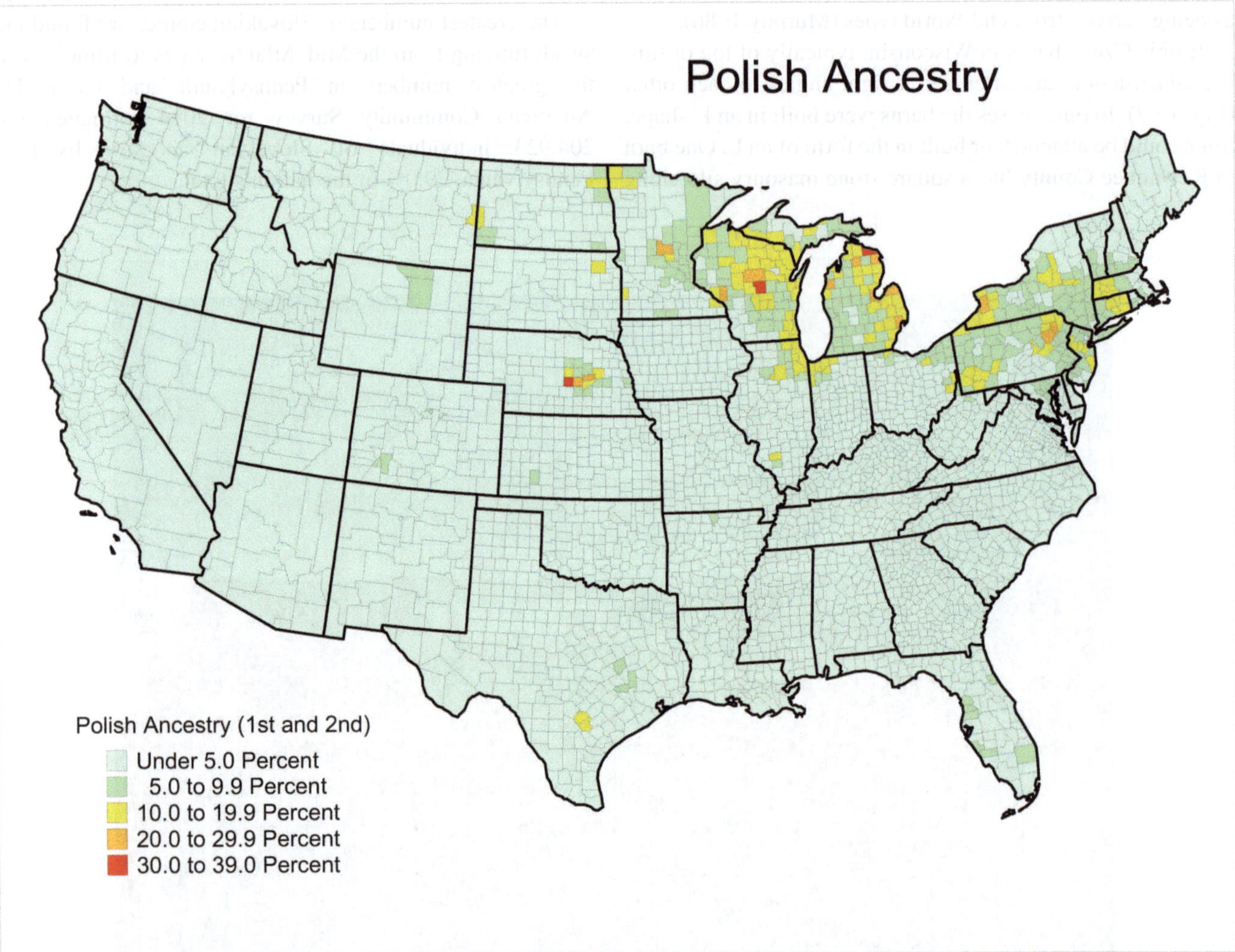

Fig. 13.8 Map showing the percentage of individuals who reported their ancestry on the 2000 Census claiming that Polish was their primary or secondary ancestry. Note the concentration of Polish-Americans in the states along the Great Lakes, plus the clusters of counties in parts of the Great Plains where over a tenth of the population reported Polish ancestry

Illinois, which includes Chicago, had 530,645 Polish ethnics—the most of any county in the nation, and it was 9.9% Polish. To the north, Milwaukee County, Wisconsin, with 124,195 Polish ethnics, was 13.2% Polish. Other Great Lakes counties with many Polish ethnics included Wayne County, Michigan, with 173,119 (8.4% of the population), Erie County, Pennsylvania (13.1%), and Erie County, New York with 181,139 (19.1%), encompassing the Detroit, Michigan; Erie, Pennsylvania; and Buffalo, New York metropolitan areas, respectively. The American Community Survey for the 5-year period ending in 2014 estimated that New York City's Polish ethnic community was similar to what the 2000 Census found, but that Chicago's Polish community had fallen to 163,137. Cook County, Illinois, still had the largest population of Polish ancestry, but its numbers had fallen to 457,575 or 8.8% of its total population. Indeed, the 2014 American Community Survey found fewer Polish ethnics in all of these metropolitan counties.

Those counties with the greatest proportion of their populations reporting Polish ancestry were scattered across the Upper Midwest and the Great Plains. Portage County, Wisconsin, which includes the city of Stevens Point, had the highest proportion of its population, 33.1%, reporting Polish ancestry of any county in the nation in 2000. It was 29.5% Polish according to the 5-year American Community Survey for 2010–14. Twenty-four additional counties in Wisconsin had populations that were between 10% and 20% Polish in 2000. Five counties within Minnesota likewise had populations that were at least 10% Polish, with Morrison County having 23.0%. Within Michigan 16 counties were at least one-tenth Polish, and Presque Isle County was 26.9% Polish ancestry. Within east central Nebraska Sherman County was 29.6% Polish, and four nearby counties were at least 10% Polish. Within eastern Pennsylvania the far larger Luzerne County was 23.9% Polish.

13.4.2 Polish-American Landscapes

With concentrations of Polish-Americans in both urban and rural communities, distinctive ethnic landscapes may be discerned in both settings. Similarities to German rural landscapes have been noticed, particularly among the earlier arriving Polish immigrants who came predominately from the western regions of Poland.

Polish communities were strongly influenced by the religious beliefs of their settlers. Indeed, the presence of a Polish-speaking priest was typically a necessary prerequisite to the establishment of Polish farming colonies, focused upon the Polish parish church. The religious landscape of the Polish communities is their most distinctive landscape element. Catholicism was intimately tied to Polish ethnicity, however not all Polish parishes are Roman Catholic. Disputes with German and Irish bishops over parish control, in particular the desires of the Polish-Americans to retain their ethnic consciousness and control of their parishes, led some individuals to leave the Roman Catholic Church and establish the Polish National Catholic Church over a century ago (Radzilowski 2005). Nevertheless, most Polish ethnics remain Roman Catholic, although many Polish communities contain overlapping parishes of each church.

13.4.3 Polish-American Churches

The church building is a conspicuous feature of Polish communities. As John Radzilowski (2005:28) explains, "Polish church buildings became the visible expressions of the community…. [T]hey channeled their hopes, dreams, and ambitions into the building of magnificent churches that expressed their new identity as Polish Americans." Geoffrey Gyrisco (1997:33) notes that "their design was intended not only to reflect but perhaps to help create a Polish-American identity that transcended earlier identities." Polish-American church buildings were the center of the community, both figuratively and spiritually (Fig. 13.9). These structures are the most prominent feature on the Polish ethnic landscape. As Radzilowski explains,

> In this era of inner-city church closing we are often confronted with the commonplace line: 'a parish is people, not bricks and mortar,' which is true enough. But this ignores the importance of buildings in the lives of our immigrant ancestors, and continuing importance they have for ethnic communities (Radzilowski 1997: 6).

This is just as true of rural communities, yet some Polish churches have been closed in recent years.

Polish ethnic churches in America were often designed and built during the late nineteenth century and early twentieth century by Polish émigré professional architects whose structures were "neither entirely east European nor entirely American" (Radzilowski 1997:6). A good example of such an architect was Victor Cordella, who "employed a vocabulary appropriated from typical elements of monumental urban churches in Poland" (Gyrisco 1997:52) in the nearly 20 churches he designed that were built in Minnesota.

Polish-American communities, and particularly those in Chicago and Detroit, invested heavily in building monumental structures that demonstrated their ethnicity and religious devotion. Described as the "Polish cathedral," these often enormous edifices contrast with the smaller Gothic or Romanesque churches erected by other Catholic populations. As explained by Peter Williams (1997:179) "the ambitious prelates and their loyal congregations in the Great Lakes Polonias often chose instead to make their

Fig. 13.9 St. Mary of the Angels Catholic Church (1914) in Chicago, Illinois. This monumental church with a central dome was erected to serve the Polish immigrant community, with a rectory and large parochial school constructed nearby

monumental statements in the Renaissance style of the mother county." Churches of the Baroque Revival style, with its "repeated round arches, shaped front gable, elaborated towers with curving forms and onion dome" (Gyrisco 1997:49), often signify a Polish-American congregation. Many of these church buildings still tower above their neighborhoods a century after they were built, although not all continue to serve Polish worshippers.

Among the eastern European architectural characteristics that often, but not always, distinguish the churches of ethnic Poles are their display of numerous domes atop the church towers, as well as a central dome. While some of the churches, such as the Basilica of St. Stanislaus Kostka in Winona, Minnesota (Fig. 13.10), have numerous dome topped towers and a small central dome, others, such as St. Josaphat Basilica in Milwaukee, Wisconsin (Fig. 13.11) have a massive central dome dominating the entire structure. The domed church not only demonstrates the influence of eastern orthodoxy upon the Polish church, but it "represents the heavenly vault and heavenly perfection" (Wolniewicz 1997:55). Even many Polish ethnic churches without a central dome have smaller domes atop each of their towers, such as Assumption of the Blessed Virgin Mary Catholic Church in Pulaski, Wisconsin. It has twin towers, each capped with a dome, flanking the front façade, while smaller towers flank the opposite end of the building.

Polish churches were often inscribed in Polish, including the word "kosciol" (Fig. 13.12), unlike so many Roman Catholic churches in the United States that were labeled only in Latin or English. Likewise, the selection of saints to honor in the naming of churches, such as St. Stanislaus, St. Kazimierza, and St. Wenceslaus, clearly displays their Polish ethnicity.

Churches were not the only way Poles displayed their religion upon the landscape. There are numerous visible expressions of their faith. The Slavic view of the sacredness of the world surrounding them found "expression in a multitude of sacred sites, which were frequently marked by shrines and crosses of almost every conceivable type… [and] were so characteristic of the Polish landscape that romantic poets liked to call Poland 'the land of graves and crosses' (Kolinski 1994:33). Such religious symbols are also visible in rural central Wisconsin, especially in eastern Portage and Marathon Counties and near Pulaski.

13.4.4 Polish Roadside/Wayside Catholic Shrines

Over 50 roadside or wayside Catholic shrines are located in the Polish ethnic region of central Wisconsin (Fig. 13.13). They are frequently located at intersections, close to the highway, although others are located along the edge of the highway right of way. These shrines are usually of brick construction, four to seven feet in height, with religious statues visible behind glass windows. Commonly displayed saints include the Virgin Mary, St. Joseph, and Our Lady of Perpetual Help. Similar shrines are sometimes located near

Fig. 13.10 St. Stanislaus Catholic Church, now the Basilica of St. Stanislaus Kostka, dedicated in 1895 in Winona, Minnesota. The selection of saints to be honored in church names also provides strong linkages to the group's ethnicity

Fig. 13.11 St. Josaphat Basilica, constructed in 1898, is located in the center of the Polish neighborhood on the South Side of Milwaukee, Wisconsin

churchyards, yet many of the roadside shrines are a mile or more from a church. In addition to the shrines, wayside crosses can be seen within Wisconsin's Polonia (Fig. 13.14). These tall, decorative crosses often have a smaller metal crucifix attached. These crosses are located along rural roads, at the edge of fields, and sometimes by cemetery entrances.

Yard shrines are more conspicuous among Polish ethnics than with other Catholic ethnics in the Midwest (Fig. 13.15).

Fig. 13.12 Polish inscription on Holy Trinity Church, Chicago, Illinois. Some of the services in this church, which was built in 1873, are still conducted in Polish

Fig. 13.13 Polish wayside shrine at intersection one mile north of Assumption of the Blessed Virgin Mary Catholic Church in Pulaski, Wisconsin. Note that the brick structure, erected in 1901, houses religious statues behind its glass windows

Some are variations of wayside shrines, built along and facing the highway. Others may be termed "bathtub shrines," given that their shape and size resembles an upright bathtub. These shrines, painted blue, shelter statues of the Virgin Mary, and are often located in the front yard near the house. In the far more crowded Polish neighborhoods of Brooklyn, New York, ethnic Poles locate their shrines to the Virgin Mary in their back yards (Krase 1997).

Fig. 13.14 Cross with mounted corpus along rural road west of Bevent, Wisconsin. Wisconsin's "Polonia" is centered in Portage County

Fig. 13.15 Polish religious shrine in yard of rural home in the hamlet of Fancher in Portage County, Wisconsin. Opposite this home is the St. Mary of Mount Carmel Catholic Church, a rectory, a parochial school, and a cemetery, which dominate this Polish ethnic community of several houses surrounded by farm fields

Grottos of rough-hewn stone and mortar, such as replications of the grotto at Lourdes, are more common among Polish Catholics than among many other groups. The grottos are sited near many Polish churches and cemeteries in the Midwest.

The presence of the roadside shrines, wayside crosses, yard shrines and grottos all contribute to the uniqueness of the Polish religious landscape in central Wisconsin. Not only did this type of prolific display of religious devotion not develop among other rural Catholic populations, but its "occurrence is so rare in the United States, that … most Americans live out their entire life without the slightest realization that such things exist within our borders" (Kolinski 1994:46).

While these shrines mark several rural Polish ethnic communities, the clustering of Polish-American urban neighborhoods around their parish churches also demonstrates the importance of religious institutions as both social and religious centers. Within Milwaukee, just as it had in partitioned Poland, the Catholic church played a critical role "as a transmitter of culture, values, and nationalism" (Kenny 1997:269). The "parish church became a social as well as a religious center for surrounding residents, as evidenced by the Polish names attached to various neighborhoods…much as a village in Poland would be named" (Kenny 1997:269–271). Within these South Side Milwaukee neighborhoods a type of housing termed "Polish flats" is commonplace.

13.4.5 Polish-American Housing

Polish flats are a duplex that represented incremental building of the structure. In essence, they are a raised cottage, with the building jacked up to permit the construction of a new lower floor, resulting in one flat or apartment on the upper floor, and one below. While nothing in the appearance of these houses is particularly Polish, this type of building was most commonly undertaken in urban Polish communities and became associated with that ethnic group. Large concentrations of such structures remain in Milwaukee and Chicago.

Within the rural Polish settlements in central Wisconsin southeast of Stevens Point architectural historians have noted a disproportionate number of two-story red brick farmhouses. These houses may be discerned by their "heavy, ornate brick hood moldings around the windows and doors" (Gyrisco 2001:29) and the placement of a potato warehouse in their basement. Considering both Portage County's rural brick houses and the Polish flats of Milwaukee, it has been suggested that "Polish-American communities took the material elements of mainstream American culture and used them in distinctive ways that set them apart from adjacent ethnic groups and the larger society" (Gyrisco 2001:32). Thus, while the styles do not directly link to what was built in Poland, they reflect "a new, distinctive, and quite special Polish-American culture" (Gyrisco 2001:32). Regardless of their origins, these characteristics distinguish the communities of Polish-Americans today.

In some rural communities Poles occasionally utilized a construction method that was quite distinctive. Stovewood walls in houses, barn foundations, and store buildings have been observed in Polish communities in Wisconsin (Fig. 13.16). In these buildings the walls are built of short logs stacked like cordwood and cemented together to form walls and foundations. While the house representing ethnic Poles in the Old World Wisconsin outdoor heritage museum has stovewood walls, the great majority of Polish ethnics did not build with stovewood.

13.5 Russian and Ukrainian Ethnic Landscapes in America

The Russian Empire, as it existed before World War I, included a variety of Slavic peoples within its most heavily populated western region, as well as many other ethnic groups along its southern and eastern frontiers. With the collapse of the czarzist government at the outbreak of the Russian Revolution, most of the former Russian Empire's Polish population resided in a newly independent Poland. Nevertheless, the Soviet Union contained a wide number of ethnic groups. Indeed, the country's official name, the Union of Soviet Socialist Republics, formally recognized this diversity by granting "Soviet Socialist Republic" status to 15 of the country's largest nationalities. The three largest groups were Russians, Ukrainians, and Belorussians. These Slavic groups are related but different. While they all use the Cyrillic alphabet, their languages are Russian, Ukrainian, and Byelo-Ruthenian, respectively. Traditionally, the majority of the population of all three nations were Orthodox Christians, although significant Jewish populations were found in the region.

13.5.1 Slavic Populations of Russia and the Ukraine

The links between the Slavic populations of Russia and the Ukraine are steeped in history. The Russian Empire annexed most of the Ukraine in the 1600s. The capital of Ukraine, Kiev, had been a Russian principality since the Ninth Century. Kiev was considered the "Mother of Cities" to Russians, being older than both Moscow and St. Petersburg, latter-day capitals of the Russians. Both Russia and Ukraine were integral parts of the Soviet Union, and at the time of the demise of the Communist Union, nearly a quarter of Ukraine's population were ethnic Russians. Continuing tensions between Russians and Ukrainians resulted in the

Fig. 13.16 Barn with stovewood foundation west of Zachow, Wisconsin. This bank barn with its forebay resembles barns erected by German-Americans, but the use of stovewood in Wisconsin is more typically associated with Polish ethnic communities. This barn collapsed in late summer 2015

Russian annexation of Crimea in 2014 and the instigation of ethnic unrest in portions of eastern Ukraine that had substantial Russian ethnic populations, which led to Russian military intervention.

Tremendous ethnic diversity marked the Russian Empire and its successor, the Soviet Union. As James Allen and Eugene Turner (1988:106) point out, individuals of Russian ancestry displayed considerable ethnic variety, possibly "more ethnic variety than any other group." They note, "Of the immigrants who arrived in the United States from Russia in the early twentieth century, only an estimated 17% were ethnic Russians or Great Russians" (Allen and Turner 1988:106). The others represented a wide variety of nationalities found within the Russian Empire. Even the Russia of today, minus the other 14 republics of the former Soviet Union, is ethnically diverse, including both many indigenous and immigrating minorities.

13.5.1.1 Religion and Russian Identity

Religion has been a key factor in motivating many Russians to migrate to the United States. It has been estimated that about 70% of the U.S. Jewish population is of Russian origin. Of the Jewish immigration to U.S. between 1881 and 1924, 75% were from the Russian Empire, fleeing various pogroms, while 18% were from Austria-Hungary, and 4% were from Romania (Goren 1980). Russian Jews so dominated this immigration, that all East European Jews were often labeled Russian. The Russian Orthodox Church was associated with many of the other Russians. Indeed, the Russian Orthodox church provided a Russian identity to many persons who were not ethnic Russians, but who came from the Russian Empire. Yet other Russian Christian minorities came to America as religious refugees, fleeing persecution from both the czarist government and the Russian Orthodox church. These included Russian Molokans, Russian Baptists, Doukhobors, and Old Believers or Staroveri, each of which settled in different areas of North America (Hardwick 1992).

While the largest numbers of Russians and Ukrainians came to the United States fleeing the anti-Jewish pogroms of the late nineteenth century and the turmoil of the Russian Revolution in the aftermath of World War I, over a half million Russians arrived in the 1980s and 1990s. Most of these were religious refugees, including both Russian Jews and Russian Baptists and Pentecostals (Hardwick 2002). Russians continued to constitute the largest share of European immigrants to the United States for the first decade and half of the twenty-first century (U.S. Dept. of Homeland Security 2014). Yet, not all Russian immigrants to the United States were or are religious refugees, nor did all arrive across the Atlantic. Furthermore, patterns of settlement of Russian Jews differed from that of non-orthodox Christians.

13.5.2 Russian Settlements in America

Russian colonization of North America brought Russians to lands that became part of the United States, even before the nation had received its independence. Russian fur traders were active along North America's Pacific coast during the eighteenth and nineteenth centuries. Russian fur trading in the Aleutian Islands of Alaska began in the 1740s, with the first permanent Russian settlement being made at Illiuliuk (now Unalaska) on Unalaska Island between 1772 and 1775. Russian activity on Kodiak Island began in the 1750s with a permanent settlement established about 1785 (Senkevitch 1987). From these sites Russian colonization spread north into the Cook Inlet area, establishing the Russian settlements of Kenai, Soldatna, and Ninilchik, among others. Novo-Arkhangel'sk, now named Sitka, was established on Baranof Island as the headquarters of the Russian-American Company in 1799. It became the colonial capital in 1808. Russia sold Alaska to the United States in 1867 and the Russian-American company was sold to a U.S. business, resulting in the departure of many Russians.

Yet not all the Russians left, and many Aleuts and Alaskan Indians, who had become highly acculturated, remained. Not only do many villages and small cities bear Russian names, but many Aleuts have Russian surnames and Russian Orthodox churches are scattered across southern Alaska into the Aleutian Islands (Figs. 13.17 and 13.18). While Sitka lost its position as territorial capital in 1906, it is the site of Sitka National Historical Park, which contains the house of the Russian bishop built in 1843, one of very few buildings surviving from the Russian colonial period. St. Michael's Cathedral with its classic Russian onion-shaped dome was built in Sitka in 1844, but it was destroyed by fire in 1966. It was rebuilt to its previous specifications, preserving the Russian ethnic landscape even though no longer with the original structure.

Russian colonists had a far smaller impact in California, where they established Fort Ross in 1812. Located north of San Francisco Bay, this post served as a Russian base for hunting sea otter and subsistence agriculture until 1841, when it was sold to John Sutter of future Gold Rush fame. One of the original Russian colonial houses remains at Fort Ross State Historic Site, while reconstructions of the blockhouses, barracks, Russian Orthodox chapel, and another house have been erected. Even farther to the south, and much farther west, on the Hawaiian island of Kauai are the earthen ruins of a Russian fort built in the late 1790s to support Russian whalers.

Fig. 13.17 St. Nicholas Russian Orthodox Church, built in 1893 in Juneau, Alaska to serve a largely Tlingit congregation. Note the characteristic onion shaped steeple topped with a cross displaying three crossbars on this small eight-sided Russian Orthodox church. While many of the oldest Russian Orthodox churches erected in Alaska were octagonal, mimicking tent-shaped churches in Russia, this is the only one remaining. The wooden iconostas and large icons displayed inside the church were brought from Russia

Fig. 13.18 Holy Transfiguration of Our Lord Russian Orthodox Church (built in 1900–01) and cemetery on bluff overlooking village of Ninilchik, Alaska. The pear- or onion-shaped domes are surmounted by crosses displaying three cross-bars. Similar crosses are placed by the graves, many of which are surrounded by picket fences, and some of which display various mementos commemorating the deceased

13.5.3 Russian-American and Ukrainian-American Population Distribution

Persons of Russian ancestry constitute a relatively small proportion of Alaska's current population, even in the areas of greatest Russian colonial settlement. For example, the Kenai Peninsula Borough reported 2.7% Russian ancestry in 2000, while the Kodiak Island Borough reported 1.9% Russian. Five-year American Community Survey estimates for 2010–14 indicate these two boroughs' populations were 2.3% and 1.9% Russian, respectively. These figures underestimate the cultural influence of the Russians because many native Aleuts and Indians, while lacking Russian ancestry, demonstrate cultures that are partly Russified.

In 2000 the county that had the greatest proportion of its population having Russian ancestry was McIntosh County, North Dakota, 19.9% (Fig. 13.19). That county was adjoined by Logan County, which was 17.9% Russian, and McPherson County, South Dakota, which was 15.9% ethnic Russian. The American Community Survey for 2010–2014 estimated that Logan County, with 19.1%, had the largest Russian proportion, while second place McIntosh County was 17.1%, and McPherson County was 12.2%. In 2000 two additional counties in South Dakota and three in North Dakota reported between 10% and 15% Russian ancestry. The county with the greatest proportion of its population reporting Ukrainian ancestry in 2000 was Billings County, North Dakota (18.4%), which was also 6.5% Russian. American Community Survey estimates for 2010–14 reduced its Ukrainian population to 15.7%. Billings County is the only county nationally in which more than 3.5% of the population claimed Ukrainian ancestry.

Nationally, 0.9% of the U.S. population, 2,652,214 persons, reported Russian ancestry in 2000, while 892,922 or 0.3% of the respondents noted Ukrainian ancestry. American Community Survey 5-year estimates for 2010–2014 put the Russian ethnic population at 2,895,000 and the Ukrainian ancestry population at 963,549. The areas with the greatest proportions of their populations having Russian ancestry are clearly in the Northern Great Plains. While five counties within North Dakota and Minnesota, plus one borough in Alaska, reported populations that were at least 2% Ukrainian, most of the Ukrainian immigrants seeking to settle within the continent's grasslands settled farther to the north, particularly within the Prairie Provinces of Saskatchewan and Alberta, where distinctive Ukrainian ethnic landscapes can be seen (Lehr 1982, 1992).

While Russians have constituted the greatest proportion of the population in rather sparsely populated rural counties, the largest numbers of persons of Russian ancestry are found in the New York City metropolitan area, with 601,984

Fig. 13.19 Map showing the percentage of individuals who reported their ancestry on the 2000 Census indicating that Russian or Ukrainian was their primary or secondary ancestry. In much of the nation the pattern of Russian and Ukrainian ethnic settlement is similar, and in parts of North and South Dakota Russian-Americans dominate the ethnic heritage of certain counties, while Ukrainian-Americans are most prominent in nearby counties. Russian and Ukrainian ethnic settlement is more dominant across the border in the Canadian Prairie Provinces

Russian ethnics counted in 2000. Within New York City proper there were 243,015 ethnic Russians, and large numbers of recent Russian immigrants, the majority of whom are of Jewish ancestry, have settled in parts of Brooklyn. American Community Survey 5-year (2010–14) estimates for New York City were lower, with 222,779 Russian ethnics and 57,178 Ukrainians reported. Kings County, New York City's borough of Brooklyn, was home to 7.7% of all Russian ethnics in the United States and 5.9% of all Ukrainian ethnics.

13.5.4 Landscape Legacy of Russians in Alaska

While few buildings constructed by Russian colonists remain in Alaska or California, the vernacular Russian dwellings were log cabins. Wood or logs were the predominant building material in both the core of Russia and in nearby Siberia, and this tradition was brought to Alaska. Anatole Senkevitch comments that "most settlements in Russian America evoked the more austere architectural style of the eastern Siberian ostrog (stockaded town), whose log buildings reflected an expedient manipulation of form and detail" (Senkevitch 1986:80–81). Most of the Russian settlements in Alaska were "simple trading outposts…[that] contained the omnipresent Russian Orthodox church or chapel and an array of dwellings for the Russian *promyshlenniki*" (Senkevitch 1987:153). The Russian settlement at Novo-Arkhangel'sk, later renamed Sitka, was better planned, and "generally echoed those of most smaller provincial Russian towns in the first part of the nineteenth century" (Senkevitch 1987:184–5). The Russian government had established a centralized planning process for new military settlements within Russia proper, and Senkevitch notes "there is a marked resemblance between its rudimentary planning format and the basic layout discernible in Novo-Arkhangel'sk" (Senkevitch 1987:185), including the existence of a parade ground, a grid system, and the orderly placement of buildings.

The most conspicuous structures are the Russian Orthodox churches, whose design best evokes aspects of Russian colonial architecture. These churches are often, but not always, topped with onion-domed steeples. They are always marked by the distinctive orthodox cross with its three bars, the lowest of which is slanted. Some, such as the St. Nicholas Russian Orthodox Church at Eklutna, Alaska are of log construction, while many others Russian Orthodox churches, such as the ones in Kenai and Ninilchik (Fig. 13.18), are wood frame buildings.

The nearly 80 Russian Orthodox churches in Alaska have several distinctive interior characteristics. They feature an east to west orientation, with their sanctuary located in the eastern end of the church, the nave in the center, and the narthex in the western end. Separating the sanctuary from the nave is the iconostas, a wall or screen of icons or religious paintings (Williams 1997). The icons face the relatively small nave, where members of the congregation stand during worship services. The Russian Orthodox churches are typically built a short distance from the center of the villages on higher ground, and they are frequently brightly painted.

The onion-shaped dome atop the church tower marks not only Russian Orthodox communities, but also those of the Ukrainian Orthodox. Both Ukrainians as well as Russian Orthodox settlers also established communities in the Upper Midwest and the northern Great Plains.

13.5.5 Ukrainian Ethnic Landscapes in the Upper Great Plains

Large scale immigration from Ukraine occurred in the two decades preceding World War I. This immigration was largely to North Dakota, Manitoba, Saskatchewan, and Alberta, with the immigration to Canada much more important numerically and proportionately. Many Ukrainian immigrants deliberately sought to locate along the northern edge of the Great Plains. In Canada the Ukrainians settled in large ethnic blocks that became almost entirely Ukrainian, however the Homestead Act made it more difficult for Ukrainians to exclusively settle such large blocks in the American Plains. As John Lehr (1982:94) explains, "their strong desire to settle close to compatriots, friends, and kinfolk led to the formation of a series of large ethnically homogeneous block settlements that eventually spanned the West from southeastern Manitoba to central Alberta." Nine such blocks were colonized from Manitoba to Alberta, plus one block in western North Dakota.

While Ukrainian Orthodox numbers were numerically larger, Ukrainian Catholics from western Ukraine (Galicia) came to North Dakota via Winnipeg in 1896. The church architecture of the Ukrainian Catholics was very similar to that of the Ukrainian Orthodox, except for the substitution of the Catholic cross for the three-barred Orthodox cross. In addition, Ukrainian Protestants from near Kiev settled in north central North Dakota.

13.5.5.1 Ukrainian Religious Landscape

The Ukrainians had a distinctive religious landscape. As with the Russian Orthodox churches, Ukrainian Orthodox churches were typified by Russified Byzantine architecture. The two major features distinguishing Ukrainian churches were their onion-shaped dome and separate bell tower. Yet, "the distinctive form of the Ukrainian church domes, more pear-shaped than rounded, became so generally associated with the Ukrainians" (Lehr 1982:99–100). Furthermore, rural Ukrainian Orthodox churches "were distinguished by an eastward orientation,.. a Byzantine dome and cupolas, a separate bell tower and a tripartite division of interior space into porch, nave and sanctuary" (Lehr 1986:164). In addition, on both the Canadian Prairies and the Dakota Plains, churches were often the most prominent building in the community, or second in stature only to the grain elevator.

Graves in Ukrainian cemeteries are often marked with decorative, blacksmith-made iron crosses. Christopher Martin (1989:97) explains that such "iron crosses usually share the trefoil cross ends and a central, symmetrical element at the cross's intersection." However, it is the shape of the cross that is more distinctive than the material, inasmuch as both Czechs and Germans also occasionally used iron crosses as grave markers.

13.5.5.2 Ukrainian Ethnic Housing

The housing in Ukrainian ethnic communities reflected long-established patterns, even though the traditional village arrangement was often difficult to replicate in the New World because of American and Canadian homestead laws requiring the prospective landowners to build their houses on their homestead. Yet in their houses, many Ukrainian immigrants replicated the house of their Old World homeland (Fig. 13.20), shaping their cultural landscape. Typical characteristics of Ukrainian settler houses include "a southward orientation in a single-storey, rectangular house…[that] had two or three rooms, a central chimney, and a gable, hipped gable, or hipped roof" (Lehr 1982:97). The diagnostic characteristics of the Ukrainian ethnic house, particularly in Canada, were its shape (rectangular and single story), its log construction, its steep roof, and its plastered or limewashed exterior walls (Lehr 1982).

Ukrainian ethnic housing in western North Dakota displayed traditional post-and-earth construction, a style common in Galicia. These dwellings are described as "built with earthfast cedar posts, each about six inches in diameter and placed in the ground 3–4 feet apart, outlining perimeter and interior walls" (Martin 1989:88), with a mixture of earth, straw, and dung held in place by a willow laths connecting the upright posts. As with the Ukrainian ethnic houses in Canada's

Fig. 13.20 Abandoned farmhouse in Ukrainian settlement southwest of Hairy Hill, Alberta. The deterioration of the plaster covering of the walls provides a good view of their structure

Prairie Provinces, the exterior walls of these dwellings were covered "with two courses of mud plaster and limewashed to a bright white color" (Martin 1989:88). There were variants that had exterior clapboard. The earthen walls were an excellent insulator in severe winters. While these folk houses were found on rural farms, many have disappeared with farm abandonment or have been replaced by new buildings. Yet as John Lehr explains, "In most areas of Ukrainian settlement the white-sided cottages may still be seen scattered across the landscape" (Lehr 1992:330). Nevertheless, most of the traditional ethnic housing have been abandoned, converted to other uses, or modified, such as by covering the walls with siding. In urban areas Ukrainian ethnic houses are more typical of American or Canadian norms.

Ukrainian ethnics also constructed post and earth barns and farm buildings. These had an earth-fast frame and horizontal wall laths. The walls were packed with flax straw and some, originally, had roofs of flax straw (Martin 1989). Few such structures remain in North Dakota.

13.5.6 Russian Jewish Immigration Since the 1880s

Beginning in the 1880s large numbers of Russian Jews migrated to the United States. While some went to rural communities, far more went to large cities, and it is in urban areas where they made their most conspicuous impact upon the cultural landscape. Jews predominated among Russian immigrants, and nearly one-third of the entire Russian Jewish population came to the United States during this period to escape anti-Semitic pogroms (Sowell 1981). Russian Jews were particularly likely to arrive from Greater Poland and the adjacent area. While Poland had initially invited the Jews, Poland was later annexed into the Russian Empire, resulting in far harsher treatment of the Jewish population. Russian Jews were more likely to come from cities or certain towns than were Russian Christians. Likewise, Russian Jews immigrated to large urban areas, while Russian Christians were more likely to be farmers, yet large numbers also settled in New York City (Fig. 13.21).

Russian Jews were especially likely to migrate to New York City. The Lower East Side of Manhattan was particularly attractive around the beginning of the twentieth century, where Jews were employed in a wide variety of trades and professions. Many Jewish women were employed in garment manufacturing at that time. The arrival of large numbers of Jews from central and eastern Europe transformed the landscape of parts of New York City. Not only did they establish over 500 synagogues in the Lower East Side between 1880 and 1915, but they "created entire communities, with slaughterhouses, ritual baths, kosher restaurants, yeshivas, and the other aspects of the built environment of the transplanted ghetto or shtetl" (Williams 1997:71). The synagogues of the Lower East Side, unlike the larger synagogues constructed by German Jews in Midtown Manhattan, were typically organized by the Russian settlement from where the immigrants had come, and they varied greatly in size and grandeur, with

Fig. 13.21 Russian Orthodox Cathedral of the Transfiguration of Our Lord built 1916–22 in Brooklyn, New York. The church follows the design of Cathedral of the Dormition in Moscow, Russia. A New York City Landmark designation plaque on the side of the cathedral explains "The Greek cross plan and the copper-covered onion-shaped domes and octagonal belfries, all crowned by patriarchal crosses, are characteristic of the Russian Byzantine style." Note that the onion-shaped domes, surmounted by the crosses with three cross-bars, are far larger and more elaborate than what was displayed on the Russian Orthodox churches shown from Alaska

some being essentially storefronts (Kaplan 2015). Among the best known and still operating synagogues are the Bialystoker and Eldridge Street Synagogues.

As with many inhabitants of urban ghettos, now much of the Jewish population has relocated to many other sections of New York City and the surrounding metropolitan area. For example, the Lower East Side of Manhattan lost nearly 30% of its Jewish population since 2002, yet several large Jewish communities in Brooklyn have experienced rapid growth. Borough Park grew by 71% between 2002 and 2011, while Williamsburg increased by 41%. The total Jewish population in New York City was estimated to be 1.027 million in 1991 (Sheskin 1993), down about 90,000 from 1981. While the city's Jewish population had fallen below a million by 2002, it was estimated at 1,086,000 in 2011 (Beck et al. 2013). It had been 1.64 million in 1920. Nevertheless, in greater New York City area, including the city's five boroughs plus New York's Nassau, Suffolk, and Westchester Counties, the Jewish population is estimated at 1,538,000 (Beck et al. 2013). New York City retains the largest Jewish population of any city in the world.

Other major urban settlements of Russian Jews developed in the Los Angeles, California metropolitan area along with Miami-Dade and Palm Beach Counties, Florida. The Jewish population of these areas was so concentrated and prominent that Ira Sheskin (1993:121) contends that "at least three locations in the U.S. may be construed as forming Jewish 'ethnic homelands': the New York metropolitan area, South Florida, and southern California." He further notes that "it is through New York that the vast majority of American Jews trace their heritage" (Sheskin 1993:122). Indeed, the Jewish settlement of Miami Beach has long been so linked to New York City that some have referred to that community as the southernmost borough of New York City.

13.5.7 Characteristic Landscapes of Russian Jewish Neighborhoods

Several geographers have described these ethnic Jewish urban neighborhoods. Ira Sheskin writes

> it is also obvious that non-Jews view New York as an area of significant Jewish presence, as a Jewish ethnic homeland. From Yiddish, Hebrew, and Anglo-Jewish newspapers on the newsstands, to kosher and kosher-style restaurants, to Jewish names for many retail establishments in the metropolitan area (Abraham and Strauss, Lerner Shops, etc.), to hundreds of synagogues and Jewish institutions, Jews have clearly left their mark on the landscape (Sheskin 1993:123).

The presence of the synagogues, whether still serving Jewish populations or now accommodating the religious needs of later arriving ethnic groups, is a key artifact contributed by

Fig. 13.22 Eldridge Street Synagogue, located in the Lower East Side neighborhood of New York City's Manhattan Island. Dedicated in 1887, this synagogue was the first that was specifically designed and erected by members of the Polish and Russian Jewish community. Featuring a Moorish design and numerous Stars of David, this synagogue now houses the Museum at Eldridge Street

the Jewish population. Yet in their architectural style, the urban synagogues, and on Manhattan Island in particular, do not always reflect the ethnic landscape of the Jewish populations that used them. Within the Lower East Side of New York's Manhattan Island many of the Russian Jewish immigrants located their synagogues in structures built to house Christian churches or German synagogues, muting the ethnic imprint of the Russian Jewish population. Nevertheless, some built synagogues illustrating "the Rundbognestil Romanesque, the Gothic, or the Moorish styles, the latter featuring bulbous domes and keyhole-shaped doors and windows" (Williams 1997:71). Thus, it was often the presence of the synagogue (Fig. 13.22), rather than the style or architecture of the structure, that marked the ethnic landscape settled by these Jewish immigrants from Eastern Europe.

While not all Jewish populations were of Russian descent—some were Sephardic Jews from Spain and Portugal and some of the Ashkenazic Jews were of German descent—the Russian Jews were by far the most numerous. While the different groups settled their own neighborhoods, in the nineteenth century Russian Jews lived in downtown New York City, the area of Lower Manhattan, while German Jews lived uptown (Sowell 1981). It was the Russian Jews who established the largest and most persistent ethnic neighborhoods. The sociologist Thomas Sowell explains, "Their religious orthodoxy restricted their ability to work in factories at a time when factories often worked on Saturdays, the Sabbath observed by Orthodox Jews…[so] they needed to live and work among other Jews, to a far greater extent than German Jews" (Sowell 1981:83). Thus, urban Jewish ghettos usually reflect the ethnic imprint of Russian Jews. Yet other Eastern European Jewish groups, including those from what are now Hungary, Poland, Lithuania, Belarus, and Ukraine, have also made a distinctive ethnic landscape.

Fig. 13.23 Former Congregation B'nai Israel Kalwarie Synagogue in the Lower East Side neighborhood of New York City's Manhattan Island. Built in 1903, this building now serves the Sung Tak Buddhist Association, given the expansion of nearby Chinatown

The rapid growth of ultraconservative Orthodox Jewish communities, such as those associated with Hasidism or Hasidic Judaism, in Brooklyn's Borough Park, Crown Heights, and Williamsburg neighborhoods and in Rockland County, New York, has resulted in a far more pronounced impact upon those communities' ethnic landscapes than that created by Russian Jews—one that will be discussed in this book's final chapter.

Many Jewish ghettos in metropolitan areas have been replaced by other ethnic groups, as many Jews have relocated to better neighborhoods or to the suburbs. Indeed, by "the 1970s, only New York City had Jews still living in the original ghetto neighborhoods" (Mesinger and Lamme 1985:155). Yet even after Jewish populations have relocated, reminders of their Jewish past remain (Fig. 13.23). Indeed, the "architectural cultural landscape created by Jewish communities in their old neighborhoods, if it survived the wrecking ball of urban renewal, was adapted by the new residents for their own needs" (Mesinger and Lamme 1985:155). Thus, today one can find Christian churches in African-American neighborhoods, or even a Buddhist temple in a Chinese district, occupying former synagogues marked by "the Star of David, the Tablets of the Ten Commandments, or some Hebrew inscriptions" (Mesinger and Lamme 1985:155). The impressive synagogues that are found in other large cities, such as Chicago and Cincinnati, herald the major Jewish populations in those cities. Yet within many smaller cities, particularly small cities within the South, abandoned or nearly abandoned synagogues are all that remain of once vibrant Jewish populations. Their Jewish populations were decimated when the children left to seek opportunities in larger metropolitan areas.

Significant growth has occurred within the Jewish communities of South Florida. While the Jewish population of Miami-Dade County, centered upon South Beach, rose rapidly in the 1950s through the mid 1970s, it has now declined as many of the first-generation Jewish-American retirees living in its high-rise apartments have died. At the same time the Jewish population of Broward and Palm Beach Counties has rapidly grown. Ira Sheskin (1992:54) explains, "by 1970, low-rise retirement complexes were being developed in Broward and Palm Beach Counties, retirement complexes that looked more suburban," in keeping with their middle-class background. Yet other than the presence of synagogues and Jewish restaurants and shops in their neighborhoods and the relative absence of churches, little else on this recently developed landscape speaks of the residents' ethnicity.

Recent Russian Jewish immigrants have generally located near already established Jewish communities within major metropolitan area, such as New York City, Chicago, Boston, Pittsburgh and the San Francisco Bay area. Several neighborhoods in the New York City have become overwhelmingly Russian, including the Brighton Beach community, with its Russian restaurants and grocery stores (Fig. 13.24), resplendent with signs and advertisements in Russian.

Fig. 13.24 Grocery store with Russian language signage along Brighton Beach Avenue in New York City's borough of Brooklyn. To the left of the grocery one can see a café with the Russian lettering "КАФЕ". Numerous shops and service providers, from bakeries, shoe stores, meat markets, to pharmacies, tout their services in Russian within this community

In contrast, Russian Baptists and Pentecostals who have arrived since the 1980s have populated other communities, with the Sacramento, California area proving attractive. Their arrival has begun to reshape the cultural landscape. As Susan Hardwick explains, "More than a dozen new Russian and Pentecostal churches have been added to the three historic churches already in place before the arrival of the newest Russians" (Hardwick 2002:261). Russian houses are far more often surrounded by fences and locked gates than those of other immigrant groups within the same neighborhoods, and they are likewise more often to be surrounded by hedges (Hardwick 2002).

13.6 Immigration of Southern Slavs

Slavic populations also immigrated from the lands of former Yugoslavia, a politically troubled region long known for conflict between its Christian Slavic and Moslem populations. While relatively few Southern Slavs immigrated to the United States until recently, the upheavals in the Balkans beginning in the 1980s has resulted in many political refugees fleeing the region, some of whom now reside in the United States. Nevertheless, the number of Southern Slaves in the United States is miniscule in comparison with the over 15 million Americans who claim Czech, Slovak, Polish, Russian, or Ukrainian ancestry. Indeed, the 2000 Census reported 113,661 Americans of Albanian ethnicity, 374,241 persons of Croatian ancestry, 140,337 individuals of Serbian ethnicity, and 328,547 of Yugoslavian heritage. The 5-year American Community Survey for 2010–2014 estimated the presence of 191,021 Albania ethnics, 412,488 Croatians, 186,122 Serbians, and 296,627 Yugoslavian ethnics, an increase of over 100,000 among these four groups. Given their sizes, relatively little has been observed or written about their ethnic landscapes in the United States. Nevertheless, recent study has occurred within a Bosnian refugee community that established itself within St. Louis, Missouri. Given that this population, which has Slavic origins, was uprooted because of its Moslem beliefs, it is discussed in Chap. 15.

13.7 Précis

Immigrants coming from the Slavic lands of Eastern Europe have created a variety of ethnic landscapes within the United States. While the material culture of some, such as the Czechs, illustrates a long association with German neighbors, others much more distinctively show the influences of eastern orthodoxy. There is considerable religious diversity amongst people from Eastern Europe, and a discussion of the Poles without mention of the strong influence of Roman Catholicism upon the religiosity of their landscape would be remiss, just as would a review of the significance of Judaism among immigrants from Russia. Religious conflict has inflamed the lands of the former Yugoslavia, creating refugees from among the Bosnian Muslims, Croatian Catholics, and Serbian Orthodox, among other religious and ethnic minorities. While this area is sending fewer refugees to the United States now than it did a decade ago, the arrival of Southern Slavs simply adds to the ethnic heterogeneity of the American population.

References

Allen, James Paul, and Eugene James Turner. 1988. *We the People: An Atlas of America's Ethnic Diversity*. New York: Macmillan Publishing Company.

Beck, Pearl, Steven M. Cohen, Jacob B. Ukeles, and Ron Miller. 2013. *Jewish Community Study of New York: 2011, Geographic Profile*. revised ed. New York City: UJA-Federation of New York.

Eckert, Eva. 2002. From Moravia to Texas: Immigrant Acculturation at the Cemetery. *Markers* 19: 174–211.

Goren, Arthur A. 1980. Jews. In *Harvard Encyclopedia of American Ethnic Groups*, ed. Stephan Thernstrom, 571–598. Cambridge (MA): Harvard University Press.

Gyrisco, Geoffrey M. 1997. Victor Cordella and the Architecture of Polish and East-Slavic Identity in America. *Polish American Studies* 54 (1): 33–52.

———. 2001. Polish Flats and Farmhouses: What Makes a House Polish? *Wisconsin Magazine of History* 84 (3): 22–33.

Hardwick, Susan W. 1992. Russian Religious Settlement along the Pacific Rim. In *Geographic Snapshots of North America*, ed. Donald G. Janelle, 223–226. New York: Guilford Press.

———. 2002. Russian acculturation in sacramento. In *Geographical Iden tities of Ethnic America: Race, Space, and Place*, ed. Kate A. Berry and Martha L. Henderson, 255–278. Reno: University of Nevada Press.

Kaplan, Paul M. 2015. *Jewish New York: A History and Guide to Neighborhoods, Synagogues, and Eateries*. Gretna: Pelican Publishing.

Kenny, Judith T. 1997. Polish Routes to Americanization: House Form and Landscape on Milwaukee's Polish South Side. In *Wisconsin Land and Life*, ed. Robert C. Ostergren and Thomas R. Vale, 263–281. Madison: University of Wisconsin Press.

Kolinski, Dennis L. 1994. Shrines and Crosses in Rural Central Wisconsin. *Polish American Studies* 21 (2): 33–47.

Krase, Jerome. 1997. Polish and Italian Vernacular Landscapes in Brooklyn. *Polish American Studies* 54 (1): 9–31.

Lehr, John C. 1982. The Landscape of Ukrainian Settlement in the Canadian West. *Great Plains Quarterly* 2 (2): 94–105.

———. 1986. Ukrainians. In *America's Architectural Roots: Ethnic Groups That Built America*, ed. Dell Upton, 160–165. Washington: Preservation Press.

———. 1992. Ukrainians in Western Canada. In *To Build in a New Land: Ethnic Landscapes in North America*, ed. Allen G. Noble, 309–330. Baltimore: Johns Hopkins University Press.

Martin, Christopher. 1989. Skeleton of Settlement: Ukrainian Folk Building in Western North Dakota. In *Perspectives in Vernacular Architecture, III*, ed. Thomas Carter and Bernard L. Herman, 86–98. Columbia: University of Missouri Press.

Mesinger, Jonathan S., and Ary J. Lamme III. 1985. American Jewish ethnicity. In *Ethnicity in Contemporary America: A Geographical Appraisal*, ed. Jesse O. McKee, 145–168. Dubuque: Kendall/Hunt Publishing Co.

Murphy, David. 1986. Czechs. In *America's Architectural Roots: Ethnic Groups That Built America*, ed. Dell Upton, 112–117. Washington: Preservation Press.

Radzilowski, John. 1997. Polish-American Architecture in Comparative Perspective. *Polish American Studies* 54: 5–8.

———. 2005. *Poles in Minnesota*. St. Paul: Minnesota Historical Society Press.

Rau, John E. 1992. Czechs in South Dakota. In *To Build in a New Land: Ethnic Landscapes in North America*, ed. Allen G. Noble, 285–306. Baltimore: Johns Hopkins University Press.

Rucker, Della G. 1995. *A Proud Heritage: History of Czech Settlements Kewaunee and Manitowoc Counties, Wisconsin, U.S.A.* Sheboygan Falls: Wisconsin's Ethnic Settlement Trail.

———. 1996. Traces of Czech Heritage Along Regional Roads. *Voyager: Northeast Wisconsin's Historical Review* 12 (2): 42–47.

Senkevitch, Anatole Jr. 1986. Russians. In *America's Architectural Roots: Ethnic Groups That Built America*, ed. Dell Upton, 80–85. Washington: Preservation Press.

———. 1987. Early Architecture and Settlements in Russian America. In *Russia's American Colony*, ed. S. Frederick Starr, 147–195. Durham: Duke University Press.

Sheskin, Ira M. 1992. The Miami Ethnic Archipelago. *The Florida Geographer* 26: 40–57.

———. 1993. Jewish Metropolitan Homelands. *Journal of Cultural Geography* 13 (2): 119–132.

Sowell, Thomas. 1981. *Ethnic America: A History*. New York: Basic Books.

Tishler, William H. 1984. Built From Tradition: Wisconsin's Rural Ethnic Folk Architecture. *Wisconsin Academy Review* 30 (March 1984): 14–18.

United States Department of Homeland Security. 2014. *Yearbook of Immigration Statistics: 2013*. Washington: U.S. Department of Homeland Security, Office of Immigration Statistics.

Williams, Peter W. 1997. *Houses of God: Region, Religion, and Architecture in the United States*. Urbana: University of Illinois Press.

Wolniewicz, Richard. 1997. Comparative Ethnic Church Architecture. *Polish American Studies* 54: 53–73.

Mediterranean Landscapes: Italians and Greeks in America

14

While an Italian sailing for Spain is heralded as the European discoverer of America, with the notable exception of the Spanish who are described in Chaps. 5 and 6, few immigrants arrived in the United States from the other lands along the Mediterranean Sea until late in the nineteenth century. In most of the United States Italian immigrants arrived after much of the urban landscape had been initially fashioned and the rural landscape dominated by earlier arriving Europeans. Thus, Italians occupied neighborhoods in Boston that had been earlier shaped by the Irish, who themselves had replaced the English. The situation was similar in New York City, but on a grander scale. In contrast, Italian immigrants within parts of California arrived sufficiently early—given that the state was settled far more recently than New England and the Mid-Atlantic states—to play a critical role in shaping the development of California's wine industry.

14.1 Cultural Influence of Mediterranean Peoples

Historically the entire area around the Mediterranean Sea was part of the Roman Empire, yet today the Mediterranean population is highly diverse, representing a variety of languages and religions. Spanish, Catalan, French, and Italian are all Romance languages, derived from Latin, while Arabic is spoken across a wide area along the southern shore of the Mediterranean, running from Morocco to Syria. The eastern Mediterranean also includes numerous other languages, particularly Greek, Turkish, and Hebrew. Spain and Italy have long been the centers of Roman Catholicism. Greece is home to Eastern Orthodox Christians, as was much of the coastal area of Turkey until the early twentieth century. Now most of Turkey is Moslem. Lebanon and western Syria are the home to followers of a variety of Christian and Islamic groups, among others, while Israel and Palestine include Jewish, Christian, and Moslem populations. North Africa, while predominately Islamic, includes a significant Coptic Christian minority in Egypt's Nile Delta, and until the middle of the twentieth century, Jewish quarters were found in most of the larger cities of North Africa, surrounded by far larger numbers of Moslems. This chapter is devoted to the discussion of the ethnic landscapes created in the United States by immigrants from those mostly Christian lands located north of the Mediterranean Sea, while Chap. 15 describes the landscapes shaped by immigrants from the predominately Moslem lands of the eastern Mediterranean and North Africa.

The architectural and landscape influence of the Roman Empire can be seen not only in many archaeological ruins scattered throughout the lands surrounding the Mediterranean, but Roman structures remain farther north in parts of England, Germany, and Hungary. Roman culture clearly had an influence on the vernacular landscape that evolved in those lands that were once under Roman rule, particularly those lands north of the Mediterranean Sea. In addition, both Classical Roman and Greek culture strongly influenced formal architectural styles that were incorporated into the design of public buildings and religious structures—and also the architect-designed homes of the wealthy—in the United States during the late eighteenth and early nineteenth centuries (Fig. 14.1). Because prominent architects sought to replicate the styles of these classical civilizations in the public buildings of the new American republic, the architecture of the United States contains numerous banks, churches, schools, and courthouses in Greek Revival style. It has been argued that its popularity can be attributed to "the often expressed sentiment that America, with its democratic ideals, was the spiritual successor of ancient Greece" (Poppeliers et al. 1983:36). Thus, architectural styles showing Mediterranean influences have been brought to America by immigrants and introduced by professional architects, reproducing both millennium-old designs as well as more contemporary Italianate styles.

J.A. Cross, *Ethnic Landscapes of America*, DOI 10.1007/978-3-319-54009-2_14

Fig. 14.1 House in Eufala, Alabama. The presence of the classical Doric columns in this nineteenth century house displays architectural links to classical Greece and Rome

14.2 Classical Greek and Roman Architecture

Early in the history of the United States, architects designed many buildings in the Early Classical Revival style, which was subsequently replaced by Greek Revival architecture. While the majority of dwellings and commercial structures continued to be built in other ethnic or vernacular styles, larger buildings and plantation houses were often designed by architects in these styles. Although some of these architects had direct training with Italian and Greek architecture and had studied classical ruins, many others were schooled in the works of English architects who had incorporated elements of Classical styles that were displayed in structures built following the Great London Fire of 1666. Thus, one can see similarities between London's domed St. Paul's Cathedral and the U.S. Capitol Building in Washington, yet these structures were not displaying British architectural traits. They were reintroducing earlier classical traditions.

Common features of Classical Revival architecture include a front porch, with the "porch roof usually supported by four simple columns [of] Roman Doric or Tuscan types" (McAlester and McAlester 1990:169). While the actual number of columns could range from two to ten, with four being most common, their conspicuous position centered at the front of the building is a key diagnostic characteristic of this style. Smooth Tuscan columns are most common (Fig. 14.2), while various styles of fluted columns are also seen.

The greatest numbers of Classical Revival buildings were constructed within the Southern states (McAlester and McAlester 1990). Thomas Jefferson was a major champion of the style, promoting its usage in the Virginia State Capitol in Richmond (Fig. 14.3), his mansion at Monticello, and at the University of Virginia, for which he served as founder and architect. As William Pierson (1970:212) explains, given that Thomas Jefferson was "consciously seeking a mode of building symbolic of the new republic, he discovered an appropriate idiom in two non-English sources, the Neoclassical architecture of contemporary France, and the architecture of ancient Rome." Drawing from this classical heritage, the University of Virginia's Rotunda was patterned after the Roman Pantheon. Yet the rigid adherence to classical Roman designs, which characterized Jefferson's buildings, differed from that of the more widely adopted Greek Revival style, which "flourished at every level of American society" (Pierson 1970:214).

Greek Revival architecture more strongly illustrated folk influences that modified classical traditions, resulting in a style that displayed significant regional variation (Pierson 1970). It has been observed that "Greek Revival was the dominant style of American domestic architecture…from about 1830 to 1850" (McAlester and McAlester 1990:182), particularly within those regions that were settled at that

Fig. 14.2 Market Hall in Charleston, South Carolina. This Greek Revival building, constructed in 1841, displays conspicuous Tuscan columns

Fig. 14.3 State Capitol building in Richmond, Virginia. Designed by Thomas Jefferson in Classical Revival style, the architecture of many public buildings reflected notions adapted from the Roman Empire and Greek ruins

Fig. 14.4 Greek Revival house in Ithaca, New York. Note the Ionic columns span the entire front of the narrow house

time. Numerous examples remain in upstate New York (Fig. 14.4) and Pennsylvania, throughout the Midwest, the Upland South, and the Mississippi Valley.

Regional variations in Greek Revival houses have been identified, particularly with respect to the size, height, and gabling of the front porch, yet these structures shared certain characteristics. These include having low-pitched gabled roofs (although occasionally hip roofs are also seen), a cornice along the edge of the roof distinguished by a wide trim, and usually the presence of a front porch, "supported by prominent square or rounded columns, typically of Doric style" (McAlester and McAlester 1990:179). Frequently, a narrow series of transom windows surround the front door. Not only were some Greek Revival houses designed by architects, but many builders relied upon a variety of published pattern books, permitting them to customize the structures to local tastes.

The front-gabled house was one of the lasting legacies of Greek Revival architecture. As McAlester and McAlester (1990) note, the "gable front and wing" house was a popular house built in rural America until the Great Depression. While a doorway entered the gabled end of these houses, the porch supported by columns might either front the gabled end of the house or be located along the side wing, yet still facing forward. While such structures displayed modified classical elements that originated along the Mediterranean, they hardly represent the type of vernacular architecture associated with ethnic communities composed of immigrants from that region.

14.3 Contemporary Italian- and Greek-American Population Distributions

Italian-Americans are by far the most numerous ethnic group from the Mediterranean when persons of Spanish ancestry are excluded. Italians were a prominent component of the last major wave of European immigrants before the United States placed severe restrictions on immigration following World War I, with Italians accounting for 25.4% of European immigrants arriving in the U.S. during the first decade of the twentieth century (Daniels 2002). The five-year American Community Survey conducted between 2010 and 2014 showed that 17.3 million Americans claimed Italian ancestry. At the time of the 2000 Census nearly 14 persons reported Italian ancestry for every individual indicating Greek ancestry. Italian-Americans also outnumbered Arab-Americans by a similar ratio, and many of the Arab ethnics came from countries that do not touch the Mediterranean Sea. With a few exceptions, ethnic communities of all of these ancestries are relatively recent. Italian-Americans are found in large numbers within only a few major metropolitan areas.

14.3.1 Italian-Americans

One quarter of all Italian-Americans are found in just two states, New York and New Jersey (Fig. 14.5). Italian-Americans comprised 8.7% of New York City's

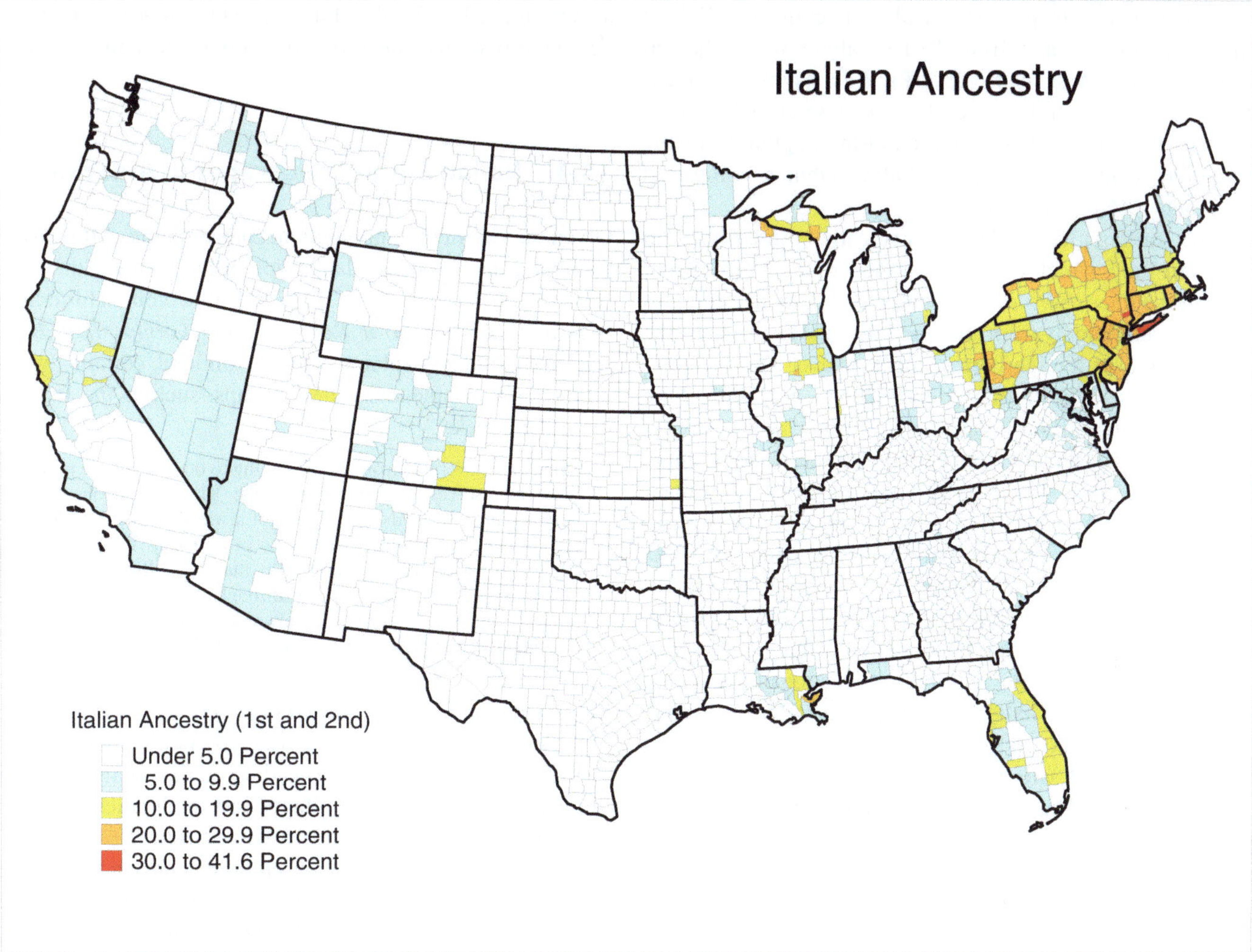

Fig. 14.5 Map showing the percentage of individuals who reported their ancestry on the 2000 Census claiming that Italian was their primary or secondary ancestry. Note that those counties with the greatest percentages of their populations being Italian are located in a broad swath from eastern Massachusetts to southern New Jersey, including the Boston and New York City metropolitan areas

population in 2000, or a total of 692,739 persons. Richmond County, which comprises New York City's Staten Island, had the greatest proportion of its population claiming Italian ancestry of any county in the nation, 37.7%. On Long Island, Nassau County and Suffolk County, New York were 23.9% and 28.8% Italian ancestry, respectively. To the north, Putnam County, New York is 31.8% Italian, while three other New York counties are over one-fifth Italian-American. In New Jersey Ocean County was 25.3% Italian ancestry and six additional counties were between 20% and 25% Italian-American in 2000. Three of Connecticut's eight counties were over 20% Italian ethnic, just as was one of Rhode Island's five counties. In Pennsylvania both Lackawanna and Lawrence Counties were over 20% Italian-Americans. Although accounting for smaller proportions of the total population, large numbers of Italian ethnics also reside within the Boston metropolitan area.

Only one other county outside New England and the Mid-Atlantic states had as many as 20% of its population claiming Italian ancestry, and that was St. Bernard Parish, Louisiana (with 20.0%, before its population exodus caused by Hurricane Katrina). Two nearby parishes were reported as being between 10% and 12% Italian ancestry in 2000. Sicilians had flocked to New Orleans beginning in the mid nineteenth century, establishing "Little Palermo" in the lower French Quarter. Like Italian-Americans in several northern cities, their wealthier descendents moved to other parts of the city and into the suburban parishes (Capanella 2006). While Italian-Americans are scattered across the nation, comprising 5.6% of all Americans, few counties outside a swath of states ranging from Ohio through Pennsylvania and New York into New England have even one-tenth of their populations being of Italian ancestry. Those areas that were proportionally more Italian in 2000 included three counties in the iron mining region of Michigan's Upper Peninsula (Dickinson, 17.3%; Gogebic, 14.2%; and Iron, 13.1%) along with adjacent Iron County, Wisconsin (19.2%). Las Animas County, Colorado (14.2% Italian ancestry) likewise attracted many Italians to work in its mines around the beginning of the twentieth century. Elsewhere, Italian ethnics comprised less than 10% of the population.

The earliest Italian immigrants came primarily from northern Italy, which developed a modern economy earlier than the southern parts of the country. A century ago foreign-born Italians, largely from northern Italy, had grown to be the largest group of foreign-born residents in over a dozen counties in northern California (Helzer 2002). Yet, with these exceptions and in San Francisco, California, where significant numbers of Italians from both the northern and southern provinces settled (Cinel 1982), the majority of Italian-Americans nationwide trace their heritage to the various provinces in southern Italy. Italy was only united in 1870, just a little over a decade before large-scale immigration of Italians began in the 1880s, thus initially most Italian immigrants identified with their home region of Italy, rather than the more unified nation. Culturally, educationally, and economically southern Italians and northern Italians differed. The better educated and more prosperous northern Italians contrasted sharply with the impoverished agricultural workers living in the villages of southern Italy (Sowell 1981).

14.3.2 Greek-Americans

Greek ethnics coming to America were more likely to immigrate from Mediterranean lands outside of Greece in which they were minorities, including Turkey, the Balkans, and Egypt, rather than from Greece proper. Greek ethnics, being far less numerous than Italians, comprised only 0.4% of the population of the United States in 2000, and a decade later they were estimated to total 1,301,000 by the American Community Survey. Greeks accounted for more than 2% of the population in only three counties in 2000, Carbon County, Utah (3.0%) and Sussex County, Massachusetts (2.2%), and Queens County, New York (2.0%), the easternmost borough of New York City. While some of their enclaves are well known for their Greek heritage, such as Tarpon Springs, Florida, where Greeks from the Greek Islands established its sponging industry (Daniels 2002), even there persons of Greek ancestry accounted for only 11.8% of the city's population. Within Pinellas County, which includes Tarpon Springs, Greek ethnics comprised only 1.4% of the population, while Greeks were 1.3% of Pasco County, immediately to the north. Cook County, Illinois, which includes Chicago, reported that 1.1% of its population was of Greek ancestry in 2000, but it had more Greeks than any other county in the nation, with 51,500 estimated by the American Community survey for 2010–2014. Thus, not surprisingly, an ethnic neighborhood just west of downtown Chicago is called Greektown.

14.4 Italian Ethnic Landscapes in America

Given the size of their population, contributions of Italians in shaping a distinct ethnic landscape in the United States are relatively underappreciated. Indeed, Dell Upton, in his edited *America's Architectural Roots: Ethnic Groups that Built America*, simply explains why he omitted certain "well-known and conspicuous ethnic groups, such as the Italians [and] the Greeks."

> "There are several reasons, some having to do with ethnic traditions themselves and more with the state of scholarship. Architecture and landscapes have not always seemed the most appropriate or desirable expressions of ethnicity. Large urban ethnic groups evidently built little that was distinctive but instead expressed their ethnicity through language, food customs, religion and social organizations" (Upton 1986: 14, Reprinted by permission of John Wiley and Son).

Most Italian immigrants arrived in America after the closing of the frontier, thus fertile well-watered lands were no longer available for farmers to homestead. Even though most Italian immigrants came from the villages and small cities of southern Italy, relatively few Italians settled farming communities, with the exception of agricultural lands in California that attracted northern Italians, although several planned agricultural settlements brought Italian fruit and vegetable farm workers to Louisiana, Texas, and Tennessee (Allen and Turner 1988). Instead, the typical Italian immigrant settled into established cities, crowding into ethnic neighborhoods.

The region of Italy from which Italians came strongly influenced where they located in the United States. As Thomas Sowell explains,

> "Italians in New Orleans were mostly Sicilians…, while Italians in San Francisco—and in California in general—were mostly from northern Italy. Neapolitans and Calabrians became open-pit miners in Minnesota, immigrants from Turin created an agricultural settlement in North Carolina, and Piedmontese settled in the town of Genoa, Wisconsin" (Sowell 1996:163).

More so than most other European immigrants, many Italian immigrants came to the United States to work as sojourners, eventually returning to their homeland. Furthermore, "Italians seldom made up a majority in any large-sized neighborhood in American cities" (Sowell 1981:111). Nevertheless, this did not prevent the establishment of "Little Italy" districts in many large cities, including New York City, Boston, Providence, and San Francisco, among others, although in certain communities, such as New Orleans, where the immigrants came overwhelmingly from Sicily, the neighborhood label named the area of Italy from which the immigrants came.

14.4.1 Urban Landscapes of Italian-Americans

Italian neighborhoods are rarely all-Italian, often including other ethnic groups. Italian-American urban areas are often marked by a higher population density than found among neighborhoods settled by other European immigrants. Italian immigrants often found "opportunities were in highly congested urban areas. Interestingly, dense environments were not unusual for Italian immigrants who left small, but highly populated villages and towns scattered throughout the countryside" (Krase 1990:177). Furthermore, their urban neighborhood had often previously been home to earlier arriving immigrants.

Because the Italian immigrants were typically not the initial occupants of their neighborhoods, occupying ghettos established by previous immigrants, they were not fully free to shape the cultural landscape, inasmuch as much of its architectural form had already been shaped. Nevertheless, they impressed their cultural values upon the community. As Jerome Krase (1997:17) explains, "Since most immigrants came to already built up urban spaces and lack the power to radically alter their environment, how they use and interpret the new environment is often more relevant to how they physically change it." Italian-American communities valued privacy for the family, and often segregated the sexes, with public spaces being predominately male. Because of their desire for privacy, Italians "build walls, not merely fences, and from the street one is greeted by drawn curtains, pulled shades, and closed venetian blinds" (Krase 1997:20). Within densely settled neighborhoods, Italian-American homes have little, if any, space for lawns or trees before their houses. Thus, in Brooklyn, New York, in keeping with "the penchant of Italians to architecturally *fa una bella figura* (to make a good appearance), the fronts of houses are frequently covered with varieties of stone or stone facing as indications of affluence and success" (Krase 1997:24). Italians also utilize decorative iron work on their houses and businesses more so than many other ethnic groups.

San Francisco has long been noted for its Italian neighborhoods, yet those of Boston (Fig. 14.6) and New York City (Fig. 14.7) are also famous. Within New York City both East Harlem and Bensonhurst have long been ethnic Italian neighborhoods, although the arrival during the past couple of decades Chinese, Russian, and Hispanic immigrants has diluted the proportion of the population that is Italian. Smaller in scale, the Hill district of St. Louis, Missouri promotes its Italian heritage, just as does the Federal Hill neighborhood in Providence, Rhode Island. Within "Little Italy" one finds both commercial and religious edifices that promote the community's Italian ethnicity. Indeed, when the Italian-American share of the population within New Haven, Connecticut's Italian-American neighborhood fell with gentrification bringing in wealthier, better-educated professionals, "the Italian Americans remade part of the neighborhood

Fig. 14.6 Italian-Americans now dominate the historic North End neighborhood of Boston, Massachusetts. Although settled during colonial times, Italians display their heritage not through the architecture of the buildings, but through the restaurant advertising signage that overwhelmingly lists Italian names

Fig. 14.7 Italian restaurants are the most conspicuous Italian ethnic feature remaining within the Little Italy neighborhood of the Lower East side area of New York City's Manhattan Island

into an 'Italian place' by a concentration of Italian businesses in one street" (Blokland 2009:1596). In both that city and Providence, Rhode Island archways were erected to draw attention to their Italian heritage.

Restaurants are one of the most common and most conspicuous draws of outsiders into Little Italys. Not only are these establishments marked by Italian names and sometimes descriptions of their offerings (Fig. 14.8), but proprietors sometimes endeavor to recreate the ambience of outdoor cafes (Fig. 14.9). While the Italian enclaves in the United States are a century old, the one in Toronto, Ontario has developed over the past half century, and it more clearly shows how the Italian-Canadians have reshaped their landscape. There "shopkeepers and proprietors could choose from a large set of Italian immigrant traditions of vernacular architecture …[including] arches, balconies, balustrades, columns, gardens, cantinas, and statuary for domestic architecture, as well as wrought iron, stucco, tile and marble that are used in both domestic and commercial buildings" (Buzzelli 2001:579–580). Their businesses primarily used "stucco, tile, and marble to change the streetscape" (Buzzelli 2001:580), in ways that are similar to what is seen in the United States.

The French Quarter of New Orleans began attracting Sicilians before the Civil War, and early in the twentieth century it was estimated that a third to a half of the district's residents were Italian, with the Italians dominating the marketing of fruit and vegetables at the French Market (Campanella 2006). While relatively few Italian ethnics live in what was "Little Palermo" today, their imprint upon the landscape remains. Italian ethnics still operate some of the stalls in the French Market, and Italian restaurants and groceries remain nearby. Furthermore, as Richard Campanella (2006:330) explains, Italian heritage is visible in the area's "beautiful tile doorway mosaics, the Italian names in pediments and palimpsests, the white tile interior and facades of certain buildings, and from the bustle of the French Market, where certain signage uses the tricolor of the Italian flag." In addition the Unione Italiana building, the mutual aid society formed in 1912, remains, as do several churches that specifically catered to the Sicilian immigrants.

In many cities Italian-American urban neighborhoods have been more persistent than those established by many other ethnic groups, who unlike the Italian-American, were more likely to relocate to suburban areas over time. Jerome Krase (1990:171) explains that Italian-Americans are better able "to defend themselves both symbolically and territorially." Strong attachments between Italian culture and specific territory were noted in both Italy and Italian-American communities. In some American cities in the aftermath of white flight resulting from the expansion of African-American neighborhoods, only Little Italy remained as a viable white community, only to be later threatened by gentrification (Krase 1990). Nevertheless, many Italian-Americans have joined the movement to the suburbs. For example, within New York City's Bensonhurst neighborhood, still far more authentically Italian than that city's Little Italy, increasing

Fig. 14.8 Italian restaurant in Providence, Rhode Island that touts its cuisine from a village a dozen miles outside of Rome, Italy. Note the mural painted on the wall of the building adjacent to the restaurant

Fig. 14.9 Italian outdoor cafe along Atwell Street in Providence, Rhode Island. Note the murals on the building walls, the decorative plantings, and the fountain in Despasquale Square. That sculpture, LaPigna or the pinecone, represents the Italian symbol of welcome, abundance and quality. A similar sculpture hangs from a archway several blocks distant

numbers of houses are sold to Chinese, Russian, or other ethnic Americans "as the Italians retire or their children marry and leave home" (Berger 2007:207). In San Francisco, "the old 'Italian quarter' with its numerous restaurants, nightclubs, bakeries, bookstores, cafes, churches, and welfare agencies still sporting Italian names has now turned into a kind of Italian-style Chinatown" (Ficheria 2011:156). Much the same has occurred within New York City's "Little Italy" on Manhattan's Lower East Side. Similarly, Chicago's Chinatown has expanded into that city's Italian neighborhood, and one can see Chinese guardian lions, or foo dogs, flanking the entrance to its St. Therese Catholic Church. Built to cater to the needs of an Italian congregation, the church now offers masses in English, Cantonese, Indonesian, and Mandarin.

14.4.2 Italians Shape California's Winery Landscape

One of the few areas in the nation where Italians played a major role in the shaping of a rural agricultural landscape is within the valleys of northern California. There Italian families helped develop the viticultural landscape. Italians brought both a familiarity with winemaking and "the inside track to a booming immigrant wine and grapes market" (Ficheria 2011:82). Italian immigrants established the largest vineyards and wineries, and they encouraged Italian peasants to immigrate, clear the land, and plant and tend the vines. There are geographic and climate similarities between California and Italy that "allegedly functioned as the basis for the successful transplantation of Old World wine culture and expertise from Langhe and Monferrato to the west coast of the United States" (Cinotto 2012:9). Nevertheless, Cinotto's recent history of the Italian dominance of the California wine industry is far more nuanced. He contends that racism provided northern Italian immigrants with advantages over southern Italians and conspired to exclude certain ethnicities from the industry while simultaneously providing employment to Italian immigrants. Indeed, it is argued that social and cultural capital associated with Italian ethnicity—particularly that from the Piedmont—enabled Italian winemakers in California to come to dominate today's industry (Cinotto 2012). Furthermore, the resemblance of California to the Mediterranean landscape had been promoted to prospective settlers since the mid-1800s. Nevertheless, that landscape which took on its Italian appearance with extensive vineyards and fields of crops introduced from Italy was not entirely natural, as the "immigrants had worked long and hard to transform this area into a suitable site for grapevine cultivation" (Cinotto 2012:53).

More recently tourism associated with the wine industry has highlighted the Italian ethnicity associated with both California's winemaking and the ethnic structures, including churches and winery buildings. While enclaves of Italian-Americans are scattered throughout northern California, many which date from the period of the Gold Rush, the wineries in Sonoma County, south of San Francisco, began to market their Italian heritage in the 1980s. As Jennifer Helzer (2001:54) explains, with this action "Sonoma County's Italian wineries were reanimated with ethnic heritage and tradition resulting in even more intense Cal-Ital landscapes." Thus, these wineries are seen promoting their Italian heritage (of their owners, more than their employees) by constructing or preserving winery buildings in Tuscan style, growing various Italian grapes, displaying olive trees and Italian flags, and offering numerous Italian foods. Indeed, as Cinotto (2012:235) concludes, "today's Piedmontese American winemakers use their immigrant heritage as symbolic capital to generate greater market value for their products." Using Italian heritage or culture to promote wineries can also be seen in other parts of the United States (Fig. 14.10).

14.4.3 Religious Landscapes in Italian-American Communities

Relative to their population, Italian immigrants built fewer churches than other ethnic groups and invested less of their earnings and labor into erecting structures that were both a physical and social focus of their communities. Nevertheless, examples, such as St. Ambrose Church in St. Louis, Missouri exist, where Italian ethnics not only contributed significant funds and labor to construct their parish church, designed to resemble a Lombard Romanesque cathedral, but vied to ensure that the interior embellishments honoring the patron saint of their region of Italy was not outdone by those immigrants from another area of Italy favoring a different saint (Mormino 1986). In California a wealthy winemaker funded the construction of the San Secondo d'Asti Church in Guasti, designed as a replica of the seventeenth century church in his home village of Asti, Italy. Stone masons from Piedmont helped in the construction, and the bells were cast in Italy. While originally "regarded as a family chapel" (Cinotto 2012:149), not unlike many that were erected in Italy, the church currently serves the entire Italian-American community.

14.4.3.1 Italian-American Churches and Cemeteries

In southern Illinois many Italian immigrants worked in the coal mines, leaving a strong imprint on small cities such as Herrin. While their ethnic legacy can still be seen in the "freiselike brickwork" on the buildings of clubs and businesses and in the grape arbor found in so many yards, religious features strongly mark the community as having Italian roots (Coggeshall and Nast 1988). Not only does Our Lady of Mt. Carmel Roman Catholic Church sport a brick Italianate style tower with a red tile roof, but in the "San Carlo Cemetery, Italian tombstones, often bearing small photographs of the deceased, preserve the native language of the newcomers" (Coggeshall and Nast 1988: 137).

Italian-Americans erected many churches in which they displayed their ethnicity. These can be found in major metropolitan areas with significant ethnic Italian populations, such

Fig. 14.10 Tuscan style architecture was incorporated in the buildings of the Villa Bellezza winery near Pepin, Wisconsin

Fig. 14.11 Roman Catholic Church of the Holy Ghost built in 1905 in Italian-American neighborhood of Providence, Rhode Island. Note the display of Italian and American flags along the street leading to the church

as Boston, Providence (Fig. 14.11) and San Francisco (Fig. 14.12), as well as in smaller communities that attracted Italian immigrants. For example, in southern Louisiana Italians settled not only in New Orleans, but within a variety of small cities upriver along the Mississippi. Thus, conspicuous St. John the Evangelist Catholic Church in Plaquemine (Fig. 14.13), built in the 1920s in the Italian Romanesque style, along with the burial vaults of the Societa M.B. Italiana

Fig. 14.12 Saints Peter and Paul Church, San Francisco, California. The present church building was completed in 1924, replacing an earlier church destroyed during the fire following San Francisco's 1906 earthquake. The church was founded by the Salesians of St. John Bosco, who were sent in the nineteenth century to minister to the city's Italian immigrants and who still serve that community

mausoleum (Fig. 14.14) in the nearby parish cemetery, illustrate the religious component of the Italian ethnic landscape.

14.4.3.2 Italian-American Yard Shrines

Religious yard shrines have been described within Italian-American neighborhoods, just as they have been noted in yards of Roman Catholic residents in certain other ethnic communities. Writing about Italian neighborhoods in New York City, Jerome Krase concludes, "Ultimately, Italian neighborhoods are 'congeries of families'...expressed in the front-yard religious shrines for patron saints of particular towns or regions" (Krase 1997:20). Nevertheless, other Roman Catholic communities show far less diversity in shrines. In his study of Italian-American neighborhoods of Kansas City, Missouri and Archbald, Pennsylvania, Joseph Manzo (1983) discovered that front-yard statues honoring the Virgin Mary are most common, although back-yard statues of St. Francis were noted in one community.

Overall, about one-tenth of the homes of Italian-Americans within the ethnic enclaves studied displayed the concrete, plaster, fiberglass, or plastic statues that are roughly two feet high. In the Columbus Park neighborhood of Kansas City, "many of the statues are adorned...around the base of the statue...[with] fresh or plastic flowers and rosary beads" (Manzo 1983:122). Some statues were free standing, while others were displayed in bathtub formed niches. Manzo (1983:119) found "a strong relationship between the existence of yard shrines and the nearness of older family members." Thus, yard shrines are typically lacking from the yards of suburban Italian-Americans. Furthermore, religious yard shrines are observed in yards of Italian-Americans from southern Italy, not those from northern Italy.

Fig. 14.13 St. John the Evangelist Catholic Church in Plaquemine, Louisiana. This church, constructed by Italian immigrants in the 1920s, displays Italian Romanesque architecture

Fig. 14.14 Mausoleum of Societa M.B. Italiana erected in 1916 in the Catholic parish cemetery in Plaquemine, Louisiana

14.4.3.3 Religious Festivals and Processions

Religious festivals highlight the ethnic heritage of many Italian-American communities. As Dominic Candeloro (1995:250) explains, "The most dramatic evidence of the retention of ethnicity among Chicago's Italian Americans has historically been demonstrated in the religious street festivals." Not only are there large festivals, with Chicago's three-day Fiesta Italiana drawing over 50,000 participants in

2014, but there are many other smaller festivals or parades in which statues of saints are carried through Italian neighborhoods. Such festivals not only feature a wide variety of Italian foods, but cater to the faithful. Thus, "the *feste* themselves, laden with symbolism, proclaim a convincing challenge to all who would dismiss the authenticity of Italian-American ethnicity in Chicago today" (Candeloro 1995:250). Similar festivals are found in many other cities. For example, in New Orleans, an annual parade celebrates St. Joseph's Day, the paramount Sicilian feast (Campanella 2006). In Providence the same feast day is celebrated, along with a three-day festival over Columbus Day weekend. The Columbus Day parade in New York City involves 35,000 participants and draws over a million spectators.

14.4.4 Italianate Architecture

Italianate style architecture, which reflected some Italian traditions, was introduced to America in the late 1830s and was subsequently widely copied by builders utilizing pattern books (McAlester and McAlester 1990). While it was very commonly displayed in both rural and urban houses build between 1860 and the 1880s, particularly within the Midwest and in the San Francisco vicinity, as well as within some of the northeastern cities, its adoption did not reflect the establishment of Italian ethnic communities. Rather, the "Italianate or Tuscan style enabled Americans to make a connection with the Italian Renaissance" (Noble 1984:141) at a time many Americans wished to display aspects of European culture in their homes.

The Italianate style, with its focus upon low pitched hipped roofs, wide eaves and distinct cornices, and tall narrow windows with arched lintels on square or slightly rectangular houses of two or three floors, was widely adopted, but it is not a vernacular building style. Similarly, Italian Renaissance style houses, with their broadly overhanging eaves, were a common architectural style between 1890 and 1935. While the style borrowed much from Italy, including their square towers, Italianate style was actually an import from England (McAlester and McAlester 1990).

14.5 Greek Ethnic Landscapes in America

Some of the most conspicuous Greek landscapes in the nation developed in large urban areas. As Constantinou and Harvey (2006:343) explain, "Greek Americans marked their communities with Greek language, signage, and other features that gave their living space a unique cultural appearance." Those urban neighborhoods in which these "visually apparent Greek-American ethnic markers" were displayed became known as Greek Towns (Constantinou and Harvey 2006:343). These were particularly well developed in Chicago, Detroit, and New York City's Astoria district. Yet over time, Greek ethnics joined other Americans in moving to the suburbs, leaving an historic core marked by Greek restaurants and coffee houses. While the Greek Orthodox Church remains a central institution for the Greek community, it may or may not be located in Greek Town. In Cleveland the original churches have been renovated, while in Chicago, they relocated nearer to the suburban parishioners. Thus, with the Greek migration to the suburbs, "[i]n many cities, there is little, if any, evidence to remind the observer of the earlier Hellenic presence in the area" (Constantinou 2002:107). In Chicago the most conspicuous ethnic legacies within Greektown are several ethnic restaurants (Fig. 14.15), a shop providing Greek Orthodox religious items, and the National Hellenic Museum.

In contrast to the Greek neighborhoods that developed in several major metropolitan areas, Greek sponge fishermen settled Tarpon Springs along the Gulf of Mexico just north of Tampa-St. Petersburg, Florida. This settlement has dominated the nation's sponge fishery for over a century, and it long "had the distinction of being the only settlement where Greeks constituted a large majority of the population" (Scourby 1984:54). Its Greek ethnicity is displayed by its "Greek church, the political clubhouse, the coffeehouses with outdoor tables, [and] the Greek signs on the stores along the street" (Scourby 1984:55). Yet, aside from the historic downtown area that promotes its Greek heritage, sponge fishing docks, and ethnic restaurants as a tourist attraction and the city's Greek Orthodox Cathedral (Fig. 14.16), much of the community appears little different from the typical suburban American landscape.

14.6 The Mediterranean's Diverse Cultural Influences

Influences from the Mediterranean upon the American landscape are highly diverse. Not only do they include imitations—not always totally correct in their renditions—of Classical Greek and Roman architecture, but they include architectural styles influenced by designs of the nineteenth and twentieth centuries. Some styles were brought to the United States directly by immigrants from the lands along the Mediterranean, while others were introduced after first having been modified and popularized elsewhere in Europe.

Mediterranean influences upon the landscape were introduced by a variety of ethnic groups from a wide geographic area, including not only the Spanish, the Italians, and the Greeks, but also various Arab and other Middle Eastern groups along the eastern and southern shores of the Mediterranean. In addition, small numbers of immigrants from the islands of Cyprus and Malta as well as from the

Fig. 14.15 Greek restaurants are one of the main attractions to the Greektown neighborhood along Halsted Avenue in Chicago. Although the National Hellenic Museum is located just a block away, now relatively few Greek-Americans in the Chicago area live in Greektown

Fig. 14.16 St. Nicholas Greek Orthodox Cathedral in Tarpon Springs, Florida. The church and shops selling Greek ethnic souvenirs and sponges gathered by Greek divers are the most conspicuous Greek features on the Tarpon Springs landscape

lands along the eastern shore of the Adriatic Sea have also contributed to America's ethnic diversity. While some of these other groups have had minimal influence on the nation's ethnic landscape—or ethnic landscapes that have not been academically described—that of the Spanish is discussed in Chaps. 5 and 6. The influences of various Arab and Middle Eastern immigrants upon the ethnic landscape of America are reviewed in the next chapter.

References

Allen, James P., and Eugene Turner. 1988. *We the People: An Atlas of America's Ethnic Diversity*. New York: Macmillan Publishing Company.

Berger, Joseph. 2007. *The World in a City: Traveling the Globe through the Neighborhoods of the New New York*. New York: Ballantine Books.

Blokland, Talja. 2009. Celebrating Local Histories and Defining Neighborhood Communities: Place-making in a Gentrified Neighborhood. *Urban Studies*. 46 (8): 1593–1610.

Buzzelli, Michael. 2001. From Little Britain to Little Italy: An Urban Ethnic Landscape Study in Toronto. *Journal of Historical Geography*. 27 (4): 573–587.

Campanella, Richard. 2006. *Geographies of New Orleans: Urban Fabrics Before the Storm*. Lafayette: Center for Louisiana Studies.

Candeloro, Dominic. 1995. Chicago's Italians: A Survey of the Ethnic Factor, 1850–1990. In Ethnic Chicago: A Multicultural Portrait< ed. Melvin G. Holli and Peter d'A. Jones, 229–259. Grand Rapids: William B. Eerdmans Publishing Company.

Cinel, Dino. 1982. *From Italy to San Francisco: The Immigrant Experience*. Stanford: Stanford University Press.

Cinotto, Simone. 2012. *Soft Soil, Black Grapes: The Birth of Italian Winemaking in California*. New York: New York University Press.

Constantinou, Stavros T. 2002. Profiles of Greek Americans. In *Geographical Identities of Ethnic America: Race, Space, and Place*, ed. Kate A. Berry and Martha L. Henderson, 92–115. Reno: University of Nevada Press.

Constantinou, Stavros T., and Milton E. Harvey. 2006. The Persistence of Greek American Ethnicity among Age Cohorts Under Changing Conditions. In *Race, Ethnicity, and Place in a Changing America*, ed. John W. Frazier and Eugene L. Tettey-Fio, 349–352. Binghamton: Global Academic Publishing.

Coggeshall, John M., and Jo Anne Nast. 1988. *Vernacular Architecture in Southern Illinois: The Ethnic Heritage*. Carbondale: Southern Illinois University Press.

Daniels, Roger. 2002. *Coming to America: A History of Immigration and Ethnicity in American Life*. 2nd ed. New York: Harper Perennial.

Fichera, Sebastian. 2011. *Italy on the Pacific: San Francisco's Italian Americans*. New York: Palgrave Macmillan.

Helzer, Jennifer J. 2001. Old Traditions, New Lifestyles: The Emergence of a Cal-Ital Landscape. *Yearbook, Association of Pacific Coast Geographers*. 63: 49–62.

———. 2002. Varieties of Ethnic Identity and Landscape among Italian Immigrants in Northern California. *The California Geographer*. 42: 25–40.

Krase, Jerome. 1990. America's Little Italies: Past, Present and Future. In *Italian Ethnics: Their Languages, Literature and Lives (Proceedings of the 20th Annual Conference of the American Italian Historical Association)*, ed. Dominic Candeloro, Fred L. Gardaphe, and Paolo A. Giordana, 169–184. Staten Island: American Italian Historical Association.

———. 1997. Polish and Italian Vernacular Landscapes in Brooklyn. *Polish American Studies*. 54 (1): 9–31.

Manzo, Joseph T. 1983. Italian-American Yard Shrines. *Journal of Cultural Geography*. 4 (2): 119–125.

McAlester, Virginia, and Lee McAlester. 1990. *A Field Guide to American Houses*. New York: Alfred A. Knopf.

Mormino, Gary Ross. 1986. *Immigrants on the Hill: Italian-Americans in St. Louis, 1882–1982*. Urbana: University of Illinois Press.

Noble, Allen G. 1984. *Wood, Brick, and Stone: The North American Settlement Landscape. Volume 1: Houses*. Amherst: University of Massachusetts Press.

Pierson, William H. Jr. 1970. *American Buildings and their Architects, Volume 1 The Colonial and Neoclassical Styles*. New York: Oxford University Press.

Poppeliers, John C., S. Allen Chambers Jr., and Nancy B. Schwartz. 1983. *What Style Is It? A Guide to American Architecture*. Washington: Preservation Press.

Scourby, Alice. 1984. *The Greek Americans*. Boston: Twayne Publishers.

Sowell, Thomas. 1981. *Ethnic America: A History*. New York: Basic Books.

———. 1996. *Migrations and Cultures: A World View*. New York: Basic Books.

Upton, Dell. 1986. Introduction. In *America's Architectural Roots: Ethnic Groups that Built America*, ed. Dell Upton, 7–15. Washington: Preservation Press.

15 Arab-American and Other Middle Eastern Landscapes

Historically the entire area around the Mediterranean Sea was once part of the Roman Empire, yet the Mediterranean population is highly diverse, representing a variety of languages and religions. Thirteen hundred years ago Moslem forces spread rapidly from the Arabian Peninsula, conquering all of North Africa and the Iberian Peninsula of Europe. The Islamic Ottoman Empire at its greatest extent spread into southeastern Europe, including Greece, lands east of the Adriatic, and Hungary. Today, Arabic is spoken across a wide area of the southern Mediterranean, running from Morocco to Syria and southeast through the Arabian Peninsula. The eastern Mediterranean also includes numerous speakers of other languages, particularly Greek and Turkish. Formerly much of coastal Turkey was settled by Eastern Orthodox Christians, while most of the rest of Turkey was Moslem. Lebanon and western Syria are the home to followers of a variety of Christian and Islamic groups, among others, while Israel and Palestine include Jewish, Christian, and Moslem populations. North Africa, while predominately Islamic, includes a significant Coptic Christian minority in Egypt's Nile Delta, and until the middle of the twentieth century, Jewish quarters were found in most of the larger cities of North Africa, surrounded by far larger numbers of Moslems.

15.1 Conflict in the Middle East and Population Movements

Religious and political conflict for over a century has generated multiple waves of Middle Eastern refugees who fled their historic homelands, and this exodus continues. Armenians underwent two decades of persecution at the hands of the Turks, culminating in a genocide that killed over a million Armenians with the dissolution of the Ottoman Empire during and immediately following World War I. While many of the political refugees at that time simply relocated to other areas of the Middle East, large numbers fled to other parts of the world. Among those who came to the United States in the largest numbers were Lebanese and Armenians, and the majority of Arab immigrants arriving in the United States through the 1950s were Christians. While the establishment of the State of Israel in 1948 led to immediate conflict and displacement of hundreds of thousands of Palestinian refugees—both Christian and Moslem, the conflict simultaneously resulted in long-established Jewish communities being displaced from throughout many of the predominately Arab nations. Yet, it was conflict in Europe and the Holocaust that caused so many European Jews to flee to British Palestine and fight to re-establish political control over their ancestral homeland, after enduring a Diaspora that had lasted over two millennia. Their actions created Palestinian refugees, resulting in the Refugee Relief Act of 1953 that authorized 2000 Palestinian families to come to the United States (Arab American National Museum n.d.).

Continuing conflict in the region, including the Six Day War between Israel and its neighbors in 1967 and the civil war in Lebanon between 1975 and 1990, generated thousands of additional immigrants. Following the staged withdrawal of American military forces from Iraq, 41,000 Iraqi refugees accounted for nearly one-quarter of all refugees admitted to the United States between 2011 and 2013 (Martin and Yankay 2014), with statistics not yet available regarding their continuing arrival. At the time of writing of this chapter, the Syrian Civil War had created a refugee crisis that was inundating Europe, with the United States agreeing to accept at least 10,000 refugees, yet Canada quickly accepted far more. Following the 2016 U.S. presidential election, it is unlikely that the United States will be accepting many more Middle Eastern refugees for the foreseeable future.

15.2 Contemporary Arab-American Population and Its Distribution

Within most of the United States, individuals of Arab ancestry comprise small minorities (Fig. 15.1). The U.S. Census Bureau categorizes individuals as being an Arab if they are

J.A. Cross, *Ethnic Landscapes of America*, DOI 10.1007/978-3-319-54009-2_15

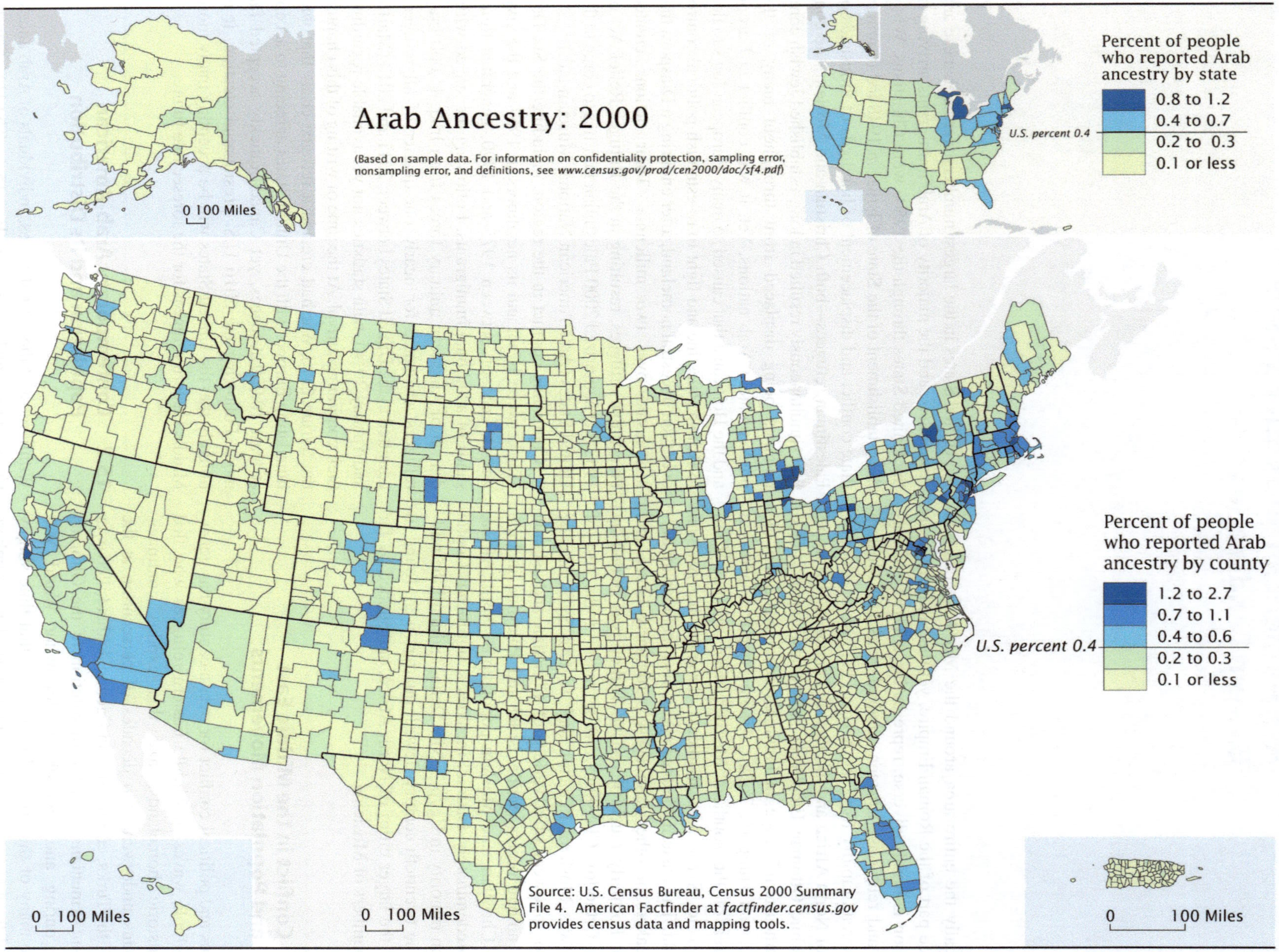

Fig. 15.1 Proportion of the population reporting Arab or Arabic ancestry in each county at the time of the 2000 Census. Note that those counties with the greatest percentages of their populations being Arabs are largely located within the southeastern Michigan (Reproduced from U.S. Census Bureau map in de la Cruz and Brittingham (2003))

"from Arabic-speaking countries or areas of the world" (de la Cruz and Brittingham 2003:1). This includes all of the countries along the southern and eastern shores of the Mediterranean from Morocco through Syria, plus Jordan, Iraq, Kuwait, Saudi Arabia, and Yemen, yet the Census publication notes that not all of the peoples living in those lands consider themselves as Arab. Even among those who view themselves as Arab, there are considerable ethnic differences, given the region's geographic expanse and the number of nationalities represented in the Arab world. Indeed, Abraham and Shryock (2000:39) argue against the notion "that Iraqi Shia, Palestinian Christians, and Yemeni Sunnis share an essential 'Arabness'." Overall, Arab-Americans accounted for 0.4% of the national population in 2000, but their numbers had grown by 38.3% since 1990 (de la Cruz and Brittingham 2003). Over a decade later the 5 year American Community Survey, based upon surveys conducted between 2010 and 2014, estimated that 1,769,645 persons, or 0.5% of the American population, had Arab ancestry. California had the largest number of Arab-Americans, an estimated 275,403, followed by Michigan with 168,478, New York with 158,893, Texas with 107,036, Florida with 109,703, and New Jersey with 96,777. Nevertheless, it has been argued that "marginalization of Arab Americans within census categories and statistics…has contributed to their overall cultural and political marginalization within the United States" (Pavlovskaya and Bier 2012:483).

With the growth in the counted numbers of Arab ethnics, the national origins of these Arabs have been changing. For example, in 1990 persons of Lebanese descent comprised 45.8% of all Arabs in the United States, yet that proportion had fallen to 37.0% in 2000 (de la Cruz and Brittingham 2003). Estimates from the American Community Survey between 2010 and 2014 indicate that the Lebanese share of the Arab-American population dropped to 27.9%. Furthermore, the majority of Arab-Americans of Lebanese ancestry were once of Maronite Christian heritage, yet Lebanese émigrés arriving since the 1980s are often Shi'a Moslems. Most Arab immigrants from other countries practice Islam, yet this population is divided between the Sunni and Shi'a. Arab-Americans are highly diverse, yet this diversity is often ignored by the majority of Americans. As Pavlovskaya and Bier explain,

> Despite being highly differentiated in terms of religion, ethnicity, class, and national origin, Arab Americans are perceived as a homogenous group, another powerful trope of marginalization. They are commonly conflated with Muslims, for example, despite comprising only one-third of US Muslims (Pavlovskaya and Bier 2012:485).

The second largest group of immigrants from the Muslim World are Iranians, a Shi'a population who view themselves as being Persians, not Arabs. Even among the Iranians, a disproportionate share of immigrants are "ethnic and religious minorities such as Jews, Armenians, Zoroastrians, and Baha'is who feared they would be marginalized and persecuted by the new fundamentalist government of the Islamic Republic of Iran" (Chacko 2016:373).

The proportion of America's Arabic population who either immigrated from Lebanon or are descended from persons from that land has declined, even though their numbers rose by 11.7% between 1990 and 2000, simply because other national groups have increased more rapidly. Several groups, including Yemenis, Kurds, Algerians, and Moroccans, more than doubled during the decade of the 1990s (de la Cruz and Brittingham 2003). Historically, the Arab-American population has been concentrated in the Midwest, where they first began arriving to work in the automobile industry a century ago. While there are now more Arab-Americans in California than in any other state, not surprising given California's size, they comprise a larger share of Michigan's population than in any other state. In addition, Michigan is home to more Lebanese Arabs than any other state.

15.2.1 Arabs in the Detroit Metropolitan Area

Wayne County, Michigan, including and surrounding Detroit, has the greatest proportion of Arabs in its population of any county in the nation, 2.7% at the time of the 2000 Census. Within the city of Dearborn, the figure was 29.9%, and that city had the largest number of Arab-Americans of any American city other than New York City. Within Macomb County, Michigan, just north of Wayne County, 2.2% of the population were Arab ethnics. Fairfax County, Virginia, is the only other county nationally where Arabs comprised as much as 2% of its population.

Differences of national origin and religion distinguish various Arab-American communities. Dearborn, Michigan is the least homogeneous of the Arab-American communities, as most are "composed of a single nationality-religious group" (Abraham 1983a:95). Yet even in Dearborn, certain Arab neighborhoods are fairly homogenous. For example, the Southend area is primarily inhabited by Yemeni Sunni Muslims with some Lebanese Muslims, while Lebanese Shi'a are the majority in the remainder of Dearborn (Schopmeyer 2000). Estimates from the American Community Survey between 2010 and 2014 showed Lebanese comprising 46.2% of Dearborn's Arab population, which was estimated at 40,713. In contrast, the nearby city of Hamtramck, which is surrounded by Detroit and is one quarter Arab-American, was estimated to have fewer than ten Lebanese residents. Its Arab-American population is primarily Yemeni, although many Bangladeshi and Albanian Moslems also reside there.

Arab-Americans of Iraqi ancestry are primarily located away from the Dearborn area, occupying several suburbs north of Detroit. As Schopmeyer (2000:72) explains, "Two-thirds of Iraqis are located in Oakland County, particularly in the areas of Southfield and Farmington Hills, with significant numbers in the more affluent Bloomfield Hills area and another segment in the vicinity of Oak Park." Other Iraqis, particularly those who came in the 1980s, "live in the traditionally Chaldean neighborhood of northeast Detroit" (Schopmeyer 2000:72). Between the Iraqi groups within Oakland County, clusters of Egyptian Coptic Christians, Syrian-Lebanese Christians, and Palestinians reside, while Ramallah Palestinians live in suburbs farther west of Detroit (Schopmeyer 2000). Highly specific geographic origins have been noted for some of the Arab immigrants. For example, in several western suburbs of Detroit Palestinian Christians primarily came from the community of Ramallah, while Chaldean Christians, mainly from Telkaif in northern Iraq, are a prominent Middle Eastern group within the Seven Mile-Woodward area (Abraham 1983a). In the Flint, Michigan area, until recently most of its Arab-American population were Christians, members of either the Lebanese Maronite Church or the Antiochian Orthodox Church. Furthermore, Flint, Michigan has been the center of the Druze in America, who came primarily from Lebanon and Syria (Hassoun 2005).

15.2.2 Arabs in the New York Metropolitan Area

New York City has the greatest urban concentration of Arab-Americans in the nation, although given its large population, Arab Americans comprise a far smaller share of its total population than within Dearborn. The five-year American Community Survey for the 2010–2014 period estimates that New York City's Arab's population totaled 93,834, of whom over half were foreign born, while 208,822 Arabs resided in the New York City-Newark-Jersey City metropolitan area.

Those Arab-Americans whose ancestors came from Lebanon and Syria early in the twentieth century remain prominent in the metropolitan area, particularly in New York City's Brooklyn and Manhattan boroughs, along with Passaic and Bergen Counties in New Jersey and Fairfield County, Connecticut. Among more recent Arab-American immigrants those from Egypt and Morocco are most numerous. The location where the Arab-Americans settled in the New York City metropolitan area is related to their ancestry. Thus, "Egyptians settled in both the traditional urban destinations such as Brooklyn and Queens and suburban counties of Middlesex, Bergen and, especially, Hudson [while] Moroccans, in contrast, avoided the suburbs while moving to Queens, Brooklyn, and Manhattan" (Pavlovskaya and Bier 2012:489). Palestinians, Jordanians, and Iraqis each have their own spatial concentrations of settlement.

15.3 Contemporary Distribution of Non-Arabs from the Middle East

A variety of non-Arab ethnic populations from the Middle East have also settled in the United States (Fig. 15.2). These include Persians or Iranians, Turks, Armenians, a relatively small number of Israelis, plus a variety of groups that are small ethnic minorities, such as the Kurds, even in the nations which incorporate their homelands. During the first decade of the twenty-first century significant numbers of Iranian, Iraqi, and Afghan refugees came to America, and more recently large numbers of Syrians and Kurds have fled their homes.

15.3.1 Armenians in California

The 2000 Census enumerated 385,488 Armenians in the United States, and of this population, 53.1% lived in California. The American Community Survey estimated their numbers had grown to 473,994 between 2010 and 2014, with California's share having risen to 55.0%. Over a century ago Armenian enclaves were established in both a rural area in Fresno County, California, where Armenians helped establish the grape and raisin industry, as well as in the New York City and Boston metropolitan areas. Later, Armenians from many countries, including Armenia, Russia, Iran, and Lebanon, established an enclave within the Glendale district of Los Angeles (Allen and Turner 1988). The Armenian Apostolic churches in these communities often display ethnic architectural links, such as an eight-sided cupola centered over the gabled roof.

15.3.2 Iranians in California

Forty-six percent of the Iranians in the United States, according to the 5 year American Community Surveys conducted between 2010 and 2014, lived in California. Nationally, the Iranian ancestry population was estimated at 462,061, just a few percent less than the Armenian population. Iranian immigrants represent a far more divergent Persian population than the Shi'a Moslems who control that nation's theocratic government. Large numbers of Iranian Jews, Armenians, Zoroastrians and Baha'i also fled Iran as political refugees (Chacko 2007). Iranians, who are not Arabs, have disproportionately settled within several upscale communities in southern California. Thirty-one percent of all Iranian-Americans lived within the Los Angeles metropolitan area in 2000, while 41.0% lived in California, thus California's share has grown since then. The Westwood section of Los Angeles has become the focus of Persian business activity in Los Angeles (Allen and Turner 2002).

Fig. 15.2 Architectural influences from the Middle East are clearly visible in the Baha'i House of Worship in Wilmette, Illinois, within the Chicago metropolitan area. The Baha'i religion evolved during the nineteenth century in Iran, and it contains elements from Judaism, Christianity, Islam, as well as Indian religions and Buddhism

15.3.3 Chaldeans in Michigan

Chaldean Christians, who are sometimes also called Assyrians, are a religious minority in the Middle East who spoke Aramaic, the ancient language of Christ. Their ancestral homeland is located in what is now northern Iraq. Their settlement in the Detroit metropolitan area, particularly Wayne and Oakland counties, has been widely estimated to number from 65,000 to 120,000, given that they are not specifically enumerated by the Census. Not only is it the only sizeable Chaldean community in the United States, it is the largest Chaldean community outside of the Middle East (Sengstock 2005). Grouped with Assyrian and Syriac populations, the American Community Survey between 2010 and 2014 estimated the national population of the three combined groups was 107,711, with 40,791 living in Michigan and 36,242 residing in California. As with many small minority populations, it is likely their numbers are underestimated, given that those surveyed may be counted in the totals of the country from which they came.

15.3.4 Turkish Ethnics in the New York City Metropolitan Area

Turkish ethnics came to the United States in only small numbers before World War I, accounting for under one-tenth of those arriving from the Ottoman Empire (Allen and Turner 1988). A century later the Arab-American population is nearly nine times greater than those reporting Turkish ancestry. Furthermore, the Turkish-American population is less than half of those having Armenian or Iranian ancestry. The 2000 Census enumerated 117,575 persons of Turkish ancestry, while the five-year American Community Survey for 2010–2014 estimated a Turkish ethnic population of 199,095.

The greatest concentration of Turkish-Americans is within the New York City-Newark-Jersey City metropolitan area. With an estimated 2010–2014 Turkish population of 41,184 within the metropolitan area, greater New York City is home to 20.7% of all Turkish-Americans.

15.3.5 Mizrahi Jews in Los Angeles and New York City

Most of the Jewish population of the United States are descendents of immigrants from central or Eastern Europe, and these individuals are termed Ashkenazi Jews. Because they had become acculturated to various degrees into the nationality of their European homelands, the ethnic landscape of this group is discussed in Chap. 13, describing Slavic ethnics, and in Chap. 18. In a similar manner, Sephardic Jews have long been associated with Hispanic cultures, even though they were expelled from Spain, or forced to convert to Catholicism, at the time of the Inquisition. Nevertheless, several smaller groups of Jews, including the Mizrahi Jews, immigrated to the United States directly from their ancient homelands in the Middle East.

Mizrahi Jews were formerly scattered throughout the Middle East, but most were expelled from their homes in Moslem dominated countries following World War II. While most went to Israel, approximately one-quarter million Mizrahi Jews now live in the United States. Persian Jews, like other Iranian-Americans, are concentrated in the Los Angeles metropolitan area, as well as Long Island, New York. Bukharian Jews, from Uzbekistan and Tajikistan, in the U.S. primarily reside within the Forest Hills neighborhood of New York City's borough of Queens, displaying their ethnicity through the presence of Bukharian restaurants and retail establishments.

15.4 Arab and Middle Eastern Ethnic Landscapes

The visual impact of Middle Eastern immigrants upon the American landscape varies depending upon their national origin and religion, the time of their arrival and their numbers, as well as their settlement intentions. For example, the majority of the Yemeni immigrants were male and fully intended to return to their native lands after earning sufficient funds to buy lands, build a house, or arrange a marriage in their homeland. Thus, in Detroit they lived in rental apartments in immigrant ghettos with fellow émigrés near their employment in the automobile industry and made minimal efforts to integrate into American society (Abraham 1983b). On the other hand Lebanese immigrants, particularly the earlier arriving ones who were predominately Christian, were far more likely to own their houses and disperse into a wider variety of suburban neighborhoods. In some communities, such as in Toledo, Ohio, as the Arab-American community became better established and more prosperous, it abandoned its former inner city neighborhood as its crime rate increased. While this enhanced the assimilation of Lebanese Catholic groups who lacked their own ethnic church, Orthodox groups had two churches to "nurture the community's cultural traditions" (Awada 2010:17). In addition, "the Muslim community was able to retain much of its cultural and religious traditions due to the continued immigration of new settlers" (Awada 2010:17), and as it outgrew their former downtown facilities, they constructed a large mosque in what had been a cornfield at the southern edge of Toledo. As with many groups, Arab-American ethnicity is best displayed by the group's religious structures and commercial buildings.

15.4.1 Religious Structures and Symbols

Religious buildings often distinguish communities settled by Middle Eastern immigrants, whether they are Arab or not, Christian or Moslem. For example, Sacred Heart Chaldean Church in Detroit, Michigan "has a façade modeled after the gates of the ancient city of Babylon, in honor of the community's Babylonian heritage" (Sengstock 2005:27). Many of the Chaldean Churches, a special rite of the Roman Catholic Church, occupy buildings with domes, just as many mosques also have domes. Some groups have carefully designed their edifices, so that "the sight is breathtaking—a gleaming gold-domed mosque with traditional twin minarets in classical Islamic style rises out of the Ohio countryside" (Adray 2010:73).

The presence of a mosque or Islamic Cultural Center (Fig. 15.3) is one of the key criteria distinguishing many Arab-American communities, yet many Arab-Americans are Christians. Furthermore, at least half of all Moslems in the United States are not from the traditional Arab countries. Instead, they are from such populous Moslem nations as Pakistan, Bangladesh, or Indonesia. In smaller Moslem communities, a single mosque may serve a diverse Islamic population. In contrast, "[t]he size of [greater] Detroit's Muslim population has allowed it to form more ethnically concentrated mosques…[with] about one-third of Detroit's mosques…attended largely by people of South Asian descent, while another 30% serve a largely Arab population" (Scheitle and Finke 2012:179). In addition, one sees considerable architectural diversity amongst the mosques, reflecting whether the structure was purpose built to serve as a mosque—and over three-quarters of the Detroit area mosques were originally built for some other function, sometimes as a church—and the ethnic ancestry of those Moslems that are served by the specific mosque (Bagby 2004). Yet, as the Building Islam in Detroit exhibit (Howell et al. 2011) explains, a mosque "can be striking for its simplicity or for its ornate designs."

For the mosque "to function as a site of communal prayer, a mosque should have certain architectural features" (Howell et al. 2011: Architecture 1). The most common of these are a qibla wall that indicates the direction of Mecca, a *haram* or sanctuary adjacent to the *qibla*, a *minbar* or pulpit, a *mida'a* or ritual cleansing area, a dome, and minarets. As James Adray (2010:85) explains, "The actual front of a mosque faces Mecca so that the correct placement of the *minbar*, the platform from which the imam addresses the congregation, is opposite the main entrance. The wall where the *mihrab*, a niche that provides the direction Muslims face when praying, is located must be without a door." Although the qibla may influence the orientation of the building, the presence of the dome and minarets are the most conspicuous features of purpose built mosques, such as one erected just south of Toledo, Ohio (Fig. 15.4). In addition, among both Sunni and Shi'a Moslems "there is a heavy reliance on calligraphy and arabesque shapes to adorn the mosque" (Howell et al. 2011).

Fig. 15.3 The Islamic Center of Milwaukee, Wisconsin. A tall minaret and arched entrance mark this center, which attends to the religious, educational, and social needs of the local Moslem community

Fig. 15.4 The Islamic Center of Greater Toledo, located at the southern edge of Perrysburg, Ohio. Erected on a spacious site on what was previously farmland, the twin minarets and prominent dome of the mosque are conspicuously visible from a nearby interstate highway. The minarets are capped by a star and crescent

Mosques are typically distinguished by the presence of tall thin rounded towers or minarets and often have arched entrances. A mosque, such as one in Dearborn, Michigan, "with its large gold dome topped by a crescent moon symbolic of the Islamic faith…[is] a constant reminder of the Southend's unique ethnic and religious character" (Abraham et al. 1983:172). The appearance of the dome varies considerably, depending upon the regional location, but its "appeal is rooted in its association with Persian and Ottoman societies" (Howell et al. 2011: Architecture 1). With growth in population, new mosques have been constructed, including one in Dearborn that can accommodate a thousand worshipers (Chacko 2007). That mosque (Fig. 15.5), the Islamic Center of America, opened in 2005 and serves a Lebanese Shi'a community. Its golden domes and towering minarets topped by the Islamic crescent are more conspicuous than several nearby churches. Its "interior is decorated with white marble, gold-leaf Arabic inscriptions and glass chandeliers—all made exclusively for the mosque and shipped from various parts of Lebanon and Saudi Arabia" (Akhtar 2015: 187–188).

From the mosque, such as in Dearborn, five times daily came "the regular call of the *muezzin* (the prayer caller)… now carried across a public address system which can be heard throughout the region" (Abraham et al. 1983:166). While the calls to worship from mosques in the Detroit area community of Hamtrammack led to protests from other ethnic groups within that city, and the minarets of some mosques today are purely decorative (Howell et al. 2011), they provide evidence of the many ways that the expansion of the Arab population alters a community's religious landscape. Islamic beliefs also influence other aspects of the ethnic landscape, including there being "a preponderance of men on the street, [and] hejab worn by women" (Chacko 2007:322). Moslem men are expected to make a pilgrimage to Mecca, a journey called the Hajj that is one of the five pillars of Islam, at some point during their life. Thus, in cities such as Dearborn, one finds businesses (Fig. 15.6) that assist in making arrangements for the Hajj, which occurs on the eighth through the twelfth days of the last month of the Islamic calendar, or the Umrah, a pilgrimage to Mecca that occurs during other times of the year.

15.4.2 Arab-American Businesses and Coffeehouses

Islamic Arab-American landscapes are identified by the presence of halal butchers and shops displaying Arabic writing and various Middle Eastern symbols. Just as Jewish communities are marked by shops selling kosher foods, Moslem communities have butchers selling halal meats (Fig. 15.7) and restaurants—and even food carts (Fig. 15.8)—touting their halal foods. Both groups have strict religious standards regarding the slaughter and handling of livestock and the meat products, as well as common prohibitions regarding the consumption of pork.

Fig. 15.5 The Islamic Center of America, Dearborn, Michigan. This mosque, which can accommodate a thousand worshipers, opened in 2005 and serves a Lebanese Shi'a community. Note the Islamic crescents atop the dome and minarets.

Fig. 15.6 Travel agency in Dearborn, Michigan that advertizes its services in arranging Hajj travel for those seeking to undertake a pilgrimage to Mecca, Saudi Arabia

Fig. 15.7 Grocery and meat market on Warren Avenue, Dearborn, Michigan advertizing its "halal" meats. The use of Arabic script, which is read from right to left, is common among Dearborn's Arab-American merchants. Note how the crescent and star design, an ancient symbol that today is often associated with Islamic countries that evolved from the former Ottoman Empire, is incorporated into the sign

Increasingly, recent Arab immigrants, who are both more likely to be devout Moslems and to have school age children, have shown an greater inclination to select homes that are in close proximity to both the mosque and the associated Islamic schools. Furthermore, the Arab Muslim concept of the home includes the "interaction between the house and its

Fig. 15.8 Food vendor selling "halal" meats in New York City. The presence of merchants selling halal meats, which require ritual killing of the animal and special handling, is indicative of a local population of Moslems who consume foods that are prepared in compliance with religious laws

immediate environment (door step, street, barber, coffeehouse, restaurants, neighbors, and others) and…its remote environment, *masjeed* (mosque), *madrasa* (school)…[and] *souk* (the market" (Amor 2008:7). Such interconnections promote a tight-knit religious community.

The Arabic coffeehouse, dispensing tea and coffee, plays a critical role as a social institution among the male ethnic population. Political conversations, card games and backgammon accompany the drinks in coffeehouses that "are distinguished not only on the basis of nationality, but also according to the village their proprietors represent" (Abraham et al. 1983:176). Thus, within Dearborn "Lebanese, Palestinians, Yemenis, and Iraqis congregate in discrete cafes. Pastry shops selling the traditional phyllo-based and other Middle Eastern-style sweets as well as French baked goods are overwhelmingly owned by Lebanese" (Chacko 2016:381).

Several communities that contain large concentrations of Arab-Americans have actually taken steps to accentuate their Arab character (Fig. 15.9). For example, Rosina Hassoun (2005:46) explains that the Southend neighborhood of Dearborn, Michigan not only includes small Arab owned businesses and a mosque, but that the "owners of these businesses cooperated in redesigning the exterior of the buildings along the main strip, through the use of façades, constructing Arabic arches and domes… to create an "Arab town" atmosphere that would bring tourists to the area." While she does not find any architectural distinctiveness in the Arab-American residential districts of Southend, she does note that some of the Yemeni ethnics wear "brightly colored traditional clothing" (Hassoun 2005:46), even though many Arab-Americans wear Western style clothing, visually marking the neighborhood as ethnic. During the 1980s Arab-American commercial activities rapidly expanded along Dearborn's Warren Avenue, prompting Barbara Aswad (1992:167) to write, "As one walks along the streets in the Lebanese Muslim community in Dearborn, one feels transplanted back to the Middle East." At that time there were over fifty Lebanese businesses, as well as eight restaurants and five bakeries, along Warren Avenue (Aswad 1992). Today, Warren Avenue remains the focus of Dearborn's vibrant Arab-American business district (Fig. 15.10). While older buildings may only display their adapted ethnicity by the Arabic writing in their signage, other more recently built structures incorporate arched windows in their architecture, enhancing their ethnic distinctiveness.

15.5 Turkish-American and Bosnian-American Landscapes

Geographers have recently published descriptions of Turkish-American (Kaya 2005) and Bosnian-American landscapes (Hume 2015). Neither of these groups are Arab, yet both of these peoples are Moslems from lands along the Mediterranean that were once part of the Ottoman Empire.

Fig. 15.9 Arab-American National Museum in Dearborn, Michigan. With a central dome, a tile mosaic centered above its entrance, and intricate carved stone panels, the building constructed in 2005 has "the look and feel of a mosque, without trying to imitate one" (Hanania 2005). The interior courtyard under the dome is designed to evoke the architecture found throughout the Arab World

Fig. 15.10 Arab-American shopping center along Warren Road in Dearborn, Michigan. Note the use of Arabic calligraphy in the signage and the arched entrances to this recently constructed business within the heart of Dearborn's Arab-American business district

Neither are well represented in the United States. The American Community Survey estimated there were 199,095 persons of Turkish ancestry living in the United States between 2010 and 2014, and Bosnians were included with others who were collectively classified as Yugoslavian, a total of 296,627 persons. Unlike Turkish ethnics, Bosnian migration to the United States has largely occurred during the past couple of decades, a consequence of the religious warfare that engulfed the former Yugoslavia as it split into separate ethnically based nations.

15.5.1 Turkish-American Landscapes in New Jersey and New York

Turks are more secular Islamic and are not Arabs. Nevertheless, Turkish-Americans are diverse, with the lower-class immigrant workers the most likely to be concentrated into Turkish ethnic communities, while middle and upper class professionals are more widely distributed. Patterson, New Jersey includes a good example of a Turkish-American blue collar community. As Ilhan Kaya explains,

> with its Turkish restaurants, grocery stores, coffee or tea houses, and mosques, Patterson is an attractive destination for lower-class Turkish immigrants. These places serve as identity-maintenance and identity preservation sites, where the customs and habits imported from traditional Turkey are kept alive (Kaya 2005:432, Reprinted with permission of the American Geographical Society).

The largest other Turkish-American communities are also in the New York City metropolitan area, including Clifton, New Jersey and the Brighton Beach neighborhood of Brooklyn (Fig. 15.11) and the Sunnyside area of Queens, both boroughs of New York City.

Turkish mosques display considerable nationalism and thus are distinguishable from those of other Muslims, not only in their propensity to provide services in Turkish and only sometimes in Arabic, but in their appearances. Thus,

> Architectural and interior designs of Turkish mosques reflect the imprints of Turkish culture and history. Most Turkish mosques in the New York City metropolitan area are named after Ottoman sultans (Fatih, Suleymaniye, Selimiye, or Osman Gazi) and usually have a Turkish flag hanging in them. Turkish American mosques with shops, grocery stores, and coffee houses next to them are not only worship places but also function as places of social interaction. (Kaya 2005:436, Reprinted with permission of the AGS).

15.5.2 Bosnian-American Landscapes in St. Louis

Bosnian refugees first came to St. Louis, Missouri in 1993, and two decades later the Bosnian-American population in metropolitan St. Louis was estimated around 70,000 (Hume 2015). Occupying a landscape created by earlier German arrivals, the Bosnian immigrants had little opportunity to create an entirely new landscape. Rather, they altered the existing landscape, forming what locals call "Little Bosnia." As Susan Hume (2015:4-5) explains, "storefronts [rather than housescapes] prove to be most distinctive in the current study of Bosnian commercial, religious, and residential landscapes."

Fig. 15.11 Turkish Kebab shop along Brighton Beach Avenue in Brooklyn, New York. Such ethnic food shops are among the most conspicuous evidence of the presence of the neighborhood's Turkish ethnic population

With a focus upon Gravois Avenue businesses in south South Louis, "Bosnian national identity is most evident in the Bevo Mill neighborhood's commercial landscape, expressed through the use of the Bosnian language and culturally significant toponyms, colors, and symbols" (Hume 2015:8–9). The national flag, or parts of it, are included in commercial signs and the national colors are featured on storefront awnings and lettering. As Hume (2015:9) explains, "the visible expression of Bosnian national identity on the commercial landscape had drawn…attention to the…neighborhood as a tourist destination to experience authentic Bosnian cuisine."

The religious landscape of the Bosnian-American community reflects both the secularization of Islam in their homeland and the adaption of existing structures to meet their needs. Thus, one of the south St. Louis mosques is a former church, which retains its steeple and is marked as a mosque only "by a small sign with its name and a gold crescent moon" (Hume 2015:13). Another mosque occupies a former bank, but it has erected a 107 foot high minaret, which "is symbolic rather than functional since it has no sound system to issue a call to prayer" (Hume 2015:13). While butcher shops and restaurants in many Arab-American communities advertize their halal meats in their signage, none of those along Gravois Avenue did when Hume carried out her fieldwork, yet halal signage was on display by summer 2016.

One of the most conspicuous reminders of the Bosnian ethnicity of the south St. Louis community was the erection (Fig. 15.12) of "an Ottoman-inspired public fountain patterned after the mid-eighteen century 'Seblij' in Bosnia and Herzegovina's capital city" (Hume 2015:15). In contrast few cultural features mark the housescapes of the Bosnian ethnic neighborhoods, except smokehouses in backyards used to smoke meats in their traditional styles.

15.6 Moslem-American Communities in the Future

While Arabs were the first Moslem group to come in relatively large numbers to the United States, many other Islamic populations have begun arriving during the past several decades. These groups include Pakistanis and Bangladeshis from the Indian subcontinent as well as large numbers of Black Africans from a broad swath of countries in West Africa. Given their different histories and the variety of their regional cultures and languages, their impacts upon the landscape can be highly diverse (Fig. 15.13). Yet at the same time, many of these individuals are dispersed among many communities, and they are occupying neighborhoods that had been long shaped by earlier ethnic groups. Thus, for example, in Chicago Moslems have purchased structures constructed by Christian groups for churches and parochial schools, and these have been converted into mosques and Islamic schools. Yet in most parts of the nation, the landscape of Middle Eastern immigrants is limited to such religious institutions (Fig. 15.14), plus a few ethnic shops and restaurants. Only within a few communities have enclaves of Arab-Americans, Turkish-Americans, Bosnian-Americans, or Iranian-Americans developed.

Fig. 15.12 The Sebilj, a Bosnian reproduction of the Sarajevo monument along Gravois Avenue, St. Louis, Missouri. The building with the green awning in the left background is the Bosnian Chamber of Commerce, in the center of the nation's largest Bosnian ethnic community

Fig. 15.13 Halal butcher shop in Flushing, Queens, New York. Conspicuously displaying the halal quality of its meats in its name, this shop and the kabab-curry restaurant next door, positioned between Chinese businesses, largely caters to a South Asian clientele, within an ethnoburb that contains large numbers of Chinese, Korean, and South Asian immigrants

Fig. 15.14 Islamic Da'wah Center in St. Paul, Minnesota. This storefront building houses an Islamic missionary outreach center serving the religious needs of the city's growing Somali immigrant population. Such structures are reminders that the Islamic religious imprint upon the landscape extends beyond the Arab-American community

References

Abraham, Nabeel, and Andrew Shryock. 2000. *Arab Detroit: From Margin to Mainstream*. Detroit: Wayne State University Press.

Abraham, Sameer Y. 1983a. Detroit's Arab-American Community: A Diversity and Commonality. In *Arabs in the New World: Studies on Arab-American Communities*, ed. Sameer Y. Abraham and Nabeel Abraham, 84–108. Detroit: Wayne State University Center for Urban Studies.

———. 1983b. The Yemeni Immigrant Community of Detroit: Background, Emigration, and Community Life. In *Arabs in the New World: Studies on Arab-American Communities*, ed. Sameer Y. Abraham and Nabeel Abraham, 109–134. Detroit: Wayne State University Center for Urban Studies.

Abraham, Saneer Y., Nabeel Abraham, and Barbara Aswad. 1983. The Southend: An Arab Muslim Working-Class Community. In *Arabs in the New World: Studies on Arab-American Communities*, ed. Sameer Y. Abraham and Nabeel Abraham, 163–184. Detroit: Wayne State University Center for Urban Studies.

Adray, James. 2010. The Islamic Center of Greater Toledo. In *Arab Americans in Toledo: Cultural Assimilation and Community Involvement*, ed. Samir Abu-Absi, 73–89. Toledo: University of Toledo Press.

Akhtar, Saima. 2015. Immigrant Island Cities in Industrial Detroit. *Journal of Urban History* 41 (2): 175–192.

Allen, James P., and Eugene Turner. 1988. *We the People: An Atlas of America's Ethnic Diversity*. New York: Macmillan Publishing Company.

———. 2002. *Changing Faces, Changing Places: Mapping Southern Californians*. Northridge: California State University Center for Geographical Studies.

Amor, Cherif M. 2008. Deciphering Immigrants' Home Environment: Arab American Muslims in the U.S. *Journal of Architectural and Planning Research* 25 (1): 6–23.

Arab American National Museum. n.d. *Arab Americans: An Integral Part of American Society*. Dearborn: Arab American National Museum (AANM Educational Series).

Aswad, Barbara C. 1992. The Lebanese Muslim Community in Dearborn, Michigan. In *The Lebanese in the World: A Century of Emigration*, ed. Albert Hourani and Nadim Shehadi, 167–187. London: Center for Lebanese Studies in association with I.B. Tauris & Co.

Awada, Hanady. 2010. Planting the Cedar Tree: The Syrian-Lebanese Community in Toledo, 1881–1960. In *Arab Americans in Toledo: Cultural Assimilation and Community Involvement*, ed. Samir Abu-Absi, 1–20. Toledo: University of Toledo Press.

Bagby, Ihsan. 2004. *A Portrait of Detroit Mosques: Muslim Views on Policy, Politics and Religion*. Clinton Township: Institute for Social Policy and Understanding.

Chacko, Elizabeth. 2007. Immigrants from the Muslim World: Lebanese and Iranians. In *Contemporary Ethnic Geographies in America*, ed. Ines M. Miyares and Christopher A. Airriess, 313–330. Lanham (MD): Rowman and Littlefield.

———. 2016. Immigrants from the Muslim World: Lebanese and Iranians. In *Contemporary Ethnic Geographies in America*, ed. Christopher A. Airriess, 2nd ed., 371–390. Lanham: Rowman and Littlefield.

de la Cruz, Patricia, and Angela Brittingham. 2003. *The Arab Population: 2000*, Census 2000 Brief C2KBR-23. Washington: U.S. Department of Commerce, U.S. Census Bureau.

Hanania, Ray. 2005. Telling Our Own Story. *Saudi Aramco World* 56 (5): 24–35.

Hassoun, Rosina J. 2005. *Arab Americans in Michigan*. East Lansing: Michigan State University Press.

Howell, Sally, Andrew Shryock, Omar Baghdadi, Mucahit Bilici, Mara Leichtman, Kate McClellan, El Safei Mohamed, Elena Godina, and Karl Longstreth. 2011. *Building Islam in Detroit*. Ann Arbor: University of Michigan. http://biid.lsa.umich.edu/panels.html. Accessed 2 Dec 2016.

Hume, Susan E. 2015. Two Decades of Bosnian Place-Making in St. Louis, Missouri. *Journal of Cultural Geography* 32 (1): 1–22.

Kaya, Ilhan. 2005. Identity and Space: The Case of Turkish Americans. *Geographical Review* 95 (3): 425–440.

Martin, Daniel C., and James E. Yankay. 2014. *Refugees and Asylees: 2013. Annual Flow Report*. Washington: U.S. Department of Homeland Security, Office of Immigration Statistics.

Pavlovskaya, Marianna, and Jess Bier. 2012. Mapping Census Data for Difference: Towards the Heterogeneous Geographies of Arab American Communities of the New York Metropolitan Area. *Geoforum* 43 (3): 483–496.

Scheitle, Christopher P., and Roger Finke. 2012. *Places of Faith: A Road Trip Across America's Religious Landscape*. New York: Oxford University Press.

Schopmeyer, Kim. 2000. A Demographic Portrait of Arab Detroit. In *Arab Detroit: From Margin to Mainstream*, ed. Nabeel Abraham and Andrew Shryock, 61–92. Detroit: Wayne State University Press.

Sengstock, Mary C. 2005. *Chaldeans in Michigan*. East Lansing: Michigan State University Press.

16 East Asians in America: Chinese, Japanese, and Southeast Asian Ethnic Landscapes

Very few Asians migrated to the United States until the mid-1800s. While Chinese workers participated in the California Gold Rush and played a major role in the construction of the Central Pacific Railroad, Asians quickly encountered racial prejudices that limited their job opportunities and restricted their residences to a few neighborhoods. By the 1880s further immigration of Asians into North America was largely prohibited, with these restrictions not being relaxed until after World War II. Nevertheless, during the last two decades of the twentieth century Asian immigration into the United States surpassed that of all other groups except Hispanics, and the United States had accepted several hundred thousand Vietnamese, Laotian, and Hmong refugees in the aftermath of the American withdrawal from the War in Vietnam.

16.1 Immigration from Asia

As the twenty-first century began, migration from East Asia continued at a rapid pace, and the Asian population in the United States increased by 43% between 2000 and 2010, four times the overall national growth rate. In total, 17.3 million Americans reported being either Asian or Asian in combination with another race in 2010, representing 5.6% of the nation's population (Hoeffel et al. 2012). Recent immigration has fueled this population growth. As of 2011, there were 11.6 million Asians in the United States who were foreign-born, representing 29% of the nation's total foreign-born population. Of this population 2,231,000 were born in China, 1,857,000 were born in India, and 1,814,000 were born in the Philippines, the three largest contributors of Asian immigrants to the United States (Gryn and Gambino 2012).

Asian immigrants from a variety of nations have left their ethnic imprint upon the American landscape, with the type of imprint varying not only by the country of origin of the immigrants, but by the time of their arrival. Put simply, the landscapes created by the Chinese immigrants of the mid-1800s and their descendents are quite different from the landscape that recent immigrants from China and Hong Kong are creating. Within most parts of the continental United States Asian immigrants are a distinct minority, only dominating a few communities or neighborhoods today (Fig. 16.1). In contrast Asians are the dominant group within Hawaii and have been for well over a century. Foreign-born Asians now comprise 14% of Hawaii's population (Gryn and Gambino 2012).

The total East Asian ethnic population in the United States in 2010 was 12.7 million, with the largest groups being the Chinese, who numbered over four million and Filipinos with a population of 3.4 million. Koreans and Vietnamese each accounted for 1.7 million individuals, and 1.3 million persons of Japanese ancestry were enumerated (Hoeffel et al. 2012). While Chinese workers participated in the California Gold Rush, Japanese immigrants began shaping the American landscape in the mid-1860s, and Filipinos began arriving when their land was an American colony, other Asian groups are far more recent arrivals. For example, before the mid-1970s, there were virtually no Vietnamese, Laotians, or Hmong in the United States. Some Asian groups are still largely unrepresented among the American people.

16.2 Chinese Ethnic Landscapes in America

The contemporary distribution of Chinese-Americans is an artifact of not only recent Chinese immigration, but of events that occurred well over a century ago. At the time of the 2000 Census, 2,734,841 persons reported being of Chinese ethnicity, with California, counting 1,111,508 Chinese ethnics, leading the list. A decade later California's Chinese population (including those from Taiwan) totaled 1,459,039, while New York state had the second largest number, 617,465. Hawaii, with 170,439 persons of Chinese ancestry in 2000 and 199,872 Chinese ethnics in 2010, had the greatest proportion of its population being ethnically Chinese of any state, 14.1%. Both California and Hawaii have a long history of Chinese settlement.

J.A. Cross, *Ethnic Landscapes of America*, DOI 10.1007/978-3-319-54009-2_16

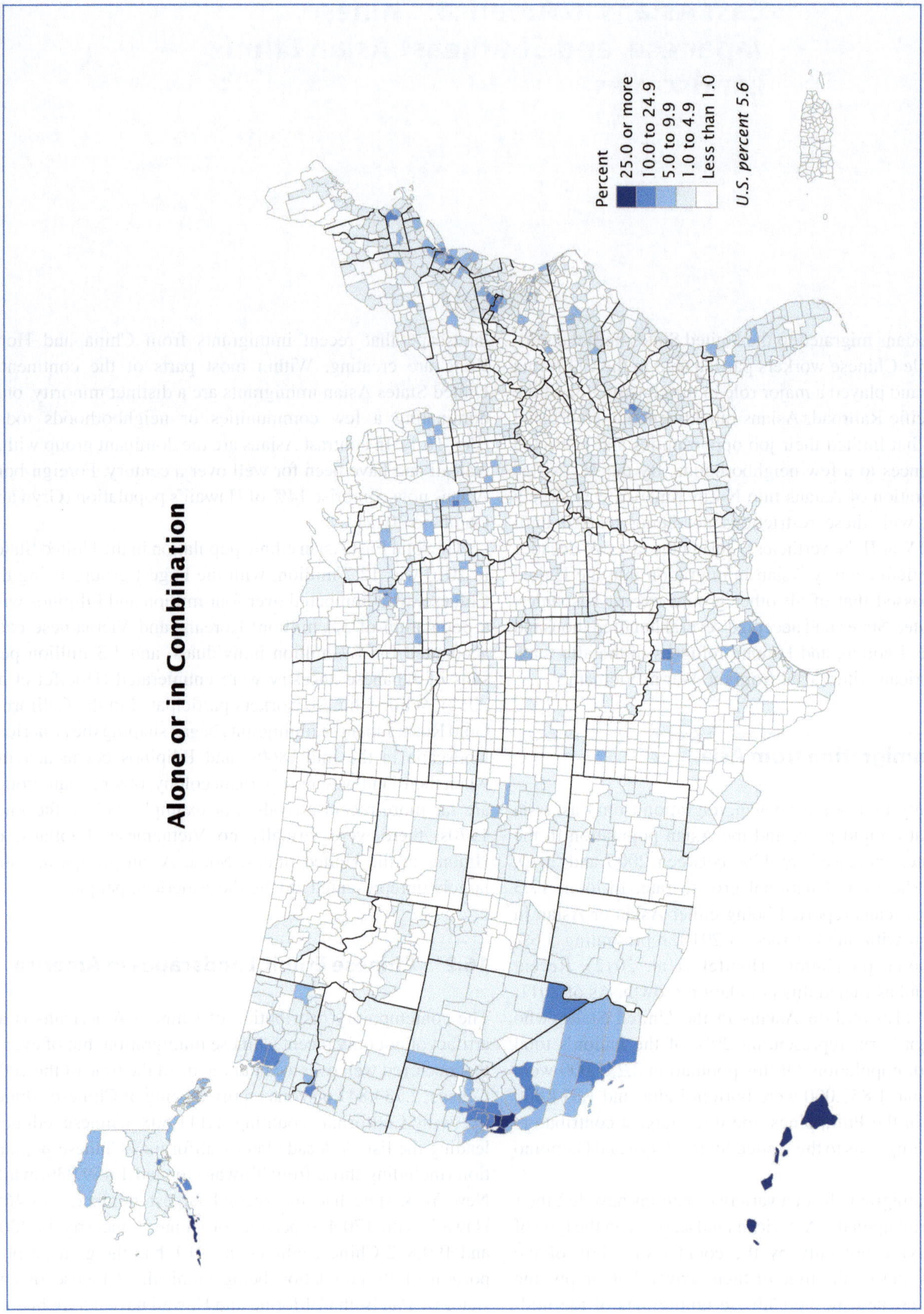

Fig. 16.1 Proportion of the population reporting Asian ancestry (either alone or in combination with another grouping) in each county at the time of the 2010 Census. Note that those counties with the greatest percentages of their populations being Asian are largely located along the Pacific coast, including Hawaii in major metropolitan areas, and in counties housing major universities (Reproduced from U.S. Census Bureau map within Hoeffel et al. (2012))

16.2.1 Historical Geography of Chinese Immigration

Chinese workers were first employed in the American West over 160 years ago. By 1851 there were 25,000 Chinese in California, participating in the California Gold Rush (Sowell 1981). The 1870 Census enumerated 49,310 Chinese in California, 8.8% of the state's population. The Chinese were the largest foreign-born group in the mining area until 1880, yet discrimination forced many off their claims, causing the Chinese to often work abandoned claims on which the easiest to obtain gold had already been removed. The Chinese also worked as miners in other mining districts, including the copper mines at Butte and Helena, Montana. Nearly 10,000 Chinese were also employed in the construction of the Central Pacific Railroad. Chinese workers likewise built the Northern Pacific Railroad in Montana and the Southern Pacific Railroad in Texas and Arizona. Chinese labor dug drainage ditches in California's Sacramento Delta, enabling the area to become a major agricultural region (Allen and Turner 1988; Takaki 1993). In fact, there were so many Chinese miners, railroad workers, and farmers that the 1870 Census found that 7% of Nevada's population was Chinese. The same census showed that Chinese accounted for 28% of the population within Idaho Territory and 9% in Montana Territory, in addition to their large numbers in California. Nevertheless, unlike so many ethnic groups, these Chinese workers very rarely brought their families to America with them. Thus the immigrant Chinese population created overwhelmingly male dominated landscapes and had far less impact upon the population than they would have had otherwise (Sowell 1981; Lai 1980).

Many thousands of Chinese were brought to Hawaii during the last half of the nineteenth century as plantation workers. Although Americans quickly gained political and economic dominance in Hawaii, the islands did not become a U.S. territory until 1898. Chinese workers were often brought under contact from southern China to work in sugarcane fields and in the sugar mills. Chinese ethnics constituted 19% of the Kingdom of Hawaii's population by 1890. After their labor contracts expired, most of these Chinese moved to Hawaiian cities (especially Honolulu) or to the cities of California. Both the Kingdom and the Republic of Hawaii were less restrictive on Chinese entry than the United States was in the nineteenth century regarding direct Chinese immigration to the American mainland.

Legal prohibitions against Chinese immigration began with the Chinese Exclusion Act of 1882, under which the only Chinese permitted to enter the U.S. were foreign-born wives and children of U.S. citizens. By the time the exclusion act went into effect, over 322,000 Chinese workers, mostly male, had come to the United States (Lai 1980). The Scott Act of 1888 was even stricter, prohibiting the entry of Chinese women, thus eliminating the possibility of the Chinese population reproducing itself within the country. The Immigration Act of 1924 banned entry of all Chinese. As the restrictions on their presence became stricter, Chinese workers left their rural worksites and concentrated into inner-city Chinatowns. With these restrictions, the Chinese ethnic population in the U.S. declined from 107,000 in 1890 to 62,000 in 1920. Because China was an ally in World War II, the Chinese exclusion was stopped in 1943.

Discrimination in employment and land ownership significantly limited the type of impacts that Chinese ethnics could make upon the landscape. White American workers prevented the Chinese from employment in many industries. Chinese were employed in jobs where they would provide less competition, such as in restaurants and laundries. Chinese restaurants were opened in both large and small communities, and they established so many laundries that the phrase "Chinese laundry" came into frequent usage (Takaki 1993). As late as the 1920s half of all Chinese workers were employed in restaurants and laundries (Sowell 1981). Some Chinese became grocers, particularly in minority communities, such as in areas of Mississippi that were predominately African-American (Allen and Turner 1988). Deed restrictions, now ruled illegal, prevented Chinese (as well as other racial minorities) from purchasing lands within many neighborhoods, resulting in densely populated Chinatowns in the small section of the city where their residency was permitted. As Wei Li (2005:31) explains, Chinatowns became a refuge for Chinese immigrants, where they "lived, worked, and sought to protect themselves from discrimination in the host society."

16.2.2 Recent Immigration from China

Within recent decades Chinese immigration has rapidly expanded. Between 1991 and 2000 the number of legal immigrants from China and Hong Kong totaled 528,893. Since 2000, the U.S. has averaged receiving over 63,000 legal immigrants annually from China, plus an additional 5000 each year from Hong Kong (U.S. Dept. of Homeland Security 2014). Only Mexico sent more immigrants than China. In addition, ethnic Chinese comprise some of the immigrants from other countries in the Orient. For example, many Vietnamese immigrants are ethnically Chinese. Ethnic Chinese immigrants to the U.S. have also come from non-Asian countries, including Cuba, Jamaica, Trinidad, and Brazil.

While most Chinese immigrants in the nineteenth century came as laborers, whether to work in the fields, the mines, or in construction, today many Chinese immigrants obtain employment in the professions or business (Fig. 16.2). While those who are less educated and less likely to speak English—as well as those who have arrived illegally—still seek refuge

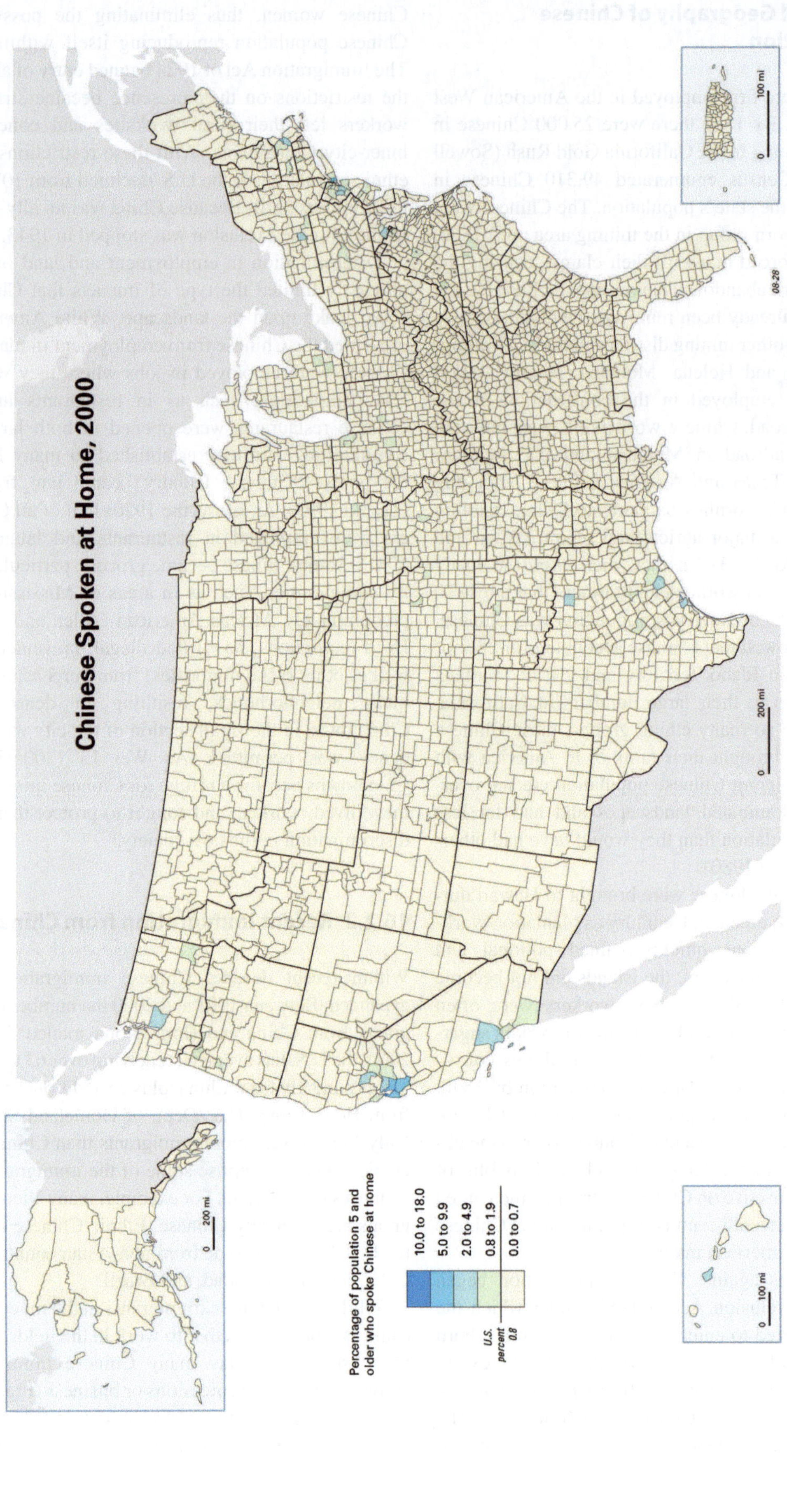

Fig. 16.2 Percent of the population speaking Chinese at home in 2000 in each county and parish of the United States. The distribution of Chinese-Americans is more highly concentrated in major metropolitan areas and university cities than is the overall Asian population (Reproduced from U.S. Census Bureau map in Suchan et al. (2007))

in the historic Chinatowns, many Chinese immigrants during the past several decades have clustered in suburban communities or revitalized declining neighborhoods, such as the Flushing area of New York City (Li 2007), which is more of an ethnoburb than a Chinatown (McGlinn 2002). Those recent middle- and upper-income immigrants typically moved into suburban areas, or into ethnoburbs, "where Chinese had significant presence [and which] continued to grow through an increasing population and a booming economy funded by Asian capital" (Li 2005:36). Ethnoburbs have developed in the San Gabriel Valley east of Los Angeles, the Fremont area of Oakland, and the Silicon Valley near San Jose. Within several suburban California neighborhoods some Chinese immigrants, particularly wealthy Hong Kong and Taiwanese businessmen who are relocating, have antagonized neighbors by building large houses that occupied disproportionate shares of their lots (Li 2005).

China is a large country, the size of the United States, and significant regional differences in culture and ethnicity can be found in the country. While many Americans have noted the broadening of Chinese cuisine to include Mandarin, Szechwan, and Hunan dishes during the past several decades, expanding upon the previous near ubiquitous Cantonese dominance of Chinese cooking, this is simply one manifestation that the source region of the Chinese immigrants has expanded. Before World War II the vast majority of Chinese immigrants to United States came from just one of the seven districts that comprise Guangdong province, the area surrounding the city of Guangzhou, formerly known to most Westerners as Canton. This region was less agricultural than the rest of China, providing more merchants and traders. Most Chinese who came to the United States and Hawaii during the nineteenth century spoke the Toishan dialect, similar to Cantonese. Since World War II most Chinese immigrants arriving in the U.S. speak Mandarin Chinese, resulting in a linguistic barrier between earlier and more recent Chinese-Americans.

16.2.3 Urban Orientation of Chinese-Americans

Ethnic Chinese accounted for over 10% of the total population within five American counties in 2010, up from three a decade earlier. San Francisco County, California, whose borders are coincident with the City of San Francisco, had the greatest proportion of Chinese in its population, 22.9% compared with 20.7% in 2000, while Honolulu County, Hawaii had the second highest proportion in 2010, 16.5%. Hawaii County was 12.2% Chinese, while Alameda and San Mateo Counties, California, were 10.8% and 10.3% Chinese (including Taiwanese), respectively. Seven additional counties, two in California, two in Hawaii, and three in New York, had populations that were between 5.0% and 9.9% Chinese.

The Chinese ethnic population in the United States is overwhelmingly urban. By 1980 97% lived in urban areas and 96% in metropolitan areas, figures that are much higher than for other ethnic groups (Allen and Turner 1988). The 2000 Census showed that 56% of all the Chinese ethnics in the United States lived in just three consolidated metropolitan areas, the San Francisco Bay Region (San Francisco, Oakland, and San Jose), Los Angeles (including Los Angeles, Orange, and Riverside Counties), and New York City (including Newark, New Jersey and Long Island, New York). The greater New York-Newark consolidated metropolitan area alone was home to 20.5% of all Chinese-Americans in 2000. By 2010, even though its Chinese-American (including Taiwanese) population had increased to 696,855, greater New York City's metropolitan area's share of the nation's Chinese ethnics had fallen to 17.3%, as Chinese ethnics dispersed into a larger number of metropolitan areas. Metropolitan Los Angeles-Long Beach-Santa Ana had 544,000 Chinese ethnics, San Francisco-Oakland-Freemont had 477,000, while nearby San Jose-Sunnyvale-Santa Clara had 173,000 in 2010 (Hoeffel et al. 2012).

Cities with at least 100,000 Chinese-Americans in 2010 within their city limits include New York City with 514,116, San Francisco with 184,513, and Honolulu (coincident with Oahu County) with 156,733. While they comprised only 6.3% of the residents in New York City, they accounted for 9.4% of the population of Queens County (New York City's borough of Queens) and 7.2% of Kings County's population (New York City's borough of Brooklyn). Other cities with substantial Chinese ethnic populations (including Taiwanese) in 2010 are Los Angeles (81,109), San Jose (73,672), Chicago (47,765), and San Diego (43,957). Within each of these large cities, a distinct downtown Chinatown either currently exists or had historically existed. Nevertheless, particularly within California and Hawaii, the Chinese population is found both within traditional Chinatowns and newly developing suburban ethnic enclaves that share many, but not all, characteristics with the long-established Chinatowns (Li 1993).

16.2.4 Chinese Urban Landscapes: Chinatown

Chinatowns are dense urban enclaves of Chinese ethnics, and they collectively comprise the most distinct Chinese ethnic landscape. White Americans played a crucial role in the development of Chinatowns. Chinatown was set apart from the rest of city and had its own spatial identity as a refuge of the Chinese (Fig. 16.3). It was an area where other city activities might not be extended, but at the same time it was often the only part of the city where the Chinese were permitted to own property. Racially restrictive real estate covenants typically kept Chinese-Americans from purchasing properties in most areas from the Nineteenth Century until well

Fig. 16.3 Gate marking the entrance to the Chinatown in Portland, Oregon. This gate, as with the gates in San Francisco and other cities, was built long after the establishment of the Chinatown, and highlights the distinct ethnicity of the enclave. The foo dogs flanking the gate with its decorative tile roof are sometimes considered as Chinese guardian lions

after World War II. The historic Chinatowns that developed, including those in San Francisco, Los Angeles, Honolulu and New York City, were settled almost exclusively by Chinese from the Canton area, now called the Guangdong province. While most surviving American Chinatowns developed in seaport cities, those established over a century ago in interior cities such as Pittsburgh, Detroit, and St. Louis were beginning to disappear by the time of World War II. Nevertheless, both Chicago and Cleveland boast Chinatowns, with that in Illinois being the nation's largest other than those in California and New York (Fig. 16.4).

Chinatowns are not limited to the United States, and indeed Chinatowns have been described in Canada, Australia, England, France, and even Singapore and Thailand, where the Chinese lived in districts separate from other Asians. Nevertheless, as Kay Anderson clearly explains, Chinatowns could not have evolved without the influence of persons of European ancestry.

> "'Chinatown' was a social construct that belonged to…[the city's] 'white' European society, who, like their contemporaries throughout North America, perceived the district of Chinese settlement according to an influential culture of race. From the vantage point of the European, Chinatown signified all those features that seemed to set the Chinese irrevocably apart—their appearance, lack of Christian faith, opium and gambling addiction, their strange eating habits, and odd graveyard practices. That is, it embodied the white Europeans' sense of difference between immigrants from China and themselves, between East and West…. The point is that 'Chinatown' was a shared characterization—one constructed and distributed by and for Europeans, who, in arbitrarily conferring outsider status on these pioneers…were affirming their own identity and privilege" (Anderson 1987, "The Idea of Chinatown: The Power of Place and Institutional Practice in the Making of a Racial Category," *Annals of the Association of American Geographers*, 594, Reprinted with permission of the American Association of Geographers).

The circumstances that led to the establishment of the historic Chinatowns in American cities no longer exist, yet these agglomerations of Chinese-Americans and their conspicuous landscapes have not only survived, but many are growing. They still are described as providing their "residents and other Chinese visitors with a 'comfort zone'" (Guan 2002:126).

The rapid rate of Chinese immigration during the past couple of decades has caused both the areal extent of Chinatowns to expand and new Chinatowns to develop. Recent immigration resulted in new Chinatowns being established in Houston and the Los Angeles suburbs, and for new middle-class Chinatowns, or ethnoburbs, to form, such as the one in Flushing, within New York City's borough of Queens (Fig. 16.5), which now rivals the population of the city's older Chinatowns in Manhattan and Brooklyn. At the same time, many of the more affluent residents have left the older Chinatowns, with their places being taken by new immigrants. Social contrasts exist between Chinatown populations

Fig. 16.4 Wentworth Avenue has long been the focus of Chicago's Chinatown, located south of the city's Loop. The Pui Tak Center Building with its pagoda shaped towers, just left of the Chinatown Gate, was constructed in the 1920s in traditional Chinese architectural style to house the On Leong Association and a religious shrine. Its ground floor now houses several retail shops

Fig. 16.5 Advertising signage in Chinese along a commercial side street within Flushing, New York. This rapidly expanding ethnoburb within New York City's borough of Queens became a destination for Chinese immigrants beginning in the 1970s, and its commercial buildings are much newer than within many long-established Chinatowns

and other Chinese-Americans. Three decades ago it was noted that "Chinatowns have more than their share of low income people, despite the above-average incomes of Chinese Americans as a group" (Sowell 1981:146). As Chuo Li (2015:1119) recently noted, within New York City's Chinatown, "commercialism and urban redevelopment, gentrification has transformed the physical and social composition of the community. The old store and traditional streetscape have been replaced with modern infrastructure that supports the role of the community as a nexus of transnational financial networks." Even though it is wealthier, a major divide remains between New York Chinatown's poorer residents and its economic elite.

16.2.4.1 City Within a City

Chinatowns basically function as a city within a city (Fig. 16.6). While his observations were based upon San Francisco's Chinatown, much of what Christopher Salter wrote over four decades ago could be said for the Chinatowns of other major cities today. He succinctly explained that "it is not necessary for a Chinese to ever leave Chinatown" given Chinatown's ability to cater to almost all of its residents' "economic, social and professional needs… within nuclear Chinatown or its immediate margins" (Salter 1978:34). Within Chinatown one could find a hospital, shops, restaurants, employment, schools, social clubs, temples, and funeral homes, just to name a few of its services. Given that so much of ethnic culture relates to dietary preferences of various populations, both the types of foodstuffs offered in groceries and butcher shops and the manner in which those products are displayed comprise a salient part of the visible ethnic landscape (Figs. 16.7, 16.8, and 16.9). Not only do such shops cater to the everyday needs of their Chinese residents, but they also are among those elements of Chinatowns that attract tourists. In a similar manner many of the newer ethnoburbs meet all of the everyday needs of their Chinese residents.

San Francisco's Chinatown is relatively unique given its intentional incorporation of Chinese architectural elements into its built environment (Fig. 16.10). At the beginning of the twentieth century guidebooks to San Francisco noted "the quarter's shops, temples, and theaters…gambling halls and hidden opium dens" (Rast 2007:42). Following its destruction from the 1906 earthquake and fire, reconstruction took place using "an exotic, 'oriental style' of architecture [that] would 'attract tourists' and 'boost business.' Hence, the Sing Chong Building at the northwest corner of California and Grant was distinguished by pagoda-like towers, upturned eaves, and a color scheme of bright red, green, and gold" (Rast 2007:54). The rebuilt Chinatown more conspicuously displayed Chinese architecture than the community it replaced. Indeed, as the Chinese-American architect and historian Philip Choy (2012:14) explains, "This new trend of a Sino-architectural vernacular, created specifically as a

Fig. 16.6 Chinese grocery along Stockton Street in San Francisco's Chinatown. Note how the displays of fruits and vegetables are overflowing onto the sidewalk

Fig. 16.7 Bok choy along with other vegetables displayed in front of a Chinese grocery along Mott Street in New York City's Chinatown

Fig. 16.8 Variety of dried fish and shellfish products displayed by entrance to food shop along Mott Street in New York City's Chinatown

response to the threat of relocation after the quake, shaped the present skyline of Chinatown." Efforts have been made in other Chinatowns to feature Oriental architectural elements. For example, in New York's Chinatown the On Leong Merchant Association building was renovated in 1950, and a "new three-roofed pagoda was added on top of the three-story brick building, as well as balconies with white rails and engaged red columns" (Li 2015:1123).

Fig. 16.9 Chinese shop displaying barbequed ducks in Flushing, New York, within New York City's borough of Queens

Fig. 16.10 Sing Chong Building at corner of Grant Avenue and California Street in San Francisco's Chinatown. Its architects "chose the most obvious representation of Chinese architecture, the pagoda…[using] multi-tiered eaves to emulate the multi-story pagoda" (Choy 2012:113)

The architecture in several Chinatowns does not necessarily fully reflect the ethnic heritage of the Chinese. Indeed, in describing part of San Francisco's Chinatown, Philip Choy (2012:45–46) notes "the illusion created is a masterful design solution, unique and indigenous, for it is neither East nor West but decidedly San Francisco." Nevertheless, Chinatowns all have their own characteristic style, even though originally they all were dominated by tenements and residential hotels that catered to male residents. These buildings often provided additional commercial and social services to their residents, as Christopher Yip explains.

> Shops would occupy the ground floors, with storage, club rooms and gambling establishments squeezed into the basements. Often, single residential hotel rooms…occupied the middle floors, while the top floor was used to house one of the important benevolent associations or a temple (Yip 1986:106, Reprinted with permission of John Wiley and Son).

Even though many of these structures resemble standard American tenements in some aspects—and may have even been constructed before the Chinese immigrants arrived, their extensive signage in Chinese characters, the exhibits of ethnic foodstuffs and other wares in shop windows and overflowing onto sidewalk displays, and the choice of bright colors and decorations all attest to Chinese ethnics remaking their urban landscape. Chinatowns frequently have higher population densities than the overall cities in which they are located. Thus, their "streets appear narrower, restaurants busier and grocery stores more crowded" (Guan 2002:130).

The expansion of New York's Chinatown, the nation's largest since the 1970s (Li 2015), into older ethnic neighborhoods creates an interesting blend of cultural landscapes. As Chinatown has grown into neighboring ethnic enclaves, "Chinese signs now overlook what used to be Little Italy; roast duck and beef in oyster sauce have taken the place of blintzes and bortsch in the old, famous Garden Cafeteria, haunt of Jewish intellectuals on East Broadway" (Pan 1994:304). A former Jewish synagogue on New York City's Pike Street, as illustrated in Chap. 13 (See Fig. 13.23), has become a Buddhist temple. Several other Chinatowns have also expanded into neighborhoods created by other ethnic groups. For example, in San Francisco and Chicago parts of their Chinatowns were originally built to accommodate Italian immigrants. Thus, while many older buildings in Chinatowns reflect American or ethnic European style, Chinese ethnicity is displayed in the material culture that is draped over these structures, and in the new ethnic usages of the buildings.

16.2.4.2 Chinatowns and Display of Ethnic Architecture

The entrances into many Chinatowns are marked by elaborate gates constructed over prominent streets leading into the communities, yet most of the Chinatowns had long existed before their construction. Chinese architecture displayed in Chinatowns is often conspicuously flaunted on some of the more recently constructed buildings, such as in the pagoda-shaped buildings along Canal Street in Lower Manhattan (Fig. 16.11) or Grant Avenue in San Francisco. One of the most distinctive types of Chinese-American structures houses various Chinese benevolent associations and temples on its top floors (Fig. 16.12). As Christopher Yip explains,

> Most benevolent associations, made up only of males, based their membership on district, village, dialect, or kinship bonds and functioned as social and quasi-political organs of the community. In large Chinatowns each association sought the most elaborate and prestigious building and often located its rooms on the top floor, whose facade would be decorated with balconies with lights, inscribed marble door surrounds and protective curving eaves above (Yip 1986:109, Reprinted with permission of John Wiley and Son).

Fig. 16.11 Modern banking and financial services building constructed in classical Chinese architectural style at intersection of Canal Street and Centre Street, just west of New York City's Chinatown-Little Italy Historic District

Fig. 16.12 Benevolence associations and temples are often placed on the top floors of buildings within Chinatowns. These buildings, which house benevolence associations and at least one has a temple, are located along Waverly Place in the historic Chinatown in San Francisco, California. Note the Nationalist Chinese flag

Describing the Tien Hou Temple in San Francisco's Chinatown, Philip Choy (2012:209) notes, "The third floor is enlivened by a deeper balcony, with two cast-iron columns and a central doorway framed by a marble surround engraved with Chinese characters and fringed by a band of egg and dart design." Sometimes the entire building housing these benevolent associations or temples displays Chinese ethnic architectural elements, at other times these are confined to the temple area on its upper floor.

A variety of common Chinese architectural features are often seen on the facade of buildings in Chinatowns (Fig. 16.13). These include a curved roof, which ranges from being slightly concave to having an upswing beginning above the side walls; a tile roof covered with large colorful tiles; lattice work displayed in doorways, windows, stairways, and courtyards; the use of plinths and columns, extending to the beams that support the roof or next floor; arches and vaults over doorways and windows; the display of a dragon or lion motif; the presence of six to eight sided Chinese lanterns, often with an ornate peak and base; and the use of vivid, brilliant colors (Salter 1978).

Colors used on buildings by the Chinese have their own unique meanings. For example, red, which is very common, symbolizes happiness and vitality; while yellow, the symbolic color of the Manchu dynasty, denotes wealth, prosperity and power. Green symbolizes life and abundance, while blue signifies heaven and tranquility. White and black are rarely used except at a time of mourning (Salter 1978). Not only are such colors vividly displayed on the exterior walls and decorations of buildings, but they along with dragons and other mythical animals are displayed in conspicuous monuments (Fig. 16.14).

Religious beliefs have strongly influenced the Chinese ethnic landscape. Temples are erected to a variety of gods and goddesses. Some temples have multiple altars, with a primary deity in the central room, while others are at the sides. Spiritual beliefs are even displayed by San Francisco's Chinatown Gate, which was built in 1970. It is aligned facing the south in accordance with the principles of Feng Shui (wind water). This ancient Chinese tradition relates to the energy and feel of a place, and it is a very important consideration in Chinese architecture and real estate transactions. Many Chinese ethnics will not occupy a structure until determining that it has been built according to the principles of Feng Shui, and experts advise builders on the appropriate locations of doors and placement of rooms. The foo dogs to the side of the Chinatown gate (in San Francisco, Portland, and Honolulu, among other cities) represent mythical creatures that ward off evil spirits.

Almost every feature found within a traditional city in China may be found in Chinatown. The city wall is the only major architectural feature of a traditional Chinese city that

Fig. 16.13 Many Chinese architectural elements are displayed on this façade along Stockton Street within San Francisco's Chinatown. Note the decorative tile awnings, the upturned fish at the corners, and the use of the bright colors and decorative designs above the windows

is missing from Chinatown, although the construction of the Chinese gate symbolically marks the entrance. Of course, many Chinese cities, including Beijing, demolished their ancient walls in their race to modernize over the past century. Furthermore, with their rush to modernize—and to accommodate their burgeoning urban populations—Beijing, Guangzhou, Shanghai, and many other Chinese cities have replaced much of their traditional one story brick courtyard housing with high rise apartments.

16.2.4.3 Tourism and Chinatown Architecture

Increasingly, many Chinese communities have flaunted their ethnicity as a means to enhance tourism (Fig. 16.15). In the case of Philadelphia, Chinatown "attracts both ethnic Chinese and American tourists" (Guan 2002:127). To attract tourists businesses have made "a more conscious effort to present an exotic image" (Yip 1986:106). To this end, they have added gateways and enhanced the usage of traditional Chinese architectural motifs in their buildings. Even in the rebuilding of San Francisco's Chinatown following the 1906 earthquake, the Chinese made a deliberate effort to "rebuild their buildings with Oriental and artistic storefronts" (Fong-Torres 1991:45). They introduced stylized pagodas such as in Chinatown Telephone Exchange, built in 1909, which now serves the Bank of Canton. In a similar manner, a recently constructed financial services building on New York City's Canal Street resembles a pagoda. As Chuo Li (2015:1118) explains, "For American Chinatowns, ethnic themes and Orientalist structures such as Chinese-style gateways and pagoda-tiered restaurants have been used to enhance their exoticism and market a sense of place."

Fig. 16.14 Nine Dragon Wall in Chinatown, Chicago, Illinois. This elaborately decorated mural is a replica of a monument in Beijing, China. Such artwork and decorative sculptures of animals and mythical creatures as well as architecture mark the ethic Chinese landscape in many Chinese-American communities

Fig. 16.15 Bright red lanterns and decorative streetlights along Grant Avenue attract tourists to San Francisco's Chinatown

16.2.5 New Chinatowns and Ethnoburbs

With large increases in Chinese immigration over the past several decades, Chinatowns in San Francisco, New York, and Chicago, as well as other communities have grown. At the same time, suburban Chinese enclaves have been established in the San Gabriel Valley of Los Angeles and elsewhere. While the older Chinatown may still be "the symbolic geographical center of the Chinese American experience" (Allen and Turner 1988:181), many Chinese ethnics now live in suburbs or ethnoburbs where new ethnic shopping centers are being built (Li 1993). The older Chinatowns are typically home to poorer residents than the new suburban enclaves, and they are far more likely to be dominated by Cantonese speaking Chinese, rather than those speaking Mandarin.

Some scholars have described the new settlements as "suburban Chinatowns," yet Wei Li (1998) argues that they are fundamentally unlike, given their demographic, socioeconomic, and economic differences. The social and legal constraints which formerly shaped the development of traditional Chinatowns have largely disappeared within contemporary American society. In contrasting these newer suburban Chinese ethnic enclaves with traditional Chinatowns, their residents benefit from "the on-going interaction with the homeland for communicational, familial, religious, and business purposes, and above all, its status as a microcosm of the Old Chinatown and the homeland" (Laguerre 2005:41).

The landscape of the traditional Chinatowns differs from that of the newly established Chinese dominated ethnoburbs, such as Monterey Park and San Gabriel. Similar suburban Chinese enclaves have also been established in Bellaire City—near Houston, Texas, and in the New York City metropolitan area (Lin and Robinson 2005). In some cases the new Chinese ethnic communities have developed in already established suburban neighborhoods, such as the Richmond district of San Francisco, while in others they are focused upon developing neighborhoods, such as Rowland Park, in suburban Los Angeles. While both Old Chinatowns and the New Chinatowns or Chinese ethnoburbs display a wide variety of businesses that display advertisements using Chinese characters (Fig. 16.16) and buildings incorporating elements of traditional Chinese architecture, the Chinese ethnoburbs are far more prosperous, less crowded, and less overwhelmingly Chinese (Li 1998). Thus, they look different from the traditional Chinatowns, differing in the density of settlement and architecture. Ethnoburbs are also the sites of both strip mall and large shopping malls (Fig. 16.17), stocked with ethnic foodstuffs and imported products sought by Chinese ethnic customers, and are replete with signage in Chinese characters and sometimes displaying Chinese-style architectural elements—such as entrances through an ethnically styled gateway (Li 2007).

In contrast with "Old Chinatowns that display a distinct architecture that set them apart from the rest of the city, New Chinatown's housing blends with the rest of the city's"

Fig. 16.16 Walk-up dim sum shop in modern Chinese ethnoburb of Monterey Park, California. Note the bright display of *yellow* and *red* colors

Fig. 16.17 Modern shopping center catering to Chinese-American shoppers in San Gabriel, California. While the architecture is not ethnically distinctive, the signage, selection of wares, and display of goods in the shops clearly showcase Chinese ethnicity

(Laguerre 2005:45). Yet in some of the newer developments, ethnic differences are visible. For example, within Los Angeles' San Gabriel Valley reside "Chinese from over thirty different countries or overseas areas" (Allen and Turner 2002:39), including China, Taiwan, Hong Kong and Southeast Asia. Some of these groups, such as those ethnic Chinese from Vietnam, identify closely with their Vietnamese heritage and utilize Vietnamese in the names advertizing their stores.

Suburban housing developers targeting the Chinese-American market have taken special efforts to accommodate their desires. As Lin and Robinson (2005:59) explain, a builder in Walnut, California "brought Buddhist monks to bless the subdivision, and instructed architects to design their homes sensitive to Chinese feng shui principles. Stairways angle away from the front door so the owner's luck won't run away." Likewise, streets are laid out to avoid having houses face a T-shaped junction, and care is taken to avoid addresses with unlucky numbers (Lin and Robinson 2005), which differ from those considered unlucky by European-Americans.

While the suburban developments may be considered as satellite Chinatowns, they have a more diverse Chinese population than within the traditional or revitalized downtown Chinatown (Aronson and Kent 2008). In some communities a modern commercial Chinatown develops, which bring together a variety of groceries, restaurants, and ethnically oriented shops catering to Chinese-American shoppers, even though many of these ethnics live at some distance from the Chinese mall or shopping center. As we will see later, some of these Chinese commercial districts, which lack contiguous ethnic residential neighborhoods, cater to a broader Asian population.

16.2.6 Chinese Landscapes in Rural America

Although the contemporary Chinese ethnic population in the United States is overwhelmingly urban in its orientation, Chinese immigrants have left some imprints on the cultural landscape in rural America. Even in communities with few Chinese residents, Chinese ethnics often operate Chinese restaurants, displaying Chinese characters in their advertising and various decorative motifs. Because Chinese labor built so many of the railroads, drainage ditches, and mines in the West during the 1800s, considerable material cultural from that period remains. For example, while remains of dugout dwellings of Chinese workers have been identified at several locations in the Pacific Northwest, even more conspicuously Chinese are the terraced gardens that they established on steep mountain slopes in Idaho to produce fresh vegetables, consistent with their ethnic dietary practices (Dubrow 2000).

Chinese labor grew fresh vegetables outside a variety of Pacific Coast communities, and with their agricultural skills brought from China, they have been credited with promoting California's emphasis upon orchard and vegetable crops rather than cereal crops (Dubrow 2000). Even in these rural agricultural environments the Chinese established villages of closely spaced two-story wooden buildings, including benevolence societies. These frame buildings with their balconies in Locke, California survive today, nearly a century after they housed a bustling ethnic Chinese farming community within the Sacramento River delta.

16.3 Japanese-American Ethnic Landscapes

The Japanese-American population has been historically concentrated along America's West Coast, although relocation and internment of both Japanese nationals and Japanese-American citizens during World War II disrupted that pattern. During the half century after the war the Pacific Coast domination slightly weakened. Nevertheless, by the time of the 2000 Census 65.1% of the entire Japanese-American population lived in just three Pacific Coast states: California, Hawaii, and Washington. A decade later the Japanese-American population totaled 1,304,286, of which 61.9% lived in those three states. Twenty-three percent of Hawaii's population reported Japanese ancestry in 2010, with Honolulu County (the island of Oahu) having the highest proportional Japanese ancestry of any county in the nation, 25.3%. The other counties in Hawaii ranged from 14.7% to 19.2% Japanese ancestry, excluding miniscule Kalawao County, comprising the leper colony. On the mainland only three counties had a 2010 population that was at least 1.5% Japanese ancestry, down from five counties a decade earlier. These were Santa Clara County, California (2.0%); San Francisco County, California (1.9%); and King County, Washington, encompassing Seattle (1.8%).

16.3.1 Japanese Immigration

Japanese immigration to the mainland United States began in the 1890s, largely to work in railroad maintenance, or to labor as agricultural workers, miners, or loggers. Like the Chinese, the Japanese encountered resentment from white American workers, and their immigration was limited in 1907, with no entry permitted for common laborers. Nevertheless, "picture brides" were granted entry for marriage to workers who had already arrived, enabling the Japanese to establish more of a family structure than was possible for many of the Chinese. Most Japanese-Americans were shop-keepers and farmers, as they rapidly left laboring jobs and used their savings to become shop-keepers or to operate their own farms (Allen and Turner 1988; Takaki 1993).

Japanese workers were also brought in large numbers to work on the sugar cane plantations in Hawaii, particularly during the period that it was the Republic of Hawaii. At the time of the 1900 Census, there were 61,111 Japanese enumerated in Hawaii, accounting for 39.7% of the territory's population.

16.3.1.1 Japanese Agricultural Legacy

As farmers, the Japanese dominated production of certain crops, such as snap beans, celery, and strawberries, among others. As Gail Dubrow explains,

> Those looking for small signs of a distinctive Asian imprint on cultural resources associated with agriculture miss the larger picture: It simply is not possible to understand the agricultural landscape of the American West without observing the central roles Asian Americans played in creating it (Dubrow 2000:156).

Not only did Japanese ethnics grow 90% of several of California's vegetable and berry crops during the second decade of the Twentieth Century, but they also raised one third of the grapes. Several all Japanese farm communities existed in California's Central Valley at that time (Dubrow 2000), yet a couple of these locales have no Japanese ethnic residents today.

Besides the Japanese farming communities near Fresno in California's central Valley and along the central California Coast near San Luis Obispo, other Japanese farmers settled in the Los Angeles Basin and near Portland and Hood River, Oregon, as well as in western Washington (Allen and Turner 1988). Malheur County in far eastern Oregon still had a population that was just under 1.5% of Japanese ancestry in 2010. In the valleys east of Seattle Japanese ethnic farmers excelled in dairying, providing half of all milk consumed in the Seattle area during the 1920s, and three-quarters of the vegetables (Dubrow 2000). There, they left a visible legacy on the agricultural landscape. For example, several dairy barns were constructed a century ago in western Washington that "incorporated the traditional *irimoya* (gable into hipped) roof form" (Dubrow 2000:154). The internment of ethnic Japanese during World War II greatly disrupted the Japanese-American farming communities, and resulted in many Japanese losing their farm lands, either being abandoned or sold in "fire sales." While some Japanese-Americans were able to resume their agricultural activities following the war, most moved to other occupations, usually in urban areas.

Little Tokyo, centered in downtown Los Angeles, provided a variety of cultural and commercial services and goods to Japanese farming communities in southern California since the beginning of the twentieth century. Like the Chinese, discrimination limited Japanese urban residents to a small downtown district.

16.3.1.2 Landscapes of Internment and Assimilation

Nearly 120,000 persons of Japanese descent were rounded up in California, Oregon, Washington, Alaska and Arizona in early 1942 and sent to concentration camps for the duration of World War II. While this disrupted or destroyed the

Japanese-American cultural landscape, within the last decade efforts have been made to preserve the physical evidence of some of the internment camps. This includes the establishment of Manzanar National Historic Site, California, where over 10,000 Japanese men, women, and children spent the war years. Some of the buildings serving the various camps still remain, as do foundations for barracks and Japanese graveyards.

Japanese-American institutions, museums, monuments, businesses, and hotels have focused upon perpetuating the memory of Little Tokyo, which the "Sansie [third-generation Japanese-Americans] have embraced…as a metaphorical, if not actual, home" (Jenks 2008:237). The site remains culturally significant, even though few of the structures date to the time when large numbers of Japanese ethnics lived in this district of downtown Los Angeles. Thus, the performance of "rituals and activities in the historically significant and emotionally meaningful place of Little Tokyo holds the increasingly diverse and geographically dispersed Japanese American community together" (Jenks 2008:238–9).

Persons of Japanese descent have undergone remarkably high rates of assimilation. Intermarriage rates exceed 50% in many areas, and many Japanese-Americans have moved into neighborhoods of white European ethnic professionals. While most Japanese-Americans now live in metropolitan areas, traces of their legacy remain upon the rural landscape. For example, "the presence of a wood-fired furo, or soaking tub, in either a freestanding outbuilding or a shed attached to a rural residence clearly marks the past presence of a Japanese household" (Dubrow 2000:157).

16.3.2 Japanese-American Religious Landscapes

Religion historically played a major role in Japanese culture and attitudes towards their environment (Fig. 16.18). For example, Donald Fellows (1972:49) writes that "there is a certain unity to all elements of Japanese culture…[that] stems from their forms of Buddhism, forms that call upon man to become an active participant in the workings of the universe." This can be seen in Japanese "architecture, painting, flower arranging or in the landscaping of gardens" (Fellows 1972:49). It is in their religious buildings that Japanese-Americans in Hawaii have most conspicuously displayed their architectural heritage (DeFrancis 1973).

16.3.2.1 Buddhist and Shinto Temples

Hongwanji are Buddhist temples, which are especially common in Hawaii (Fig. 16.19), and although their numbers have declined, 33 remained open in Oahu in 2013 (Hill 2013). These temples display a mixture of Japanese and local building styles, such as pews in the temples. The hongwanji typically have pitched roofs. Many of these religious buildings are wooden, sometime built without nails. Within the American setting, these temples hold Sunday services, being more scheduled and structured than in the Japanese homeland. Pagodas and the Golden Pavilion in Hawaii are replicas of monuments in Japan. For example, the Honolulu replica of Nara pagoda is 116 feet high, consisting of three tiers, topped with a ringed copper spire and a turquoise colored roof. Additional pagodas are located at various memorial parks with temples in Hawaii, especially on the islands of Oahu and Maui.

Fig. 16.18 Kauai Soto Zen Temple in Hanapepe, Hawaii. This Buddhist temple was constructed by immigrants from Japan. Note the curved roof

Fig. 16.19 Lahaina Hongwanji is located on the Hawaiian island of Maui. It was erected in 1927 to serve Japanese-American Buddhists residing in Lahaina

Other structures associated with these temples include entrance gates and a large bell with a wooden log to strike its side. Statues of Buddha patterned after those in Japan, being either wooden or cast bronze, are also to be seen.

16.3.2.2 Shinto Shrines in Hawaii

Many Japanese practice a mixture of Buddhism and Shintoism. The design of Shinto shrines in Hawaii is described as being "representative of the oldest architecture in Japan" (DeFrancis 1973:88). Shinto shrines are typically square and made of unpainted wood. Diagnostic characteristics include "torii gates that mark the approaches to the temples..., stone lions and lanterns and places for ritual washing of the hands near the entrances, the prayer ropes hung near the doorways of the shrines, and the exposed roof beams crossed at the ends" (DeFrancis 1973:88). The religious heritage of Japan is also displayed in churches erected by Christian converts. For example, the Makiki Christian Church is modeled after the medieval Tosa Castle on Shikoku Island, Japan (Fig. 16.20).

16.3.2.3 Japanese Houses and Gardens

While the Japanese ethnic influence upon the architecture of religious buildings such as hongwanji is far more impressive, Japanese immigrants have influenced residential architecture in both Hawaii and California. For example, "the influence of Japanese Buddhism on American architecture can be observed in the number of 'bungalows' erected during the early 1920's which displayed the curving roof lines and ornate entrance arches of Buddhist temples" (Fellows 1972:49). It is noted that those neighborhoods and inner suburbs of metropolitan Los Angeles dating from that time "abound with these old residences which gained inspiration from earlier Japanese participation in expositions and world fairs" (Fellows 1972:49). Japanese style houses in Hawaii typically had a wood supporting frame and lacked load bearing walls. Sliding screens and doors were used instead of windows, and movable screens took the place of walls. Architectural historians have argued that Frank Lloyd Wright's ideas of continuous internal space were influenced by or paralleled Japanese concepts (Lee 1986). Japanese influenced residences commonly have pitched roofs. It has been proposed that Japanese styles blended with those of other Pacific groups and Europeans into the Hawaiian style, which subsequently influenced building on the West Coast.

The Japanese garden is very commonly placed in yards of Japanese-Americans. Not only does this garden add beauty, but it is "a complex work of art" that has Buddhist religious significance. As Fellows (1972:50) explains, "The Buddhist religious garden is a universe in miniature…[that] provides the Buddhist with a place for contemplation and for communion with nature and one's own soul." Common symbolic elements of the Japanese-Buddhist garden include a guardian stone, a black pine, and a lantern or pagoda. The guardian

Fig. 16.20 Makiki Christian Church located in Honolulu, Hawaii was designed to resemble a medieval Japanese castle

Fig. 16.21 Japanese ethnic cemetery near Wailua on the Hawaiian island of Kauai. Note the small stone Buddha figure along with other sacred rocks

stone is large, and is viewed as a home of gods, representing the great mountain (Fellows 1972). The black pine, or other trees symbolizes long life. In addition, a variety of other symbolic stones are displayed.

Cemeteries are another location where Japanese-Americans display their ethnicity (Fig. 16.21). Not only are tombstones inscribed in Japanese characters, but stone statues of Buddha are displayed along with flowers and lanterns.

16.4 Vietnamese and Laotian Hmong Landscapes in America

Southeast Asians represent a recent immigrant group, as there were virtually none in the U.S. at the time of the 1970 Census. Beginning with the Communist victory in Vietnam in 1975, over three-quarters of a million refugees came to the United States. By the time of the 2000 Census, these individuals and their offspring accounted for nearly two million residents. The 2000 Census reported 1,223,736 persons of Vietnamese descent, 206,052 residents of Cambodian ancestry, 198,203 of Laotian heritage, and 186,310 Hmong. The numbers of Vietnamese and Hmong had more than doubled since the 1990 Census. In addition, many ethnic Chinese also came as refugees from Southeast Asia. The 2010 Census found that the Vietnamese ethnic population had risen to 1,737,433, those of Cambodian ancestry numbered 276,667, the Hmong had grown to 260,073, and Laotians totaled 232,130. Thus, the Southeast Asian population had grown by nearly 40%.

16.4.1 Vietnamese-American Population

California has been the major destination for Vietnamese refugees, and it was the home to 37.3% of all U.S. Vietnamese at the time of the 2010 Census, yet it had a smaller proportion than two decades earlier. Santa Clara County, California had the highest percentage of its population reporting Vietnamese ancestry of any county in the nation in 2010, 7.6%. Orange County, California is 6.5% Vietnamese, and it numerically includes more Vietnamese than any other U.S. county.

Many Vietnamese settled along the Gulf of Mexico where they engaged in off-shore fishing. In 2010 Vietnamese-Americans accounted for between 1.8% and 2.0% of the population of three Louisiana parishes along the Lower Mississippi River, including Jefferson, Orleans, and Plaquemines Parishes, establishing a "Little Saigon" in the Versailles neighborhood in northeastern New Orleans. Vietnamese comprised nearly 1.5% of the residents of Harrison and Jackson counties, Mississippi, across Lake Pontchartrain.

16.4.2 Hmong Ethnic Population

The Hmong, a people who had occupied the highlands of Laos since the early nineteenth century, are largely concentrated in California, where 35.1% of the nation's Hmong population was located in 2010, down from 38.5% a decade earlier, and in two upper Great Lakes states, Minnesota and Wisconsin, where together are found 44.4% of the U.S. Hmong. Indeed, over half of the nation's entire Hmong population is concentrated into just six counties, of which three are in California, two in Minnesota, and one in Wisconsin. Ramsey County, Minnesota, where the Hmong comprise 7.0% of the total population, has a greater Hmong population than any other American county. It is part of the Minneapolis-St. Paul metropolitan area, which is the home of one-quarter of the nation's Hmong. The second largest concentration is in Fresno County, California.

The distribution of the Hmong is quite different from that of many other Asian immigrant groups in the United States. Although initial federal efforts to resettle the Hmong scattered them across the country, "Hmong clan leaders designated a distinct set of cities to which the Hmong secondarily migrated, forming enclaves to preserve culture" (Miyares 1997:217). While the physical climate of the Midwest hardly represented anything that would have attracted the Hmong, "Minnesota's strong faith-based, voluntary community was instrumental in the initial resettlement of Hmong refugees" (Vang 2008:11–12). This led to chain migration from other locations, bringing the Hmong to places where few minority populations were present that might compete for employment.

Within many of the West Coast's metropolitan counties where Asians comprise a significant share of the population, the Hmong are barely represented. Two-thirds of California's Hmong population are found within just three counties, all noted for their agricultural production. Within many counties where the Hmong are numerous, they are clearly the largest Asian group. In some of the counties of Minnesota and Wisconsin with significant Hmong populations, they comprise the most conspicuous racial minority. In three Wisconsin counties the Hmong now comprise over 2% of the total population, with Marathon County having the second largest proportion nationwide of its population being of Hmong ancestry, 4.4%. Nine Wisconsin counties each contain at least 1% of the nation's Hmong population.

16.4.3 Vietnamese and Hmong Ethnic Landscapes

Given the short time that many Vietnamese and Hmong have resided in the United States and the fact that many reside in dwellings that were previously built for Americans of other ethnicities, the most conspicuous visual impacts that they have had upon the landscape are their market gardens, ethnic food shops, and restaurants. The market-garden landscape of Vietnamese immigrants is particularly noteworthy. Although most Vietnamese came primarily from rural backgrounds, being even poor by Vietnamese standards, they have primarily settled in urban areas, bringing their agrarian traditions to urban areas. Before Hurricane Katrina devastated New Orleans, that city's Versailles neighborhood included the largest single concentration of Vietnamese in the United States, approximately 12,000 persons. The rate of return and reconstruction by the Vietnamese ethnics immediately following Hurricane Katrina considerably exceeded that by the city's African-American population (Airriess 2008). New Orleans counted 6214 Vietnamese residents in 2010, comprising 1.8% of its population.

Vietnamese immigrants in New Orleans created a distinctive rural focused landscape within an urban environment. As Christopher Airriess and David L. Clawson (1994:16) note, "Perhaps the most conspicuous evidence of the persistence of rural traditions in the enclave is the vegetable and herb gardens." The back-yard gardens of the Vietnamese immigrants in New Orleans typically occupied the area behind their house or duplex that most other residents would have devoted to grass. These gardens often extended onto utility easements or to the edge of a canal, as the area of the garden was increased by squatting. Sometimes, the gardens were on levees, but these gardens are not located in front yards (Airriess and Clawson 1994). As the Vietnamese ethnic population became increasingly dominated by younger or second-generation families and culturally familiar vegetables become more commercially available, fewer gardens are being developed (Airriess 2008). Nevertheless, "market-garden areas in Versailles show a patchwork quilt of plots, paths, lattices, shades, and other elements of traditional, low-technology market gardening. It is a cultural landscape that is not only transplanted from a different place, but a different time" (Campanella 2006:362).

The gardening of the Vietnamese supplied traditional foodstuffs and provided medicinal herbs. A wide variety of crops are grown in a small area, including tubers, cucurbits, leafy greens, cassava, taro, sweet potatoes, legumes, and various herbs, among others. The garden boundaries are typically a narrow path of bare soil (Airriess and Clawson 1994). Such market gardens also found elsewhere among Southeast Asian immigrants. For example, in Hmong residential areas of Fresno, California, gardens are planted near the entrances to their apartments. Growing a variety of "fruits, vegetables, and spices native to Southeast Asia" they have created what "has become a dominant 'Hmong signature' on Fresno's landscape" (Miyares 1996:104). The Hmong have established extensive community gardens near St. Paul, Minnesota, as well as around the mid-sized cities in Wisconsin where they settled. Before the arrival of the Hmong refugees, these gardens did not exist. The produce of both Hmong and Vietnamese gardeners is frequently displayed at local farmers' markets. Within certain areas of Wisconsin, Hmong vegetable stands occupy nearly half of the spaces in farmers' markets, where the Hmong sell both traditional American produce as well as distinctive Oriental vegetables. The selection of crops grown varies regionally, and most closely resembles crops of their homeland in New Orleans because of its similar climate.

The Hmong have had minimal impact upon their built environment in the United States. Indeed as a somewhat migratory people without a clearly defined homeland in Southeast Asia, "permanent structures were rare in villages" (Miyares 1996:105) they had historically occupied in Laos. Hmong settlements in California's Central Valley, Minnesota, and Wisconsin continue growing, with the Hmong occupying houses built for earlier residents, with the strongest visible ethnic feature being their extensive gardens. In contrast, among the largely Catholic Vietnamese population of New Orleans who occupy housing similar to that of other Americans, religious statuary and shrines are found in many front yards, fences enclose yards, and the free Vietnam flag is flown (Campanella 2006).

The Vietnamese enclave near Westminster and Garden Grove in California's Orange County is called Little Saigon. Allen and Turner (2002:43) describe how this area "slowly built up with new Vietnamese shopping centers [which] contained numerous restaurants and noodle houses and bakeries, doctors' and lawyers' offices, grocery and shoe stores, and travel agents—all patronized by the Vietnamese community." Nearby residential neighborhoods are increasingly dominated by Vietnamese residents, and Westminister is now described as an ethnoburb, in which Vietnamese account for 34% of the suburban city's population (Airriess 2016).

Vietnamese retail shopping centers have also developed at several locations in northern Virginia within the Greater Washington metropolitan area. With signage in Vietnamese and a "tower and pagoda to resemble Saigon's Ben Thanh Public Market," these retail centers have come to "resemble Vietnamese market towns" (Frazier 2006:283). Developed by ethnic Chinese Vietnamese, northern Virginia's "Little Saigon" not only caters to ethnic Vietnamese and other Asians, but seeks to attract tourists by its large array of ethnic shops that are constructed in Asian and French Colonial styles (Airriess 2007). In a variety of communities, including San Jose and Los Angeles, California, Houston and Dallas, Texas, as well as in Northern Virginia, Vietnamese entrepreneurs have acquired older strip shopping centers and created ethnically oriented "Little Saigons." In his study of Vietnamese ethnic shopping centers Joseph Wood (2006) notices similarities in the long linear arrangements of retail establishments and the clustering of shops selling similar items in these shopping centers with the pattern of shops in the Old Quarter of Hanoi. Thus, the clustering of similar shops and signage display Vietnamese ethnicity, while the basic architecture of the buildings is generic American. He writes that it appears as if "American locales were pre-adapted for use by Vietnamese, and a pattern repeats itself in every 'Little Saigon' one encounters in the United States" (Wood 2006:29). To more conspicuously display the ethnicity of the shopping center, some, such as Eden Center in Northern Virginia, are entered through "a pagoda entrance gate with crouching lions astride it" (Wood 2006:34).

16.5 Other Asian-American Landscapes in the Making

Asian immigrants to the United States are not just confined to the Chinese, Japanese, Vietnamese, and Laotian Hmong. While the Chinese community is the largest and best known, locally other Asian immigrant groups have created ethnic enclaves and contributed to shaping the local cultural landscape. Filipino neighborhoods have long been recognized in the Los Angeles area, and Koreans have long functioned as merchants in the African-American communities of that city, as well as establishing their own enclaves. Indeed, Filipinos are the largest Asian population within most western states (Fig. 16.22), while South Asian Indians are the most numerous in many southern states (Hoeffel et al. 2012).

16.5.1 Koreans

The largest Korean community in the nation is located west of Los Angeles' central business district. Conflict between Korean shopkeepers and African-American customers was a contributing factor to destruction of parts of Los Angeles' "Koreatown" by rioters in 1992. As a result many Korea-Americans moved into suburban Los Angeles, with Latinos and Bangladeshi Muslims occupying their former neighborhoods, such that ethnic Koreans account for only one-fifth of the residents of Koreatown (Lee and Park 2008). Yet at the same time the spatial extent of Koreatown has expanded with the investment by additional Korean transnational corporations, resulting in significant social stratification between the business elite and small scale business operators. Koreans have frequently operated dry cleaners, nail salons, or green groceries (Reimer 2007). The focus of the Korean community remains upon Koreatown, best shown by a wide variety of Korean-language signage on many local businesses (Lee and Park 2008). Thus while Los Angeles' Koreatown presents an Asian ethnicity, in reality it is highly multicultural and multiracial. A sizeable Koreatown has also developed in New York City's Flushing community, immediately east of its Chinatown, with the change in language of advertizing signage one of the conspicuous indicators of the Korean presence.

Because the majority of Korean immigrants are members of Korean Christian churches, these institutions play major spiritual, social, and economic roles within their communities. Religion often helps define the ethnic identity of various immigrants, however the proportion of Koreans who are Protestants in the United States considerably exceeds that in Korea, where a far greater share of the population is Buddhist (Lee 2016).

Koreans are concentrated on the east and west coast of the United States. During the first 8 years of the twenty-first century, the United States averaged receiving 19,753 Korean immigrants a year (US Dept. of Homeland Security 2008). Between 2012 and 2014, that number had grown to 21,558 annually (US Dept. of Homeland Security 2015).

16.5.2 Filipinos

Even larger numbers of Filipinos have been arriving, yet this group, whose migration began in earnest when The Philippines was a U.S. commonwealth, has been termed an "invisible ethnic community" (Tyner 2007). Since 2000, an average of 54,679 Filipino immigrants have arrived annually, more than any group except Mexicans, Chinese, and Indians (U.S. Dept. of Homeland Security 2014). Given such rates of immigration, these groups might be expected to leave a larger mark upon the cultural landscape in the future. Filipinos have been immigrating for over a century, with many of the recent immigrants being female health care workers. Nevertheless, with the exception of several areas of the Hawaiian Islands of Molokai and Lanai as well as neighborhoods of Honolulu and Los Angeles where Filipino communities developed, many Filipino immigrants widely dispersed throughout their new communities, particularly within the suburbs (Tyner 2007). Thus, Filipinos "have not established many easily recognized ethnic enclaves comparable to Chinatowns, Little Tokyos, or Little Saigons" (Tyner 2016:309). Forty-three percent of all Filipinos in the United States in 2010 resided in California, with an additional 10.0% in Hawaii. The proportion in both states had declined slightly over the previous decade, as Filipinos had become the largest Asian group in all of the states from the Rocky Mountains westward with the exceptions of Colorado, Utah, and Oregon, where Chinese were most numerous (Hoeffel et al. 2012).

Filipinos and Vietnamese immigrants bring a cultural heritage that simultaneously reflects their Asian roots and their long history of European colonization. Before the American acquisition of The Philippines in 1898, it had been a Spanish colony since the sixteenth century. The French had acquired the lands that comprised Vietnam in the middle of the nineteenth century. Thus, Vietnamese ethnics have opened French restaurants in the United States and many Filipino and Vietnamese communities are focused upon their Roman Catholic churches.

16.5.3 South Asians: Asian Indians

Large numbers of Asian Indians have immigrated over the past decade. While the 2000 Census enumerated 1,899,599 Asian Indians, a decade later the census counted 3,183,063 (Barnes and Bennett 2002; Hoeffel et al. 2012). Since the beginning of 2000, the United States has averaged receiving 60,843 Indians from Asia annually (US Dept. Homeland 2014). Asian Indians have dispersed widely across the

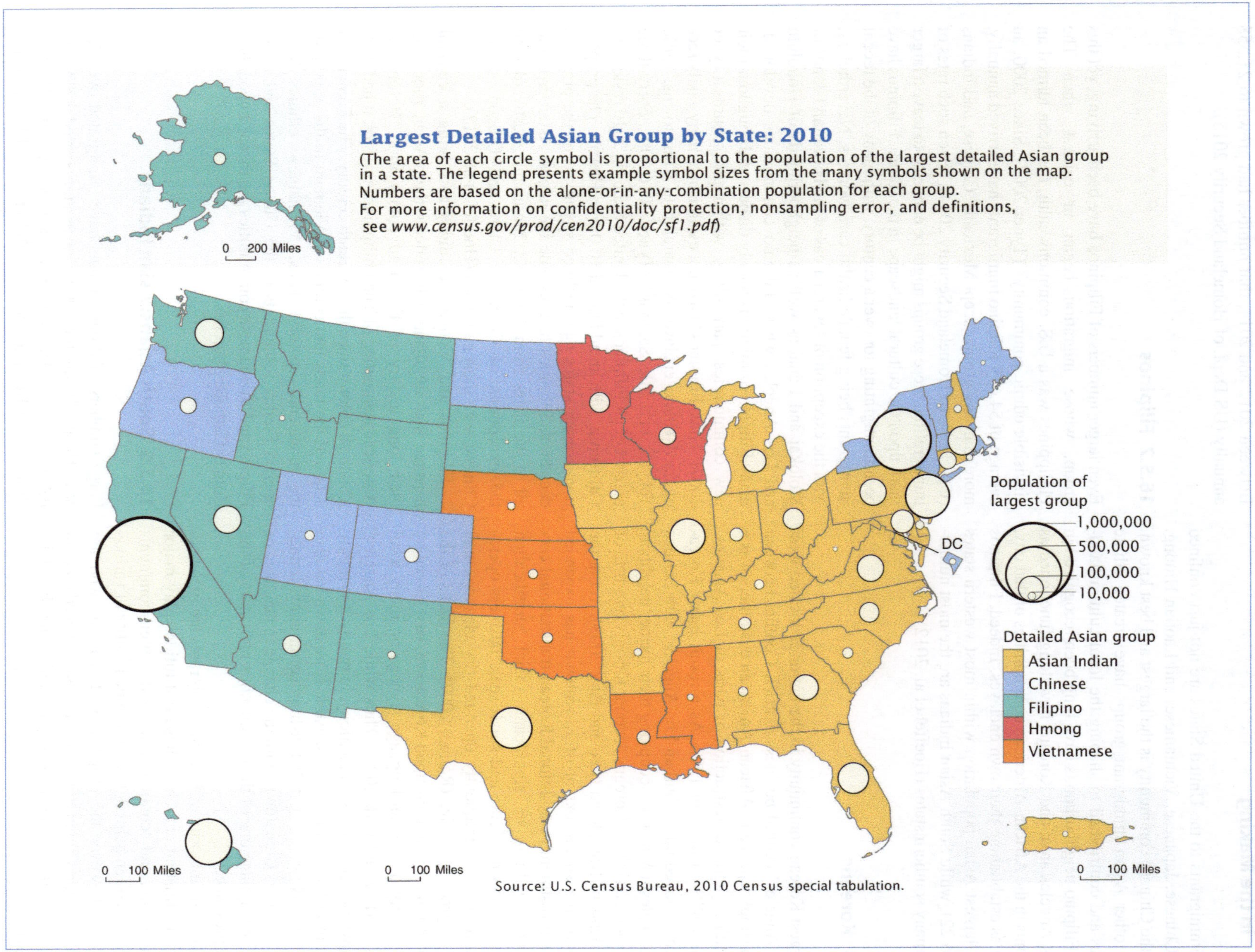

Fig. 16.22 Largest Asian ancestry group within the various states. Given that Chinese and Japanese ethnics are highly concentrated within just a few states, in many states where Asians comprise small proportions of their total populations either Asian Indians or Filipinos are the largest Asian groups (Figure reproduced from U.S. Census Bureau map in Hoeffel et al. (2012))

United States. However, their concentration in the Jackson Heights neighborhood of New York City's borough of Queens and subsequent spread into the more affluent neighborhoods of Queens Village and Bellerose is displayed by ethnic businesses and alteration of housescapes (McGovern and Frazier 2015). This expansion of upscale Asian Indian neighborhoods has resulted in larger housing replacing former smaller houses, which were sometimes "modified to display 'Indian-ness'...include[ing] soft pastel colors, marble stairs, ornate fencing, and the replacement of front lawns with brick or concrete" (McGovern and Frazier 2015:23).

Not only are Asian Indians well represented among a variety of professionals and educators, they are frequently the owners and managers of motels throughout the nation. In fact Indians have become so closely associated with this occupation that the writer V.S. Naipaul (1989) devoted a portion of his travelogue, *A Turn In the South*, to discussing their appearance upon the landscape. Several decades later, Paul Theroux (2015), another astute traveler and observer as well as Naipaul critic, commented upon their presence in his *Deep South*.

The rate of immigration of Asian Indians is comparable to that of the Chinese. A greater share of the Asiatic Indians has dispersed into the smaller cities of the United States than have Chinese immigrants, and they are less concentrated in the Western U.S. Indeed, Indians are the largest Asian population group within 24 states, of which Texas is the westernmost (Hoeffel et al. 2012). While small clusters of Asian Indians have focused upon university cities and suburban areas of various metropolitan centers, along with larger concentrations within the New York City and Chicago areas, Asian Indians are often termed "Invisible Immigrants" because they have developed few neighborhoods dominated by Asian Indians (Skop and Altman 2006). However, "Hinduism has become a centralizing and 'public' feature in many newly established immigrant communities" (Skop 2007:282). As with many ethnic groups, religious structures are often the most conspicuous—and sometimes among the only—distinctive aspect of their ethnic landscape.

16.6 Asians Add Their Ethnic Imprint

While their numbers are smaller and their ethnic landscapes less likely to have attracted academic notice, other Asian groups are also adding their ethnic imprint to the American landscape. While many of the smaller Asian ethnic groups are attracted to California, just as with the most numerous Asian groups, others have settled in areas where they are obvious minorities. In the rural countryside south of St. Paul, Minnesota one can now see a Buddhist temple complex, with Cambodian lettering on the entrance gateway (Fig. 16.23). A simple web search located over 200 Hindu temples scattered across the United States, found in 43 states, with many

Fig. 16.23 Newly constructed Buddhist temple east of Farmington, Minnesota. The Watt Munisotarm temple was constructed by Cambodian ethnic Buddhists and opened in 2007, following 5 years of construction. A 60 foot high pagoda was subsequently added nearby. Note the seated golden statue of Buddha in the temple

of them displaying ornate carved stone construction closely mirroring what would be seen in India. Indeed, when a new Hindu temple was recently constructed in Chicago, the carved stones were imported from India, and Indian workers assisted in its assembly.

References

Airriess, Christopher A. 2007. Conflict Migrants from Mainland Southeast Asia. In *Contemporary Ethnic Geographies in America*, ed. Ines M. Miyares and Christopher A. Airriess, 291–312. Lanham: Rowman and Littlefield.

———. 2008. Spaces and Places of Adaptation in an Ethnic Vietnamese Cluster in New Orleans, Louisiana. In *Immigrants Outside Megalopolis: Ethnic Transformation in the Heartland*, ed. Richard C. Jones, 163–188. Lanham: Lexington Books.

———. 2016. Mainland Southeast Asia: Diversity of Conflict Migrants. In *Contemporary Ethnic Geographies in America*, ed. Christopher A. Airriess, 2nd ed., 347–370. Lanham: Rowman and Littlefield.

Airriess, Christopher A., and David L. Clawson. 1994. Vietnamese Market Gardens in New Orleans. *The Geographical Review* 84 (1): 16–31.

Allen, James P., and Eugene Turner. 1988. *We The People: An Atlas of America's Ethnic Diversity*. New York: Macmillan Publishing.

———. 2002. *Changing Faces, Changing Places: Mapping Southern Californians*. Northridge: California State University Center for Geographical Studies.

Anderson, Kay J. 1987. The Idea of Chinatown: The Power of Place and Institutional Practice in the Making of a Racial Category. *Annals of the Association of American Geographers* 77 (4): 580–598.

Aronson, Emily, and Robert B. Kent. 2008. A Midwestern Chinatown? Cleveland, Ohio in North American Context, 1900-2005. *Journal of Cultural Geography* 25 (3): 305–329.

Barnes, Jessica S., and Claudette E. Bennett. 2002. *The Asian Population: 2000*, Census 2000 Brief C2KBR/01-16. Washington: U.S. Department of Commerce.

Campanella, Richard. 2006. *Geographies of New Orleans: Urban Fabrics Before the Storm*. Lafayette: Center for Louisiana Studies.

Choy, Philip P. 2012. *San Francisco: A Guide to Its History and Architecture*. San Francisco: City Lights.

DeFrancis, John. 1973. *Things Japanese in Hawaii*. Honolulu: University of Hawaii Press.

Dubrow, Gail Lee. 2000. Asian American Imprints on the Western Landscape. In *Preserving Cultural Landscapes in America*, ed. Arnold R. Alanen and Robert Z. Melnick, 143–168. Baltimore: The Johns Hopkins University Press.

Fellows, Donald K. 1972. Japanese Buddhism—Its Imprint on a California Landscape. *California Geographer* 13: 49–57.

Fong-Torres, Shirley. 1991. *San Francisco Chinatown: A Walking Tour*. San Francisco: China Books and Periodicals.

Frazier, John W. 2006. Asians in the United States: Historical and Contemporary Settlement Patterns. In *Race, Ethnicity, and Place in a Changing America*, ed. John W. Frazier and Eugene L. Tettey-Fio, 265–285. Binghamton: Global Academic Publishing.

Gryn, Thomas, and Christine Gambino. 2012. *The Foreign Born from Asia: 2011*, American Community Survey Brief ACSBR/11-06. Washington: U.S. Department of Commerce.

Guan, Jian. 2002. Ethnic Consciousness Arises on Facing Spatial Threats to Philadelphia Chinatown. In *Urban Ethnic Encounters: The spatial Consequences*, ed. Aygen Erdentug and Freek Colombijn, 126–141. London: Routledge.

Hill, Tiffany. 2013. Buddhism in Hawaii: Fading Tradition. *Honolulu Magazine*. http://www.honolulumagazine.com/Honolulu-Magazine/July-2013/Buddhism-in-Hawaii-Fading-Tradition/.

Hoeffel, Elizabeth M., Sonya Rastogi, Myoung Ouk Kim, and Hasan Shahid. 2012. *The Asian Population: 2010*, 2010 Census Brief C2010BR-11. Washington: U.S. Department of Commerce.

Jenks, Hillary. 2008. Urban Space, Ethnic Community, and National Belonging: The Political Landscape of Memory in Little Tokyo. *GeoJournal* 73 (3): 231–244.

Laguerre, Michel S. 2005. The Globalization of a Panethnopolis: Richmond District as the New Chinatown in San Francisco. *GeoJournal* 64 (1): 41–49.

Lai, H.M. 1980. Chinese. In *Harvard Encyclopedia of American Ethnic Groups*, ed. Stephan Thernstrom, 217–234. Cambridge (MA): Harvard University Press.

Lee, Ronald K.K. 1986. Japanese. In *America's Architectural Roots: Ethnic Groups that Built America*, ed. Dell Upton, 136–141. Washington: Preservation Press.

Lee, Jane Yeonjae. 2016. Korean Americans: Entrepreneurship and Religion. In *Contemporary Ethnic Geographies in America*, ed. Christopher A. Airriess, 2nd ed., 285–301. Lanham: Rowman and Littlefield.

Lee, Youngmin, and Kyonghwan Park. 2008. Negotiating Hybridity: Transnational Reconstruction of Migrant Subjectivity in Koreatown, Los Angeles. *Journal of Cultural Geography* 25 (3): 245–262.

Li, Wei. 1993. Geographical Study of Ethnicity: Comparison Between Downtown and Suburban Chinese in Metropolitan Los Angeles. *GeoJournal* 30 (3): 317–322.

———. 1998. Ethnoburb Versus Chinatown: Two Types of Urban Ethnic Communities in Los Angeles. *Cybergeo* (70). http://193.55.107.45/culture/weili/weili.htm.

———. 2005. Beyond Chinatown, Beyond Enclave: Reconceptualizing Contemporary Chinese Settlements in the United States. *GeoJournal* 64 (1): 31–40.

———. 2007. Chinese Americans: Community Formation in Time and Space. In *Contemporary Ethnic Geographies in America*, ed. Ines M. Miyares and Christopher A. Airriess, 213–232. Lanham: Rowman and Littlefield.

Li, Chuo. 2015. Commercialism and Identity: Politics in New York's Chinatown. *Journal of Urban History* 41 (6): 1118–1134.

Lin, Jan, and Paul Robinson. 2005. Spatial Disparities in the Expansion of the Chinese Ethnoburb of Los Angeles. *GeoJournal* 64 (1): 51–61.

McGlinn, Lawrence A. 2002. Beyond Chinatown: Dual Immigration and the Chinese Population of Metropolitan New York City, 2000. *Middle States Geographer* 35: 110–119.

McGovern, Brendan, and John W. Frazier. 2015. Evolving Ethnic Settlements in Queens: Historical and Current Forces Reshaping Human Geography. *Focus on Geography* 58 (1): 11–26.

Miyares, Ines M. 1996. To Be Hmong in America: Settlement Patterns and Early Adaption of Hmong Refugees in the United States. In *Human Geography in North America: New Perspectives and Trends in Research Innsbrucker Geographische Studien Band 26*, ed. Klaus Krantz, 97–113. Innsbruck: Universität Innsbruck.

———. 1997. Changing Perceptions of Space and Place as Measures of Hmong Acculturation. *Professional Geographer* 49 (2): 214–224.

Naipaul, V.S. 1989. *A Turn in the South*. New York: Vintage International.

Pan, Lynn. 1994. *Sons of the Yellow Emperor: A History of the Chinese Diaspora*. New York: Kodansha International.

Rast, Raymond W. 2007. The Cultural Politics of Tourism in San Francisco's Chinatown, 1882–1917. *Pacific Historical Review* 76 (1): 29–60.

Reimer, Dana G. 2007. Korean Culture and Entrepreneurship. In *Contemporary Ethnic Geographies in America*, ed. Ines M. Miyares and Christopher A. Airriess, 233–250. Lanham: Rowman and Littlefield.

Salter, Christopher L. 1978. *San Francisco's Chinatown: How Chinese a Town?* San Francisco: R & E Research Assoc., Inc.

Skop, Emily. 2007. Asian Indians and the Construction of Community and Identity. In *Contemporary Ethnic Geographies in America*, ed. Ines M. Miyares and Christopher A. Airriess, 271–290. Lanham: Rowman and Littlefield.

Skop, Emily H., and Claire E. Altman. 2006. The Invisible Immigrants: Asian Indian Settlement Patterns and Racial/Ethnic Identities. In *Race, Ethnicity, and Place in a Changing America*, ed. John W. Frazier and Eugene L. Tettey-Fio, 309–316. Binghamton: Global Academic Publishing.

Sowell, Thomas. 1981. *Ethnic America: A History*. New York: Basic Books.

Suchan, Trudy A., Marc J. Perry, James D. Fitzsimmons, Anika E. Juhn, Alexander M. Tait, and Cynthia A. Brewer. 2007. *Census Atlas of the Unites States, Series CENSR-29*. Washington: U.S. Census Bureau.

Takaki, Ronald. 1993. *A Different Mirror: A History of Multicultural America*. Boston: Little, Brown and Company.

Theroux, Paul. 2015. *Deep South: Four Seasons on Back Roads*. Boston: Houghton Mifflin Harcourt.

Tyner, James A. 2007. Filipinos: The Invisible Ethnic Community. In *Contemporary Ethnic Geographies in America*, ed. Ines M. Miyares and Christopher A. Airriess, 251–270. Lanham: Rowman and Littlefield.

———. 2016. Filipinos: The Invisible Ethnic Community. In *Contemporary Ethnic Geographies in America*, ed. Christopher A. Airriess, 2nd ed., 303–324. Lanham: Rowman and Littlefield.

United States Department of Homeland Security. 2008. *Yearbook of Immigration Statistics: 2007*. Washington: U.S. Department of Homeland Security, Office of Immigration Statistics.

———. 2014. *Yearbook of Immigration Statistics: 2013*. Washington: U.S. Department of Homeland Security, Office of Immigration Statistics.

———. 2015. *Yearbook of Immigration Statistics: 2014*. Washington: U.S. Department of Homeland Security, Office of Immigration Statistics.

Vang, Chia Youyee. 2008. *Hmong in Minnesota*. St. Paul: Minnesota Historical Society Press.

Wood, Joseph S. 2006. Making America at Eden Center. In *From Urban Enclave to Ethnic Suburb: New Asian Communities in Pacific Rim Countries*, ed. Wei Li, 23–40. Honolulu: University of Hawai'i Press.

Yip, Christopher Lee. 1986. Chinese. In *America's Architectural Roots: Ethnic Groups That Built America*, ed. Dell Upton, 106–111. Washington: Preservation Press.

Native Hawaiian Landscapes

17

Hawaii is the most multi-ethnic state of the nation. Forty-two percent of its population reported being of Asian ancestry on the 2000 Census, 24.3% indicated they were Caucasian, and 9.4% claimed Native Hawaiian or other Pacific Islander heritage. A full 21.4% of the population reported being of two or more races. A decade later the 2010 Census indicated that Caucasian population had slightly increased—to 24.7%, and the Native Hawaiian and other Pacific Islander numbers had marginally increased to 10.0%. While the influence of several Asian ethnic groups was discussed in the previous chapter, here we consider the indigenous peoples, the Native Hawaiians.

17.1 Native Hawaiian Population

On the occasion of the 2000 Census, 80,137 Hawaiians, 6.6% of the state's population, reported being solely of Native Hawaiian ancestry. That number had slightly risen to 80,337 by the time of the 2010 Census. When considering those persons whose ancestry included both Native Hawaiian and other groups, they numbered 239,655 in 2000, or 19.8% of Hawaii's population. The five-year American Community Survey for 2010–2014 estimated 84,584 Native Hawaiians resided in Hawaii, with an additional 86,324 Native Hawaiians living in the continental United States. California was home to 27.4% of these Native Hawaiians. In addition, Hawaii was home to 54,151 other Pacific Islanders, according to the 2014 American Community Survey, including 27,023 Micronesians and 16,393 Samoans.

Native Hawaiians have left a lasting legacy upon the state's cultural landscape, a landscape that also reflects the ethnicity of the outsiders. Native Hawaiians comprised 10.1% of the population of Hawaii County according to the 2014 five-year American Community Survey, the highest among Hawaii's counties. While they comprised 6.4% of Maui County's population, they accounted for 33.2% of the population on the island of Molokai, which is part of Maui County. While Native Hawaiians comprise only 5.0% of Honolulu County's population, the state's largest, they constitute a greater proportion of the population in some of the more rural communities in the northern and western parts of Oahu, all of which is coexistent with Honolulu County. Not only do the state and county names display the Native Hawaiian place names, but so do many local toponyms. Nevertheless, as RDK Herman (1999:76) explains, "Hawaiian names themselves remain relatively intact while their use, meaning, and context has changed."

17.1.1 Outsiders Influence Hawaiian Landscape

Hawaii's history has been profoundly influenced by contact with both European Americans and Asians. The earliest outsiders to overwhelming influence the Hawaiian landscape were whalers and missionaries, largely from the United States. New England missionaries first arrived in Hawaii in 1820, just 1 year after its old religion had been discredited after being disavowed by the Hawaiian monarchy. The old temple houses and oracle towers were destroyed, and there was a widespread adoption of Congregationalism by the Hawaiian population. Thus, many of the older churches in Hawaii reflect a New England style with tall central steeples (Williams 1997). Missionaries and their descendents gained control of the land and the island's government. Thus Castle & Cooke, operators of Dole Pineapple, bore the names of three missionary families, and President Dole was the first ruler of the Republic of Hawaii, following the toppling of the monarchy. While many Americans associate pineapple production with Hawaii, it was whaling that first brought Americans to Hawaii from the 1790s to the late 1800s. Lahaina, later followed by Honolulu, was the primary location for resupplying the whaling ships. Sugar plantations obtained huge acreages, disrupting the indigenous landholding system, and they imported thousands of plantation workers, primarily from China, Japan, and the Philippines, but

J.A. Cross, *Ethnic Landscapes of America*, DOI 10.1007/978-3-319-54009-2_17

also from Portugal. The sugar and pineapple growers wielded tremendous political influence, ousted the monarchy, established the Republic of Hawaii, and ultimately urged the U.S. acquisition of the territory.

The heritage of the Kingdom and Republic of Hawaii can be seen in the Iolani Palace, located near the state capitol building, as well as the seasonal palace preserved on the island of Hawaii. The Bishop Trust, controlling one-sixth of the state's lands that were the former royal family holdings, operates schools for native Hawaiians. Indigenous land claims have long been controversial in Hawaii (Creighton 1978), just as they are in parts of Mainland America. While farmlands were allocated via Hawaiian Homesteads to many Native Hawaiians resident on Molokai under the Hawaiian Home Lands Act of 1920, the promises of the legislation fell short. The intention to restore lands to native Hawaiians was largely unsuccessful, except in Molokai, and the promised lands were often never given. At the time of the 2010 Census, 30,858 persons lived on Hawaiian home lands, as those lands held in trust for Native Hawaiians are now termed (Hixson et al. 2012). Thus, only 16% of the Native Hawaiian population (those reporting only Native Hawaiian ancestry) live on the home lands. There are insufficient lands in the state to fully implement the program, yet Native Hawaiian groups are pushing for both lands and sovereignty. While efforts have been made to pass legislation to give Native Hawaiians rights similar to those granted to Native Americans, this is yet to occur.

17.2 Native Hawaiian Landscapes

Many archaeological sites and ruins dot the Hawaiian Islands (Fig. 17.1). While some of these sites preserve the legacies of villages and others display artifices related to farming and fishing activities, a great preponderance of the features are religious sites. Even though the indigenous religion lost its primacy nearly 200 years ago, the relics of the former temples and sacred sites are receiving a resurgence of interest today. As Kana'iaupuni and Malone (2006:287) explain, "Place is intertwined with identity and self-determination of today's Native Hawaiians in complex and intimate ways." Not only are the Hawaiian Islands the homeland of these indigenous peoples, but the lands figure in their spirituality. Their gods created the islands and the Hawaiians and gave them the taro plant. Having occupied the islands for over a 1000 years, this spirituality ties the Native Hawaiians and their history to this land and the seas that surround it.

17.2.1 Religious Features on the Landscape

Numerous religious artifacts remain upon the Hawaiian landscape. Charles Heatwole explains,

> The ancient Hawaiians [believed]...[e]very object or phenomenon was associated with a deity. ... Because the gods were everywhere, the rules of behavior were many and complex. Numerous actions and places were kapu, that is, taboo, forbidden, or marked off (Heatwole 1988:6).

Fig. 17.1 Ruins of stone walls of 600 year old native village on the island of Hawaii. This site is preserved in Lapakahi State Historic Park, Hawaii

Because of a resurgence in sacred devotion to these sites today among Native Hawaiians—many who respect both the old gods and the Christian God, offerings of flowers, fruit, and other items can frequently be spotted upon the old stone walls and ruins.

17.2.1.1 Heiaus

Heiaus are stone temple foundations that are widely dispersed across the Hawaiian landscape, what remained after the wooden temples were thrown down two centuries ago (Fig. 17.2). King Kamehameha II ordered the wooden temples atop the heiaus destroyed when he abolished the kapu system. Heiaus are found in various configurations, including elevated stone mounds, terraces, and walls (Fig. 17.3). Peter Buck (1957:517-518) provides a typical classification of their structural details: terraces or courts; terraced temples, with a paved open court, and with enlargements on different terraces; and walled temples, consisting of an elaboration of terraced temple with stone wall on all sides and a raised platform at one end of the court. Heiaus on Kauai averaged 101 by 162 feet and had walls six feet thick and were five feet high.

While all that remains at most sites today are the heiaus, it is clear that many varieties of temples once existed. On Oahu composite temples, which are a combination of walled temples and terraced temples, were common. Today these can be seen as two terraces, with walls on three sides. In addition, a stone wall often surrounded the religious site, which along with the heiau remained when the temple was destroyed.

Heiaus were often located at prominent or strategic places, "such as a hill top, the seaward end of a range, or a promontory of higher land which overlooked valleys, villages, or the sea" (Buck 1957:516). Most heiaus faced the sea. A prominent example includes the Puu o Mahuka Heiau, overlooking an inlet near Waimea on the island of Oahu.

Because the indigenous Hawaiian religion recognized many deities, each with its particular role or activity, many different types of heiaus were constructed. The various purposes of heiaus included peace, war, good fishing, and agricultural success, among other activities. In addition, not all heiaus were built for the same users. Some heiaus were erected for commoners, while others served the kings. Likewise, the diverse temples required different sacrifices, ranging from easy offerings to human sacrifice.

Pu'uhonua O Hōnaunau (City of Refuge) National Historical Park on the island of Hawaii was a refuge for violators of kapus and for war captives. As a refuge, it was a site of forgiveness and safety. Its priests could purify, as kapu violators would be killed if they did not seek refuge. The site, along the shore, is surrounded with stone walls, and it includes both the temple grounds and heiaus, along with several reconstructed wooden structures. Another prominent stone heiau is Pu'ukoholā, which covers an area of nearly 25,000 square feet (Kirch 1996). Located on a slope overlooking the Pacific Ocean, it is preserved in Pu'ukoholā Heiau National Historic Site farther north along the west coast of Hawaii.

Fig. 17.2 Pu'ukohola Heiau towers over Mailekini Heiau on the dry western side of the island of Hawaii. Such large stone platforms once supported wooden religious structures. These heiaus are preserved in Puukohola Heiau National Historic Site, Hawaii

Fig. 17.3 Stone walls from a heiau located near the northern shore of the island of Oahu. Note that a religious offering of pineapples has been placed atop the wall of Puu O Mahuka Heiau, located near Waimea, Hawaii

17.2.1.2 Petroglyphs and Rock Cairns

Numerous sites were sacred to the Native Hawaiians besides heiaus. Sacred sites include rock cairns, which conspicuously mark the landscape in parts of northwestern Maui, and petroglyphs carved in lava rock, such as those preserved in Hawaii Volcanoes National Park (Fig. 17.4). There the Pu'uloa petroglyph field includes over 15,000 carvings of circles, depressions, and human figures, scratched onto the surface of the pahoehoe lava (Kirch 1996).

Not all the rock cairns visible today are ancient religious monuments, even though their inspirations are (Fig. 17.5). Peter Buck (1957:529) explains that these road shrines "were marked by an upright stone or small stone platform and stood besides roads or tracks to well-known valleys." A wide variety of offerings, such as leaves, stone, among other items were added to the shrine by travelers "as a placatory offering to the guardian spirit to ward off any mishap that might otherwise befall them" (Buck 1957:529). He notes that visitors continued adding to these shrines, and this author can attest to seeing that such practices continue to the present.

Religious offerings to this day can be seen at various sacred sites, whether that be on the ancient stone heiaus or prominent places, such as volcanic crater rims (Fig. 17.6), sea cliffs, or the southernmost point on the island of Hawaii. The offerings, which traditionally include rocks wrapped in ti leaves, are commonly placed on a lele, or an offering stand. The leles range from a single pole to a four pole platform with wicker shelf, although small stone platforms can also be seen (Fig. 17.7). At some sites today one sees offerings of fruits, vegetables, or flowers, left by either Native Hawaiians or by recent Vietnamese immigrants.

Signs bearing the word "Kapu" are commonly displayed to this day in Hawaii. Very common around both religious and secular properties, the word "Kapu" equates with the words "No Trespassing" that are used elsewhere. This is a reminder that many sites are considered sacred, and that it was simply "kapu", or forbidden, for commoners to enter.

Even the largest landscape features are sacred, and the indigenous religion had a volcano goddess, Pele. Protests by Native Hawaiians beginning in 2014 have blocked construction of a new 30 meter telescope atop Mauna Kea on the island of Hawaii, viewing it as "yet another eyesore despoiling an ancient sacred landscape" (Overbye 2016). The removal of volcanic rock from Hawaii is believed to bring bad luck to those who take it.

17.2.2 Native Hawaiian Housing

The built cultural landscape displaying Native Hawaiian ethnicity has largely disappeared with the exception of the archaeological sites and ruins. As Ines Miyares (2008:522) explains, "remnants of plantation-era worker housing remain on O'ahu in rural communities…, but they are threatened with extinction." Virtually all Native Hawaiian style houses have long disappeared from the Hawaiian landscape,

Fig. 17.4 Pu'u Loa Petroglyphs carved in volcanic pahoehoe lava within Hawaii Volcanoes National Park, Hawaii

Fig. 17.5 Rock cairns near Nakalde Point east of Honokohau, Hawaii. This large assemblage of rock cairns erected by Native Hawaiians overlooks the Pacific Ocean on the island of Maui

Fig. 17.6 Religious offerings to the volcano goddess Pele placed at the rim of Kilauea Caldera. This site, within Hawaii Volcanoes National Park, has been the location of recent volcanic activity

Fig. 17.7 Stone platform of heiau topped with a lele near the southernmost tip on the island of Hawaii. Offerings of ti leaves and small stones are often left on the leles

Fig. 17.8 Thatched house reconstruction in Pu'uhohua o Honaunau National Historic Park, Hawaii. At one time this was the typical Hawaiian house, but now it is only seen in museums and historical recreations

although many remained during the early part of the Twentieth Century (Fig. 17.8). Today, all that is left are various archaeological sites and a few cultural centers where a few structures have been reconstructed. The best remaining example is at the Bishop Museum. Native Hawaiian houses were of wood, of a pole post and frame design, with thatched roofs and walls. A variety of house types were built, with the main differences related to the nature of the walls and roofs (Buck 1957). Some houses were constructed without walls, with the roof resting on ground. Others had stone walls or a roof on loose stone walls that were about three feet high. Both gabled and hipped roofs with grass thatched walls were used.

The legacy of the Hawaiian houses today can be seen, as Ray Morris (1986:41) explains, in "the continued preference for one-story houses in Hawaii [that] may grow out of memories of the thatched house tradition." He writes that "Hawaiian architects have introduced mixtures of Asian and Hawaiian forms into their designs for resorts and other new buildings as more explicit references to the islands' traditional architecture" (Morris 1986:41).

17.2.3 Native Hawaiian Agricultural Landscapes

The ethnic heritage of Native Hawaiians can be seen in how they divided their lands and in the crops that are grown. While much of the land of Hawaii is no longer owned by the indigenous population, the legacy of the traditional land tenure system is displayed in the shapes of many landholdings, the boundaries of districts, and the arrangements of fences. The Bishop Trust, which owns one-sixth of the state, manages the former lands of the royal family, with the revenues underwriting an exclusive school system dedicated to educating ethnic Hawaiians.

17.2.3.1 Kuleana

Kuleana were land grants given to commoners by the Hawaiian monarchy in the mid 1800s (Kelly 1983), during the Great Mahele between 1846 and 1855. These plots of farmlands were derived from an earlier system in which native Hawaiians were given cultivation lands for their use by the local chiefs. A part of the chief's ahupua'a, which extended from the sea to the mountain, the kuleana typically covered two to four acres. While the holder of a kuleana could pass the lands on to his heirs, the lands reverted to the ahupua'a should its occupier die without an heir. The kuleana was in essence a landholding within a holding, with the kuleana owner having gathering rights and water rights on the ahupua'a.

Kuleanas were surveyed by metes and bounds, resulting in their irregular shapes and boundaries. The legacy of the system can be seen today, both in property lines and in the political subdivisions or districts within Hawaii's counties that approximate the old ahupua'a boundaries. Only 19% of these land parcels remained as recently as 1980 (Kelly 1983), as the others have been consolidated into the sugar plantations or other large landholdings. However, they are still important in those regions where the native economy is best preserved, such as on Molokai, Nihua, and parts of several of the other islands. The 175 mile long Ala Kahakai National Historic Trail, established in 2000, connects over 200 ahupua'a in the state.

17.2.3.2 Taro Fields

Taro fields are one of the most conspicuous Native Hawaiian agricultural features (Fig. 17.9). Taro was the most important staple crop that was used to make poi, a paste-like foodstuff. Taro, a high carbohydrate root with large leaves, is grown in several settings. Dry taro plantings are found in forest clearings and on upland slopes. In contrast, wet taro plantings are frequently positioned along stream banks or in swamps. Irrigation of taro fields is quite common, and the fields are often seen with terracing. Rock walls frequently separate taro fields. In places, the irrigation works can be quite extensive, such as the extensive diversion canals seen in Kauai. Today taro fields can be seen near centers of Native Hawaiian populations, particularly on Molokai, where the dry plantings are more common, and along the Hana Coast of Maui and in Kauai, where extensive plantings of taro can be seen in waterlogged fields.

Extensive systems of stone walls demarking prehistoric agricultural fields were developed in the Kona area of western Hawaii and Kohala area of northwestern Hawaii where dryland cultivation of taro and sweet potatoes was widespread. As Patrick Kirch (1996:81) notes, such "field walls, called *kuaiwi* ("backbone") in Hawaiian, probably had several functions, including the delineation of separate garden plots and soil retention, as well as for stone clearance." Portions of these field systems with their long low stone walls are still visible.

17.2.3.3 Fish Ponds and Salt Ponds

Fishponds built along the coast where the waters were shallow enabled the indigenous Hawaiians to raise and net fish. Built of rock walls rising a couple of feet above high tide to separate shallow waters along the shore from the open sea, the southern coast of Molokai had the most. There, 73 fishponds have been identified, a few which are still usable. These ponds were used to raise mullet and milkfish, as well as shrimp, and ranged in size from one to over 500 acres (Kirch 1996). Although most fishponds were constructed along the coast, a stone-faced earthen wall extends over a half mile along the Hulē'ia River in Kauai, separating the Menehune Fishpond, which occupies a former river meander, from the river proper.

Salt production ponds (or evaporation ponds) can also be spotted along some of the drier coasts, where very shallow areas could be flooded at high tide and then sealed off from the sea, permitting the water to evaporate, precipitating the salt.

Fig. 17.9 Taro fields in the Hanalei Valley on the island of Kauai. Note the irregular shape of the kuleanas, which must the flooded during part of the period that the taro is growing

17.3 Native Hawaiians and Other Pacific Islanders on the Mainland

Native Hawaiians left their mark throughout the landscape of the state of Hawaii, seen in ruins of numerous religious structures, the stone walls separating fields, and the shape of properties. Increasing vigor of those seeking to express their sovereignty, combined with the demise of agricultural enterprises which grew vast acreages of sugar cane and pineapples, has raised many questions about who should control the future development and use of lands across much of the state. Just as many Native Americans have left their reservations to migrate to other parts of the nation, many Native Hawaiians have left the traditional communities in which they comprised the majority of the population and moved to the larger cities of the state. Many have traveled farther, as only 55.0% of all Native Hawaiians in the United States lived in Hawaii in 2010. California is home to 14.2% of the Native Hawaiian population, while 3.8% live in Washington (Hixon et al. 2012).

Native Hawaiians are not the only Polynesian peoples who migrated to the mainland of the United States, with the 2000 Census reporting that 221,458 residents of California reported Native Hawaiian or other Pacific Island ancestry (Fig. 17.10). Their number had risen to 286,145 by the time of the 2010 Census (Hixson et al. 2012). California has a quarter as many persons of Native Hawaiian ancestry as Hawaii, but one and one-half times as many Samoans as Hawaii and far greater proportions of Tongans, Guamanians, and Fijians.

The population of Pacific Islanders other than Native Hawaiians has grown faster than the Native Hawaiian population in the United States. The 2010 Census showed that California was home to 30.1% of the Guamanians or Chamorro in the United States, 33.0% of the Samoans, 40.0% of the Tongans, and 74.5% of the Fijians (Hixson et al. 2012). Four years later, 36.9% of the nation's residents of Samoan ancestry resided in California, according to the 2014 American Community Survey, as did 33.0% of the Guamanians and 80.9% of the Fijians. Of the "Other Pacific Islanders" enumerated by the Census, only the Marshallese are found in greater numbers in Hawaii than within some other state.

Those Native Hawaiians and Other Pacific Islanders who settled on the mainland focused upon the western states. Fourteen mainland counties—all located in states west of the Mississippi River—reported Native Hawaiian and Pacific Islander populations that exceeded 1% of their total populations in 2000, yet in no case did their proportion exceed 3% (Greico 2001). By the time of the 2010 Census five mainland counties reported Native Hawaiian and Pacific Islander populations that equaled or exceeded 2% of their total population. Anchorage Municipality in Alaska led the list, with Native Hawaiians and Pacific Islanders accounting for 2.8% of its population. While two of these mainland counties were located in the along the Pacific coast, Pierce County, Washington and San Mateo County, California, one was Salt Lake County, Utah while Washington County, Arkansas had a Native Hawaiian and Pacific Islander population that comprised 2.2% of its total population, the sixth highest ranking in the nation (Hixson et al. 2012). Of the Polynesian groups, Native Hawaiians outnumbered Samoans, the second largest group, by a three to one margin.

Geographic scholarship describing the ethnic contributions of Native Hawaiians upon the landscape of their communities on the North American continent is lacking. In his exploration of the impact of Samoan immigrants upon the landscape of southern California, Stephen Koletty (2002:136) writes that the "Samoan community is almost invisible because of their (low) population density." Thus, he explains that Samoan residential areas "do not really stand out from the rest of the neighborhood." Nevertheless, "the assemblage of exotic plants found around the home exterior…reflects a culture literally transplanted" (Koletty 2002:136). These plants include taro, which is also integral to the Native Hawaiian diet. As with many immigrant groups, close affiliations with particular churches defines many of the Polynesian ethnic communities on the mainland.

17.4 Envoi

It is only fitting that the final chapter discussing specific ethnic landscapes in America considers an indigenous First Nations group, just as did the first chapters in the book. The United States has expanded beyond its continental boundaries to include a state whose population and ethnic history strongly diverge from that of most of the rest of the country, yet whose landscape also shows a strong American ethnic imprint, brought half way across the Pacific Ocean. While the Polynesians and American Indians are culturally quite different and occupy highly dissimilar environments, they are similar in that they both were the first human group to shape many of their regional environments, they were overrun by settlers of various European heritages, and they are similar in that they are striving to both maintain their sovereignty and contemporary ethnic identity. Likewise, both groups left a rich archaeological heritage of ruins that conspicuously mark their homelands. Members of both groups have migrated in large numbers away from their traditional lands to metropolitan centers, where their impact upon the urban environment is constrained by their poverty and competition with multiple competing ethnic Americans. Indeed, it remains to be seen what will be their contributions to the American ethnic landscape of the future.

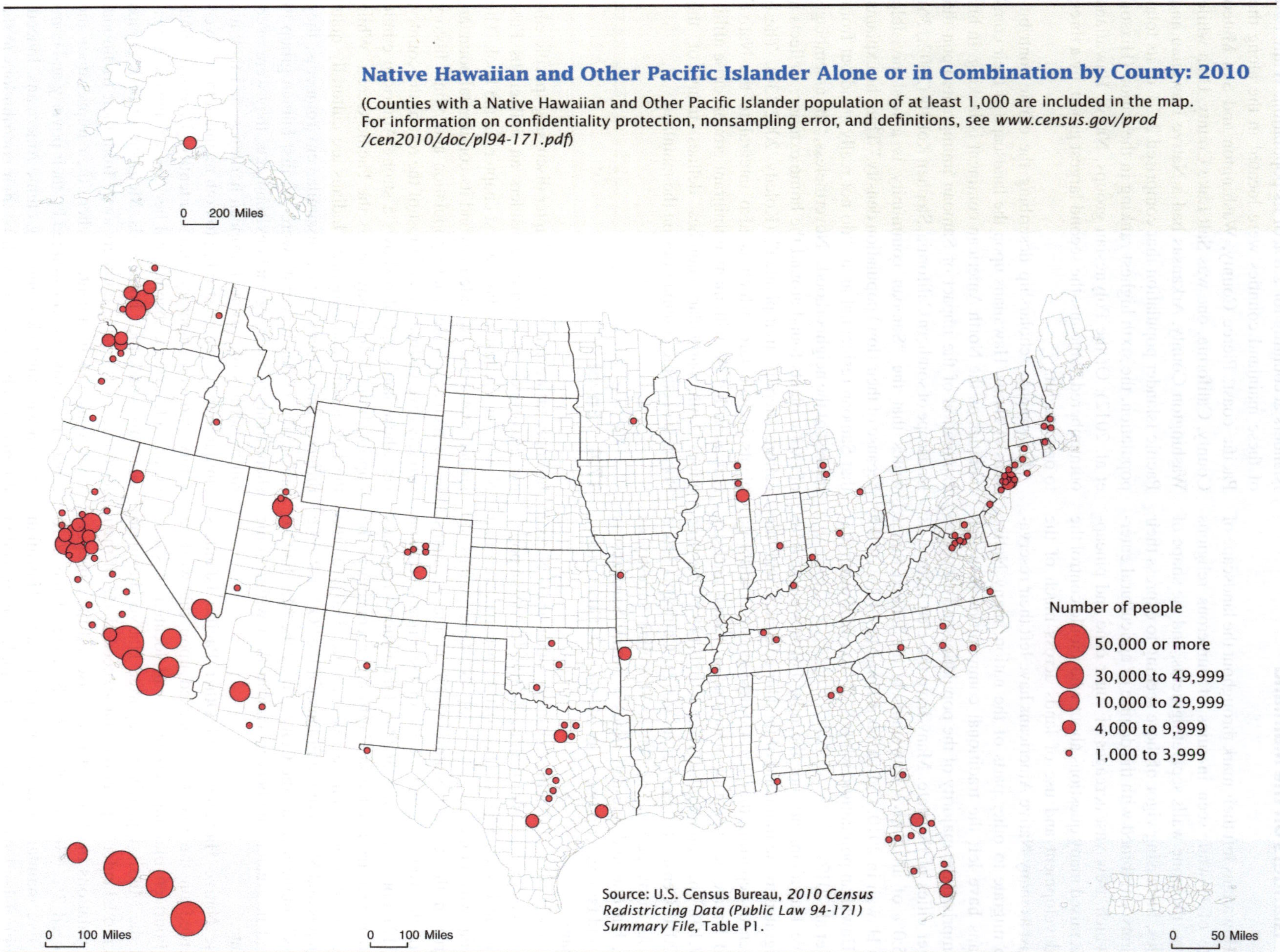

Fig. 17.10 Distribution of Native Hawaiians and Other Pacific Islanders among the counties of the United States in 2010. Note that besides Hawaii, the population of these islanders are heavily focused upon counties in California, and to a lesser extent, elsewhere in the far western United States (Reproduced from U.S. Census Bureau map within Hixson et al. (2012))

References

Buck, Peter H. 1957. *Arts and Crafts of Hawaii*, Special Publication 45. Honolulu: Bishop Museum Press.

Creighton, Thomas H. 1978. *The Lands of Hawaii: Their Use and Misuse*. Honolulu: The University Press of Hawaii.

Grieco, Elizabeth M. 2001. *The Native Hawaiian and Other Pacific Islander Population: 2000*, Census 2000 Brief C2KBR/01-14. Washington: U.S. Department of Commerce, U.S. Census Bureau.

Heatwole, Charles A. 1988. In Pele's Realm: Reflections on Traditional Hawaiian Religion. *The American Geographical Society's Focus* 38 (3): 6–9. 36.

Herman, R.D.K. 1999. The Aloha State: Place Names and the Anti-conquest of Hawai'i. *Annals of the Association of American Geographers* 89 (1): 76–102.

Hixson, Lindsay, Bradford B. Helper, and Myoung Ouk Kim. 2012. *The Native Hawaiian and Other Pacific Islander Population: 2010*, Census 2010 Brief C2010BR-12. Washington: U.S. Department of Commerce, U.S. Census Bureau.

Kana'iaupuni, Shawn Malia, and Nolan Malone. 2006. This Land is My Land: The Role of Place in Native Hawaiian Identity. In *Race, Ethnicity, and Place in a Changing America*, ed. John W. Frazier and Eugene L. Tettey-Fio, 287–300. Binghamton: Global Academic Publishing.

Kelly, James L. 1983. The Loss and Survival of 'Kuleana' in Hawaii. *Journal of Cultural Geography* 3 (2 (Spring/Summer 1983)): 27–39.

Kirch, Patrick Vinton. 1996. *Legacy of the Landscape: An Illustrated Guide to Hawaiian Archaeological Sites*. Honolulu: University of Hawai'i Press.

Koletty, Stephen R. 2002. The Samoan Archipelago in Urban America. In *Geographical Identities of Ethnic America*, ed. Kate A. Berry and Martha L. Henderson, 130–148. Reno: University of Nevada Press.

Miyares, Ines M. 2008. Expressing 'Local Culture' in Hawai'i. *The Geographical Review* 98 (4): 513–531.

Morris, Ray. 1986. Native Hawaiians. In *America's Architectural Roots: Ethnic Groups that Built America*, ed. Dell Upton, 36–41. Washington: Preservation Press.

Overbye, Dennis. 2016. Under Hawaii's Starriest Skies, a Fight Over Sacred Ground. *New York Times*. October 3, 2016: D1.

Williams, Peter W. 1997. *Houses of God: Region, Religion, and Architecture in the United States*. Urbana: University of Illinois Press.

18 The Future Ethnic Landscape of America

The ethnic landscape of the future will be different from what we currently see. Indeed, the cultural landscape is constantly evolving, and some of the buildings that the author photographed only a few years ago no longer exist. At the same time, new types of structures have appeared, testifying to the increasing prominence of ethnic groups who were hardly represented in the United States just a decade ago. Likewise, other immigrant groups, such as Mexicans, who more recently have been joined by a variety of Central Americans, are increasingly settling into areas, such as the Deep South, which had not previously been associated with any Hispanic culture (Fig. 18.1), creating a patina of Mexican, Guatemalan, or Salvadorian culture upon an ethnic landscape generally associated with the dominant ancestry of those who preceded them. Yet it is not just change in the ethnic composition of the population that alters the landscape. New technology and economic activities may render certain activities, which may or may not have strong ethnic or cultural connections, obsolete or redundant, while new processes or operations may be introduced. In describing the tremendous changes in the landscape of southeastern England, a British historical geographer notes that landscapes are "always moving on; not static but always in flux, with different aspects and areas changing faster or more slowly than others" (Short 2006:11). This observation is just as true in North America, yet the speed of such change varies from region to region and from community to community, as well as over time.

In projecting what will be most visible in the cultural landscape of the future, several critical questions must be asked and answered. These questions are centered around predicting what ethnic populations will dominate the immigration to the United States in the future—as well as projecting whether or not current trends will be permitted to continue, forecasting different rates of natural increase and assimilation among various ethnic groups, identifying groups whose actions resist or outright oppose assimilation, considering what efforts are made to preserve those ethnic structures and neighborhoods that currently exist, and projecting what future changes in technology and public policy may have upon the preservation of obsolete or redundant structures from the past. In addition, just as the revelation of the *Book of Mormon* in upstate New York eventually led the followers of the Church of Jesus Christ of Latter Day Saints to undertake their exodus to Utah in the 1840s where they established their new "ethnic" homeland (Bennion 2001), complete with a highly distinctive physical imprint upon the cultural landscape (Meinig 1965; Jackson 2003), other domestic groups may rise to strongly shape, or reshape, the cultural landscape in specific regions of the country.

18.1 Immigration in the Future

The United States has been built by waves of immigrants, first from eastern Siberia via Beringia nearly fifteen thousand years ago, and then from Europe. The initial European immigrants came largely from the British Isles, with much smaller numbers of French, Dutch, German, and Spanish settlers arriving in colonial times. Several of these European groups were responsible for bringing African slaves to America. With the huge onslaught of European immigrants that began around 1840 and continued until the outbreak of World War I, the nation first saw large numbers of Irish and German immigrants, and then the greatest numbers came from eastern and southern Europe. While the nation received large numbers of immigrants who altered its ethnic mix, the country also grew by adding territory and populations that had previously been foreign. New Orleans was a sizeable foreign city at the time of Louisiana's annexation. The Hispanos in New Mexico and Tejanos in Texas had occupied their homelands for one to two centuries before their lands were acquired from Mexico following the Mexican-American War. Hawaii was not acquired until after the massive waves of immigrants from northern and central Europe had largely taken up settlement within continental United States, yet its annexation added not only Pacific Islanders, but also several Asian groups to the American population. Although small

J.A. Cross, *Ethnic Landscapes of America*, DOI 10.1007/978-3-319-54009-2_18

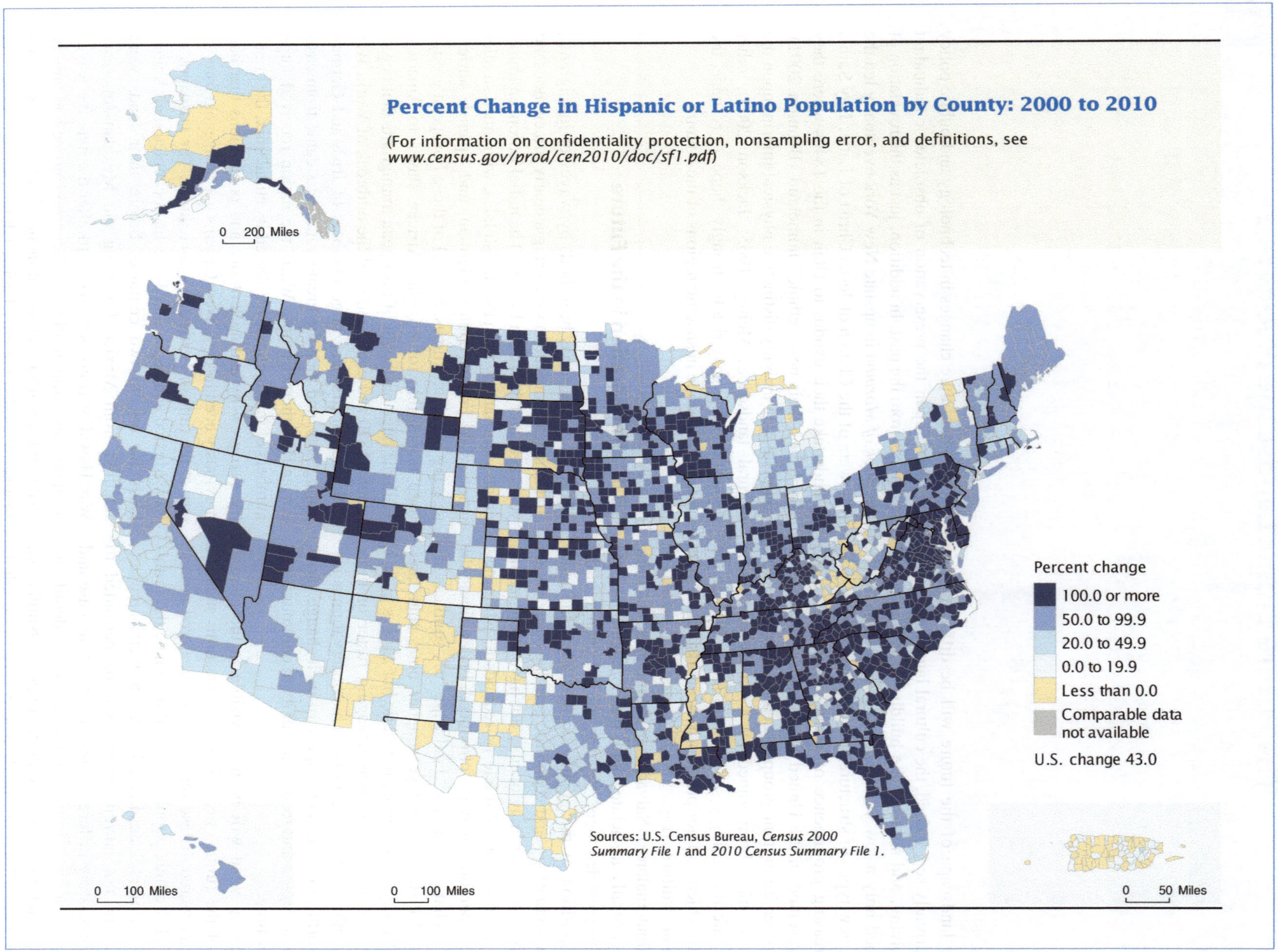

Fig. 18.1 Percent change in the Hispanic or Latino population by county between 2000 and 2010 censuses. The spread of the Hispanic population into many rural counties of the Great Plains, the Midwest, and the South since 1990 has introduced considerable ethnic diversity within many regions whose populations had been dominated by other ethnic groups for well over a century. Although some of those counties that experienced the greatest percentage increases had very small Hispanic populations a decade or two ago, there are now many rural counties in these regions where Hispanics now comprise 5 to 10% of the population, up from very few several decades ago (Reproduced from U.S. Census Bureau maps in Ennis et al. (2011))

numbers of Asians had arrived in the mid 1800s, they did not constitute a major segment of American immigrants until the last third of the twentieth century, by which time European immigration had significantly waned. Over the past half century immigration from Latin America, particularly from Mexico, Central America, and several islands of the West Indies, had risen to its current preeminent place.

Not only have the sources of immigrants changed dramatically over time, but the total numbers of immigrants arriving have also varied. Between 1820 and the outbreak of the First World War, America received approximately 34 million immigrants (Daniels 2002). Yet even during this time immigration rates were not constant, exemplified by the huge dip during the American Civil War. Following World War I rates of immigration were more stringently controlled, and during the Great Depression relatively few immigrants arrived. With the enactment of the Immigration Act of 1965, which opened immigration to a wide diversity of ethnicities, and the opening of the border to refugees from Cuba, Vietnam, Somalia, and elsewhere, America welcomed its largest numbers of immigrants, now numbering well over one million annually.

In projecting future immigration, three broad questions need to be answered. First, how many individuals will be moving to America? That number is determined both by domestic policy and economic and political circumstances in the United States, acting as pull factors—and at times the exact opposite, and by conditions in other nations, tending to push various ethnic populations to flee their homelands, whether for economic, political, religious, or environmental reasons. These conditions are constantly changing. Indeed, during the severe economic downturn beginning in 2008 not only did the rate of arrival of new immigrant workers from Latin America and Eastern Europe decline, but unemployed and discouraged workers left the United States to return home. Second, what nations will dominate the immigration stream in the future? Given America's history, that clearly has changed over time and is likely to continue evolving. As the rhetoric of the 2016 presidential campaign demonstrated, political sentiment to exclude or expel certain ethnic groups, such as has previously marked American history, remains an undercurrent that has not gone away. Thus, not only do political, economic, and environmental conditions abroad determine those who seek to immigrate, but domestic conditions may determine who is welcomed or sent home. Third, what areas of the United States are most attractive to immigrants? Not only does this change regionally with time, but are immigrants attracted to areas that other Americans are leaving (Barcus and Simmons 2013), or are they simply responding to circumstances that are attracting both domestic and international workers? Certain regions have jobs and are welcoming of outsiders, others are dominated by nativists who will not accept immigrants—or at least those who physically look different or worship in unfamiliar ways, which influences interstate migration of certain ethnic populations (Ellis et al. 2016). Finally, populations within America are constantly moving. These internal movements of populations whose identities are not linked to an Old World ethnicity but to an American region have had, and will continue to have, profound impacts upon reshaping—or in some locales averting the reshaping—of the cultural landscape.

Many, but not all, immigrant groups display far higher rates of natural population increase than ethnic and racial populations that are already well established in the United States. Indeed, the larger family sizes displayed by Mexican-Americans, Hasidic Jews, and Hmong refugees are clearly recognized within the communities where they have settled. Yet, within some, such as the Hmong, the population pyramid has rapidly begun to narrow, as family sizes are shrinking among the second generation. In addition, some long-established domestic groups, such as the Mormons and the Amish, have population doubling rates that equal or exceed those of many immigrant groups. Disparities in reproduction rates and high rates of immigration prompted the U.S. Census Bureau to predict that by 2042 that non-Hispanic whites will have become a minority of the United States population (Perez and Hirschman 2009). Thus, not only will some neighborhoods transition from predominately white European residents to Hispanic or Asian, but entirely new immigrant communities are likely to become established that will considerably differ from those established in earlier times by previous immigrants. Indeed, recent controversies have arisen in Los Angeles as newly arrived Bangladeshis have attempted to have part of Koreatown officially designated as Little Bangladesh (Jang 2009).

At the same time that ethnic populations are becoming more conspicuous in new gateway cities, certain long established ethnic communities have witnessed an exodus of some of their "longtime residents who have provided a distinctive sense of place" (Godfrey 2016:59). Thus, certain communities not previously known for the presence of particular ethnic groups have developed new, sometimes highly visible ethnic landscapes, while other ethnic neighborhoods have seen a dilution of the cultural group for which they had been well known.

18.2 Environmental Change and Landscapes

Changes to the physical environment have the potential to force huge changes upon the cultural landscape. While river valleys have been inundated by the construction of giant dams impounding water for irrigation or hydro-electric generation for well over a century, tens of thousands of giant windmills, whose towers may be 100 meters high and whose blades may extend 50 meters from the towers, have been erected across many areas of rural America during the past decade, and many more such wind power projects are planned. Thus, when looking at a German ethnic barn on a farmstead whose shape and dispersal of structures illustrate the American public land survey system, one cannot help to see giant windmills dotting the landscape in many parts of the American Midwest and Great Plains.

Climate experts have identified many dire consequences of global warming, including a significant increase in the severity of certain atmospheric hazards (van Aalst 2006). Rising sea levels and spreading drought will undoubtedly not only change land use and settlement patterns in those affected parts of the United States (Melillo et al. 2014), but such environmental degradation elsewhere will undoubtedly increase the number of environmental refugees who arrive in the United States (Reuveny 2007). Many of these immigrants will arrive from areas of the world that had not previously sent large numbers of people to America. Furthermore, many political analysts project political instability in some parts of the world promoted by the stress of climate change, with warfare and related political strife adding political refugees to the numbers of environmental refugees. Altering the mix of immigrants will undoubtedly influence the ethnic landscape.

Climate change is now expected to result in a rise of sea level of at least two to three feet by the end of the twenty-first century (Kopp et al. 2016). Such an increase will subject many coastal communities to heightened risk of inundation and erosion, risks that are exacerbated when combined with exposure to hurricane hazards (Hauer et al. 2016). Such a rise in sea level threatens destruction of a number of historic ethnic communities, including French Cajun, Native American, and African-American communities in southern Louisiana, Cuban settlements in southern Florida, and English settlements along the coast of the Carolinas and Virginia. The recent decision to move the Native American residents of Isle de Jean Charles, Louisiana was touted as simply the first such relocation (Davenport and Robertson 2016), yet several small Inuit villages in northern Alaska are also facing annihilation from a combination of sea level rise and thawing permafrost (Marino 2012). Climate change resulting in a hotter and drier climate will also seriously threaten the livelihood, and thus the long-term economic viability, of various Native American communities within the American Southwest (Cook et al. 2015; MacDonald 2010). It may also accelerate the abandonment of already threatened rural communities in portions of the Great Plains. Thus, certain ethnic landscapes will be lost, yet environmental refugees will contribute to the reshaping of the communities that provide them haven.

18.3 Preservation and Loss of the Past

America's attitudes towards preserving its past have shifted over time, while at the same time fires, floods, earthquakes, windstorms, and other calamities have removed structures that were recognized as worthy of preservation. Even buildings recognized by preservationists as deserving safeguarding have been felled by the wrecking ball to make way for urban renewal, highway construction, or industrial or commercial expansion. Others have succumbed to benign neglect, as funds could not be raised to begin essential maintenance to forestall collapse or condemnation. The passage of time favors the preservation of structures built of stone or brick over those constructed of logs or lumber, and it is biased towards those elements of material culture located in dry versus wet environments, as well as cold versus hot.

18.3.1 What Gets Preserved?

Even when communities have made conscious efforts to preserve fragments of their past, the choices made have often focused upon certain monumental structures, such as churches or the houses of prominent (and often wealthy) citizens, ignoring the more typical and commonplace vernacular houses, shops, or farm buildings erected by the majority of early ethnic settlers. Often those communities that display the greatest assemblages of structures from the past retained them not by deliberate action, but by default. Not only were many mining communities in the American West largely abandoned when ore bodies were exhausted and mines closed, but many rural farming settlements have seen a steady out-migration of their youth, leaving a declining population with little motivation to invest in new buildings or to even raze those that have been abandoned.

While many rural areas provide some of the best preserved examples of ethnic landscapes that were shaped in the late nineteenth century, even in these areas there has been a demise of rural churches and numerous rural communities. In many counties of the Great Plains that were settled by ethnic immigrants between the late 1880s and the outbreak of the First World War, the population reached its maximum around a century ago—at the time the immigrant population was also at its maximum—and has steadily declined ever since. With the loss of population has come the desire for many rural churches to consolidate, and many church buildings no longer house active congregations (Brown 2002). Indeed, some churches or abandoned rural schools are about all that is left of settlements that have died. While some states, such as North Dakota, have made a concerted effort to preserve these landmarks of their past, the "Spruce Up South Dakota" project in 2000 leveled many abandoned buildings in an effort to beautify that state. In contrast, New Mexico's government has assisted in protecting its historic landscape by providing funding for church restoration, enhancing the preservation of its already distinct Hispano homeland. Nevertheless, for many communities that have gone beyond the declining stage, all that is left are the rural cemeteries without their previous churches. While some of these are well maintained, and historical markers are sometimes erected to commemorate the now-departed church and settlement, others are overgrown by vegetation.

Grave markers, particularly those of stone or metal, are all that remain at the site of some rural churches, or even of some abandoned rural settlements, yet these stones provide lasting evidence of many ethnic populations. Not only can one see the ethnic surnames and the various languages that adorn the tombstones, but there are tremendous ethnic and religious variations between the types and amount of artwork and religious symbolism displayed. In contrast, some groups were less likely to erect permanent markers, or to even make burials. The loss of rural population, less involvement with organized religion, and the increased tendency for burials to occur in multi-ethnic cemeteries and for many Americans to be cremated may result in today's contemporary cemeteries conveying a different ethnic message in the future than those from the eighteenth or nineteenth centuries.

Not all items of material culture are equally likely to get preserved. As John Lehr (1990) has noted, certain structures are more resistant to loss, both because of their physical makeup and because of their cultural or religious significance. Cemetery markers are a classic example. On the other hand, many vernacular buildings—particularly those associated with outdated modes of living or obsolete economic activities are disappearing. A half century ago the lowly outhouse was spotted behind a large share of rural farmhouses; now they are rarely seen. Large barns—many of which represented a far greater financial investment than their owner's residences—were still the norm on most farms even a quarter century ago. Today, very few farmers have horses. Furthermore, the nation's number of farms with milk cows fell by 98% during the twentieth century, resulting in a huge number of redundant barns. While many such farm structures displayed aspects of their builder's ethnicity, today most have either been relegated to other uses or have fallen into ruin. In some locations they have been demolished to simultaneously salvage their timbers and to reduce the assessed value of the structures subject to property taxes. In parts of rural New England, the fields have also been abandoned and overgrown, only marked by the stone fences that delineate their former edges.

Within many highly urbanized areas the ethnic imprint of various populations that arrived several centuries ago has become highly muted. As discussed in Chap. 10, the visible architectural imprint of the Dutch colonial settlement on Manhattan Island could still be seen in the city's buildings after a century of English colonial rule. Yet after extensive destruction of the city by fire during the American Revolution, followed by other fires and several cycles of urban renewal that resulted in New York City's downtown becoming dominated by high rise buildings, little of the colonial Dutch imprint remains other than place names and the location of certain city streets. At the same time, many recently arrived ethnic groups, new to the American scene (Fig. 18.2), are reshaping their neighborhoods and establishing ethnoburbs

Fig. 18.2 Hindu temple in Flushing, New York City's borough of Queens, New York. The Hindu Temple Society of North America was, at the time of its construction in 1977, the only Hindu temple on the continent. It remains one of the most prominent Hindu temples in the New York City area, with its prominent rajagopuram or grand entrance tower. Similar Hindu temples have now been built in other American cities, including near Chicago, yet many Asiatic Indians live dispersed throughout the metropolitan regions

Fig. 18.3 Ethnic shops in the Jackson Heights neighborhood in New York City's borough of Queens. Note that the pharmacy, which is catering to a Bangladeshi population, also notes its identity in Spanish. Jackson Heights now has a highly mixed ethnic population, including many Hispanics from northwestern South America and South Asians from Bangladesh and Pakistan

composed of several different groups of ethnic immigrants (Fig. 18.3). Other ethnic populations arrive and disperse across neighborhoods rather than concentrating their numbers, muting their visible impact.

18.3.2 Changes to the Built Environment

The replacement of structures suitable to smaller populations by huge modern buildings—as well as highways, railroad yards, wharf facilities and airports—constantly alters, diminishes, or obliterates the ethnic imprint of earlier settlers. This process is on-going. Chicago recently struggled to remove St. Johannes Cemetery, established by German Evangelical Protestants a century and half ago in suburban Bensonville, so that it could expand the runways of O'Hare Airport. New construction, regardless of how carefully the graves might be removed or populations relocated, inevitably alters the cultural landscape. In urban centers throughout the country, many cemeteries from the eighteenth and nineteenth century have simply been paved over.

Buildings need not be replaced to diverge significantly from their original purpose and form. Over time, as buildings are renovated, upgraded, and repaired, their appearance changes. Thus, over time their original cladding may be replaced, such as a brick veneer covering the original frame walls, the roof may be replaced by a new one of a different pitch and roofing material, windows are added or subtracted, and various decorative elements are installed or removed. For example, Italianate decorations may be added or the original cupola or towers removed. Vinyl windows with double or triple pane glass may replace older wooden sash windows, whose replacement with authentic materials would be not only more expensive, but less energy efficient. Even greater alterations to a building's appearance occur when it undergoes expansion, whether by jacking it up to add a basement or new ground floor, adding additional floors on top, adding one or more new wings, or having all of these actions occur. Many old buildings have undergone a series of remodeling and upgrades, such that their appearance increasingly diverges from their original image and multiple ethnicities or generic American styles are displayed (Brand 1995). Thus, old buildings may not faithfully display the ethnic heritage of their original builders.

18.3.3 Threats to Material Culture of Several Ethnic Groups

In many urban centers, as well as in rapidly growing suburbia, where new or different ethnic populations make-over existing structures into their own, the material culture that shaped the landscape of earlier ethnic groups may be lost. The new inhabitants as well as developers may lack an appreciation of the significance of structures erected by the previous ethnic populations. Gentrification may rehabilitate neighborhood structures to accommodate more prosperous residents who replace the original tenants, or ghetto formation may result in townhouses being converted into tenements housing many families. Within the African-American ghettos of both Chicago and Milwaukee are found many churches and synagogues built by European immigrants during the nineteenth century, yet many of these buildings have been relegated to various secular uses or have been adapted to meet the religious needs of denominations whose traditions and artistic heritage are quite different from those of the original congregations. Even when structures are faithfully maintained and appreciated, such as the Pilgrim Baptist Church that was destroyed by fire in Chicago in 2006, they come to represent two cultures—the one that built them and then the one that used them, but with a different set of adornments and purpose. If such a destroyed cultural landmark is to be rebuilt, not only can the question be asked about whether it will display a distinct ethnic imprint, but if so, whose?

The relative poverty and powerlessness of certain racial and ethnic populations hinders the preservation of their ethnic imprint upon the landscape. Urban renewal has often focused upon areas inhabited by racial or ethnic minorities. Numerous cities in the 1950s and 1960s deliberately chose to run their expressways through low-income minority neighborhoods, demolishing many vernacular structures and destroying the cohesiveness of many ethnic communities. Examples include Chicago, Milwaukee, Los Angeles, and New Orleans. Later, during the inundation of New Orleans in the aftermath of Hurricane Katrina, the poorer and lower-lying African-American neighborhoods bore the brunt of the destruction. The wealthier neighborhoods of the upper-class white population around the Garden District, along with the historic French Quarter, were largely unscathed. Thus, much of the Black Creole city has been lost, the buildings are in ruins with many properties having become vacant lots, and the population is scattered across the nation.

It has been argued that "the preservation of traditional built forms and spaces not only reflect ethnic identity, but also serve to reinforce group identity and influence the retention of culture" (Manger 2000:4). Indeed, such survival of pueblo structures within the American Southwest only strengthens the identity of various Native American groups occupying the few remaining historic communities. In a like manner, the presence of the San Antonio Missions in Texas and over 20 Spanish missions in California provides a constant reminder of the deep roots that those areas' Hispanic populations have in their community. Yet we again need to ask the question about how representative are these legacy buildings. As we saw in Chap. 8, there are many examples of antebellum mansions to remind the viewer of the Southern plantation system, but there are few surviving slave quarters to illustrate the labor of those whose work facilitated the plantations' economic success.

In some locales structures that have little current utility add to the assessed property value, thus simple avoidance of taxation results in the demolition of the buildings, both in rural and urban areas. Indeed, solidly built masonry buildings in the downtown section of the former mining city of Rhyolite, Nevada were dynamited shortly after the closing of its mines to avoid taxes. Today, over three-quarters of a century later, the ruined walls of the downtown buildings are a reminder of the former existence of the small city. Sometimes entire downtown blocks within cities have been allowed to fall into ruin, such as what one sees today in Cairo, Illinois, where the small residual population neither shares the prosperity nor ethnicity of the city's founders.

Where groups of people have had either the strength in numbers or sufficient financial wealth and political clout, much of their ethnic heritage has been preserved. Yet, such unequal influence introduces biases into the preservation of the cultural landscape, where the eye is drawn not only towards the palatial houses of the wealthy, who may or may not be the most numerous ethnic population, but towards the displayed ethnic heritage that a community wants to exhibit or showcase. Furthermore, does the present or former population have either an understanding of or respect for the heritage of the past?

18.4 Landscape Losses with Preservation

Losses and landscape change can occur even with preservation. Consider what has happened in the historic district that comprises the core of Spanish colonial Saint Augustine, Florida. Given the city's existence for four and a half centuries, not only are most of the structures built during the two periods of Spanish colonial rule gone, but so are many of those built during the periods of British rule and early American control. While many buildings have been reconstructed along the original narrow streets in a manner to at least mimic the colonial architecture in keeping with local historic district ordinances, the functional landscape is in sharp contrast with the original. Electric lights glow in the buildings and motor vehicles convey both passengers and goods. Many shops in the historic district display trinkets, candies and ice cream, among other items for sale to tourists, hardly the type of wares that would have characterized the retail landscape in a seventeenth or eighteenth century Spanish colonial outpost. Thus, only a portion of what we see today reflects the genuine historic ethnic heritage.

Many communities throughout the United States have attempted to preserve collections of historic and ethnic buildings, with varying degrees of success. Some of these, such as the magnificent selection of structures in Historic

Colonial Williamsburg, Virginia, contain buildings that have been either preserved or reconstructed on their original sites, with those structures representing different time periods being removed from the site, yet many collections do little to create the authentic spatial arrangements seen on the original ethnic landscape. For example, throughout North America most farmers lived dispersed on their own farmsteads, yet as John Lehr (1990) notes, their structures have been gathered for preservation in various village-like clusters. Through such clustering, not only are the buildings of several farmsteads brought together, but these selected buildings are also arranged in close proximity rather than in the original dispersed manner. While this provides for a better arrangement for tourists and may be the best option for preservation, it not only ignores, but misrepresents, important spatial components on the landscape.

Selective preservation in such historic villages is always problematic. Oftentimes those buildings that are relocated to such sites were those that had had both survived and those that their owners wanted removed. Thus, buildings in the way of a highway expansion, suburban development, or a reservoir project might be available, but these might not be truly representative of the variety of structures in the original ethnic community. Houses might be relocated without the appropriate associated outbuildings, and places of work may be missing. Thus, as Richard Francaviglia (2000) notes, there are many different types of preservation of heritage landscapes, ranging from passively to actively preserved, with some being restored and others assembled. Yet, each variety conveys a different type of landscape image to the viewer.

18.5 Ethnic Tourism

Many communities have capitalized upon ethnic tourism. A listing of just a few examples would include Spanish Santa Fe and Taos, New Mexico, San Antonio, Texas, and St. Augustine, Florida; the French Quarter of New Orleans, Louisiana; English Williamsburg, Virginia, and Plymouth Plantation, Massachusetts; German New Ulm, Minnesota, Fredericksburg, Texas, Frankenmuth, Michigan and Amana, Iowa; Amish Holmes County, Ohio and Lancaster County, Pennsylvania; Swiss New Glarus, Wisconsin; Danish Solvang, California; Dutch Holland, Michigan and Pella, Iowa; Greek Tarpon Springs, Florida; the Cuban Ybor City district of Tampa, Florida and "Little Havana" in Miami, Florida; and Chinatowns in San Francisco, California, Chicago, Illinois, and New York City, New York. Some cities simply promote their ethnic neighborhoods as one of many attractions available to tourists. Commercial services offering guided tours to a variety of ethnic neighborhoods are available within several large cities, including New York City and Chicago. Wisconsin has created an Ethnic Heritage Trail linking many ethnic European settlements along Lake Michigan, extending from the Illinois border to Green Bay, while Louisiana recently began promoting its African-American Heritage Trail that linked ethnic communities, museums, and heritage sites across the state.

18.5.1 Creation of Ethnic Tourism

Ethnic tourism does not necessarily involve the showcasing of a preserved ethnic landscape. Sometimes, neither buildings nor much other material culture that evoke the ethnic heritage of the group remain. For example, while Milwaukee hosts German Fest annually, drawing thousands to partake in German music, drink, and food in a city that is rich with German architecture and monuments, within other communities it is the ethnic festival that underpins their ethnic tourism. While the ethnic Swiss population of New Glarus, Wisconsin had long performed several Swiss festivals to help maintain its identity, its original Reformed church, described as "the only structure in town that bore a distinctive Old World stamp" (Hoelscher 1998:62) was demolished a century ago. In the mid-twentieth century in an effort to further promote its ethnic tourism, efforts were made to add Swiss influenced architectural embellishments to its commercial district and then professional architects were engaged to construct Swiss style buildings—both commercial and residential, such that New Glarus has a stronger Swiss appearance today than it did a century ago. Yet questions can be raised about the authenticity of such ethnic landscapes. As Steven Hoelscher explains regarding New Glarus's Swiss ethnicity, "the meaning of that ethnicity, its social and economic implications, and its symbolic representation—in short the construction of Swissness—have changed radically over time" (Hoelscher 1998:232). Yet things have also changed in Switzerland, and "people cross the Atlantic to find 'genuine' Swiss culture in New Glarus" (Hoelscher 1998:232).

Another example of the creation of ethnic tourism is the case of Lindsborg, Kansas, which promotes itself as "Little Sweden, U.S.A." Lindsborg's ethnicity is showcased during festivals when many residents, including some who have no Swedish ancestry, wear ethnic costumes designed to portray the areas of Sweden which their ancestors left, participate in Swedish dance groups, partake of Swedish foods, display Swedish crafts, and tell Swedish stories (Schnell 2003a). The red Dala horse, patterned after Swedish painted wooden toy horses, has become a city symbol, painted on sign-posts and name-plates on two-thirds of its houses, and displayed as statues within the business district. Swedish-themed stores selling gifts, crafts, and food have been established. To enhance their ethnic appearance, local businesses have remodeled their storefronts "with faux Swedish half-timbering, ironwork, and peaked rooflines, giving the old

Victorian-era storefronts a decidedly playful and festive air" (Schnell 2003b:43). This construction to evoke ethnicity, of course, has not gone as far as what is seen atop Al Johnson's Swedish Restaurant in Sister Bay, Wisconsin, which is famous for its sod roof being actively grazed by several goats.

Similarly, Caldwell, Texas has few outward architectural manifestations of its Czech heritage, yet it promotes itself as "The Kolache Capital of Texas." These Czech pastries are sold at several of its bakeries and restaurants and the city hosts an annual kolache festival. Verdigre, Nebraska claims to be "The Kolache Capital of the World," and similarly hosts an annual "Kolache Days." Hillsboro, Wisconsin also touts its Czech heritage, even though it has few architectural reminders of that ethnicity.

Business interests in Wenatchee, Washington have added architectural elements and signage to their buildings in an effort to enhance their contrived Bavarian ethnicity, as we saw in Chap. 11. Indeed, the communities described above are simply a small snippet of those that that have tried to capitalize on ethnicity tourism, even if it is at a small scale. On a much larger scale, Disney World's Epcot Center outside Orlando, Florida has constructed village neighborhoods that feature shops and restaurants that entice tourists to sample a variety of European and Asian ethnic cultures.

18.5.2 Ethnic Tourism and Its Impact upon the Ethnic Group

In some areas of the United States, ethnic tourism capitalizes upon vibrant communities that represent ethnic islands or even full homelands, comprised by a locally dominant ethnic group. Tourism within northern New Mexico, focused upon both its vibrant Hispano and Native American communities, has made Santa Fe and Taos popular tourist destinations. Similarly, many tourists are drawn to the bayous of southern Louisiana to partake of the Cajun cuisine and music, as well as to encounter the swamp, whose "heritage ecology" has been described as "a Cajun tourism commodity" (Keul 2014:506). Amish settlements in Holmes County, Ohio and Lancaster County, Pennsylvania—just to name the largest two—attract many tourists to view plain clad residents riding horse-drawn buggies and to purchase Amish cheeses, candy, and vegetables, yet numerous questions may be asked regarding whether it is the old-fashioned nostalgia or the ethnic or religious heritage that piques the visitors' interests. In addition, the interactions with tourists influence Amish culture (Trollinger 2012), and some of the most conservative Amish avoid such commercial and social interactions.

Chinatowns in several cities draw many tourists, some from considerable distances, while others come from different neighborhoods or suburbs of the same city. Santos and colleagues (2008:1008) note that guides escorting tourists through Chicago's Chinatown often "construct Chinatown as an exotic and different place through references to unique elements of Chinese heritage, typically by offering discussions of Chinese holidays, their meaning, and the ways they are celebrated," even when the tours occur at times of no holidays. Yet they note, "touristification of Chinatown has policy consequences, since the ethnic identity of the neighborhood must be preserved, if tourism to the neighborhood is to flourish" (Santos et al. 2008:1011). Yet at the same time, because of tourism, "ethnic identities are increasingly subject to renegotiation, rediscovery and re-evaluation" (Santos and Yan 2008:881). Indeed, the more that Chinatown—or any ethnic neighborhood—relies upon tourism to underpin its economy, the greater the risk that its ethnicity will be represented in a way to fulfill the tourists' expectations.

Visiting an ethnic homeland or community does not necessarily result in the visitor gaining an informative or unbiased education about the ethnicity of that group. For example, the construction of casinos by many Native American Nations has drawn large number of tourists to many reservations or off-reservation sites, yet many of these visitors see little of the ethnic heritage of the host group. While some of the casinos house well-displayed museums or gift shops featuring aspects of the indigenous heritage, in others nothing is displayed or offered for sale that promotes the Native American culture. Some Native American groups do not even want visitors, placing signs along roads leading to certain pueblos or sacred sites that inform outsiders to stay out.

18.6 Creation of New Ethnic Landscapes by Resisting Assimilation

Without doubt, new ethnic enclaves and neighborhoods will be created by some of the new immigrant groups who arrive in the future, as described earlier in this chapter. Yet, many newly arrived ethnic populations will join previous groups who have become increasingly assimilated into American culture. Indeed, through the assimilation process many populations, such as the Irish, Germans, Italians, and Mexicans, have reshaped what one thinks is the generic American culture, as new terms are added to everyday language, building styles evolve, and the American diet incorporates new dishes. Yet at the same time, not all groups have assimilated. It may well be that the groups that most vigorously resist assimilation will provide some of the most interesting and growing ethnic imprints upon the future American landscape.

Based upon work in the Canadian prairies, which were settled by Europeans similar to those who flocked to the Dakotas, Oliver and Edwald (2016:204) found that "those groups who resisted assimilative forces were those that placed significant emphasis upon religion, evident in the building and maintenance of ethnic churches." Through the

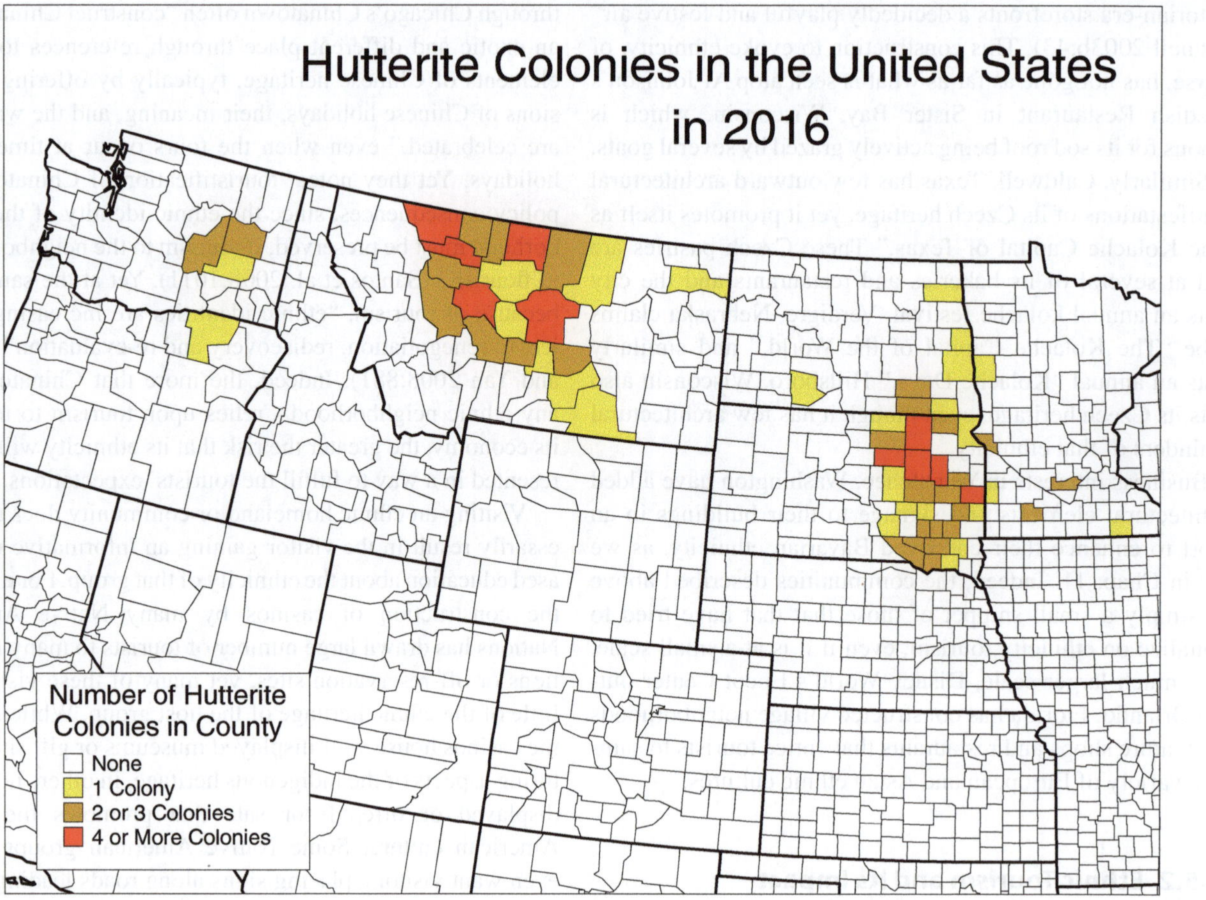

Fig. 18.4 Hutterite colonies per county in the United States in 2016. While far more Hutterite colonies are located within Canada's Prairie Provinces, the majority of the Hutterite colonies in the United States are located in the northern Great Plains

loss of many rural settlements and their ethnic churches that was noted earlier, the vibrancy of their Scandinavian and Slavic communities that were established over a century ago has been substantially muted. Yet, both the Canadian Prairies and the Dakotas are the home of one religious group that has not only resisted assimilation, but which has rapidly grown in numbers within its enclaves that are located apart from other populations within the sparsely settled region.

18.6.1 The Hutterites

The Hutterites, an Anabaptist group of German Russians who migrated to North America in 1874, live in nearly 500 rural colonies, of which over two-thirds are in Canada's three Prairie Provinces, yet in 2016 South Dakota was home to 62 colonies and Montana had 52, while North Dakota and Minnesota each had nine, with smaller numbers in Oregon and eastern Washington (Hutterian Brethren 2016). Largely separated geographically from other populations and worshiping in Hutterisch, a dialect derived from Bavarian German, this group—and particularly its most conservative Lehrerleut and Dariusleut subgroups—maintains its separation from mainstream American and Canadian society. With over 10,000 residents in the northern Great Plains of the United States (Fig. 18.4), the Hutterites "live in self-contained communities, practice a community of goods, maintain traditional dress and customs, and attempt to replicate the ideal of the early Christian communities" (Zimmerman 2015:74). They have been described as being "the most successful communal society in modern history" (Janzen and Stanton 2010:302). Largely employed in farming, the Hutterites account for large shares of the statewide livestock and egg production in both Montana and South Dakota. They cultivate thousands of acres of grain and have extensive vegetable gardens in colonies that may own 20,000 acres of farmland.

While Hutterites comprise slightly over half of 1% of the total populations of Montana and South Dakota, the concentration of their colonies, each averaging 80 individuals and typically ranging from 50 to 120 residents (Hutterian Brethern 2016), results in considerably higher percentages of the population in several rural counties. Indeed, Hutterites comprise over 15% of the population of Faulk County, South Dakota, and 12% of Clark County (ARDA n.d.). Furthermore, given their exceptionally high birthrates, pressure to establish new colonies is continuing.

Fig. 18.5 Aerial view, looking towards the south, of Blumengard Hutterite colony in Faulk County, South Dakota. The geometric arrangement of the residential buildings, linked by sidewalks to the church and dining hall, is clearly visible. Larger buildings to house livestock, particularly pigs, are located to the right

Unlike the Amish, their Anabaptist brethren, the Hutterites have accepted many modern technologies, including highly mechanized agriculture, trucks and vans, electricity, and computers (including maintaining their own website) that are used for business by the colony, but live in family quarters that are part of communal housing, eat their meals communally, own all property communally with income going to the colony rather than the individual, and wear clothing—that while colorful—bespeaks to it being sewn at home (Janzen and Stanton 2010).

The Hutterite colony has a distinct imprint upon the cultural landscape, one that is identifiable from the air (Fig. 18.5), as their "relatively large and complex footprint is in marked contrast to that of neighboring family farms" (Evans and Peller 2016:52). Furthermore, their colonies are planned, and most "colonies also conform to simple geometric grid patterns" (Evans and Peller 2016:74). Nevertheless, in contrast to their conspicuous footprint when viewed from above, "many Hutterite colonies appear hidden even when they are located near major roads" (Janzen and Stanton 2010:202). Similarly, Kraybill and Bowman (2001:22-23) observe that "colony buildings, clustered like a small village, are often invisible from major roads."

Several scholars have described the structures that are found in Hutterite colonies. At the center of the colony's grid is the kitchen, dining hall, and church building, which are surrounded by apartment buildings. Sidewalks are often placed in front of residential dwellings that link them to the dining hall and church and trees are planted, yet ornamental flowers and lawns are lacking among the Lehrerleut and Dariusleut colonies (Janzen and Stanton 2010).

The location of their farm buildings helps to hide the residential core of the Hutterite colony. Within their colony, "Hutterites usually place barns, shops, and other farm buildings on the outskirts of their communities [which] surround and protect the more private environs of dwellings, dining hall, school, and church (Janzen and Stanton 2010:202). Overall, "each colony is physically cloistered from the outside world, [yet] linked to the economy of the region, but residential segregation shields colonists from daily interaction with outsiders" (Kraybill and Bowman 2001:22). Thus, the Hutterites "have turned away from contemporary America more single-mindedly than even their spiritual cousins, the Amish and the Mennonites" (Wilson 2000:2).

18.6.2 Hasidic Jews

In contrast to the Hutterites and Amish who have resisted assimilation in the rural countryside, the Hasidic Jews have done so in the nation's largest metropolitan area. They are best seen in New York City's borough of Brooklyn.

Of Brooklyn's total 2.5 million persons in 2011, the overall Jewish population was estimated at 561,100, with 24% of the borough's population living in a Jewish household (Beck et al. 2013). Hasidic Jews account for about a third of Brooklyn's total Jewish population, and they are concentrated within its Borough Park, Crown Heights, and Williamsburg neighborhoods. The Jewish population of all three of these neighborhoods is dominated by Haredi Jews, which includes the ultraorthodox Hasidic Jews, with Haredi Jews accounting for 80% of Williamsburg's Jewish population and 76% of Borough Park's (Beck et al. 2013). Thus, within just these two neighborhoods reside over 150,000 Hasidic Jews, and over 78% of the population in the Borough Park area is Jewish. Indeed, "Hasidic Jews have established entire neighborhoods within the city that allow them to live apart from a surrounding culture that seems to threaten many of their beliefs and practices" (Scheitle and Finke 2012:197–198.) Thus, they resist assimilation into the overall culture and into the larger, mainstream Jewish population.

Highly pietistic with "an intense devotion to religious practice and the laws of the Jewish faith" (Scheitle and Finke 2012:202), the Hasidic community, or as translated the "pious ones," are divided into approximately 25 groups, each with its own rebbe or spiritual leader, who often inherits this position from his father. Not only is the cultural landscape conspicuously marked by the presence of synagogues, yeshivas or Jewish religious schools (Fig. 18.6), and school busses labeled in Yiddish, using Hebrew letters (Fig. 18.7), but with many shops to cater to their needs keeping in compliance with religious laws. Thus, one sees kosher bakeries, kosher meat markets, and shops displaying Shtreimel or Spodik hats made with fox tail fur and long black coats. Advertising signs promote the care of the Torah, the responsibility of the Sofrim, and the provision of custom wigs, to cover the heads of women, both of which cater to Hasidic religious needs. Maternity shops abound.

Many shops and institutions are only labeled in Yiddish, as are many of the advertizing placards and billboards (Fig. 18.8). In describing the Borough Park Hasidic neighborhood of Brooklyn, New York, Scheitle and Finke (2012:214) note, "The dress of the pedestrians was the first and most obvious sign that we had entered a Hasidic neighborhood." The Hasidim rely upon their own buses to commute within the city, labeled with Hebrew letters and segregating the sexes, with women sitting in the rear. Certain buildings have separate entrances for women. On sidewalks leading to Jewish schools warning signs caution visitors about carrying objects on Shabbat, the Sabbath day beginning at sundown on Friday and lasting until the next evening. Indeed, on the day of worship no work is to be performed and no driving, nor riding in a vehicle—whether car, bus, or subway—is to be done.

Such rules keep the congregation living close to the synagogue and have prompted residents in certain Orthodox communities to erect eruvs, in which wires or string mark the

Fig. 18.6 Yiddish signage, using Hebrew lettering, dominates the arched brickwork façade of Yeshiva Beth Josef Zvi in the Hasidic Jewish Williamsburg neighborhood of Brooklyn, New York. Note the distinctive black cloak being worn by the man

Fig. 18.7 Bais Rachel, a Hasidic Jewish school in a building that predates the establishment of the Williamsburg Hasidic community. Note the Yiddish language, using Hebrew lettering, on both the school building and the school busses. Numerous busses with Hebrew lettering are parked in the neighborhood on school days, with children being bussed to its many Orthodox Jewish from other parts of New York City

Fig. 18.8 Billboard, showing a rabbi holding scrolls captioned in the Yiddish language, advertising religious gifts store in Williamsburg Hasidic community of Brooklyn, New York

boundary of ritual space in which objects may be carried on Shabbat. Yet in certain communities the use of eruvs are prohibited, and notices highlight the controversy surrounding them. As a local rebbe explained, "the density of the population is what makes a strong Jewish community possible. Orthodox Jews can walk to synagogue, and all Jews can be close to kosher markets, Hebrew schools, and other institutions that are distinctly Jewish" (quoted in Scheitle and Finke 2012: 214). The Hasidic communities also have their own ambulance service, bearing the label "Hatzalah," to provide emergency services to the ultraconservative Yiddish-speaking Orthodox Jewish community, and an organized neighborhood watch security force, or "Shomrim." Their activities are largely limited to Haredi Jewish neighborhoods.

18.6.3 Resistance to Assimilation and Population Growth

Clearly, neither the Hutterites, the Old Order Amish (who are discussed in Chap. 11), nor the Hasidic Jewish population have any desire to assimilate with others. Not only are all three groups ultraconservative in their religious practices, all avoid television, but their communities are close knit, and all three speak dialects that enhance their separation from the broader society. Furthermore, among both the Hutterites and the Hasidic Jewish population, even intermarriage between members of different Hutterite leuts or Hasidic communities, respectively, is strongly discouraged. Similarly, intermarriage among the Amish and outsiders does not occur—unless one is leaving their community, and most Amish marry an individual from the same or similar Amish affiliation, even if that individual comes from a different Amish settlement.

All three groups are growing rapidly in population. Indeed, the Hutterites displayed the highest fertility rate of any group when studied in the mid-twentieth century, and while their birth rate dropped 23% between 1950 and 1990, it is three times higher than that for other South Dakota rural population groups (Kraybill and Bowen 2001). The Hasidic Jewish populations in Williamsburg and Borough Park likewise have very high birth rates, with 50% of Williamsburg's Jewish population being no more than 17 years of age, while Borough Park's proportion of children was 49% (Beck et al. 2013). The birth rate among the Amish is also quite high, resulting in a doubling of their population every 20 years (Donnermeyer 2015).

Unlike so many ethnic groups who over time experience a drop in their birth rates as they assimilate into greater American society, these groups not only promote their separation and maintain their distinctive culture, but they are growing in population so rapidly that new settlements or new colonies must be continually created. If anything, the Amish, the Hutterites, and the Hasidic Jews will have a much stronger and geographically widespread impact upon the nation's future ethnic landscape. Between 1992 and 2016, the number of Amish settlements grew from 227 to 509 (Young Center 2016). The Hasidic Jewish population not only continues to rapidly increase in Brooklyn, but its growth has spawned the spread of Hasidic settlement into counties north of New York City. Indeed, the Satmar Hasidic village of Kiryas Joel in Orange County and the Skverer Hasidic village of New Square in Rockland County are virtually entirely Hasidic and are among the most rapidly growing communities in the state (Chan and McGinty 2007), with the five-year American Community Survey estimates for 2011–15 indicating their median ages are 12.9 and 14.1 years, respectively.

Within this book's opening chapter Karl Raitz (1979:88) was quoted, explaining how these groups established "a way of life similar to that of their homeland." Yet, as they have grown in numbers they have become increasingly separated from the norms of American society. They are establishing new homelands in America—different from those that they left behind. Indeed, neither the Amish nor the Hutterites have any physical or cultural presence remaining in Europe, and the Jewish diaspora has resulted in the largest Hasidic Jewish communities being in the United States, Israel, and the United Kingdom, not within their Eastern European homelands that they fled following World War II. Much of the ethnic landscape we see in their communities today reflects a new homeland, displaying how their culture has responded to the American environment, and how the American culture has accommodated and adjusted to their presence.

18.7 Globalization and Landscapes

Modernization and the increased urbanization of society have resulted in many changes to the landscape. Globalization is now creating fundamental changes. Not only are the economies of nations increasingly linked, with an increasing share of the world's products grown or manufactured far from their consumers, but facilities to accommodate growing numbers of tourists are reshaping many parts of the world. In his essay about globalization in the countryside, James McCarthy (2008:129) writes, "We see that the countryside is increasingly a global commodity, one produced to standardized and exacting specifications at a growing number of locations." The shaping of the landscape is becoming increasingly homogenized, not only in the United States, but in other parts of the world. Americans eyed the German autobahns with envy, built their huge interstate highway system, and have been exporting their affinity for highway transportation to many parts of the world. American style suburban development has come to parts of Europe and the Middle East, as have the American-style supermarket and shopping mall. Indeed, the world's largest shopping malls are now in China.

The flow of ideas goes both directions, with numerous permutations and feedbacks influencing the ultimate built landscape. For instance, English architects strongly influenced urban building in colonial and post-Revolutionary America, resulting in Georgian architecture and evolving into the Federalist and later styles. A century later, Americans championed the construction of skyscrapers, which now mark the skyline in parts of London, Paris, Frankfort, Shanghai, Beijing, and hundreds of other world cities. Prominent contemporary architects have designed high rise buildings in both the United States and other nations, as styles become far less ethnic but trans-national in the globalizing society. As urbanites have established exurban vacation homes and suburban development has stretched farther from the urban core, new modern architectural styles, which then may be copied by other residents, modify the original rural landscape, while at the same time introduce rural attitudes about space into the urban scene. All of these modify the landscape, often reducing the intelligibility of the earlier developed ethnic, or multiple-ethnic, landscape and increasing the dominance of a homogenizing American imprint upon the United States landscape.

Yet at the same time such effects of globalization and increased urban and suburban dominance diminish the explicit imprint of many older ethnic landscapes, these same forces bring many newcomers. They often come from parts of the world different from those that provided the participants in the earlier waves of immigrants, and thus the new arrivals begin to reshape the landscape in keeping with their ethnic traditions. Thus, we see newly opened Ethiopian or Somalian restaurants and shops within strip shopping malls, and Middle Eastern mosques and Hindu temples being sited in suburban landscapes dominated by tract housing. Globalization both homogenizes and introduces new ideas and new forms to the landscape, with some communities being dominated by homogenizing forces, while others experience greater heterogeneity.

What we have observed on the United States landscape is not unique to the American setting. While the particular mix of ethnic landscapes and the survival of several distinctive ethnic homelands within the United States is unique for its specific blend of ethnicities and their relative prominences and histories, migrations of peoples have created similarities in other lands (see Sowell 1996). For example, while Canada shares parts of its French and British roots with the United States and the cultural landscape of the Maritime Provinces shows many similarities with New England from where many of its Loyalist settlers came following the American Revolution (Jasanoff 2012), Canada lacks a prominent Hispanic influence along its southern border. Yet it too experienced waves of immigrants from central and eastern Europe who settled its Prairies, just north of the American Great Plains. Americans also flooded across Canada's border, not only as Loyalists fleeing persecution in the aftermath of the American Revolution to profoundly shape the settlement patterns in Ontario, New Brunswick, and Nova Scotia, but over a century later American farmers played a major role in settling the Prairie Provinces, where land was partitioned into one mile squares similar to those created by the survey used south of the 49th parallel. Yet, throughout much of the nineteenth century and well into the twentieth century, Canadians from the Maritimes and Quebec moved to New England and the Mid-Atlantic states for employment. Canadians from Ontario and Quebec have established vacation and retirement communities in Florida and elsewhere.

Like the United States and Canada, Australia saw its indigenous people become overwhelmed by British settlers, who were later supplemented by large numbers of central and southern Europeans, creating ethnic landscapes that differed from those of the predominately British settlers. All three countries have now developed Chinatowns, along with many non-Anglophone nations that also have become home to the Chinese Diaspora. With globalization, the migration of Americans to many areas of the world, whether for business purposes, retirement, or military assignment, has created communities of Americans who occupy single family detached houses or apartments and shop in American style retail centers in parts of Latin America, the Caribbean, Europe, the Middle East, Africa, and Asia.

18.8 Persistence of Ethnic Landscapes

Change will continue to occur to the American landscape, rapidly in some locales and slowly in others, caused by movements of population, growth in populations that resist assimilation, new technologies, economic shifts, changing environmental conditions, and the passage of time. While the ethnic mix that comprises the American population will continue to evolve, just as it has been changing for the half millennia since the first Europeans established their settlements along the Atlantic coast, the imprint of previous ethnic settlements remains on the landscape. While the prominence of such historic contributions to the current cultural landscape clearly varies with both the passage of time and the number and character of landscape alterations brought by subsequent populations, the legacy of many groups upon the landscape remains long after a group has departed or become relegated to obscurity. As we noted in Chap. 2, those Native American peoples who built the thousands of mounds that still dot the landscape of much of the eastern half of the nation have been gone for nearly a thousand years, yet their mounds remain highly visible features on the landscape. In a like manner, the hundreds of well-preserved cliff dwellings along many canyons in the Southwest—as well as many abandoned pueblos—date from the thirteenth century. Landscape alterations made by the earliest European settlers persist, even though the first European groups were subsequently replaced by new waves of immigrants, coming first from different regions of Europe, and then later from parts of Latin America, the Caribbean, Africa, or Asia. Thus, the arrangement of streets and the governance of Boston were shaped by the English, even though certain neighborhoods today have been occupied by Italians for over a century and, more recently, a Chinatown has developed. The naming of places, whether rivers, settlements, or counties, often demonstrates the influence of ethnic groups who no longer occupy a particular landscape.

Many ethnic impacts upon the landscape have a high degree of persistence. Not only were German immigrants prominent among those waves of Europeans who arrived in the Midwest a century and a half ago, but the descendants of those German ethnics still own over half of the farmland within the Cornbelt region. In areas where long-established ethnic populations have remained prominent for many generations, such as Hispanos, Tejanos, and Native Americans in the American Southwest, vibrant ethnic homelands remain. Yet even where a group did not enjoy such prominence, at least trace elements of its ethnic landscape remain. Consider that the naming of certain neighborhoods and the layout of streets in the southern part of Manhattan Island remain a legacy to Dutch colonization four centuries ago. The way the nation's land was surveyed and the location of its roads persists over the generations. While change does come to the ethnic landscape of the United States, reminders of past ethnic landscapes remain etched on the ground. Undoubtedly, the arrival of new ethnic populations will alter the American landscape, just as earlier groups shaped and reconfigured the multi-ethnic society that currently occupies the American landscape. Geographers are already describing the visible impacts of many newly arrived Asian, African, and Middle Eastern groups upon the American landscapes (Airriess 2016), noting how the arrival of many such groups results in the formation of not only ethnic enclaves and ethnic economies (Kaplan and Li 2006), but also ethnic gentrification of formerly run-down neighborhoods.

America will remain a multi-ethnic society, but the flavor of the stew today is different than it was in the past, and it will be different in other ways in the future. While this blending of cultures results in many Americans simply reporting an "American" ancestry to Census takers—the creation of a new ethnicity that represents an amalgamation of numerous heritages—the continuing arrival of newcomers, the development of ethnic settlements that defy assimilation, and the stirring of the pot results in a landscape that is simultaneously both American and ethnic, changing over both time and space.

References

Airriess, Christopher A. 2016. *Contemporary Ethnic Geographies in America*. Lanham: Rowman and Littlefield.

Association of Religious Data Archives (ARDA). n.d. Hutterian Brethren Counties. http://www.thearda.com/ql2010/QL_C_ALL_2_1089c.asp. Accessed 1 Aug 2016.

Barcus, Holly R., and Laura Simmons. 2013. Ethnic Restructuring in Rural America: Migration and the Changing Faces of Rural Communities in the Great Plains. *The Professional Geographer* 65 (1): 130–152.

Beck, Pearl, Steven M. Cohen, Jacob B. Ukeles, and Ron Miller. 2013. *Jewish Community Study of New York: 2011, Geographic Profile*. revised ed. New York City: UJA-Federation of New York.

Bennion, Lowell C. Ben. 2001. Mormondom's Deseret Homeland. In *Homelands: A Geography of Culture and Place Across America*, ed. Richard L. Nostrand and Lawrence E. Estaville, 184–209. Baltimore: Johns Hopkins University Press.

Brand, Stewart. 1995. *How Buildings Learn: What Happens After They're Built*. New York: Penguin Books.

Brown, Patricia Leigh. 2002. Beacons of Faith are Dimming on the Prairie. *The New York Times*. July 7, 2002.

Chan, Sewell, and Jo Craven McGinty. 2007. Explosive Growth Since 2000 in State's Hasidic Enclaves. *New York Times*. June 29, 2007.

Cook, B.I., T.R. Ault, and J.E. Smerdon. 2015. Unprecedented 21st Century Drought Risk in the American Southwest and Central Plains. *Science Advances* 1: e1400082. doi:10.1126/sciadv.1400082.

Daniels, Roger. 2002. *Coming to America: A History of Immigration and Ethnicity in American Life*. 2nd ed. New York: Harper Perennial.

Davenport, Coral, and Campbell Robertson. 2016. Resettling the First American 'Climate Refugees'. *New York Times*. May 3, 2016, A1.

Donnermeyer, Joseph. 2015. Doubling Time and Population Increase of the Amish. *Journal of Amish and Plain Anabaptist Studies*. 3 (1): 94–109.

Ellis, Mark, Richard Wright, and Matthew Townley. 2016. State-Scale Immigration Enforcement and Latino Interstate Migration in the United States. *Annals of the Association of American Geographers* 106 (4): 891–908.

Ennis, Sharon R., Merarys Rios-Vargas, and Nora G. Albert. 2011. *The Hispanic Population: 2010*, Census 2010 Brief C2010BR/4. Washington: U.S. Census Bureau.

Evans, Simon M., and Peter Peller. 2016. Hutterite Colonies and the Cultural Landscape: An Inventory of Selected Site Characteristics. *Journal of Amish and Plain Anabaptist Studies*. 4 (1): 51–81.

Francaviglia, Richard. 2000. Selling Heritage Landscapes. In *Preserving Cultural Landscapes in America*, ed. Arnold R. Alanen and Robert Z. Melnick, 44–69. Baltimore: The Johns Hopkins University Press.

Godfrey, Brian J. 2016. New Ethnic Landscapes: Place Making in Urban America. In *Contemporary Ethnic Geographies in America*, ed. Christopher A. Airriess, 2nd ed., 59–89. Lanham: Rowman & Littlefield.

Hauer, Mathew E., Jason M. Evans, and Deepak R. Mishra. 2016. Millions Projected to be at Risk from Sea-Level Rise in the Continental United States. *Nature Climate Change* 6: 691. doi:10.1038/NCLIMATE2961.

Hoelscher, Steven D. 1998. *Heritage of Stage: The Invention of Ethnic Place in America's Little Switzerland*. Madison: University of Wisconsin Press.

Hutterian Brethren. 2016. Geographic Location. http:/ http://www.hutterites.org/the-leut/distribution/. Accessed 10 Sept 2016.

Jackson, Richard H. 2003. Mormon Wests: The Creation and Evolution of an American Region. In *Western Places, American Myths: How We Think About the West*, ed. Gary J. Hausladen, 135–165. Reno: University of Nevada Press.

Jang, Mira. 2009. Koreans and Bangladeshis Vie in Los Angeles District. *New York Times* (April 7).

Janzen, Rod, and Max Stanton. 2010. *The Hutterites in North America*. Baltimore: Johns Hopkins University Press.

Jasanoff, Maya. 2012. *Liberty's Exiles: American Loyalists in the Revolutionary World*. New York: Vintage Books.

Kaplan, David H., and Wei Li. 2006. *Landscapes of the Ethnic Economy*. Lanham: Rowman and Littlefield Publishers.

Keul, Adam. 2014. Reliving off the Swamp: A Cajun Tourism Commodity. *Geographical Review* 104 (4): 506–522.

Kopp, Robert E., Andrew C. Kemp, Klaus Bittermann, Benjamin P. Horton, Jeffrey P. Donnelly, W. Roland Gehrels, Carling C. Hay, Jerry X. Mitrovica, Eric D. Morrow, and Stefan Rahmstorf. 2016. Temperature-Driven Global Sea-Level Variability in the Common Era. *Proceedings of the National Academy of Sciences* 113: E1434. doi:10.1073/pnas.1517056113.

Kraybill, Donald B., and Carl F. Bowman. 2001. *On the Backroad to Heaven: Old Order Hutterites, Mennonites, Amish, and Brethren*. Baltimore: Johns Hopkins University Press.

Lehr, John C. 1990. Preservation of the Ethnic Landscape in Western Canada. *Prairie Forum* 15 (2): 263–276.

Manger, William F. 2000. The 'Idealized' Mexican American Housescape. *Material Culture* 32 (1): 1–36.

Marino, E. 2012. The Long History of Environmental Migration: Assessing Vulnerability Construction and Obstacles to Successful Relocation in Shishmaref, Alaska. *Global Environmental Change* 22 (2): 374–381.

MacDonald, Glen M. 2010. Water, Climate Change, and Sustainability in the Southwest. *Proceedings of the National Academy of Sciences* 107 (50): 21256–21262.

McCarthy, James. 2008. Rural Geography: Globalizing the Countryside. *Progress in Human Geography* 32 (1): 129–137.

Meinig, Donald W. 1965. The Mormon Culture Region: Strategies and Patterns in the Geography of the American West, 1847-1964. *Annals of the Association of American Geographers* 55 (2): 191–220.

Melillo, Jerry M., Terese (T.C.) Richmond, and Gary W. Yohe, eds. 2014. *Climate Change Impacts in the United States: The Third National Climate Assessment*. Washington: U.S. Global Change Research Program. doi:10.7930/J0Z31WJ2.

Oliver, Jeff, and Ágústa Edwald. 2016. Between Islands of Ethnicity and Shared Landscapes: Rethinking Settler Society, Cultural Landscapes and the Study of the Canadian West. *Cultural Geographies* 23 (2): 199–219.

Perez, Anthony Daniel, and Charles Hirschman. 2009. The Changing Racial and Ethnic Composition of the US Population: Emerging American Identities. *Population and Development Review* 35 (1): 1–51.

Raitz, Karl B. 1979. Themes in the Cultural Geography of European Ethnic Groups in the United States. *The Geographical Review*. 69 (1): 79–94.

Reuveny, Rafael. 2007. Climate Change-Induced Migration and Violent Conflict. *Political Geography* 26: 656–673.

Santos, Carla Almeida, Yaniv Belhassen, and Kellee Caton. 2008. Reimagining Chinatown: An Analysis of Tourism Discourse. *Tourism Management* 29: 1002–1012.

Santos, Carla Almeida, and Grace Yan. 2008. Representational Politics in Chinatown: The Ethnic Other. *Annals of Tourism Research* 35 (4): 879–899.

Scheitle, Christopher P., and Roger Finke. 2012. *Places of Faith: A Road Trip Across America's Religious Landscape*. New York: Oxford University Press.

Schnell, Steven M. 2003a. Creating Narratives of Place and Identity in 'Little Sweden, U.S.A. *The Geographical Review*. 93 (1): 1–29.

———. 2003b. The Ambiguities of Authenticity in Little Sweden, U.S.A. *Journal of Cultural Geography* 20 (2): 43–68.

Short, Brian. 2006. *England's Landscape: The South East*. London: Collins.

Sowell, Thomas. 1996. *Migrations and Cultures: A World View*. New York: Basic Books.

Trollinger, Susan L. 2012. *Selling the Amish: The Tourism of Nostalgia*. Baltimore: The Johns Hopkins University Press.

van Aalst, Maarten K. 2006. The Impacts of Climate Change on the Risk of Natural Disasters. *Disasters* 30 (1): 5–18.

Wilson, Laura. 2000. *Hutterites of Montana*. New Haven: Yale University Press.

Young Center for Anabaptist and Pietist Studies. 2016. Amish Population Change 1992–2016. Elizabethtown College. https://groups.etown.edu/amishstudies/files/2016/06/Population_change_1992-2016.pdf. Accessed 10 Sept 2016.

Zimmerman, Janelle M. 2015. Practicality and Identity as Functions of Education in Old Order Mennonite and Hutterite Communities. *Journal of Amish and Plain Anabaptist Studies* 3 (1): 71–93.

MIX
Papier aus verantwortungsvollen Quellen
Paper from responsible sources
FSC® C105338

If you have any concerns about our products,
you can contact us on
ProductSafety@springernature.com

In case Publisher is established outside the EU,
the EU authorized representative is:
Springer Nature Customer Service Center GmbH
Europaplatz 3, 69115 Heidelberg, Germany

Printed by Libri Plureos GmbH
in Hamburg, Germany